Higher National Engineering

Higher National Engineering

Mike Tooley and Lloyd Dingle

 **Newnes**

OXFORD   AUCKLAND   BOSTON   JOHANNESBURG   MELBOURNE   NEW DELHI

Newnes
An imprint of Butterworth-Heinemann
Linacre House, Jordan Hill, Oxford OX2 8DP
225 Wildwood Avenue, Woburn, MA 01801-2041
A division of Reed Educational and Professional Publishing Ltd

 A member of the Reed Elsevier plc group

First published 1998
Reprinted with amendments 1999
Reprinted 2000, 2001

British Library Cataloguing in Publication Data
A catalogue record for this book is available from the British Library

ISBN 0 7506 4629 2

Edexcel

Success through qualifications
*Edexcel recommends this book
as a suitable text for its Higher National
Engineering programmes*

FOR EVERY TITLE THAT WE PUBLISH, BUTTERWORTH-HEINEMANN
WILL PAY FOR BTCV TO PLANT AND CARE FOR A TREE.

Printed and bound in Great Britain

Printed on acid free paper

Contents

Introduction

This book has been written to help you achieve the learning outcomes of the core units of BTEC's new Higher National Engineering programme and the two authors have many years experience of teaching HNC and HND engineering students. In producing the book, the authors' principal aim has been that of capturing, within a single volume, the core knowledge required of all engineering students at HNC/HND level.

The six core units covered in the book are: Business Management Techniques, Engineering Design, Engineering Principles, Electrical and Electronic Principles, Mechanical Principles, and Analytical Methods. The book has been organized on a logical basis with each chapter devoted to a single core unit. We have, however, attempted to reduce duplication and some material is appropriate to more than one of the core units. Furthermore, to put mathematical concepts into context, we have developed a number of mathematical topics within the individual chapters. This material appears as **Mathematics in Action**, and features throughout the text. In addition, these mathematical topics are summarized in the chapter dealing with Analytical Methods. You will also find that, where difficult concepts are introduced, we have included notes in the margin headed **Another View**. These will provide you with an alternative way of understanding them.

This book has been designed to provide you with a thorough introduction to each of the core units. Despite this, you are advised to make use of other reference books and materials wherever and whenever possible. You should also get into the habit of using all of the resources that are available to you. These include your tutor, your college or university library, computer centre, engineering laboratories and other learning resources, such as the Internet (see Appendix). You should also become familiar with selecting materials that are appropriate to the topics that you are studying. In particular, you may find it useful to refer to materials that will provide you with several different views of a particular topic.

Throughout the book we have provided worked examples that show how the ideas introduced in the text can be put into practice. We have also included problems and questions at various stages in the text. Depending on the nature of the topic, these questions take a variety of forms, from simple problems requiring short numerical answers to those that may require some additional research or that may require the use of an analytical software package in their solution. Your tutor may well ask you to provide answers to these questions as coursework or homework but they can also be used to help you with revision for course assessments.

Business Management Techniques are introduced in Chapter 1. This chapter will provide you with an introduction to standard costing techniques as well as an insight into the key functions that underpin financial planning and control, project planning and scheduling. Chapter 2, **Engineering Design**, deals with the thought processes and procedural activities concerned with the design of engineering artifacts and systems. At this level the mathematical rigour often associated with 'designing' has been omitted. Instead, we have emphasized the production of the design specification, management report and the use of computer technology as a design aid.

The scientific principles that underpin the design and operation of modern engineering systems are introduced in Chapter 3, **Engineering Science**. This chapter provides essential preparation to the two 'principles' units that follow. It also provides a valuable introduction to engineering science for anyone who has not studied engineering before.

Electrical and Electronic Principles are covered in Chapter 4. This chapter introduces electrical circuit theory, networks and complex waveforms. **Mechanical Principles** are introduced in Chapter 5, which explains the principles that underpin the design and operation of mechanical engineering systems and deals with complex loading systems, loaded cylinders and beams, power transmission and rotational systems.

Chapter 6, **Analytical Methods**, covers the essential mathematical principles, methods and applications that underpin a study of engineering at HNC/HND level. Each topic is presented in three parts: *formulae*, *methods* and *applications*. This novel method of presentation is particularly effective because it will allow you to develop analytical and mathematical skills *alongside* the relevant engineering concepts.

Finally, we would like to offer a few words of practical advice to students. At the beginning of your HNC or HND course you will undoubtedly find that some topics appear to be more difficult than others. Sometimes you may find the basic concepts difficult to grasp (perhaps you haven't met them before), you may find the analytical methods daunting, or you might have difficulty with things that you cannot immediately visualize.

No matter what the cause of your temporary learning block, it is important to remember two things: you won't be the first person to encounter the problem, and there is plenty of material available to you that will help you overcome it. All that you need to do is to recognize that it *is* a problem and then set about doing something about it. A regular study pattern and a clearly defined set of learning goals will help you get started. In any event, don't give up – engineering is a challenging and demanding career and your first challenge along the road to becoming a practising engineer is to master the core knowledge that engineers use in their everyday work. And that is what you will find in this book.

May we wish you every success with your Higher National studies!

Mike Tooley and Lloyd Dingle

Acknowledgements

The authors would like to thank a number of people who have helped in producing this book. In particular, we would like to thank Richard Tooley for taking our rough drawings and turning them into 'real' artwork, Gerry Wood for comments and advice on the chapter dealing with Business Management Techniques, Matthew Deans, Diane Chandler and all members of the team at Newnes for their patience and perseverance. Last, but by no means least, we would like to say a big 'thank you' to Wendy and Yvonne. But for your support and understanding this book would never have been finished!

1 Business management techniques

Summary

This unit is designed to give students an appreciation of the application of standard costing techniques as well as an insight into the key functions underpinning financial planning and control. The unit also aims to expand students' knowledge and interest in managerial and supervisory techniques by introducing and applying the fundamental concepts of project planning and scheduling.

1.1 Costing systems and techniques

To meet the requirements of this unit, you need to be able to identify and describe appropriate costing systems and techniques for specific engineering business functions. You also need to be able to measure and evaluate the impact of changing activity levels on engineering business performance. We will start by introducing some common costing systems used in engineering.

Costing systems

Any modern business enterprise needs to have in place an effective costing system that take into account the real cost of manufacturing the product or delivering the service that it provides. Without such a system in place it is impossible to control costs and determine the overall profitability of the business operation.

Cost accounting is necessary for a company to be able to exercise control over the actual costs incurred compared with planned expenditure. From the point of view of cost control, a costing system should not only be able to identify any costs that are running out of control but should also provide a tool that can assist in determining the action that is required to put things right.

Job costing

Job costing is a very simple costing technique. It usually applies to a unique operation, such as fitting a part or carrying out a modification to a product. Typical operations in which job costing is commonly used include:

- supplying a unique or 'one-off' item;
- painting and decorating a building;
- converting or adapting a product to meet a particular customer's requirements.

Job costing always has at least three elements: direct labour; direct materials, and absorbed overheads. Sometimes there is an additional direct cost – direct expenses. Let's take an example:

A *jobbing builder*, John Smith, has been asked to supply a quotation for supplying a 'one-off' shipping container for a diesel generator. The parts and materials required to build the shipping container are as follows:

Item	Quantity	Price per unit	Cost (£)
½" chipboard	4 m × 2 m	@ £1.75/m^2	14.00
2" × 1" timber	10m	@ £0.80/m	8.00
Panel pins	50	@ £0.01/ea.	0.50
Countersunk screws	30	@ £0.02/ea.	1.60
Adhesive	1	@ £1.28/ea.	1.28
Joints	16	@ £0.45/ea.	7.20
Total for parts and materials:			**£32.58**

To this should be added the cost of labour. Let's assume that this amounts to 3 hours at £20.00 per hour (this figure includes the overheads associated with employment, such as National Insurance contributions). Hence the cost of labour is:

Item	Quantity	Price per unit	Cost (£)
Labour	3 hours	@ £20.00/hour	60.00
Total for labour:			**£60.00**

We can add the cost of labour to the total *bill of materials* to arrive at the final cost for the job which amounts to £92.58. Note that parts and materials *may* be supplied 'at cost' or 'marked up' by a percentage which can often range from 10% to 50% (and sometimes more).

Example 1.1.1

John Smith has decided to mark-up the costs of timber by 25% and all other sundry items by 10%. Determine the amount that he will charge for the transit container.

The revised bill of materials is as follows:

Item	Quantity	Price per unit	Cost (£)	Charge(£)
½" chipboard	4 m × 2 m	@ £1.75/m^2	14.00	17.50
2" × 1" timber	10m	@ £0.80/m	8.00	10.00
Panel pins	50	@ £0.01/ea.	0.50	0.55
Countersunk screws	30	@ £0.02/ea.	1.60	1.76
Adhesive	1	@ £1.28/ea.	1.28	1.41
Joints	16	@ £0.45/ea.	7.20	7.92
Labour	3 hours	@ £20.00/hour	60.00	60.00
Total amount charged:				**£99.14**

Large companies also use job costing when they produce a variety of different, and often unique, products. These products are often referred to as *custom built* and each is separately costed as a 'job' in its own right. This type of production is described as *intermittent* (and traditionally referred to as *job shop* production) to distinguish it from the *continuous* or *assembly-line* production associated with the manufacture of a large number of identical units.

In jobbing production, individual manufactured units are normally produced to meet an individual customer's requirements and production is not normally speculative. Costs are agreed before manufacturing starts and form the basis of a contract between the manufacturer and the customer.

Contract costing

Contract costing relates to larger jobs (so is conceptually the same as job costing) and is longer lasting. Contract costing is usually used for things like civil engineering, ship building and defence. Contract costing is more complex than job costing.

Parts costing

Parts costing is straightforward and is simply a question of determining the cost of all of the physical parts and components used in a manufactured or engineered product. Parts costing works from the 'bottom up' – in other words, the cost of each individual component (i.e. the *per unit cost*) is determined on the basis of the given *standard supply multiple*. As an example of parts costing, consider the following example:

Centralux is a small engineering company that specialises in the manufacture of domestic central heating controllers. Their latest product uses the following parts:

Component	Qty.	Cost ea.	Total
Bridge rectifier	1	0.18	0.18
Capacitor ceramic disk 20%	4	0.04	0.16
Capacitor electrolytic 20%	1	0.21	0.21
Capacitor polyester 10%	4	0.12	0.48
Clips – plastic	2	0.04	0.08
Connector – mains	1	0.25	0.25
Connector – PCB type	2	0.08	0.16
Connector – solder terminal	2	0.02	0.04
Display – LED	1	0.45	0.45
Fascia trim	1	0.15	0.15
Fuse – 20mm	1	0.12	0.12
Fuse holder – 20mm	1	0.13	0.13
Keypad – membrane type	1	0.89	0.89
Miniature PCB transformer	1	1.99	1.99
Nuts – M3	8	0.02	0.16
Opto-isolator	2	0.22	0.44
Pillars – plastic	4	0.03	0.12
Plastic enclosure	1	0.89	0.89
Printed circuit board	1	1.45	1.45
Programmed controller chip	1	1.05	1.05
Resistor 0.25W 5%	5	0.01	0.05
Resistor 0.5W 5%	2	0.02	0.04
Screws – M3	8	0.03	0.24
Switch – mains	1	0.55	0.55
Temperature sensor	1	0.25	0.25
Transducer – piezoelectric	1	0.33	0.33
Triac	2	0.42	0.84
Voltage regulator	1	0.21	0.21
Washers – M3	8	0.01	0.08
Total cost:			**£11.99**

It is often useful to group together individual component parts under groupings of similar items. The reason for this is that such groupings tend to be subject to the same fluctuation in cost. We can thus quickly determine the effect of market fluctuations by examining the effect of changes on particular groups of parts.

Example 1.1.2

Group together the parts used in the Centralux domestic central heating controller under the following headings; *Hardware*, *Semiconductors*, *Passive components*, and *Miscellaneous*. Determine the proportion of the total cost by part category.

Hardware

Item	Qty.	Cost ea.	Total
Clips – plastic	2	0.04	0.08
Fascia trim	1	0.15	0.15
Nuts – M3	8	0.02	0.16
Pillars – plastic	4	0.03	0.12
Plastic enclosure	1	0.89	0.89
Printed circuit board	1	1.45	1.45
Screws – M3	8	0.03	0.24
Washers – M3	8	0.01	0.08
Sub total:			**£3.17**

Semiconductors

Item	Qty.	Cost ea.	Total
Bridge rectifier	1	0.18	0.18
Display – LED	1	0.45	0.45
Opto-isolator	2	0.22	0.44
Programmed controller chip	1	1.05	1.05
Triac	2	0.42	0.84
Voltage regulator	1	0.21	0.21
Sub total:			**£3.17**

Passive components

Item	Qty.	Cost ea.	Total
Capacitor ceramic disk 20%	4	0.04	0.16
Capacitor electrolytic 20%	1	0.21	0.21
Capacitor polyester 10%	4	0.12	0.48
Miniature PCB transformer	1	1.99	1.99
Resistor 0.25W 5%	5	0.01	0.05
Resistor 0.5W 5%	2	0.02	0.04
Temperature sensor	1	0.25	0.25
Transducer – piezoelectric	1	0.33	0.33
Sub total:			**£3.51**

Miscellaneous

Item	Qty.	Cost ea.	Total
Connector – mains	1	0.25	0.25
Connector – PCB type	2	0.08	0.16
Connector – solder terminal	2	0.02	0.04
Fuse – 20mm	1	0.12	0.12
Fuse holder – 20mm	1	0.13	0.13
Keypad – membrane type	1	0.89	0.89
Switch – mains	1	0.55	0.55
Sub total:			**£2.14**

| **Total cost:** | | | **£11.99** |

The proportion of costs by part category is shown in the pie-chart of Figure 1.1.1.

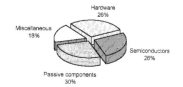

Figure 1.1.1 *Proportion of costs by part category for Centralux's domestic heating controller*

Question 1.1.1

Determine the effect on the total cost of the Centralux domestic central heating controller when the cost of semiconductors increases by 5% and the cost of hardware falls by 10%.

Process costing

Process costing takes into account the cost of a continuous manufacturing process and apportions part of the cost of each process to an individual product. Typical processes might be:

- forming, bending or machining of metal and plastic parts;
- flow soldering of printed circuit boards;
- heat treatment of metal parts;
- paint spraying and finishing.

Process costing is used in industries that operate on a continuous basis, such as chemical plants, petroleum or food production. In order to carry out process costing it is necessary to show how the flow of products is costed at each stage of the process; Process 1, Process 2, Process 3, and so on to the finished product.

The following example illustrates one stage of process costing. Note that, when determining the total cost of manufacturing a product, it is essential to take into account the notional cost of *all* of the processes involved.

Example 1.1.3

Centralux has invested in a flow soldering plant in order to partly automate the manufacture of their domestic central heating controller. The flow soldering plant operates at a rate of 50 units per hour and its operating cost (including capital cost recovery calculated over a nominal 8-year asset life) amounts to £6000/week plus £10 material costs per hour. Determine the unit cost of the flow soldering process based on: (a) 70 hours operation per week; and (b) 84 hours per week.

(a) Based on 70 hours operation per week, the total cost of the flow soldering process will be given by:

Total cost = £6000 + (70 × £10) = £6700.

At 50 units per hour, the total weekly production will be given by:

Total production = 70 × 50 = 3500.

The cost, per unit, will thus be given by:

Cost per unit = £6700/3500 = £1.91.

(b) Based on 84 hours operation per week, the total cost of the flow soldering process will be given by:

Total cost = £6000 + (84 × £10) = £6840.

At 50 units per hour, the total weekly production will be given by:

Total production = 84 × 50 = 4200.

The cost, per unit, will thus be given by:

Cost per unit = £6840/4200 = £1.63.

Costing techniques

Any engineering business is liable to incur a variety of costs. These will typically include:

- rent for factory and office premises;
- rates;
- energy costs (including heating and lighting);
- material costs;
- costs associated with production equipment (purchase and maintenance);
- salaries and National Insurance;
- transport costs;
- postage and telephone charges;
- insurance premiums.

Given the wide range of costs above, it is often useful to classify costs under various headings, including fixed and variable costs, overhead and direct costs, average and marginal costs, and so on.

In order to be able to control costs, it is, of course, vital to ensure that all of the costs incurred are known. Indeed, the consequences of not being fully aware of the costs of a business operation can be dire!

This section examines a number of different methods used by businesses to determine the total cost of the product or service that they deliver. The prime objective of these techniques is that of informing commercial decisions such as:

- How many units have to be produced in order to make a profit?
- Is it cheaper to make or buy an item?
- What happens to our profits if the cost of production changes?
- What happens to our profits if the cost of parts changes?

Absorption costing

One method of determining the total cost of a given product or service is that of adding the costs of overheads to the direct costs by a process of *allocation*, *apportionment* and *absorption*. Since *overheads* (or *indirect costs*) can be allocated as whole items to production departments, it is possible to arrive at a notional amount that must be added to the cost of each product in order to cover the production overheads.

$$Mark\text{-}up = \frac{Total\ of\ fixed\ and\ variable\ costs\ attributable\ to\ the\ product}{Total\ number\ of\ units\ produced}.$$

Another view

In absorption costing, each product manufactured is made (at least in theory) to cover all of its costs. This is achieved by adding a notional amount to the total unit cost of each product. We sometimes refer to this as *cost-plus*.

Marginal costing

Marginal costing provides us with an alternative way of looking at costs that provides an insight into the way costs behave by allowing us to observe the interaction between costs, volumes, and profits. The marginal cost of a product is equal to the cost of producing one more unit of output (we shall return to this later).

There are a number of advantages of using marginal costing, notably:

- Marginal costing systems are simpler to operate than absorption costing systems because they avoid the problems associated with overhead apportionment and recovery.

- It is easier to make decisions on the basis of marginal cost calculations. Where several products are being produced, marginal costing can show which products are making a contribution and which are failing to cover their variable costs.

The disadvantages of marginal costing include:

- The effect of time (and its effect on *true cost*) tends to be overlooked in marginal costing.
- There is a temptation to spread fixed costs (or to neglect these in favour of more easily quantified variable costs).

Marginal costing is more useful for management decision-making than absorption costing because it avoids using estimation to determine overheads. The choice of whether to use absorption costing or marginal costing is usually governed by factors such as:

- The system of financial control used within a company (e.g. *responsibility accounting* is consistent with absorption costing).
- The production methods used (e.g. marginal costing is easier to operate in simple processing applications, whereas absorption costing is usually preferred when several different products require different plant and processing techniques).
- The significance of the prevailing level of overhead costs.

Activity based costing

Activity based costing is an attempt to assess the 'true' cost of providing a product or service. Knowledge of the 'true' cost is not only important in helping us to identify opportunities for cost improvement but it also helps us to make strategic decisions that are better informed.

Activity based costing focuses on indirect costs (*overheads*). It does this by making costs that would traditionally be considered indirect into direct costs. In effect, it traces costs and overhead expenses to an individual *cost object*. The basic principles of activity based costing are shown in Figure 1.1.2.

Activity based costing is particularly useful when the overhead costs associated with a particular product are significant and where a number of products are manufactured in different volumes. Activity based costing is particularly applicable where competition is severe and the margin of selling price over manufacturing cost has to be precisely determined.

The steps required to carry out activity based costing are:

(1) Identify the activities.
(2) Determine the cost of each activity.
(3) Determine the factors that drive costs.
(4) Collect the activity data.
(5) Calculate the product cost.

The use of activity based costing is best illustrated by taking an example. A small manufacturing company, EzBild, has decided to carry out activity based costing of its two products: an aluminium folding ladder and a modular work platform. The following table summarises the activity required for these two products:

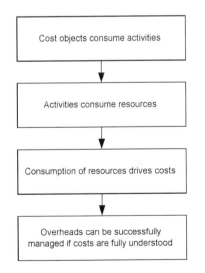

Cost objects consume activities

↓

Activities consume resources

↓

Consumption of resources drives costs

↓

Overheads can be successfully managed if costs are fully understood

Figure 1.1.2 *Principles of activity based costing*

Activity	Ladders (per unit)	Cost (£)	Platforms (per unit)	Cost (£)	Total (£)
Set-up	1 @ £25 000	25 000	1 @ £35 000	35 000	60 000
Manufacture	1500 @ £6	9 000	500 @ £30	15 000	24 000
Assembly	1500 @ £2	3 000	500 @ £10	5 000	8 000
Inspection	1500 @ £1	1 500	500 @ £2	1 000	2 500
Packaging	1500 @ £1	1 500	500 @ £6	3 000	4 500
Total		40 000		59 000	99 000

The activity based product cost for each ladder thus amounts to £40 000/1500 = £26.67, whilst the activity based product cost for each platform amounts to £59 000/500 = £118. To this should be added the direct (material) costs of each product. Assuming that this amounts to £20 for the ladder and £80 for the platform, we would arrive at a cost of £46.67 for the ladder and £198 for the platform.

Traditional cost accounting would have arrived at two rather different figures. Let's assume that 3300 hours of direct labour are used in the manufacturing plant. Dividing the total overhead cost of £99 000 by this figure will give us the hourly direct labour cost of £30 per hour. If ladders require 1 hour of direct labour and platforms require 3.6 hours of direct labour the allocation of costs would be £30 per ladder and £108 per platform. Adding the same direct (material) costs to this yields a cost of £50 for the ladder and £188 for the platform!

Question 1.1.2

DataSwitch Inc. specialises in the production of switches that can be used to link several personal computers to a shared printer. The company currently manufactures a low-cost manually operated data switch and a more expensive automatic data switch. Both types of switch are packed in multiples of five before they are despatched to retail outlets. An analysis of the company's production reveals the following:

	Manual data switch	Automatic data switch
Production volume	1000	250
Direct materials cost	£15/unit	£25/unit
Activity based production analysis:		
Set-up costs	1 @ £10 000	1 @ £10 000
Manufacturing costs	1000 @ £4/unit	250 @ £10/unit
Assembly costs	1000 @ £2/unit	250 @ £4/unit
Packaging	200 @ £10/unit	50 @ £10/unit
Despatch/delivery	200 @ £20/unit	50 @ £20/unit

Determine the cost of each product using activity based costing.

Engineering business functions

Within an engineering company there are a number of discrete business functions. These include design, manufacturing and engineering services. We shall briefly examine each of these essential functions and the effect that they have on costs.

Design

By definition, an engineered product cannot be manufactured until it has been designed. Design is thus an essential engineering function. Design is itself a complex activity requiring inputs from a team of people with differing, but complementary, skills. To be effective, this team needs to undertake a variety of activities as part of the design process. These activities include liaison with clients and customers, concept design, specification, layout and detail design, and liaison with those responsible for manufacturing, sales, service and customer support. With most engineering projects, design costs may be significant, furthermore these costs are normally incurred *before* manufacturing starts and income (attributable to the product or service being designed) is received.

Manufacturing

Manufacturing involves having the right components and materials available and being able to apply appropriate processes to them in order to produce the end product. In this context, 'right' must not only be taken to mean appropriate in terms of the design specification but also the most cost-effective solution in every case. Costs of manufacturing are appreciable. These costs can be attributed to a number of sources including material and component costs and the *added value* inherent in the manufacturing process resulting from labour, energy, and other overheads. Later we shall examine this in greater detail.

Engineering services

Engineering services can be described as any engineering activity that is not directly concerned with manufacturing. Thus maintenance, sales, and customer support can all be described as *engineering services*. These functions may also represent significant costs which normally have to be recovered from manufacturing income.

Measures and evaluation

An engineering company will normally employ a number of different control methods to ensure that its operation is profitable. These control methods include making forecasts of overall profitability, determining the contribution made by each individual activity towards overheads and fixed costs, and performing 'what-if' analysis to determine the effects of variations in cost and selling price. We shall start by describing the most simple method, *break-even analysis*.

Break-even charts

Break-even charts provide a simple (and relatively unsophisticated) method for determining the minimum level of sales that a company must achieve in order for the business to be profitable. Consider the simple relationship illustrated in Figure 1.1.3. Here total income has been plotted against total costs using the same scale for each axis. At point A, total costs exceed total income and the operation is not profitable, i.e. it makes a *loss*. If we charge more for the product, whilst keeping the costs fixed, we would move from point A to B. At a certain point, total income exceeds total costs and we move into *profit*. Finally, let's assume that

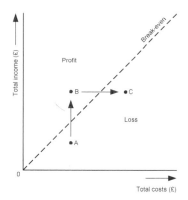

Figure 1.1.3   *Total income plotted against total costs showing profit and loss regions*

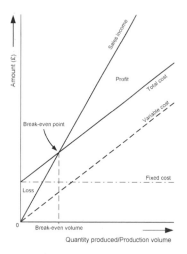

Figure 1.1.4   *Fixed and variable costs plotted against production volume showing break-even*

our total costs increase whilst the total income from sales remains unchanged. We would then move from profit (point B) to loss (point C).

The *break-even point* is the volume of sales at which the operation becomes profitable and it marks the transition from loss into profit. A *break-even chart* takes the form of a graph of costs plotted against volume of product sold. At this point, it is important to recall that the total costs of the business operation are the sum of the fixed and overhead costs with the variable costs of production. Thus:

Total cost = fixed cost + overhead cost + variable cost.

The income derived from the sale of the product (assuming a constant pricing structure) will simply be the product of the quantity sold (i.e. the *volume* of product sold) and the price at which it is sold (i.e. the per unit selling price). This relationship (a straight line) can be superimposed on the break-even chart and the point of intersection with the total cost line can be identified. This is the *break-even point* and the corresponding production volume can be determined from the horizontal axis, see Figure 1.1.4.

The *break-even quantity* can be determined from:

$$Break\text{-}even\ quantity = \frac{fixed\ cost}{selling\ price - variable\ cost}$$

(Note that, in the above formula, *selling price* and *variable cost* are per unit).

It is also possible to use the break-even chart to determine the profit that would result from a particular production quantity.

Profit can be determined from:

Profit =

(selling price × quantity sold) − (fixed cost + (variable cost × quantity sold))

(Note that, in the above formula, *selling price* and *variable cost* are again per unit).

Example 1.1.4

Centralux has analysed its fixed and overhead costs which together amount to £250 000 whilst the variable costs of its domestic central heating controller amount to £15 per unit manufactured.

Construct a break-even chart and use this to determine:

(a)   The break-even production volume when the controller is sold at:
    (i)   £25 per unit;
    (ii)   £30 per unit;
    (iii)   £35 per unit.
(b)   The profit for a production quantity of 25 000 units if the selling price is £35 per unit.

The break-even chart for the Centralux business operation is shown in Figure 1.1.5.

(a)   From Figure 1.1.5 the break-even points are:
    (i)   25 000 units;
    (ii)   16 667 units;
    (iii)   12 500 units.
(b)   Also from Figure 1.1.5 the profit based on 25 000 units at a selling price of £35 per unit is £250 000.

Figure 1.1.5 *Break-even chart for Centralux's domestic central heating controller*

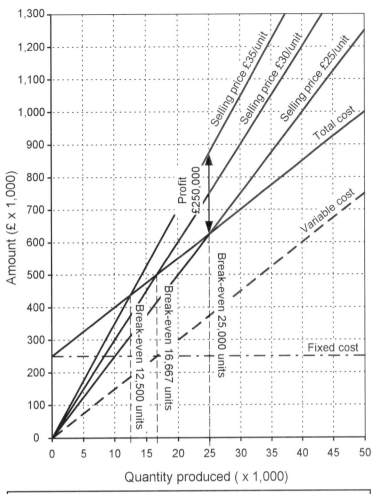

Quantity produced (× 1,000)

Example 1.1.5

A small manufacturing company, KarKare, manufactures a trolley jack for the DIY motor enthusiast. The fixed cost of the company's jack manufacturing operation is £175 000 and the variable costs of producing the trolley jack amount to £21 per unit.

If the jack is to be sold for £45, determine the break-even quantity and the profit that would be returned from sales of 5000 units.

Now *Break-even quantity* $= \dfrac{fixed\ cost}{selling\ price - variable\ cost}$.

Thus *Break-even quantity* = £175 000/(£45 − £21) = 7292.

The profit based on sales of 10 000 units will be given by:

Profit = (*selling price* × *quantity sold*)−(*fixed cost* + (*variable cost* × *quantity sold*))
Profit = (£45 × 10 000) − (£175 000 + (£21 × 10 000)) = £450 000 − £385 000.

Thus the profit on 10 000 units will be £65 000.

It is important to realise that simple break-even analysis has a number of serious shortcomings. These may be summarised as follows:

- The sales income line (i.e. the product of the volume produced and its selling price) takes no account of the effect of price on the volume of sales. This is important as it is likely that the demand for the product will fall progressively as the selling price increases and the product becomes less competitive in the open market.
- The assumption that fixed costs remain fixed and variable costs increase linearly with production are somewhat dangerous. The reality is that both of these will change!

For the foregoing reasons it is important to regard break-even analysis as a 'rule-of-thumb' method for evaluating product pricing. Before making any business decisions relating to pricing and targets for production volume it is important to undertake further research into effect of pricing on potential sales as well as the pricing of competitive products.

Average cost

The fixed costs associated with production have to be shared between the entire volume produced. Hence, a proportion of the final cost of a product will be attributable to the fixed costs of manufacture. The larger the quantity produced, the smaller this proportion will be. In other words, the *average cost* of the product will fall as the volume increases. We can illustrate this in the form of a graph (see Figure 1.1.6). Note that:

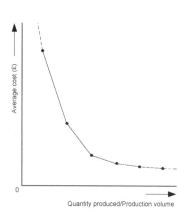

Figure 1.1.6 *Average cost plotted against production volume*

$$Average\ cost = \frac{total\ cost}{quantity\ produced}.$$

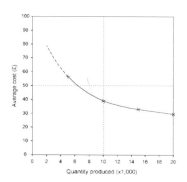

Figure 1.1.7 *Average cost plotted against production volume for KarKare's trolley jack*

Example 1.1.6

Determine the average costs of KarKare's trolley jack based on fixed costs of £175 000 and variable costs of £21 per unit for production levels of 5000, 10 000, 15 000 and 20 000 units.

Now *Average cost* $= \dfrac{total\ cost}{quantity\ produced}$

and *Total cost = fixed cost + variable cost.*

Thus:

Quantity produced:	5000	10 000	15 000	20 000
Fixed cost (£):	175 000	175 000	175 000	175 000
Variable cost (£):	105 000	210 000	315 000	420 000
Total cost (£):	280 000	385 000	490 000	595 000
Average cost (£):	**56**	**38.5**	**32.67**	**29.75**

Marginal cost

Once we have established a particular volume of production, the cost of producing one more unit is referred to as the *marginal cost*. The marginal cost of a product is that cost of the unit that results only from changes in those costs that do not vary with the amount produced. Marginal cost is not the same as average cost – the reason for knowing the marginal cost of a product is that it can help us decide whether or not to increase production from an existing level. This is best illustrated with the use of an example:

Example 1.1.7

KarKare has established a production level of 12 000 units for its trolley jack. As before, the fixed cost of the company's jack manufacturing operation is £175 000 and the variable costs of producing the trolley jack amount to £21 per unit. The jack is normally sold for £45 but a large high-street chain store has offered to take an additional 2000 units at a non-negotiable price of £30 per unit. Use marginal costing to determine whether this proposition is financially sound.

The total cost associated with a production volume of 12 000 units is found from:

Total cost = fixed cost + variable cost

Total cost = £175 000 + (12 000 × £21) = £175 000 + £252 000 = £427 000.

Now the average cost (based on 12 000 units) will be given by:

$$\textit{Average cost} = \frac{\text{total cost}}{\text{quantity produced}} = £427\,000/12\,000 = £35.58.$$

Based on an average cost of £35.58, a selling price of £30 per unit does not appear to be sound business sense. However, if we consider the marginal cost of the trolley jack based on an existing production level of 12 000 units, we arrive at a different view. The rationale is as follows:

Let's assume a scenario in which we sell 12 000 trolley jacks at £45 and 2000 trolley jacks at £30. The total income produced will be given by:

Total income = (12 000 × £45) + (2000 × £30) = £540 000 + £60 000 = £600 000.

The total cost associated with producing 14 000 trolley jacks will be:

Total cost = £175 000 + (14 000 × £21) = £175 000 + £294 000 = £469 000.

Thus the resulting profit will be given by:

Profit = total income − total cost = £600 000 − £469 000 = £131 000.

Had we decided not to accept the order for the extra 2000 units, we would have generated a profit given by:

Profit = total income − total cost = £540 000 − £427 000 = £113 000.

Thus, meeting the order for an additional 2000 units at £30 has helped to increase our profits by £18 000. The important thing to note here is that, although the selling price of £30 per unit is less than the average cost per unit of £35.58, it is actually greater than the marginal cost of £21!

Another view

In marginal costing, we consider the cost of a product when all of the variable costs are removed. We arrive at this figure by calculating the cost of producing just one more unit – the difference in the cost of this unit and the previously manufactured one is the variable cost attributable to just one unit (i.e. the variable cost *per unit*). This assumes that the variable cost per unit is the same for all volumes of production output. This will usually be true for significant production volumes (note how the average cost tends towards a fixed value as the quantity increases in Figure 1.1.6).

Profitability

Profit, or *return on capital employed* (ROCE), is the expressed or implied goal for every business. Being able to make a realistic forecast of profits is an essential prerequisite to making a financial case for investment. It is also an essential ingredient in any business plan.

The need to maximise profits should be an important factor in decision making. Traditional theory assumes that a company will invest in the most profitable projects first, and then choose projects of descending profitability, until the return on the last project just covers the funding of that project (this occurs when the *marginal revenue* is equal to the *marginal cost*).

The process of choosing projects is, however, much more complex. It may, for example, involve strategic issues (such as the need to maintain a presence in a particular market or the need to developing expertise in a particular technology with the aim of improving profits at some later date). Furthermore, many companies do not have sufficient funds available to reach the marginal position. Instead, they will rely on one or two 'hurdle' rates of return for projects. Projects that do not reach these rates of return will be abandoned in favour of those that are considered 'profitable'.

Contribution analysis

In marginal costing, the excess of sales revenue over variable cost is known as the *contribution margin*. This margin represents the contribution made by the item in question to the fixed costs and profit. In Figure 1.1.4 the contribution margin is equivalent to the distance from the variable cost curve and the sales income line. We can thus say that:

Contribution = sales revenue – variable costs.

But since:

Sales revenue = fixed costs + variable costs + profit.

Thus:
Contribution = fixed costs + profit.

Contribution margin can be easily calculated. For example, if the selling price of a Centralux central heating controller (see Example 1.1.4) is £25 and its variable costs amount to £15 then the contribution margin is £(25 – 15) = £10.

1.2 Financial planning and control

By now, you should have begun to understand how costing systems and techniques are applied in a typical engineering company. This next topic introduces you to the financial planning process. To satisfy the requirements of this unit you need to be able to describe the factors influencing the decision-making process during financial planning, examine the budgetary planning process and its application to financial planning decisions. You also need to be able to apply the standard costing techniques that we met in the previous section, analysing deviations from the planned outcome.

Financial planning and control

Adequate financial planning is essential if a business is to achieve its objectives and profit targets. The basic procedure required to formulate a financial plan is as follows:

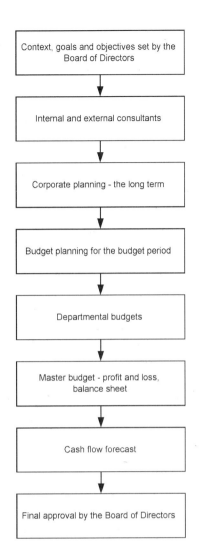

Context, goals and objectives set by the Board of Directors

↓

Internal and external consultants

↓

Corporate planning - the long term

↓

Budget planning for the budget period

↓

Departmental budgets

↓

Master budget - profit and loss, balance sheet

↓

Cash flow forecast

↓

Final approval by the Board of Directors

Figure 1.2.1 *Process of financial planning and control*

(a) Formulate company policy, profit targets and long term plans.
(b) Prepare forecasts for sales, production, stocks, costs, capital expenditure and cash.
(c) Compile these separate forecasts into a master forecast.
(d) Consider all the alternatives available and select the plan which gives the best results, for example, in terms of profit and long term financial stability.
(e) Review limiting factors and the principal budget factor. This process takes place concurrently with (d) and enables work to begin on the framing of the budgets in (f).
(f) Prepare individual budgets and finally the master budget which includes a forecasted profit and loss account and balance sheet.

This process is illustrated in Figure 1.2.1.

Budgetary control

The starting point of budgetary control is with the Board of Directors who determine the scale and nature of the activities of the company. This policy and objective setting is done within the constraints which exist at the time. For example, plans may have to be made within the current capacities and capabilities of the company, since making changes to the location of operations, the size and composition of the work force and the product range are usually long term matters. The budget is essentially for the short term, usually for one year, and created within the framework of long term corporate planning. Successive budgets will be influenced by the preparation of long term plans, but will always relate to the current period.

Some organisations prepare outline budgets over much longer periods perhaps for a 5–10 year horizon, but such budgets are really part of the long term corporate planning activity and subject to major revision before being used as a basis for current period budgetary planning.

External factors will exercise considerable effects on the company in preparing its forecasts and budgets. Government policy, the proximity of a general election, taxation, inflation, world economic conditions and technological development will all combine to constrain or influence the budget planning process. Once the Board of Directors has settled on a policy within the prevailing situation, then the process of turning the policy into detailed quantitative statements can begin.

We normally assume a budget period of one year, which is usual for most industries. It is therefore recognised that the budget period is fixed in relation to the needs of the organisation concerned and could be any period ranging from three months to five years. The shorter the period the more accurate the forecasts, and that is why most companies find that an annual budgeting procedure is a satisfactory compromise.

Preparing a business plan

On occasions, it is necessary to provide a detailed business plan in order to make a case for a particular business venture or project. Before a business plan is written, it is necessary to:

• clearly define the target audience for the business plan;

- determine the plan's requirements in relation to the contents and levels of detail;
- map out the plan's structure (contents);
- decide on the likely length of the plan;
- identify all the main issues to be addressed (including the financial aspects).

Shortcomings in the concept and gaps in supporting evidence and proposals need to be identified. This will facilitate an assessment of research to be undertaken before any drafting commences. It is also important to bear in mind that a business plan should be the end result of a careful and extensive research and development project which must be completed before any serious writing is started.

A typical business plan comprises the following main elements:

- An *introduction* which sets out the background and structure of the plan.
- A *summary* consisting of a few pages which highlight the main issues and proposals.
- A *main body* containing sections or chapters divided into numbered sections and subsections. The main body should include financial information (including a *profitability forecast*).
- *Market and sales projections* should be supported by valid market research. It is particularly important to ensure that there is a direct relationship between market analysis, sales forecasts and financial projections. It may also be important to make an assessment of competitors' positions and their possible response to the appearance of a rival product.
- *Appendices* should be used for additional information, tabulated data, and other background material. These (and their sources) should be clearly referenced in the text.

The financial section of the plan is of crucial importance and, since it is likely to be read in some detail, it needs to be realistic about sales expectations, profit margins and funding requirements ensuring that financial ratios are in line with industry norms. It is also essential to make realistic estimates of the cost and time required for product development, market entry, and the need to secure external sources of funding.

When preparing a plan it is often useful to include a number of 'what-if' scenarios. These can help you to plan for the effects of escalating costs, reduction in sales, or essential resources becoming scarce. During a *what-if analysis*, you may also wish to consider the *halve–double* scenario in which you examine the financial viability of the project in the event that sales projections are halved and costs and time are doubled. The results can be sobering!

When writing a business plan it is necessary to:

- avoid unnecessary jargon;
- economise on words;
- use short crisp sentences and bullet points;
- check spelling, punctuation and grammar;
- concentrate on relevant and significant issues;
- break the text into numbered paragraphs, sections, etc.;
- relegate detail to appendices;
- provide a contents page and number the pages;
- write the summary last.

Finally, it can be useful to ask a consultant or other qualified outsider to review your plan in draft form and be prepared to adjust the plan in the light of comments obtained and experiences gained.

Budgetary planning

Budgets are used as a means of achieving planning and control objectives in most businesses and in many non-commercial organisations. A budget has been defined as:

A financial or quantitative statement prepared and approved, prior to a defined period of time, of the policy to be pursued during that period for the purpose of attaining given objectives.

The benefits that derive from budgetary control arise from the ability to co-ordinate policy, plans and action and to be able to monitor the financial consequences of carrying out the plans.

An engineering company will prepare a number of budgets, each corresponding with a particular functional area. Each budget will normally be controlled by a specified manager, although some managers may control several budgets according to the particular management organisation employed within the company. In a typical engineering business you will find the following:

- marketing budget;
- manufacturing budget;
- research and development budget;
- administration budget;
- capital expenditure budget;
- cash budget.

Each of these budgets may be sub-divided into further budgets. For example, the manufacturing budget may be sub-divided into a budget for direct materials, a budget for direct labour, and a budget for factory overheads (heating, lighting and other energy costs).

Each functional manager will forecast his/her own budget however, there is a need for managers and departments to co-ordinate their budget activities. For example, the capital expenditure budget may reflect the purchase of major items of capital equipment (such as a fork-lift truck or an overhead crane) that will be shared by several departments.

Budget centres

The concept of *budget centres* (we have used this term in preference to *cost centre* or *profit centre*) is central to the process of budgetary control because it provides the means by which it is possible to identify and control costs at the point at which they are incurred. The number and selection of budget centres varies according to the size and complexity of the company. The following are typical:

- sales and marketing;
- production;
- personnel;
- research and development;
- finance and administration.

Example 1.2.1

The Sales Manager at KarKare has been asked to produce a quarterly budget forecast for next year's sales. In order to set about preparing the budget, he has discussed this with the Production Manager and they have agreed the following production volumes and likely levels of sales in each quarter of the year (note that Battery Charger production is not due to start until the third quarter).

First quarter

Product	Units	Selling price
Battery charger	nil	£20
Jack	4000	£30
Warning triangle	2500	£8

Second quarter

Product	Units	Selling price
Battery charger	nil	£20
Jack	4000	£30
Warning triangle	3000	£8

Third quarter

Product	Units	Selling price
Battery charger	1000	£20
Jack	3500	£30
Warning triangle	2500	£8

Fourth quarter

Product	Units	Selling price
Battery charger	2000	£20
Jack	3000	£30
Warning triangle	2250	£8

Complete the quarterly sales budget showing the revenue by product line and total income on sales for each quarter. Include summary columns for the 12-month period. (Note that this task is greatly simplified by making use of a spreadsheet package!). Table 1.2.1 shows how the quarterly sales budget can be presented.

Table 1.2.1 *KarKare sales budget 1999*

Product line	Quarter 1			Quarter 2			Quarter 3			Quarter 4			12 months	
	Units	Price (£)	Revenue	Units	Price (£)	Revenue	Units	Price (£)	Revenue	Units	Price (£)	Revenue	Units	Revenue
Battery charger	0	20	0	0	20	0	1000	20	20 000	2000	20	40 000	3000	60 000
Trolley jack	4000	30	120 000	4000	30	120 000	3500	30	105 000	3000	30	90 000	14 500	435 000
Warning triangle	2500	8	20 000	3000	8	24 000	2500	8	20 000	2250	8	18 000	10 250	82 000
Total			140 000			144 000			145 000			148 000		577 000

Question 1.2.1

Using the data and spreadsheet model of Table 1.2.1, determine the effect of each of the following 'what-if' scenarios:

(a) A 10% increase in all prices with effect from the start of the second quarter.

(b) Price increases of 15% with effect from the start of the third quarter for the battery charger and the jack (no price increase for the warning triangle).

(c) Production of all three items achieving only 50% of forecast in the fourth quarter.

(d) A 50p reduction in selling price of all three items with effect from the start of the third quarter.

Budgets for materials

The direct materials costs of production depends upon knowing the quantity of materials and component parts purchased for use in the manufacture of an individual item. Once again, the process is most easily explained by the use of an example:

Example 1.2.2

Having agreed KarKare's quarterly sales budget, the Managing Director has asked the Production Manager to produce a direct materials budget for the year. The *bill of materials* for each item is as follows:

Battery charger

Material/part	Units	Price/unit
Transformer	6000	2.75
Fuse panel	6000	0.95
Indicator	6000	0.25
Ammeter	6000	0.75
Wire	90	15.95
Solder	30	39.00
Mains cable	60	25.50
Mains plug	6000	0.45
Plastic sheet	1500	15.50
Crocodile clips	6000	0.20

Trolley jack

Material/part	Units	Price/unit
Steel bar	1500	15.75
Steel rod	2500	21.00
Steel sheet	1750	19.75
Spray paint	1500	5.00

Warning triangle

Material/part	Units	Price/unit
Coated aluminium strip	10 250	0.75
Plastic mouldings	30 750	0.95
Reflector panels	21 500	1.50
Fixings	102 500	0.05

Complete the direct materials budget showing the production value of each material/component part and the total material costs for each product. Also determine the KarKare's total direct

cont.

materials cost. (One again, this task is greatly simplified by making use of a spreadsheet package). Table 1.2.2 shows how the direct materials budget can be presented.

Table 1.2.2 *KarKare direct materials budget 1999*

Item	Units	Price/unit (£)	Value (£)	Total
Battery charger				
Transformer	3000	1.95	5850.00	
Fuse panel	3000	0.55	1650.00	
Indicator	3000	0.25	750.00	
Ammeter	3000	0.65	1950.00	
Wire	90	9.50	855.00	
Solder	30	22.50	675.00	
Mains cable	60	19.90	1194.00	
Mains plug	3000	0.39	1170.00	
Plastic sheet	750	9.95	7462.50	
Crocodile clips	6000	0.15	900.00	
				£22 456.50
Trolley jack				
Steel bar	1500	15.75	23 625.00	
Steel rod	2500	21.00	52 500.00	
Steel sheet	900	19.75	17 775.00	
Spray paint	11 500	5.00	7500.00	
Bearing	29 000	1.75	50 750.00	
				£152 150.00
Warning triangle				
Coated aluminium	10 250	0.65	6662.50	
Plastic mouldings	30 750	0.42	12 915.00	
Reflector panels	21 500	0.35	7525.00	
Fixings	102 500	0.05	5125.00	
				£32 227.50
Total				£206 834.00

Question 1.2.2

Using the data and spreadsheet model of Table 1.2.2, determine the effect of each of the following 'what-if' scenarios:
(a) The cost of steel increases by 10% over the entire period.
(b) The design of the battery charger is improved. The ammeter is eliminated and the fuse panel is replaced by a thermal trip costing 35p when purchased in quantities of 1000 or more.
(c) A cheaper source of plastic material is located. The new supplier offers a 15% discount on plastic sheet and 50% discount on moulded plastic parts.
(d) Finally, determine the effect of all three scenarios, (a), (b) and (c) applying at the same time.

Direct labour budgets

The direct labour budget can be produced from a knowledge of the work time required to produce an item even though that time might be divided between several workers, each responsible for a different manufacturing operation or process. For each product we must determine an average time (i.e. the *standard time*) to manufacture a single unit. We can then multiply this figure by the mean wage rate for workers in that area.

If the manufacturing operation requires a variety of different skills and competency, wage rates may vary significantly (less skilled workers will command lower pay rates). In such cases, a more accurate determination of labour costs will require a more detailed analysis based on the time contribution made by workers on each different pay rate. Once again, the process is most easily explained by the use of an example:

Example 1.2.3

KarKare's Production Manager has been asked to produce a labour budget for the next 12-month budget period. He has consulted the Personnel Manager on current wage rates and determined the standard work time required for each product. The results of these calculations are as follows:

Battery charger	*Units*
Standard hours/item	1.10
Wage rate in this area	£4.75 (this product requires soldering/wiring skills)

Trolley jack	
Standard hours/item	1.25
Wage rate in this area	£6.25 (this product requires skilled operatives)

Warning triangle	
Standard hours/item	0.25
Wage rate in this area	£4.25 (this product uses unskilled labour)

Using the production volumes agreed with the Sales Manager (see Table 1.2.1), determine the labour costs per product over the next 12 months. Also determine KarKare's total direct labour costs for the next 12 months.

Once again, a spreadsheet model is recommended, and Table 1.2.3 shows one way of presenting this.

Table 1.2.3   *KarKare direct labour budget 1999*

Item	Battery charger	Trolley jack	Warning triangle
Total units	3000	14 500	10 250
Standard hours/item	1.10	1.25	0.25
Total hours	3300	18 125	2562.5
Wage rate	£4.75	£6.25	£4.25
Total labour cost	£15 675.00	£113 281.25	£10 890.63

Question 1.2.3

Using the data and spreadsheet model of Table 1.2.3, determine the effect of each of the following 'what-if' scenarios:

(a)   Minimum wage rates for production workers are set at £4.50 per hour for the entire period.

(b)   The improvements to the design of the battery charger

are instrumental in reducing its standard assembly time to 0.95 hours.

(c) The design of the trolley jack is to be improved in order to increase the load that can be placed on it. This requires an additional six minutes of welding time.

(d) Finally, determine the effect of all three scenarios, (a), (b) and (c) applying at the same time.

Manufacturing overhead budgets

We mentioned earlier that manufacturing overheads are incurred regardless of the volume of production. This is a rather simplistic view. Some overhead costs can truly be regarded as *fixed*, others can more correctly be referred to as *semi-variable*.

Semi-variable costs are those that have both a fixed element (this cost will be incurred regardless of production volume) and a variable element (this cost will increase as the production volume increases). A good example of a semi-variable cost is the supply of electrical energy. Some electrical energy will be required to support non-production activities (e.g. heating, lighting, office equipment, security equipment, etc.). Electrical energy will also be required to supply machine tools, soldering plants, and many other manufacturing processes (including standard factory plants such as conveyors, lifts, hoists, etc.). This use of electrical energy will clearly increase according to the level of production (note that activity based costing would actually apportion these costs to individual products). Let's take a further example to show how this works:

Example 1.2.4

KarKare's Production Manager has been asked to produce a manufacturing overhead budget for the next 12-month budget period. He has established the fixed costs attributable to business rates, rent and salaries (including his own):

Fixed costs	£
Business rates	6500
Rent	12 000
Salaries	68 000 (including costs of employment)
Electricity	500
Water	450
Gas	650

Semi-variable costs	
Electricity	15p per hour for all three products
Water	5p per hour for the battery charger and trolley jack only
Gas	20p per hour for the trolley jack only

Using the production volumes agreed with the Sales Manager (see Table 1.2.1) and the estimated production hours (see Table 1.2.3), determine the total fixed overhead and the semi-variable overhead for each product. Finally, determine the total manufacturing overhead for the next 12 months.

A spreadsheet model is again recommended and Table 1.2.4 shows one way of presenting this.

Table 1.2.4 *KarKare manufacturing overhead budget 1999*

	Fixed	Var./hour	Battery charger	Trolley jack	Warning triangle
Total units:		3000	14 500	10 250	
Standard hours/item:		1.10	1.25	0.25	
Production hours:		3300	18 125	2562.5	
Semi-variable costs:					
Electricity	£500.00	0.15	£495.00	£2718.75	£384.38
Water	£450.00	0.05	£165.00	£906.25	£0.00
Gas	£650.00	0.20	£0.00	£3625.00	£0.00
Fixed costs:					
Business rates	£ 6500.00				
Rent	£12 000.00				
Salaries	£68 000.00				
Totals:	£88 100.00		£660.00	£7250.00	£384.38
Total overhead					£96 394.38

Question 1.2.4

Using the data and spreadsheet model of Table 1.2.4, determine the effect of each of the following 'what-if' scenarios:

(a) A 5% increase in salaries for six months of the 12-month period.
(b) A £2500 increase in business rates over the entire 12-month period.
(c) A 30% increase in the cost of electricity over the entire 12-month period.
(d) Finally, determine the effect of all three scenarios, (a), (b) and (c) applying at the same time.

Question 1.2.5

Unfortunately, KarKare's Production Manager has not taken into account all of the fixed costs in his model (see Table 1.2.4). Suggest ONE additional semi-variable cost and ONE additional fixed cost that he should have taken into account.

Cash flow budgets

Cash flow budgets are important to all businesses. Even a business that is highly profitable can get into serious difficulty if it has a shortage of cash. This may seem surprising but you must not forget that cash flows and profits may occur at quite different times. In the final analysis, if a company cannot pay employees and creditors when payments are due, it may fail, regardless of whether profits may be significant at some point in the future.

Cash budgets normally start from the opening cash balance and the flow of expected revenues from sales. However, as cash flows occur later than the sales which generate them, the budget has to account this time phasing. We will illustrate this with a simple example.

EzBild's cash flow budget is shown in Table 1.2.5. This budget shows income from sales over the six-month period, January to June.

Take a good look at Table 1.2.5. There are a number of important things to notice:

Table 1.2.5 *EzBild cash flow budget January to June 1999*

	Dec	*Jan*	*Feb*	*Mar*	*Apr*	*May*	*June*
Monthly sales	177	399	502	779	499	381	401
Income							
Balance (b/f)		13 400	7915	2403	(17 202)	8786	19 139
Receipts		17 700	39 900	50 205	77 012	49 933	38 381
		31 100	47 815	52 608	59 810	58 719	57 520
Expenditure							
Payments		23 185	45 412	69 810	51 204	39 580	31 224
Balance (c/f)	13 400	7915	2403	(17 202)	8786	19 139	26 296

(1) The volume of sales each month is shown in the top row. The highest volume of sales is achieved in March (for EzBild this is the start of the highest season for sales). December is a quiet month for sales (perhaps nobody wants a ladder or a platform as a Christmas present!).

(2) The balance *carried forward* (c/f) in the bottom line is *brought forward* (b/f) into the income stream for the next month.

(3) Negative sums are shown enclosed in brackets. Note how March is a particularly poor month – production is high (thus outgoings are significant) but income from sales has not yet been received!

Temporary shortfalls of cash can be met in several ways. The company might liquidate some investments such as bank deposits, stocks and shares, or it might arrange a bank overdraft. The production of a cash flow forecast enables the company to do proper financial planning. Banks are happy to provide short-term overdrafts if they are planned, but are much less happy with businesses who fail to plan for these eventualities, for obvious reasons.

EzBild's cash flow budget has been greatly simplified. In practice, much more detail is required. Receipts have been assumed to be for sales only. However, receipts may include interest on investments, disposals of fixed assets, etc. Similarly, payments will represent purchase of supplies, wages, purchase of equipment, overdraft interest, auditors fees, payment of loan interest or dividend paid to shareholders.

Master budgets

The overall budget planning process shown in Figure 1.2.1 indicates the production of a master budget. This comprises a forecasted *profit and loss statement* and a *balance sheet*. Budgets are often redrafted if the initial master budget they generate is unsatisfactory. It may not contain sufficient profit or some costs may be considered too high. This is an iterative process, whereby the information is refined and operating problems solved.

EzBild's master budget is shown in Table 1.2.6. This balance sheet summarises the financial position as at the end of the budget year. In effect, it is a snapshot of the position at 31 December 1999 (after the events of the year have taken place) whereas the profit and loss account is a period statement that provides us with a summary of the year's activities.

EzBild's balance sheet provides us with some important information about the company. In particular, it shows us where the money came from to run the business and what was done with the money obtained.

Table 1.2.6   *EzBild master budget*

Profit and loss account – January to December 1999

	£	£
Sale of goods		601 293
Less cost of goods sold:		
Opening stock of finished goods	19 501	
Cost of finished goods	315 249	
	334 750	
Less closing stock of goods:	27 866	
		306 884
Gross profit		294 409
Selling and distribution costs	88 730	
Administration and finance costs	76 211	
Design and development costs	45 629	
		210 570
Net profit		83 839

Balance sheet as at 31st December 1999
Assets employed:
Net fixed assets

	£	£
Premises	95 800	
Equipment	75 222	
	171 022	
Current assets		
Stock	6950	
Debtors	28 565	
	12 771	
Cash	48 286	
Less current liabilities		
Creditors	(9545)	
Overdraft	(8000)	
Net current assets		30 741
		201 763

As before, there are a number of important things to be aware of. In particular, you should be able to see that:

- EzBild sold goods worth approximately £600 000 during the year. The *net profit* from these sales amounted to approximately £84 000.
- EzBild has a number of *fixed assets*. These include their office premises and various items of production equipment.
- EzBild's current assets amount to around £200 000. The company is financed by some share capital (valued at £140 000) and the accumulated profit and loss position.
- A loss of approximately £22 000 was made on the previous year's trading. This has been carried forward into the current year. Fortunately, this year's operations show a profit which adds to the available funds.
- The funds from shareholders and profits are the sole source of long-term funding for EzBild's business. The use of funds (or capital employed) is shown in the fixed assets and current assets totals.

Fixed assets are the long-term property of the business and current assets are the circulating assets of the business. The latter are so called because they constantly change through time, reflecting day-to-day business operations. The net current asset figure is an important figure because it shows what is left of current assets after current liabilities are met. Since current liabilities are sums of money owed to creditors (usually the company's suppliers) or to employees for wages, they have to be met out of current assets. Current assets are usually stock, debtors and cash.

The net current assets figure is also called *working capital*. It must be enough to support the company's day-to-day business operations. If it is too small, the company may have difficulty in meeting its commitments.

1.3 Project planning and scheduling

A project can be defined as a series of activities with a definite beginning and ending, and with a series of actions that will lead to the achievement of a clearly defined goal. You need to be able to apply basic project planning and scheduling methods, including establishing timescales and resource requirements and the relationships that exist between the various activities that make up a project.

Project planning is different from other forms of planning and scheduling simply because the set of activities that constitute a project are unique and only occur once. Production planning and scheduling relate to a set of activities that may be performed a large number of times.

Project planning involves considering the full set of activities necessary to achieve the project goal. Typical of these activities are:

* appointing consultants;
* appointing suppliers;
* forming a team to be responsible for carrying out the project;
* preparing a budget for the project;
* preparing a detailed costing;
* producing drawings;
* producing itemised parts lists;
* producing specifications;
* obtaining management approval;
* obtaining planning permission;
* scheduling the phases of the project.

Project goals must be clearly defined at the outset. If these are not clear, not properly understood or agreed by all members of the project team, then the chances of a successful outcome (or any outcome at all, for that matter) can be significantly reduced. In addition, you need to be able to establish the project resources and requirements, produce a plan with an appropriate time-scale for completing the project, identify human resource needs, and identify approximate costs associated with each stage of the project.

Programme Evaluation and Review Technique (PERT)

Programme Evaluation and Review Technique (PERT) was developed for the US Navy in 1958 for planning and control of the Polaris nuclear submarine project. This project involved around 3000 contractors and the use of PERT was instrumental in reducing the project completion time by two years. PERT is widely used today as a fundamental project management tool both by governments and in industry.

PERT requires that project activities should be discrete and have definite start and end points. The technique provides most benefit when projects have a very large number of interrelated activities where it can be very effective in helping to identify the most effective sequence of activities from a variety of possibilities.

One important aspect of PERT is that it allows us to identify the path through the network for which the total activity times are the greatest. This is the *critical path*.

Critical Path Method (CPM)

Critical Path Method (CPM) is also widely used by both government departments and industry. CPM and PERT are very similar and the critical path is important for several reasons:

(1)   Since it represents the most time-critical set of activities, the total time to reach the project goal can *only* be reduced by reducing time spent in one or more of the activities along the critical path. In other words, the critical path highlights those activities that should be critically reviewed to see whether they can be shortened in any way. Putting extra resources into one or more of the critical path activities can usually be instrumental in reducing the overall project time.

(2)   The critical path is unique for a particular set of activities and timings. If any of these are changed (e.g. by directing extra resources into them) a new critical path will be revealed. We can then apply PERT evaluation to this new critical path, critically reviewing the activities that it points us to. This process is iterative – in a large project we can continue to reduce overall project time making changes as the project develops.

(3)   The critical path shows us where the most risky and potentially time threatening activities occur. Since any problems or delays with activities on the critical path may jeopardise the entire project it is in our interests to focus particular attention on these tasks.

Applying PERT

PERT is straightforward to apply. It comprises:

(1)   Identifying all of the activities that make up the project.
(2)   Identifying the sequence of the activities in step 1 (and, in particular, the order of precedence of these activities).
(3)   Estimating the timing of the activities.
(4)   Constructing a diagram that illustrates steps 1, 2 and 3.
(5)   Evaluating the network and, in particular, identifying (and clearly marking) the *critical path*.
(6)   Monitoring actual performance as the project is carried out against the schedule produced in 5, revising and re-evaluating the network as appropriate.

(a)

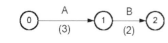

(b)

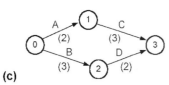

(c)

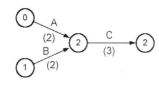

(d)

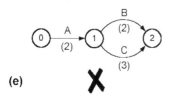

(e)

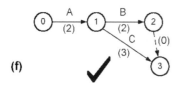

(f)

Figure 1.3.1 *Some simple network diagrams*

Network diagrams

The network diagram used in PERT is comprised of a series of events which form the nodes in the network. Events are linked together by arrows which denote the activities, as shown in Figure 1.3.1.

Figure 1.3.1(a) shows two events, 0 and 1 linked by a single activity A. The expected time for the activity is 3 units of time (e.g. days, weeks, months, etc.).

Figure 1.3.1(b) shows three events, 0, 1, and 2 linked by two activities A and B. Activity A precedes activity B. The expected times for activities A and B are respectively 3 and 2 units of time.

Figure 1.3.1(c) shows four events, 0, 1, 2, and 3, linked by four activities, A, B, C and D. In this network, activity A precedes activity C, whilst activity B precedes activity D. Event 3 is not reached until activities C and D have both been completed. The expected times for activities A, B, C, and D are, respectively, 2, 3, 3, and 2 units of time. There are several other things to note about this network:

- the events occur in the following order; 0, 1, 2, 3;
- activity B is performed at the same time as activity A;
- activity D is performed at the same time as activity C;
- the total time to reach event 3, whichever route is chosen, amounts to 5 units of time.

Figure 1.3.1(d) also shows four events, 0, 1, 2, and 3, linked by three activities, A, B and C. In this network, activities A and B both precede activity C (in other words, activity C cannot start until both activities A and B have been completed). The total time to reach event 3, whichever route is chosen, amounts to 5 units of time.

Within a network diagram, the activities that link two events must be unique. Consider Figure 1.3.1(e). This shows that event 2 can be reached via activities B and C (where the expected time for activity B is 2 units whilst the expected time for activity C is 3 units). To avoid potential confusion, we introduce a *dummy activity* between event 2 and event 3. This activity requires no time for completion and thus its expected time is 0 (note that we have adopted the convention that dummy activities are shown as a dashed line). Finally, you should see that event 2 is reached before event 3 and that the total expected time through the network amounts to 5 units and that there is *slack time* associated with activity B amounting to 1 unit of time (in other words, activity B can be performed up to 1 unit of time late without affecting the expected time through the network).

Critical path

Within a network diagram, the *critical path* is the path that links the activities which have the greatest expected time. Consider Figure 1.3.2. This diagram shows six events linked by six real activities plus one dummy activity. The relationship between the activities and their expected times can be illustrated in the form of a table:

Activity	Preceding activity	Expected time
A_1	none	4
A_2	A_1	3
B	none	3
C	B	5
D	B	2
E	A_2, C, D	1

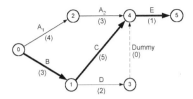

Figure 1.3.2 *Simple network diagram showing the critical path*

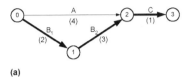

(a)

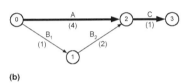

(b)

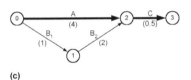

(c)

Figure 1.3.3 *Using the network diagram to reduce overall project time*

The critical path constitutes activities B, C and E which produce a total expected time, between event 0 (the start of the project) and event 5 (the completion of the project) of 9 units.

The critical path allows us to identify those activities that are critical. By reducing the time spent on activities along the critical path we can reduce the expected time for the complete project. Next consider the network diagram shown in Figure 1.3.3.

Figure 1.3.3(a) shows a network diagram in which there are four events, 0–3, and four activities, A, B_1, B_2 and C. The critical path links together activities B_1, B_2 and C resulting in an expected completion time of 6 units. Let's assume that we can put some additional resources into activities B_1 and B_2 and that this reduces the time spent on these tasks to 1 and 2 units, respectively. The critical path will move and it will now link activities A and C. The expected completion time will be reduced to 5 units. Our next task is to review tasks A and C to see if there is any way of reducing the time spent on these activities. Let's assume that we cannot reduce the time spent on activity A but we are able to reduce activity C by 0.5 unit. This leaves the critical path unchanged but the expected time for completion will now be 4.5 units. This process is typical of that used on larger much more complex projects. It is, however, important to note that there is a trade-off between time and cost. We shall examine this next.

Project costs

Projects involve two types of cost; *indirect project costs* and *direct activity costs*. Indirect costs include items such as administrative overheads and facilities costs (heating, lighting, etc.). Direct costs are concerned with additional labour costs, equipment leasing, etc. We can spend extra money to reduce the time taken on the project however, this only makes sense up to the point where further direct cost expenditure (such as the cost of employing additional contract staff) becomes equal to the savings in indirect project costs (heating, lighting and other overheads).

To examine the trade-off between project time and costs, we need to have the following information:

(1)   A network diagram for the project showing expected times and indicating the initial critical path. We also need to know the minimum time for each activity when there are no resource constraints (this is known as the *crash time*).

(2)   Cost estimates for each project activity expressed in terms of indirect expenditure per unit time.

(3)   The costs of providing additional resources for each project activity and the consequent time saving expressed in terms of expenditure per unit time reduction.

With the above information we can reduce the critical path activity times, beginning with the activity that offers the least expenditure per unit time reduction. We can then continue with the second least costly, and continue until we are left with the most costly until either we reach the target minimum time for the project or the additional direct cost expenditure becomes equal to the savings in indirect costs.

Next we shall look at time analysis in a little more detail.

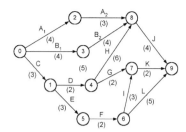

Figure 1.3.4 *See Question 1.3.1*

Problems 1.3.1

(1) The following data refers to the activities that make up a project. Use this information to construct a network diagram and identify the critical path.

Activity	Preceding activity	Expected time
A_1	none	1
A_2	A_1	2
B_1	none	2
B_2	B_1	3
C	A_2, B_2	1
D	C	5
E	C	3
F	D, E	1

(2) Determine the critical path through the network shown in Figure 1.3.4.

Determine the expected time to complete the project.

Project time analysis using PERT

PERT defines a number of important times in the project life cycle. These are as follows:

Expected time	t_e	The expected time for an activity is simply the average time for the activity.
Optimistic time	t_o	This is fastest time for the completion of the activity. This time will rarely be achieved and will only be bettered under exceptionally favourable circumstances.
Pessimistic time	t_p	This is slowest time for the completion of the activity. This time will nearly always be bettered and will only be exceeded under exceptionally unfavourable circumstances.
Most likely time	t_m	This time represents the 'best guess' time for the completion of the activity. This time is the statistical mode of the distribution of the times for the activity.
Earliest expected time	T_E	The earliest expected time for a particular event is the sum of all of the expected times (t_e) that lead up to the event in question.
Latest allowable time	T_L	The latest allowable time for a particular event is the latest time that the event can take place yet still allow the project to be completed on time.
Slack time	T_S	The slack time is the difference between the earliest expected time (T_E) and the latest allowable time (T_L). In effect, it is the amount of time that can elapse after completing one activity and starting another whilst still allowing the project to be completed on time. Note that, by definition, there is no slack time when following the critical path through a network.

Estimates of project times are often based on previous experience of performing similar tasks and activities. The expected time, t_e, is often calculated from the formula:

$$t_e = \frac{(t_o + 4t_m + t_p)}{6}$$

whilst the variance of an activity time can be determined from:

$$\sigma^2 = \left(\frac{t_p - t_o}{6}\right)^2.$$

Detailed time analysis using PERT can best be illustrated by taking an example:

Example 1.3.1

KarKare has engaged a consultant to advise on the construction of a new production facility for its trolley jack. The consultant has identified the following sequence of activities:

Ref.	Activity	Preceding activity	Time estimate (weeks)		
			t_o	t_m	t_p
A	Outline design	none	4	5	6
B	Prepare cost estimates	A	2	2	3
C	Client review	B	2	3	4
D	Planning application	C	8	12	14
E	Site investigation	C	2	3	4
F	Detailed plans	C	4	5	6
G	Building regulations	F	6	7	8
H	Tendering	F	4	5	6
I	Client approval	H	1	2	4
J	Construction	I	30	34	44
K	Fitting out	J	1	2	3
L	Client handover	K	1	2	3

(i) Draw the network diagram for the project.
(ii) Determine the expected time for each activity and mark this on the network diagram.
(iii) Determine the critical path and mark this on the network diagram.
(iv) Determine the earliest expected time and latest permissible time at each node of the network and use this to determine the slack time for each event.

Figure 1.3.5 shows the network diagram and the critical path for KarKare's new facility.

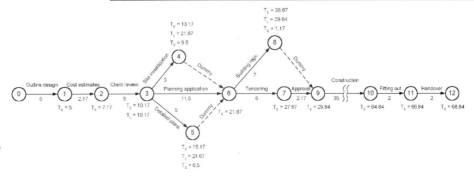

Figure 1.3.5 *See Example 1.3.1*

Question 1.3.2

The Board of Directors at Centralux has decided to replace the company's ageing telephone system. The following events and activities have been identified as forming part of this project:

Ref.	Activity	Preceding activity	Time estimate (weeks) t_o	t_m	t_p
A	Prepare a specification	none	1	1	2
B	Identify potential suppliers	A	1	1	2
C	Plan new system location	A	2	3	4
D	Tendering	B	4	5	6
E	Board approval	C, D	1	1	2
F	New system built by supplier	E	5	6	8
G	New system delivered and installed	F	1	2	3
H	New system commissioned	G	1	2	3
I	Training	G	1	1	2
J	Handover to Centralux	H, I	1	1	2
K	Allocate new numbers	E	1	2	4
L	Prepare new telephone directory	K	1	2	4

(i) Draw the network diagram for the project.
(ii) Determine the expected time for each activity and mark this on the network diagram.
(iii) Determine the critical path and mark this in the network diagram.
(iv) Determine the earliest expected time and latest permissible time at each node in the network and use this to determine the slack time for each event.

Symbol	Meaning
[	Start of an activity
]	End of an activity
[———]	Actual progress of an activity
▨▨▨▨□	(alternative representation)
V	Time now

Figure 1.3.6 *Symbols used in Gantt charts*

Gantt charts

A Gantt chart is simply a bar chart that shows the relationship of activities over a period of time. When constructing a Gantt chart, activities are listed down the page whilst time runs along the horizontal axis. The standard symbols used to denote the start and end of activities, and the progress towards their completion, are shown in Figure 1.3.6.

A simple Gantt chart is shown in Figure 1.3.7. This chart depicts the relationship between activities A and F. The horizontal scale is marked off in intervals of 1 day, with the whole project completed by day 18. At the start of the eighth day (see *time now*) the following situation is evident:

- Activity A has been completed;
- Activity B has been partly completed but is running behind schedule by two days;
- Activity C has been partly completed and is running ahead of schedule by one day;
- Activity D is yet to start;
- Activity E has started and is on schedule;
- Activity F is yet to start.

Let's move on to a more realistic example. The Board of Directors at Centralux has decided to relocate their main office. The Gantt chart

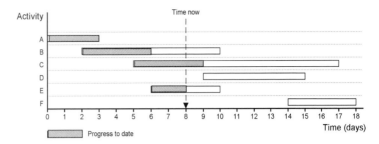

Figure 1.3.7 *A simple Gantt chart*

for this project is shown in Figure 1.3.8. This chart has been produced by a specialist project management software package. Note the following:

- The entire project is expected to be completed in a time period of eight months;
- The chart identifies key personnel involved in the project;
- Arrows have been added to indicate the flow of activities;
- The first part of project involved searches, visits and surveys (equivalent to the design phase of a manufacturing project);
- The Board of Directors take the initial decision to engage in the project and later authorise the next (legal/financial) phase of the project which has major financial implications.

Figure 1.3.8 *Gantt chart for Centralux's head office relocation*

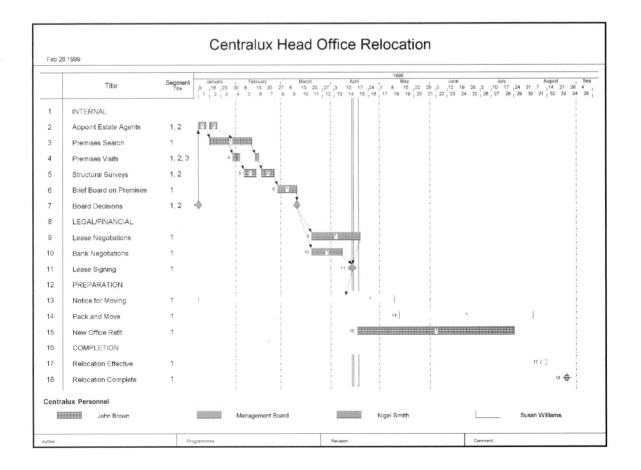

Figure 1.3.9 *See Question 1.3.3*

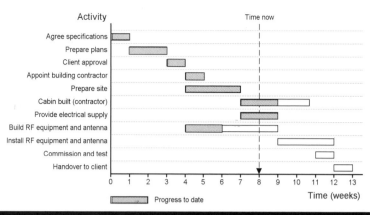

Question 1.3.3

Centralux has decided to install a site radio communication system. The company has selected a supplier, Bizcom, who will carry out the work of building, installing and commissioning the radio equipment. At the beginning of the eighth week of the project, Centralux's Managing Director has asked Bizcom to supply a Gantt chart showing progress to date (see Figure 1.3.9). Use the Gantt chart to identify:

(a) the number of weeks required to complete the project;
(b) activities that have been completed;
(c) activities that have been completed ahead of schedule;
(d) activities that are behind schedule (and how far each is behind).

Question 1.3.4

KarKare is developing a car roof box mounted on a metal rack. This new product will go into production at the end of the current year (1998). The Gantt chart for this project is shown in Figure 1.3.10. Use the Gantt chart to identify:

(a) the number of weeks required to complete the project;
(b) activities that are due to be completed at the end of the current month (July);
(c) activities that are ahead of schedule but not yet completed;
(d) activities that are more than two weeks behind schedule.

Figure 1.3.10 *See Question 1.3.4*

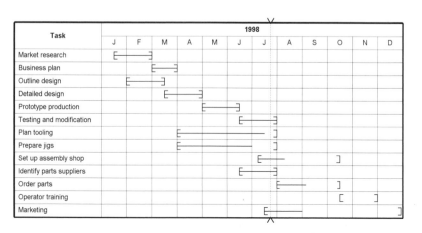

2 Engineering design

Summary

The aim of this unit is to give students an opportunity to experience the various phases of carrying out a design project, including the preparation of a detailed design specification and the production of a comprehensive design report. Computer technology is used extensively in engineering design and students are expected to demonstrate an ability to use appropriate hardware and software as part of the engineering design process.

Introduction

Design itself is a complicated activity requiring individual inputs from a whole group of people, who go to make up the *design team*. The total design process requires the design team to consider activities that include; market forces, design management, design specification, concept design, layout design, and detail design, materials technology, manufacturing methods, financial and legal aspects, publicity and sales. The total design process requires us to first analyse the design requirements, and then, to synthesise the many diverse design parameters so that, apart from the simplest of design tasks, the whole design process requires the combined skills and knowledge of a design team.

In this chapter we will be concentrating on those aspects of engineering design, which enable us not only to produce engineering components, but also to consider some of the problems associated with the design of engineering systems. We will be very much concerned with looking at the overall design process and the production of a design report. The use of computers in engineering design, will also be considered.

We start by examining the essential components of a *design specification*, since it is this document that forms the essential reference point for all the activities concerned with the engineering design process.

Having established the nature of the design specification, another essential task for engineers, is to present their ideas to company management. It is no use achieving design perfection, if we are unable to persuade those in authority that our ideas are viable! To help us achieve this aim, we look at how to prepare and present the *design report*.

Computer technology has been an integral part of engineering design for many years. So our final topic is concerned with the use of computers in the overall design process and in particular, the use of computer packages for drafting, project scheduling and mathematical analysis.

2.1 The design specification

Introduction

As an introduction to specification writing consider the following modified paragraphs taken from BS7373 *Guide to the preparation of specifications*. Note that we are primarily concerned here with the product design specification (PDS), that is the specification which conveys the designers description of the product.

A specification is essentially a means of communicating the needs or intentions of one party to another. It may be a user's description, to a designer (a brief), detailing requirements for purpose or duty; or it may be a designer's detailed description to the operator, indicating manufacturing detail, materials and manufacturing tolerances; or it may be a statement, by a sales person, describing fitness for purpose to fulfil the need of a user or possible user. It may, of course, be some or all of these in one.

The contents of the specification will, therefore, vary according to whether it is primarily from the using, designing, manufacturing or selling aspect. Specifications will also vary according to the type of material, or component being considered, ranging from a brief specification for a simple component to a comprehensive specification for a complex assembly or engineering system. We are primarily concerned here, in preparing a design specification from the point of view of the designer attempting to meet customer needs.

The requirements of the specification should be written in terms of describing the optimum quality for the job, not necessarily the highest quality. It is usually unwise to over specify requirements beyond those for a known purpose. It is costly and restrictive to seek more refinements than those necessary for the function required. The aim, should therefore be, to produce a minimum statement of optimum quality in order not to increase costs unnecessarily; not to restrict processes of manufacture and; not to limit the use of possible materials.

As already mentioned the design specification has to take into account parameters such as function, performance, cost, aesthetics and production problems, all of these issues are concerned with *customer requirements* and this has to be the major consideration when producing a design specification. The major *design parameters* such as; layout, materials, erection methods, transportation, safety, manufacture, fabrication and legal implications; must also be considered when producing the design specification. Finally, all *design information* must be extracted from appropriate sources, such as British Standards and International Standards, and all legislative requirements concerning processes, quality assurance and the use of new technologies must be applied.

Customer requirements

In order to provide a successful winning design, it is essential that the requirements of the customer are met whenever possible. Because we wish to produce a specification which expresses our customer requirements it is most important that we spend time in consultation with them, to ensure that customer needs are well understood and, if necessary, to agree amendments or reach a compromise dependent on circumstances.

Remember that a specification is essentially a listing of all the parameters essential to the design. Then, generally, a customer will list essential values as part of his/her requirements, but each value must be examined before transfer to the specification. Thus the design specification must always be formulated by the designer.

To illustrate the customer requirements for a particular engineering design, and the role of the designer in interpreting such requirements, consider the following example:

Example 2.1.1

A potential customer approaches you, as a designer, with a *brief* for the design of an electric drill. In order to secure the job, you need to produce a comprehensive specification, which takes into account all of your customer's requirements which are listed below:

A. B. Brown Engineering
Outline specification for electric drill

Performance: Capable of taking drill bits up to 0.75 inch diameter.
Operate from 240 volt, 50 Hz power supply.
Capable of two speed operation.
Have hammer action.
Operate continuously for long periods of time.
Suitable for soft and hard drilling.
Eccentricity of drilling action must be limited to 0.01 inches, for drill bits up to 12 inches in length.
Have a minimum cable reach of 5 metres.

Environment: Able to operate internally and externally, within a temperature range of -20 to $+40$°C.
Have no adverse effects from dirt, dust or ingress of oil or grease.
Capable of operation in wet conditions.
Capable of operation where combustible dusts are present.

Maintenance: Capable of being dismantled into component parts, for ease of maintenance.
Require no special tools, for dismantling/assembly operations.
Component parts to last a minimum of two years, before requiring replacement or rectification.

Costs: To cost a maximum of £60.00

Quantity: Two thousand required from first production run.

Aesthetics/ ergonomics: Polymer body shell with two colour finish.
Pistol grip lower body, and upper body steady handle.
Metal chuck assembly, with chrome finish.

Size/weight: Maximum weight of 3 kg.
Overall length not to exceed 30 cm.

cont.

Safety: Complies with all relevant British Standards.

As designers we must ensure that the customer's requirements are unambiguous, complete and attainable. This is where the dialogue with the customer begins! For the purpose of this example, let us re-visit each of the requirements.

The title for the specification needs clarification, is electric drill, a suitable title? We know that A. B. Brown Engineering have specified information about type of grip and body design, so perhaps a more accurate description might be: *electric hand drill (mains operated)*, since cable length and supply conditions have been mentioned.

Performance

If we consider the *performance* requirements, we note that in certain areas they are ambiguous and generally incomplete. Is the electric hand drill to be made available for export? The metric equivalent of 0.75 inches may be necessary, in any event it is a good marketing ploy to ensure that all dimensions are available in Imperial and SI units, this appeals to both the European and USA markets. Thus to ensure ease of production of the 'chuck' a metric equivalent should be given.

Are the supply details and power requirements for the drill sufficient? There is, for example, no indication of the power requirements for the drill motor, this must be given together with details of the supply, that is, alternating current. The rpm of the two speed operation must also be given, this will depend on service loading, time in use and materials to be drilled.

Statements such as 'operate continuously for long periods of time' and 'suitable for soft and hard drilling' should be avoided. What periods of time? What type of materials are required to be drilled?

Although not directly obvious, the eccentricity requirements will have an effect on the quality of the gearing and the type of bearing required for the drive spindle. High quality bearings are expensive and this will need to be taken into account when designing to a maximum cost.

Environment

The criteria for the operating environment are quite clear, however, they do have quite serious consequences for the design. To be able to operate the drill in quite harsh external conditions, the insulation for the plug and cable assembly will need to meet stringent standards. Motor insulation and protection, will be required to insure that sparking and arcing does not occur when the drill is being used in a combustible gas atmosphere (refer to BS 4999, BS 5000, BS 5501 and BS 6467). Suitable motor caging will be required to ensure that the ingress of dirt, dust, oil and grease do not adversely affect the performance of the drill motor.

Maintenance

The number and nature of component parts needs to be established, prior to any detail design being carried out. There are obvious cost implications if all component parts are to last a

cont.

minimum of two years. This needs clarification to ascertain whether we are only talking about major mechanical parts or genuinely all parts. The likelihood of failure dependent on service use also needs to be carefully established to determine the feasibility of this requirement.

Costs
The viability of this figure needs to be determined by taking into account the costs of component parts, tooling requirements, fabrication costs, machining costs based on required tolerances and materials finishing and environmental protection costs.

Quantity
This will determine the type of manufacturing process, and whether or not it is necessary to lay-on additional tooling, or buy-in standard parts and concentrate only on assembly and test facilities, for the production run. Future component numbers will need to be established in order to make predictions about the most cost-effective production process.

Aesthetics/ergonomics
Is the two colour finish absolutely necessary? This will depend on target market and results of consumer research, which will need to be known by the designer, prior to determining the unit cost. Chrome finishing is an expensive process and not altogether suitable for a drill chuck, which will be subject to harsh treatment in a hostile environment. Knocks, dents, scratches and pollution in the work environment would quickly affect the chrome protective coating. Consideration needs to be given to alternative materials, which provide good corrosion protection and durability, as well as looking aesthetically pleasing.

Size/weight
The weight and size criteria are not overly restrictive and allow the designer some room for manoeuvre. Light alloys and polymers may be used to help keep weight down, providing the performance criteria are not compromised.

Safety
There is a need to establish whether or not the product is intended for the European and/or World market. European legislation already has a major influence on safety standards, particularly relating to electrical goods. So ISO standards, European legislation and other relevant quality standards would need to be followed. Is it, for instance, the intention of A. B. Brown Engineering to provide an electric hand drill which is capable of operation from the continental 220V supply?

Having considered the customer's requirements and obtained answers to the questions posed, we need to ensure that all design parameters have been covered, prior to the production of the specification. For our example there is a need to include more design requirements on materials specification, component function, testing, prototype production, and timescales.

Question 2.1.1

Based on the information provided in Example 2.1.1, re-write the specification for A. B. Brown Engineering, avoiding ambiguous information and adding design requirements for materials specification, component function and timescales. For the purpose of this exercise, avoid reference to British Standards or other sources of information. Consider only the argument offered in the example and information supplied in text books. You should assume that the electric hand drill is to be designed for the home and European market.

Since the design requirements, their associated design parameters and the requirements of the customer, are inextricably linked, we next look in more detail at the design requirements and parameters necessary to produce a comprehensive product design specification (PDS). At this point, it is worth remembering that it may not always be necessary to include *all* requirements since these will be dependent on the complexity (or otherwise) of the engineering component or system.

The PDS and design requirements

Product design specification (PDS)

The designer's description of the product is presented in the form of a PDS. The structure of the PDS is again mentioned in PD 6112 and this should be referred to for full details. A brief summary of what might be included, based on the information provided in PD 6112, is given below, this list is neither exclusive nor exhaustive, but it does act as a useful guide for those new to product design specification writing.

- Title: This should provide an informative unambiguous description of the product.
- Contents: A list of contents, which acts as a useful introduction to the document for the reader.
- Foreword: This sets the scene and provides the reader with useful background information concerned with the project and the customer's brief.
- Scope of specification: This section provides the reader with details on the extent of the coverage and the limitations imposed on the information provided. It also gives details on the function of the product or system under consideration.
- Consultation: Information on any authorities who must be consulted concerning the product's design and use. These might include the health and safety executive (HSE), fire service, patent office or other interested parties.
- Design: Main body of text detailing the design requirements/parameters for the product or system, such as; performance, ergonomics, materials, manufacture, maintenance, safety, packaging, transportation, etc.
- Appendices   Containing definitions that might include: complex terminology; abbreviations; symbols and units. Related information and references such as statutory regulations, British Standards, ISO standards, design journals, codes of practice etc.

Figure 2.1.1

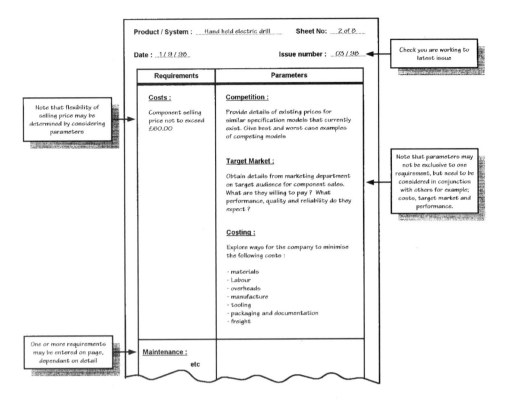

Figure 2.1.1 shows a typical layout for the 'design' element of a PDS, detailing the customer requirements and the parameters which might affect these requirements.

Design requirements

Information on some of the major design requirements, in the form of a list, are detailed below. You should ensure that you understand the significance of each of these and, in particular, you are aware of the important parameters associated with each.

Performance

You will already be aware of some typical performance parameters, when studying Example 2.1.1. The performance specified by customers, must be attainable and clearly defined. The performance required by A. B. Brown Engineering in our electric hand drill example, shows how easy it is to produce ambiguous performance criteria and leave out essential information.

A balance needs to be established between performance and costs, ultimate performance demanded by a customer is likely to be prohibitively expensive. Designers must be aware of the economic viability of meeting performance requirements and, where these are not feasible, dialogue must be entered into to seek a more cost-effective solution.

Over specification of performance is more likely to occur in specialist 'one-off' products where operational information is limited. Designers should draw on the knowledge and experience gained from similar designs, or seek the advice of more experienced colleagues. In any case, the urge to over design should be avoided, computer simulation, scale models or the use of prototypes, may be a way of establishing appropriate performance data, for very large one-off products, when trying to avoid over specification.

Ergonomics (or human factors)

The word ergonomics originates from the Greek, *ergos* – work, and *nomics* – natural laws, thus ergonomics literally means '*the natural laws of work*'. It first came into prevalence during the Second World War, when aircraft pilots confused, for example, the landing gear lever with the flap lever, which on occasions, resulted in disaster. Prior to the Second World War, little interest had been paid, by engineers, to ergonomic design. We only have to look at the arrangement of controls and displays used in old steam engines, power stations and heavy process equipment, to realise that little consideration had been given to the needs of the operator.

Ergonomic design has risen in prominence over the past decade and the needs of the user, have taken on much more importance. Consideration should now be given to one or more of the following ergonomic design parameters: controls and displays; instruments and tools; workspace arrangement; safety aspects; anthropometrics (measurement of physical characteristics of humans, in particular human dimensions); environment – visual, acoustic and thermal.

No matter how complex, sophisticated or ingenious the product, if human operation is involved, the ergonomics of the situation need careful consideration by the designer.

Environment

All aspects of the product or system's environment need to be considered. In our electric drill example, the operating environment was to include, dirt, dust, oil and grease, as well as combustible dust. This has a significant effect on the robustness of the design and the environmental protection required for safe and efficient use.

So factors such as, dirt, dust, oil, grease, chemical spillage, temperature, pressure, humidity, corrosives and other pollutants, animal infestation, vibration and noise should be considered. Apart from the effects on the designed product or system we, as engineers, should also take into account the likelihood of our design polluting the natural environment. For example, when considering the design of nuclear reactors, the vast majority of the design effort is focused on ensuring that fail-safe systems and back-up facilities exist. Thus minimising the possibility of a nuclear accident, which might result in an ecological disaster. Filtration systems to prevent the leakage of dangerous substances from, plant, machinery and vehicles, should also be introduced into the design, as a matter of routine.

Maintenance

When purchasing a domestic appliance like our electric drill, the ease with which it can be maintained and serviced is of importance to the average DIY enthusiast. Therefore, the ease with which parts can be obtained and the drill can be assembled/dismantled are important when considering design for maintenance. When deciding whether or not to design-in a significant amount of maintainability, there are several factors that need to determined. For example, we will need to know the likely market and establish their philosophy on servicing, repair and rectification. Estimates of component life will need to be found in order to assess the economic viability of repair, reconditioning, or, repair by replacement.

So when designing for ease of maintenance, we need to consider the extra costs involved in using more sophisticated manufacturing processes and component parts, and balance these against customer satisfaction with the finished product or system. The service life of the product and the life of component parts, also needs to be established in order to make informed decisions about the need for maintenance.

Costs

A realistic product cost needs to be established as early as possible in the negotiating process. Estimates for products and systems are often set lower than reality dictates, because of the need to gain the competitive edge. However, it is no use accepting or giving unrealistic estimates which are likely to put a business into debt. Costing has become a science and design engineers need to be aware of how to accurately estimate costs and financially evaluate the viability of new ventures. Because of the importance of finance, more in-depth information on costs and costing is given later in this chapter.

Transportation

Think of the consequences of designing and assembling a very large 100 ton transformer in Newcastle which is required in Penzance! The cost and practicalities of transporting such a monster, would make the task prohibitively expensive, if not impossible. Transportation becomes an issue when products that are very large and very heavy need to be moved. Thus when faced with these problems as a design engineer due consideration needs to be given to ease of transportation and packaging. Small items such as our electric drill can be packaged and transported by rail, road, sea or air, with relative ease. Size and weight restrictions must always be pre-determined, to ensure that the product will fit into the space allocated by the customer.

If, for example, we are designing, installing and commissioning an air-conditioning system for a large hotel in Cairo. Then due consideration must be given, at the design stage, to ease of transportation and assembly. In fact, large structures and systems are designed in kit form and dry-assembled, this ensures that all component parts are available, that they fit together, and that the installation does not exceed required dimensions, prior to shipment. Thus enabling the product or system to be easily installed and commissioned on sight and preventing any unnecessary transportation costs from being incurred.

Manufacture

Due consideration needs to be given at the design stage, to the ease of manufacture of products and their associated parts. The cost implications of 'over-engineering' must be remembered when designing, particularly with respect to design detail. Component parts should be designed with the over-riding thought of saving costs. All non-functional features and trimmings from part should be omitted. For example, do not design-in radii if a square corner will do and do not waste money on extra machining operations, if stock-size material is available and acceptable.

The cost and complexity of production methods also need to be considered, at the design stage. One-off items are likely to require 'jobbing production', which is expensive in time and requires a high degree of skill. Batch production and mass flow production may need to be considered, always remembering the facilities available on customer premises.

Should items be bought-in, assembled using standard parts, or manufactured on-site? This will again depend on the philosophy adopted by the customer and the manufacturing facilities and tooling available.

Aesthetics

Once all the primary requirements regarding function, safety, use and economy have been fulfilled. The aim of designers is to create products that appeal to customers, thus industrial design lies somewhere between engineering and art. Consideration needs to be given to shape or form, the use of colour and surface texture.

The aesthetic design of an engineering product or system, may be outside the remit of the engineering designer. Specialists such as graphic artists, could very easily be seconded to the design team to assist with product aesthetics, packaging, labelling and graphics, for a particular market.

Legal implications

Designers will need to be aware of the legal constraints involved with any particular product or system. With the advent of more and more EEC legislation concerning product liability, disposal of toxic waste, COSHH and general Health and Safety. All legal aspects must be considered during the early stages of the design process.

The fundamentals of the law regarding patents and copyright should be understood by the engineering designer. A patent, for example, does not stop anybody using your idea, it merely provides a channel for redress. The incentive gained by patenting a product normally results in the item being made and marketed, knowing that legal protection is offered. New industrial designs are protected under the provisions of the Copyright Act (1968) in the UK. This protection is offered without any form of registration and is valid for 15 years from the date of manufacture (in quantity) or when first marketed.

Some knowledge of the law of contract is also very useful for the engineering designer. A contract is a formal written agreement between two parties. Both parties agree to abide by the conditions that are laid down, in all respects. However, certain 'let-out' clauses may be included in order to accommodate unforeseen difficulties.

Safety

Safety requirements vary according to the product or system being designed and the use to which they will be put. Many areas of engineering exist within highly regulated industries, where safety is of paramount importance, for example, nuclear power, petro-chemical, aircraft operation, hospitals and the emergency services. For these industries and many others, adherence to HASAWA, COSHH regulations, and BS for product liability will form an essential part of the design process.

Quality

To ensure quality in design, the design methods adopted, technical documentation, review processes, testing and close co-operation with the customer; must be such that: the performance, reliability, maintainability, safety, produceability, standardisation, interchange-ability and cost are those required.

Many companies have achieved the Total Quality Management System kite mark, in that they have gained BS5750 or the ISO 9000 series equivalent. To obtain such a standard requires a company to establish documented methods of control of quality-affecting activities, training personnel in these methods, implementing these methods, verifying implementation, measuring effectiveness and identifying and correcting problems to prevent reoccurrence, in short good business practice.

Quality control (QC) is the final part of the total quality (TQ) process, if the TQ process is effective QC will be less and less needed, until eventually the ultimate *right first time* is achieved.

We have spent some time explaining the nature of the design parameters, although rather tedious, this should be treated as essential learning.

Question 2.1.2

Your company receives the undermentioned customer's description (design brief) for a 'hydraulic hose connector'. From the customer's brief, using appropriate sources of reference, produce a product design specification (PDS) which meets all essential requirements and constraints. Your specification should be written following the format given earlier. Under the 'design' heading you should include all design requirements you think necessary. The style of the PDS should enable it to be used for communication between your company (via you, the design engineer) and your customer who wishes to produce the new hose connectors.

At this time, and for the purpose of this question, you do not need to include estimates of costs, timescales or sketches of alternative design solutions. You will be asked to complete these exercises later!

Customer's brief

We currently produce portable hydraulic power packs used on a range of agricultural machinery. Our current design of flexible hose fittings used with the pack, have caused difficulties when removing/fitting the power pack from/to the machines during maintenance.

Hose end connectors are required that will allow the attachment and removal of hoses to be carried out quickly and safely, without undue leakage of hydraulic fluid. The fittings will need to accommodate 10 mm, 14 mm, 20 mm and 30 mm flexible wire reinforced rubber hoses, capable of withstanding pressures up to 50 bar. The fittings must be robust enough to withstand harsh agricultural environmental conditions.

2.2 The design report

Introduction

When a professional engineer carries out a design project, a note book is kept in which are recorded the initial specification, design requirements, design parameters, alternative solutions, ideas, test requirements and results, calculations, general schemes, references, contacts and a host of other related information. The contents of the note book are primarily for the design engineers use and serve as an essential reference during the lifetime of the project.

When the project is completed, the designer will generally convey his ideas by producing a report, which contains the all important 'description of the final design' together with layout drawings. The information required for the report being obtained and transferred from the designer's note book. The reason for writing a report is to present findings in a readily understood form which would allow any reader to appreciate the nature of a particular design and, to allow people the opportunity to comment on and make suggestions for possible design improvements, should the specification change or the need arise.

The production of the design report is the culmination of the design process, and it forms an important part of the final design documentation. It requires a high degree of intellectual ability in order to analyse all the major design parameters, draw conclusions, make judgements and synthesise all the parts into a coherent, logical and effective design solution, which is to be presented in the form of a design report. Some of the techniques, which have been designed to assist these thought processes, are described in the following section.

The design process

Before considering the design report itself, it will be useful to look at the overall design process and determine some logical order and structure, in which to proceed from the design problem to a possible design solution.

Figure 2.2.1 provides an overview of the design process showing one possible way in which to reach a satisfactory solution. We start by considering *the design problem*, this may be presented in the form of a customer's brief, or it may come as an idea for improving an existing in-house artefact. Tackling the design problem requires us to establish exact customer requirements, determine the major design parameters, research and obtain design information from appropriate sources and prepare and produce *the design specification*.

Using the design specification as our guide, we next need to prepare an analysis of possible design solutions, produce concept designs, evaluate alternative concepts and select an *optimum design solution*. From the *conceptual design* phase, where engineering principles are established, we move to the *layout design* or general arrangement design. Here we are concerned with the selection of appropriate materials, determining the design of the preliminary form, checking for errors, disturbances and minimum costs and producing the *definitive layout and design report*. The final stage of the design process, is the *detail design*, where we are concerned with the arrangement, dimensions, tolerances, surface finish, materials specification, detail drawings, assembly drawings and production costs, of the individual parts of the product or system being designed.

Determining possible design solutions

In our search for an optimum solution to an engineering design problem, there are many methods available, which will help us to find such a solution. Which method should be used in which situation, will depend upon the nature of the problem, the magnitude of the task, the information available and the skill, knowledge and experience of the designers.

All solution finding methods are designed to encourage lateral thinking and foster an open-minded approach to problem solving. These methods may be conveniently divided into *general methods* and *problem specific* methods. The former are not linked to a specific part of the design process or to a particular product or system. They do, however, enable us to search for solutions to general problems that arise throughout the design process.

Figure 2.2.1

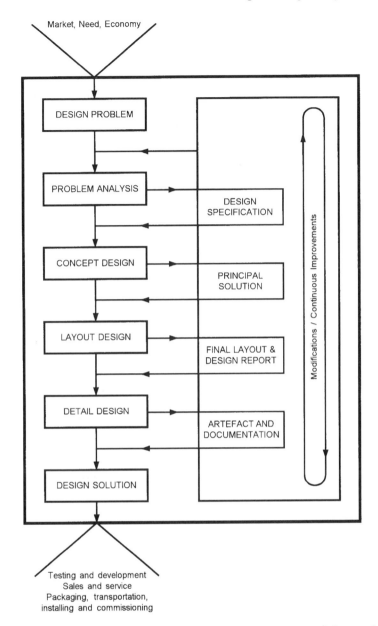

Problem specific methods, as their name suggests, can only be used for specific tasks, for example to estimate costs or determine buckling capability of specified materials. In this section we will consider two examples of general methods; *brainstorming* and the *systematic search method.*

Brainstorming

The object of brainstorming is to generate a flood of new ideas, it is often used when there is a feeling that matters are becoming desperate. It involves using a multidisciplinary team of people, with diverse backgrounds, who are brought together to offer differing perspectives to the generation of ideas. This method is particularly useful where no feasible solution principle has been found; where a radical departure from the conventional approach is considered necessary or, where deadlock has been reached.

The group will normally consist of between 6 and 15 people drawn from a diverse variety of backgrounds, and must not be limited to specialists. They are formally asked to focus their attention on a specific problem, or group of problems, in order to rapidly generate ideas for a solution. This technique was originally suggested by Alex Osborn, during his time in the advertising industry. He devised the set of criteria given below, in order to ensure that all participants were given equal opportunity to express themselves freely without inhibition.

- The leader of the group should be responsible for dealing with organisational issues and outlining the problem, prior to the start of the brainstorming session. The leader should also ensure that all new ideas are encouraged and that no one criticises the ideas of other group members.
- All ideas are to be accepted by all participants, no matter how absurd, frivolous or bizarre, they may seem.
- All ideas should be written down, sketched out, or recorded for future reference.
- Building on the ideas of others to create a group chain-reaction should be encouraged.
- The practicality of any suggestion, should be ignored at first, and judged later.
- Sessions should be limited to less than one hour. Longer sessions tend to cause participant fatigue and the repetition of ideas.
- The results should be reviewed and evaluated by experts, to find potential solutions to design problem/s.
- After classification and grading of the ideas, one or more suggested final solutions should be presented again to the group for interpretation, comment and feedback.

Systematic search method

This method relies on a mechanistic approach to the generation of ideas, through the systematic presentation of data. Data is often presented in the form of a *classification scheme*, which enables the designer to identify and combine criteria to aid the design solution. The choice of classifying criteria and their associated parameters requires careful thought, since they are of crucial importance. In order to illustrate this method, consider the following example:

Example 2.2.1

An engineering system is required to operate the ailerons of a light aircraft. The ailerons must be capable of being operated by the pilot from the cockpit. Use the systematic search method to produce design ideas for possible motion converters for the required aileron system. Figure 2.2.2 illustrates the required output motion for the system.

Possible motion parameters might include, linear to linear, linear to rotary, rotary to linear, rotary to rotary and oscillatory, all of these forms of motion are sub-sets of translational (linear) and rotational motion. In order to provide design ideas for possible motion converters, a classification scheme needs to be produced. The first attempt is shown below in Figure 2.2.3 the column headings indicate the motion parameters and the solution proposals are entered in the rows.

cont.

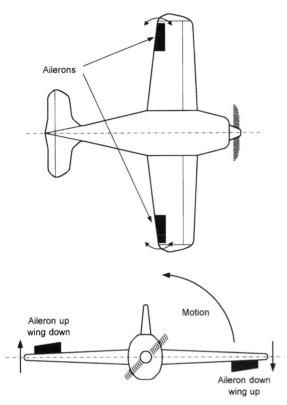

Motion parameters / Solution proposals	Translational and Rotary			
	Linear to Linear	Linear/Rotary Rotary/Linear	Rotary to Rotary	Rotary/Linear to Oscillatory or Intermittent
1. Gears		Worm and wheel		Driven / Driver
2. Cables/Belts/Pulleys	Cable Conduit Fixed			
3. Cams/Followers				
4. Chains/Sprockets				Geneva wheel
5. Actuators/Rams	Mass			
6. Screws				
7. Linkages/Rods/Levers			Universal coupling	

Figure 2.2.2 *(above left)*

Figure 2.2.3 *(above right)*

Ailerons

Aileron up wing down

Motion

Aileron down wing up

In my classification scheme illustrated above, you will note that gears are included as a possible motion converter. In fact improvements to our scheme could be made by sub-dividing the solution proposals. For example, gears, could be sub-divided into spur and bevel gears, epicyclic gears, harmonic drives, worm gears, helical gears. Note that our motion parameters have already been sub-divided from translational and rotary motion. This process of layering our classification scheme, enables us to produce many varied design proposals. No mention has been made of energy sources for the system, this has been left for you as an exercise.

Question 2.2.1

For Example 2.2.1, sub-divide cables/belts/pulleys, actuators/rams and linkages/rods/levers into as many different variants as you can. Reference to standard engineering design text books and specialist texts on mechanisms, systems and machine design, should prove useful.

Question 2.2.2

List the possible energy sources that might be used to provide the input power for the aileron system described earlier. Also give solution proposals for the possible types of energy provider, that might be used to power the system.

There are many other solution finding methods available, which should be considered in addition to those mentioned. These include: literature search, analysis of natural and existing technical systems, model testing, galley method and many more. As design engineers, all these methods should be familiar to you, time does not permit a full study of them all in this chapter, but further information is available from the reference material given at the end of this book.

Selecting and evaluating possible design solutions

Until now, we have been concerned with generating ideas for possible design proposals and we have looked in detail at one or two methods which help us to produce design concepts. It is now time to consider ways in which we can evaluate and so select the best of these solution variants.

The evaluation matrix

One of the most common methods is to produce an *evaluation matrix*, where each solution concept is set against a list of selection criteria. For each criteria some kind of scoring system is used to indicate how the individual design concept compares with an agreed norm. This process is illustrated in Figure 2.2.4.

Prior to inclusion in the evaluation matrix, if there are a large number of solution proposals, the design engineer should produce some form of pre-selection procedure, in order to reduce the proposals to a manageable size. The pre-selection process being based on fundamental criteria which the design proposal must meet. Such criteria might include: compatibility with required task, meets the demands of the design specification, feasible in respect of performance, etc. meets mandatory safety requirements and expected to be within agreed costs.

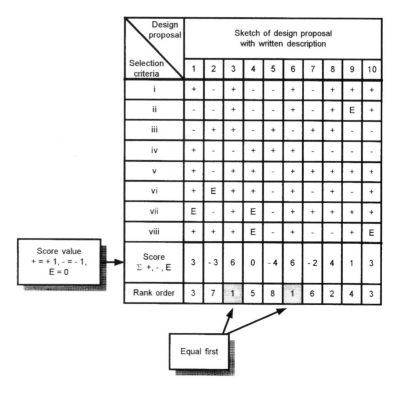

Design proposal / Selection criteria	Sketch of design proposal with written description										
	1	2	3	4	5	6	7	8	9	10
i	+	-	+	-	-	+	-	+	+	+
ii	-	-	+	-	-	+	-	+	E	+
iii	-	+	+	-	+	-	+	+	-	-
iv	+	-	-	+	+	+	-	-	-	-
v	+	-	+	+	-	+	+	+	+	+
vi	+	E	+	+	-	+	-	+	-	+
vii	E	-	+	E	-	+	+	+	+	+
viii	+	+	+	E	-	+	-	-	+	E
Score Σ +, -, E	3	-3	6	0	-4	6	-2	4	1	3
Rank order	3	7	1	5	8	1	6	2	4	3

Score value
+ = + 1, - = - 1,
E = 0

Equal first

Figure 2.2.4

The concept proposals should be in the form of sketches together with a short written explanation. Equality with the agreed norm is shown in the skeleton matrix by the letter 'E', if the design solution is considered better than the norm, in some way, then a + sign is used, conversely a − sign is used, if the design solution is worse than the norm, in some way. A score may be obtained by allocating a +1 to the positives, − 1 to the negatives and 0 to the Es. More sophisticated scoring systems may be used involving 'weightings', when the selection criteria are not considered to be of equal importance.

Example 2.2.2

Assume that you are to manufacture a large diameter flywheel, for a heavy pressing machine, at minimum cost.

(i) Write a short list of selection criteria against which the given design solutions can be evaluated.
(ii) Produce an evaluation matrix using the 'casting method' of manufacture as your norm and, rank all the remaining design solutions.

cont.

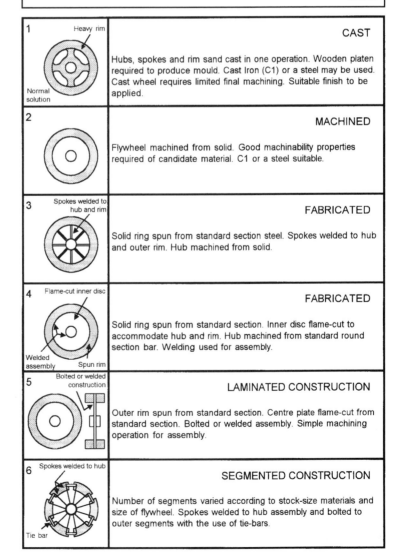

1	Heavy rim / Normal solution	CAST
		Hubs, spokes and rim sand cast in one operation. Wooden platen required to produce mould. Cast Iron (C1) or a steel may be used. Cast wheel requires limited final machining. Suitable finish to be applied.

2		MACHINED
		Flywheel machined from solid. Good machinability properties required of candidate material. C1 or a steel suitable.

3	Spokes welded to hub and rim	FABRICATED
		Solid ring spun from standard section steel. Spokes welded to hub and outer rim. Hub machined from solid.

4	Flame-cut inner disc / Welded assembly / Spun rim	FABRICATED
		Solid ring spun from standard section. Inner disc flame-cut to accommodate hub and rim. Hub machined from standard round section bar. Welding used for assembly.

5	Bolted or welded construction	LAMINATED CONSTRUCTION
		Outer rim spun from standard section. Centre plate flame-cut from standard section. Bolted or welded assembly. Simple machining operation for assembly.

6	Spokes welded to hub / Tie bar	SEGMENTED CONSTRUCTION
		Number of segments varied according to stock-size materials and size of flywheel. Spokes welded to hub assembly and bolted to outer segments with the use of tie-bars.

Figure 2.2.5

Figure 2.2.5 shows the given design solutions, we now need to produce our evaluation criteria. We could use one or more of the previous methods, to generate ideas. However, for the purpose of this example, since cost is of paramount importance, we will just look closely at the manufacturing methods which minimise cost. We will assume that the requirements of the specification have been met and that all design alternatives are compatible for use with the pressing machine under all operating conditions.

Then for each of the design options, we need to consider:

- materials costs;
- skill and amount of labour required;
- complexity of construction;
- tooling costs;
- machining and finishing costs;
- safety (this will be related to the integrity of the design solution assuming it is chosen);
- amount of waste generated;
- company preference – knowledge, skills, equipment.

The above list of criteria is not exhaustive, but should enable us to select one or two preferred design alternatives. Further refinements/criteria may be necessary if two or more concepts are closely ranked.

Scoring

Figure 2.2.6 shows the completed evaluation matrix for this problem. You will note that the company preference, immediately skews the scores. The company does not have, or does not wish to use foundry facilities, in any case, the production of the mould would be prohibitively expensive for what appears to be a 'one-off' job. Obviously you, as the design

cont.

Selection criteria \ Design proposal	See Figure 2.2.5 for design proposal sketch and written description					
	1	2	3	4	5	6
Materials	+	E	–	–	E	+
Labour	–	–	+	+	+	–
Complexity of constraint	+	+	E	E	E	–
Tooling	–	–	E	+	+	+
Machining/finish	+	–	E	–	+	+
Safety	+	+	E	E	E	–
Waste generation	+	–	+	–	E	–
Company preference	–	+	+	+	+	+
Score Σ +, –, E	2	–1	2	0	4	3
Rank order	NORM	5	3	3	1	2

Figure 2.2.6

engineer, would be aware of these facts before evaluating the options. Note that options 2–6 all involve some form of fabrication, assembly or machining, which we will assume is the company preference.

Proposal 2: Machining parts for a heavy flywheel, will require several machining operations and the use of elaborate fixtures, not to mention operator skill for an object of such size, so labour, tooling and machining costs are relatively high. This process also involves a large amount of material waste.

Proposal 3: The major advantage of this method is that standard stock materials can be used. Difficulties include, the use of jigs and fixtures, weld decay and possibility of complicated heat treatments.

Proposal 4: Similar advantages and disadvantages to option 3.

Proposal 5: Advantages include use of standard stock materials, little machining required after assembly, relatively easy to assemble. Disadvantages include necessity for positive locking of bolts after assembly and outer rim would require skimming after spinning.

Proposal 6: Labour intensive fabrication and assembly, complex assembly and integrity of construction would raise safety issue. No specialist tooling required, finishing relatively easy and cheap, minimal waste from each machining operation, company preference.

Note that if company preference had been for casting, then option 1 would probably have been preferable, providing it met the cost requirements. Options 5 and 6 appear next to favourite, although option 3 might also be worth looking at again, dependent on the skills of the labour force.

If there was insufficient evidence on which to make a decision, then more selection criteria would need to be considered. For example, do the options just meet or exceed the design specification, bursting speeds and other safety criteria might have to be further investigated. This process would need to be adopted no matter what the artefact, an *iterative approach* being adopted, in an attempt to get ever closer to the optimum design solution.

Costing

The cost of an engineering component or system, is of paramount importance. Engineering designs require the specification to be met, the artefact to be produced on-time and *at the right cost*, if the design solution is to be successful. Thus an understanding of costs and costing procedures is something that every design engineer needs to achieve. A detailed exposition on costs and costing methods is given in Chapter 1 (Business Management). Set out below are one or two important points concerning costing, directly related to the production of an engineering artefact.

Importance of costing and pricing

The importance of producing a successful tender cannot be over-emphasised, in fact the future of jobs within the company may depend upon it. To ensure that a commercial contract to design, manufacture and supply on time, is won, there must be an effective costing and pricing policy.

Not only must the contract be won, against competition, but a profit margin needs to be shown. Price fixing needs careful planning, clearly too high a price may not result in a successful tender and too low a price may cause financial loss to the company, particularly if there are unforeseen difficulties.

For profit, and as a 'useful rule of thumb', we should know our costs to within plus or minus 2%. The tolerable margin between maximum and minimum prices is small, thus the necessity for *design* and *costing* accuracy.

Some important general reasons for costing are to:

- determine the viability of a proposed business venture;
- monitor company performance;
- forecast future prospects of a business deal;
- price, products and/or services;
- meet legal requirements to produce records of company viability, for public scrutiny as required.

Below are some useful definitions concerned with cost and price.

Price: money paid for products or services.
Value: the amount of money someone is prepared to pay for products or services.
Cost: all money spent by a supplier to produce goods and services.
Material cost: (volume × density × cost/kg) plus an amount for wastage.
Labour cost: (operational time × labour rate) plus wasted labour time which is not directly related to the task.

Standard costing sheets

These are used to ensure that all parameters are considered when costing a product or service. Some standard costing sheet headings together with their definition, are given here:

A: Direct material cost: raw material and bought-in costs.
B: Direct material scrap: materials subsequently scrapped (typically 3–5% of A).
C: Direct labour cost: wages of production operations, including all incentive payments.
D: Direct labour scrap: time spent and paid for on artefacts, which are subsequently scrapped, this would include the costs of machine breakdown or other reasons for stoppages to production (typically 3–5% of C).
E: Prime cost: the sum of all material and labour costs that is A + B + C + D.
F: Variable overheads: cost of overheads which vary with rate of production, these might include; fuels costs, cost of power supplied, consumables, etc. (typically 75–80% of C).
G: Manufacturing cost: this is the sum of prime costs and variable overheads (E + F).

H, I, J:These are packaging, tooling and freight costs, respectively.

K: Variable cost (VC) this is the sum of the previous costs, G + H + I + J.

L: Fixed overheads (FO): overheads which do not vary with production output, these include all indirect personnel not involved with production, marketing costs, research and development costs, equipment depreciation, premises costs, (typically 30–40% of K).

M: Total cost (TC): the sum of all direct variable costs (K) plus indirect costs (L).

Thus: Total costs (TC) = direct variable costs (VC) + indirect costs (FO).

Example 2.2.3

A company has been commissioned to produce 2000 high quality metal braided shower hoses, complete with fixtures and fittings. Assuming that:

(i)   Direct material cost per item is £1.50 and material scrap is estimated to be 3% of material costs.

(ii)   Direct labour costs total £8000 and the labour scrap rate is 4% of direct labour costs.

(iii) Variable overheads are 75% of direct labour costs.

(iv)   Fixed overheads are 30% of variable costs.

(v)   Packaging, tooling and freight costs are 10% of manufacturing costs.

Estimate the selling price of the shower hose, if the company wish to make a 30% profit.

   This problem is easily solved by laying out the costing sheet as shown below, and totalling the amounts.

Cost		Amount (£)
Direct material cost	(1.5 × 2000)	3000
Direct material scrap	(3% of 3000)	90
Direct labour cost		8000
Direct labour scrap	(4% of 8000)	320
Prime cost		1 410
Variable overheads	(75% of 8000)	6000
Manufacturing cost	(prime + variable)	17 410
Packaging, tooling and freight cost	(10% of 17 410)	1741
Variable cost (VC)	(manufacturing + packaging, tooling freight)	19 151
Fixed overheads (FO)	(30% of 19 151)	5745
Total cost	(variable + fixed overheads)	
£24 896		

Now company are required to make 30% profit.

So selling price per item is $\dfrac{24\,896 + 7469}{2000} = £16.18$.

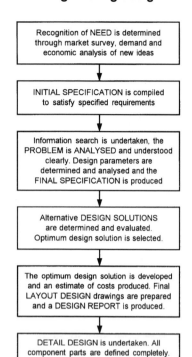

Recognition of NEED is determined through market survey, demand and economic analysis of new ideas

INITIAL SPECIFICATION is compiled to satisfy specified requirements

Information search is undertaken, the PROBLEM is ANALYSED and understood clearly. Design parameters are determined and analysed and the FINAL SPECIFICATION is produced

Alternative DESIGN SOLUTIONS are determined and evaluated. Optimum design solution is selected.

The optimum design solution is developed and an estimate of costs produced. Final LAYOUT DESIGN drawings are prepared and a DESIGN REPORT is produced.

DETAIL DESIGN is undertaken. All component parts are defined completely. Assembly and detail drawings are prepared ready for production. Modularisation or other design features may be incorporated

Figure 2.2.7

Here is something to remember when considering costs: *always design parts with the over-riding thought of saving money and do not forget that omitting all non-functional features and trimmings from parts saves production time!*

Summary of the design process

We leave this section with a summary of the design process (Figure 2.2.7). If the process is followed it will ensure that an orderly and logical approach to engineering design is achieved.

Format of the design report

In the last section we looked at the design process, including a number of ways in which to identify and evaluate design solutions. These evaluation methods are equally suitable for determining solutions during the concept, general arrangement (layout) and detail phase of the design process. Information on the layout design, as well as conceptual design should appear in the report. At the end of this chapter you will find several questions, which have been designed to help you improve your ability to think laterally, throughout all phases of the design process. Here, we are concerned with report writing, the layout of the design report and the detail expected within each section.

The following general information is given for guidance only. The report content and layout may differ slightly from that given, depending on the nature and requirements of the design task. More specific information on report writing may be found in *BS 4811 The presentation of research and development reports BSI (1972)*.

Title page: This should include a clear and precise title for the design and contain the designer's name and company details as appropriate.

Acknowledgements: These should always appear at the front of the report. They should include individuals, companies, or any associated body who has provided the design engineer with help and advice. This may include assistance with regard to literature, materials, information, finance or any form of resource.

Summary: This should provide a brief statement of the design problem, its solution and any further recommendations with respect to development and testing. References may be made to other areas of the report, in order to clarify the design description.

List of contents: This should contain a list, which provides the page number of all the main headings as they appear in the report. A separate list of all diagrams, sketches, drawings, illustrations and photographs should be provided, indicating figure numbers, page numbers, plate numbers and drawing numbers, as appropriate.

Introduction: This should provide all background detail to the project and, give an indication to the reader as to why the design was undertaken.

Specification: This section should include the design requirements in the form of a statement of the initial design specification.

Design parameters: A description of all the design parameters related specifically to the design in question, should be given. The design parameters will include those concerned with the engineering aspects of the product or system being considered, as well as organisational factors. Any modifications to the original design specification should be given, stating all assumptions made and, giving reasons for such decisions.

Description of design: This is the most important section within the report. It should contain an explicit, succinct description of the final design solution, indicating clearly its function and operation. Sketches should be provided to clarify specific areas of the design solution and references to formal drawings, in particular, the general arrangement drawing should be made.

Design evaluation: This section should contain a critical appraisal and appreciation of the final design solution. Recommendations for further development and testing should also be given, to enable improvements to be made to specific features of the design, as required.

References: The reference list should contain only those references that are mentioned in the text. They are normally numbered in the same order in which they appear in the text.

Appendices: These contain all supporting material necessary for the report which is not essential for inclusion or appropriate for inclusion into the main body of the report. The material contained in the appendices should be referred to in the text of the main report. Appendices are often identified using a Roman numeral. The following list gives a typical selection of appendix material for a design report.

- Evaluation of alternative design solutions, including sketches and description of alternatives.
- Details of decision making processes, such as evaluation matrices, decision trees, etc.
- Theoretical calculations, mathematical derivations, formulae and repetitive calculations.
- Evaluation of materials selection, for all phases of the design.
- Evaluation of appropriate manufacturing processes.
- Consideration of human factors.
- Costing considerations and pricing policy.
- Details of correspondence, associated with the design.
- Description of specialist test and development equipment.
- Details of experimentation and record of associated data.
- Computer programs and evaluation of computer printouts.

General layout: The design report should follow a recognised hierarchy for headings. There are several ways in which the relative headings can be laid out these include; numbering, indenting, use of capital and lower case letters, emboldening and underlining. Below is an example of the decimal numbering system, which has the advantage of easy and accurate cross-referencing, when required.

2. **FIRST LEVEL HEADING**
2.1 SECOND LEVEL HEADING
2.1.1 **Third level heading**
2.1.1.1 Fourth level heading

The report should be typed or word-processed on one side of the paper, with appropriate margins. The text is normally double-spaced or similar, to provide the opportunity for specific comments/ suggestions, for amendment.

Pages before the table of contents are numbered using lower case Roman numerals, pages following the table of contents should be numbered using Arabic numerals. The style and placement of page numbers should be consistent.

In order to assist with the design process and the design report, the design engineer needs to be familiar with some important aspects of computer technology, this is detailed in Section 2.3.

2.3 Computer technology and the design process

Introduction

The computer has become a very important and powerful tool in engineering design and manufacture. Computer aided design (CAD) is a design process which uses sophisticated user friendly computer graphic techniques together with computer software packages, which assist in solving the visualisation, analytical, development, economic and management problems associated with engineering design work.

The key features of a CAD system include: 2D and 3D drafting and modelling, parts and materials storage and retrieval, provision for engineering calculations, engineering circuit design and layout and circuit and logic simulation and analysis.

Computers may be applied directly to the design process in a number of areas, these include: *geometric modelling*, where structured mathematics is used to describe the form or geometry of an object. The *analysis* of engineering situations where forces, motion parameters, endurance and other variables may be investigated, for individual design situations. *Reviewing and evaluating* the design is made easier using computer graphics, size dimensions and tolerances may easily be checked for accuracy and, minute detail can be closely scrutinised using the magnification facility of the graphics system. *Automated drafting* has greatly improved the efficiency of producing hard copy drawings, which can be easily amended, as the design evolves (Groover and Zimmers, 1984).

There is a plethora of software packages associated with the engineering design process. 2D drafting packages enable the production of single part, layout, general arrangement, assembly, sub-assembly, installation, schematic and system drawings. 3D drafting and modelling packages, in addition to engineering drawing, enable us to visualise and model, product aesthetics, packaging, ergonomics and the differing effects of colour change, textures and surface finish.

Design analysis packages enable us to perform calculations involving area, volume, mass, fluid flow, pressure and heat loss as well as the analysis of stress and, the modelling and analysis of static, kinematic and dynamic engineering problems. Circuit and system modelling packages enable us to simulate and modify the layout, design and operation of electronic, fluid, control and other engineering systems.

Many other packages exist which enable design engineers to:

- program CNC machinery;
- design jig and fixtures;
- project plan;
- store, retrieve and modify technical information, publications and company literature;
- prepare tenders and estimates;
- generate and maintain materials stock lists;
- perform other engineering management functions.

The computer and its associated software, has thus become a very important part of a design engineer's armoury.

Computer Aided Design (CAD)

The use of a CAD system has many advantages over the more traditional design systems it has replaced. Improvements in productivity have obvious benefits for the company. Lead times from conception to embodiment, detail design and manufacture may be significantly reduced, dependent on the type and sophistication of the CAD system adopted.

Drafting packages range from simple 2D systems to semi-automated highly sophisticated 3D drafting and modelling systems, which contain large libraries of commonly used utilities. *Macros* may be provided which enable the user to define a sequence of commands to be executed in one instruction, this is a useful aid to speeding up repetitive routines (Lock, 1992). Drawing visualisation is easier with 3D modelling systems, orthographic projection may unknowingly be misinterpreted. However, being able to produce the equivalent isometric or perspective projection, which may be further enhanced with shading or colour, makes the CAD drawing much easier to interpret.

Design analysis using a CAD system, has already been mentioned, but it is worth emphasising here the versatility, power and accuracy of these systems. Greater accuracy may be achieved in design calculations, with correspondingly fewer errors. Since errors cost time and money to rectify, the employment of a CAD analysis system, at the right scale for company needs, is advantageous.

The classification of engineering information into specific categories is necessary in order to be able to establish and implement document handling procedures. A wealth of historical data which may be needed by the company for future projects, needs to be stored in such a manner that it is easily retrieved for use, the modern CAD system is ideal for this purpose. Drawings, planning documents, correspondence, tenders and other useful information is easily stored on magnetic tape, floppy disk, CD ROM or some other electronic form, which is easily catalogued for quick access.

The remainder of this section is concerned with the computer hardware needed for a typical CAD workstation and, a more in-depth look at software packages for drafting, modelling and analysis.

Computer hardware

A typical engineering design computer work-station will contain the following items of hardware:

- Input devices;
- Microprocessor and interface unit;
- Visual display unit (VDU);
- Output devices;
- Storage devices.

Figure 2.3.1 shows the components of a typical CAD workstation.

Input devices

Typical input devices include the *alphanumeric keyboard* which is used to input numbers, text or instructions. The *electronic digitiser* which is used for entering information on co-ordinates from existing graphical images with the help of a hand-held cursor. A *menu tablet* may be used as a low-resolution digitiser for accessing standard large scale details. The *mouse* which is a hand held device with a roller in

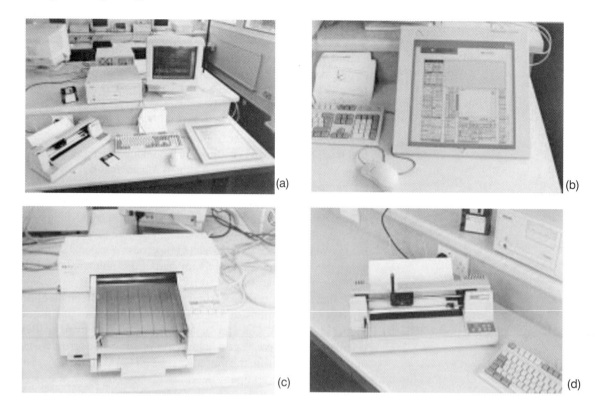

Figure 2.3.1 a–d

its base which controls the screen cursor by quickly following the input movements made by the mouse. The *joystick* is another cursor control device which functions in the same way as a mouse. The *light pen*, which looks similar to an ordinary writing pen except that it has a stylus tip, it is operated by placing the tip on the screen and moving as required.

Although not strictly hardware information for use in the computer can be stored, loaded and down-loaded to or from the machine using floppy disks, CD ROMs, scanners and now, the electronically modelled information from a digital camera. The latter with the ability to alter detail and other features of the digital photograph may prove useful for designers in the future.

Microprocessor

This is often thought of as the 'brain' of the computer, the microprocessor temporarily stores and executes program instructions on data. A microprocessor is a miniature version of the central processing unit (CPU) of a digital computer. Its versatility is due to the fact that it is program controlled, simply by changing the program it can be used as the 'brain' of not only a microcomputer but a calculator, washing machine, industrial robot, for sequencing traffic signals and many other applications.

The operations within the microprocessor dictate its architecture, which essentially consists of seven pieces of hardware (Figure 2.3.2). The *arithmetic and logic unit* (ALU) which handles the arithmetic calculations and logic decision making processes. *Registers* which temporarily store data, address codes and operating instructions. One of them, called the *accumulator*, contains the data actually being processed at any one time. Some type of *control unit* must be provided for the control and timing of operations within the entire microprocessor chip. The *clock* circuit generates timing pulses at a frequency of several megahertz to synchronise and control operations. The *bus controller*,

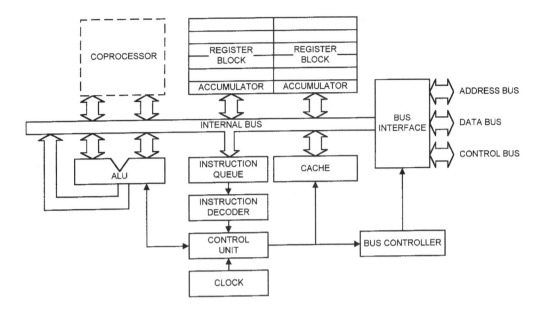

Figure 2.3.2

connects and controls the data, address and control lines between the system bus and the processor. The *coprocessor* is activated when complex mathematics is processed, for example, when using spreadsheet or CAD software packages. The internal *cache* improves the data throughput of the microprocessor by allowing frequently used data and instructions to be stored internally within the chip, thus reducing the need for access to the external bus.

Visual display unit

The VDU screen is used to display information and, where only alphanumeric data is displayed, low resolution monitors are adequate. However, when graphics are used high resolution screens are usually employed, as they give a clearer picture. The high resolution monitors are normally larger, permitting a greater area to be viewed without loss of definition.

When the design engineer wishes to display graphics there are a variety of graphics monitors, which may be chosen. Considerations such as resolution, computer power requirements, speed of operation and animation requirements dictate which type of VDU to use. The three most common types of monitor, that all use the cathode ray tube (CRT) as the display device are; the direct view storage tube (DVST), the vector refresh system and the raster refresh system. *Refresh* is a system whereby the image has to be constantly regenerated in order to avoid it visibly flickering on screen.

The CRT is a vacuum tube in which an image is made visible on a phosphorescent viewing screen that glows when a narrow beam of electrons strikes the phosphor. There are three basic elements to a CRT these are:

- the electron gun which produces and focuses the beam of electrons;
- some means of deflecting the beam according to the signals present; and
- the screen which converts the screen into visible light.

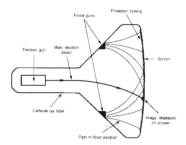

Figure 2.3.3

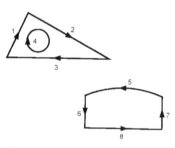

Figure 2.3.4

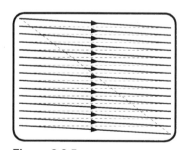

Figure 2.3.5

The DVST system

In this system, information is sent to the screen once only, by a flood of electrons which converge onto the main electron beam and light the required area even after the beam has moved on to another position (Figure 2.3.3). This system produces excellent line quality and is able to retain the projected image for a period of time, but this system has several disadvantages. These disadvantages include; the requirement to redraw the entire screen if just a single line is edited, thus slow in operation; animation is difficult to achieve; colour images are not usually available; the system is unable to create filled-in areas on the drawing and; the system needs to be used in dimly lit areas since flood lighting is insufficiently bright under normal lighting conditions.

Vector refresh

This system was developed to overcome many of the disadvantages of the DVST system. The screen is regularly updated within a fraction of a second, under CPU control. There are no flood guns as with the DVST system. Editing of single elements is easily achieved. This system has excellent line quality, high drawing speed and can accommodate 2D and 3D animation.

Disadvantages include high costs, although with improving technology, prices are now almost comparable with other graphic display systems. Screen flicker can present problems with complex drawings, if the refresh rate slows sufficiently to become less than the flicker threshold of the human eye. Difficulties may also be experienced when trying to in-fill areas on the screen.

The generation of images using this system relies on vector drawing in sequential steps, the refresh system also uses a vectored line approach (see Figure 2.3.4).

Raster refresh system

This system is similar in operation to that of a normal television, where thousands of picture element dots (pixels) may be illuminated and the brightness of the pixels controlled by the intensity and direction of a narrow electron beam. The refresh mechanism zig-zags from the top right to the bottom left of the screen approximately 50 times a second (Figure 2.3.5). The picture resolution may be varied according to how many pixels are available, these typically vary from 320×240 low resolution to 1024×1024 high resolution. Advantages include bright and clear monochrome and colour picture quality, no flickering problems, image in-fill easily achieved and, low costs, making this a widely used system.

Disadvantages include 'stair-casing' effect on lines and curves, the fact that the pixel display must be calculated for each refresh which makes heavy demands on computer memory and animation is only available when extensive computing facilities exist.

Output devices

The output devices most relevant to a CAD workstation are pen plotters, electrostatic plotters, and printers. There are essentially two types of pen plotter, flat bed and drum (see Figure 2.3.1(d)). Both types of plotter use an ink pen which produces lines on the paper, as the pen and paper move relative to one another, when signalled. Multi-colour production is achieved by using a numbers of pens. Drum plotters are generally less expensive than their flat bed counterpart. Drawings of virtually any length can be produced on the drum plotter, although the width is limited by the length of the drum. Drum and flat bed plotters have relatively slow plotting speeds, although the drum plotter is normally faster than the flat bed, however, both types of plotter are highly accurate.

Electrostatic plotters have fast plotting speeds but low accuracy, they are ideal for producing draft copies quickly. These plotters can only process data that is presented in the raster format, unless some form of conversion equipment is added. They are, like the drum plotter, capable of producing drawings of any length, the width being limited to around 1.5 m.

Many CAD systems can now output to printers or onto photographic film. There are three main types of printer that are used commercially, the dot matrix, laser and ink jet. Laser printers have high quality reproduction and can now print in colour. Ink jet printer have slightly poorer definition, when compared with the laser, but are cheaper. Dot matrix printers, have until recently, given relatively poor reproduction but with the advent of 24 pin systems, the quality of reproduction has improved. All of these printing devices are still limited to relatively small paper sizes, when compared to conventional plotters.

Storage devices

Mention has already been made of magnetic tape and floppy disk storage devices. These storage devices and others, are required for the files generated by the CAD system. For personal computers and workstations, there is typically storage available near to the processor, or a common storage device provided through a network. Local storage can be achieved by saving to the hard disk or a floppy disk, whereas storage for a network might be saved to a common magnetic tape system, with back-up on floppy disks. On central systems, the storage devices are normally attached to the central processor.

The use of networked or central storage provides greater flexibility and the rapid transfer of information between users. When information is available to all users, the question of accessibility and confidentiality, becomes a problem. Systems need to be introduced which safeguard the integrity of stored information, this is most easily achieved through the use of networked or centralised storage systems, where individual control is maintained through the use of access codes.

Computer aided drafting and modelling

Computer graphics

At the heart of any computer aided design system is the software which enables the artefact to be graphically represented and described. This software enables the computer to understand geometric shapes, when the operator inputs commands. A computer aided design and drafting package needs to be able to perform the following operations:

- Generate graphic elements such as points, lines, arcs, circles, tangents.
- Allow objects to be transformed by scaling, moving, mirroring or rotating them.
- Display components by windowing, providing alternative views, layering, etc.

- Allow common shapes to be drawn and stored for future use in a component library.
- Ability to program the computer to generate a component drawing or model automatically by specifying as a set of dimensions (user defined parametric macros).
- Provide standard macros.
- Allow non-graphical information to be stored with the drawing or model (attributes).
- Ability to program package to suit individual needs.

In Figure 2.3.6 is an illustration of a TurboCAD computer drafting package window showing a pull-down menu which is accessed from the main menu, using the cursor or keyboard. The tools menu, for example, is used for pinpointing the ends of lines, centres of circles, etc. Note also the icon menus, typical examples of screen icons used with a graphics package are shown in Figure 2.3.7. Macros, you may remember, allow a sequence of operations to be executed in one instruction. Two macro programs are given below, one is a 'script file' used with TurboCAD, whilst the other is an AutoSketch 'macro'. Both result in the production of the nine-sided polygon shown in Figure 2.3.8!

TurboCAD script file:

```
DrawLinePolygon(9)    ; Polygon will have nine sides
ClickAt(5,5)          ; Centre it on 5,5
ClickAt(5,7)          ; Starting point for drawing
SetTextJustify(5)     ; Justify label text
SetTextCSize(0.5)     ; Set up text size and
DrawText(POLYGON)     ; place the text in
ClickAt(5,5)          ; the centre of the polygon
EditMode
```

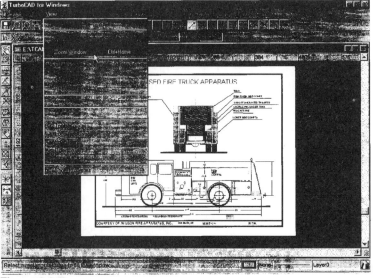

Figure 2.3.7

Figure 2.3.6

Figure 2.3.8

Autosketch macro:
```
DRAWPOLYGON
SETPOLYGON
SET POLYGONSIDES 9
DialogBoxReturn 1
DRAWPOLYGON
POINT 5,5
POINT 5,7
DRAWQUICKTEXT
SETTEXT
SET TEXTHEIGHT 0.500000
SET TEXTJUSTLEFT 0
SET TEXTJUSTCENTER 1
DialogBoxReturn 1
DRAWQUICKTEXT
POINT 4.808734,4.896243
STRING POLYGON\013
```

Computer modelling

There are three basic ways in which we may graphically model an object. *Two-dimensional* representation, which most closely resembles the method of working on the drawing board. This representation has limitations, only the plane of the screen or paper is used (the x–y plane), so the object has to be drawn using different elevations (see Figure 2.3.9). The points that go to make up a view are separately produced and so are not inter-related, thus changing features in one view will not automatically produces changes in the other views.

Two and a half dimensional representation, portrays a three-dimensional object, but with the third dimension being constant in section or having a constant section which is allowed to revolve around its axis, for example as in turning operations (Lock, 1992).

Three-dimensional representation which can completely define the shape of the component (Figure 2.3.10). Three-dimensional systems are sub-divided into; *wire-frame*, *surface* and *solid modelling* which grow, respectively, in sophistication and so require a corresponding increase in available computer power. 3D software packages store the finished component in the computer's memory as a real shape, even though the CAD system is only able to display it on a 2D screen.

Three-dimensional wire-frame modelling

This is considered to be the lowest level of 3D modelling, where the component is described in terms of points and lines only (see Figure 2.3.11). This system was originally developed to model simple machining operations. It has a number of limitations which include; no detail or shading of faces; the inability to distinguish between the interior and exterior of a solid object, thus causing problems in visualising different orientations; inability to detect interference between component parts and; unreliable calculations with respect to physical properties, such as mass, surface area, centre of mass, etc.

Surface modelling

In this system the points, lines and faces are defined, this system is more powerful than wire-frame modelling and as a consequence requires more computer power, such as that available from a 32-bit minicomputer. This system is able to recognise and display complex curves and profiles (see Figure 2.3.12). Since faces are recognised, shading is possible to help enhance the computer image. Holes are also easily identified.

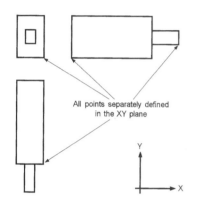

All points separately defined in the XY plane

Figure 2.3.9

Figure 2.3.10

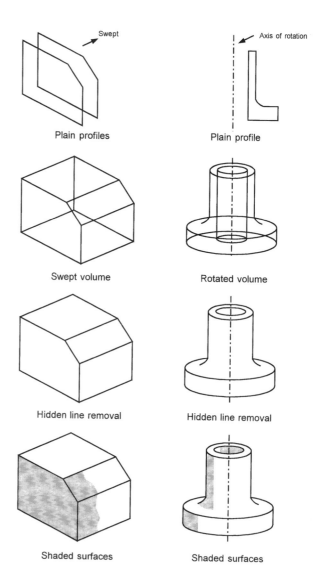

Applications of the system include complex tool path simulation and robot simulation. Complex curved surfaces such as automobile body panels, plant ducting, cowlings and fairings may be designed using this system.

Limitations of the system include; the inability to comprehend solid volumes and so subsequent volume data may be unreliable; hidden lines cannot be easily removed and; internal sections are difficult to display.

Solid modelling

This is the most sophisticated of all computer aided modelling systems requiring a large amount of computing power, from at least a 32-bit computer. Solid modelling enables the user to describe fully and unambiguously three-dimensional shapes (Figure 2.3.13).

The advantages of this system are numerous. All three-dimensional features are easily defined, including internal and external detail. There is a facility for automatic line removal. Shading and variable colour graphics are available. Finite element analysis is easily carried out (see later). Component sectioning is easily and effectively achieved. Physical parameters may be accurately calculated. The system provides excellent simulation facilities for motion parameters such as, mechanism dynamics and robot configurations.

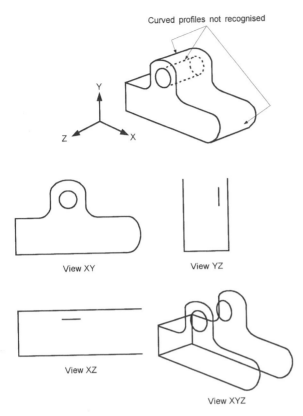

Figure 2.3.11

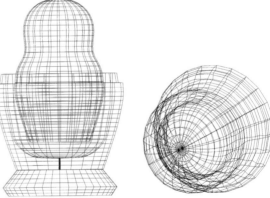

Figure 2.3.12

Figure 2.3.13

The obvious disadvantage has already been mentioned, a large amount of computer power is required so the system is very expensive. However, as technology continues to advance, cheaper hardware and even more sophisticated software will become available.

Computer aided analysis

We have discussed a variety of tasks that may be undertaken by typical design analysis packages, to illustrate this area of computer aided design we will discuss a method known as *finite element analysis* (FEA). This widely used technique predicts the mechanical characteristics of a design under load. These loads may result from imposed stresses, temperature and pressure changes, or other forces acting on the system.

Before a finite element analysis can be carried out, the finite element model (FEM) of the design must be determined. This is then used as an input to the FEA program. The purpose of an FEM is to represent the proposed design in a testable form.

Finite element modelling (FEM) is the process of simplifying the design, for analysis by sub-dividing the entire object into a large number of *finite elements* usually rectangular or triangular for 2D work, and cubic or tetrahedral for 3D work. These elements are *discrete* (individually recognisable) and as such form an interconnecting network of concentrated *nodes*, as shown in Figure 2.3.14.

Finite element analysis consists of software programs that use an FEM to perform a variety of tests when given a set of parameters. FEA software programs are separate from CAD programs but use CAD generated models to perform the specific tests. In addition to FEM these programs require the input of other information variables such as the material specification, the loads to be imposed on the structure, dimensions and other geometric factors, heat flow characteristics, and so on. FEA is thus able to analyse the entire component for stress, strain, heat flow, etc. by calculating the interrelating behaviour of each node in the system.

Today, modern advanced CAD systems have the capability to automatically model the network (meshes and nodes), without the design engineer having to specify the FEM. This saves valuable time and ensures the relevance of the model selected. The output of the FEA provides an indication as to whether or not the design conditions have been met. The results also include a graphical output which is displayed on the workstation monitor. Such outputs may be in the form of contour plots for stress, strain, or heat flow, among other factors (see Figure 2.3.15).

Computer aided design and manufacture

Little has been said about the relationship between computer aided design (CAD) and computer aided manufacture (CAM), where in a CAD/CAM system data is exchanged between each. CAM systems cover activities which convert the design of the product into instructions that define how the component will be produced. CAM is thus any automated manufacturing process which is controlled by computer. These will include computer numerical controlled (CNC) lathes, millers and punches, plasma flame cutters, process systems and robotic cells, to name but a few. Other more sophisticated computer controlled manufacturing techniques such as flexible manufacturing systems (FMS) and computer-integrated manufacturing systems (CIMS), all come under the umbrella of CAM.

CAM systems are also used for generating machine tool programs, which vary in sophistication from simple two-dimensional applications

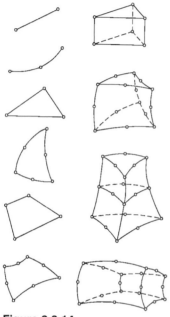

Figure 2.3.14

Figure 2.3.15

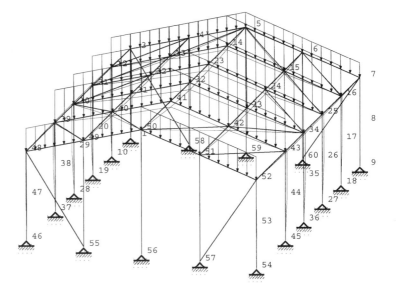

such as turning or drilling, to more complex software programs for multi-axis operations.

A CAD/CAM system would include the following features; an interactive graphics system and the associated software for design; software packages for manufacture; and a common CAD/CAM database organised to serve both design and manufacture. The CAD interactive graphics system is likely to have facilities for engineering analysis, automatic drafting, design review and evaluation and modelling. The CAM software might include packages for, tool and fixture design, NC part programming, automated process planning and production planning (Groover, 1984). Figure 2.3.16 shows the desired features of a typical CAD/CAM system.

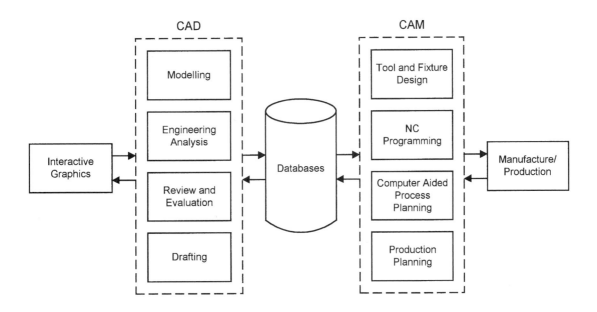

Figure 2.3.16

Figure 2.3.17

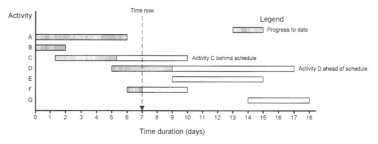

Computer aided planning and resource scheduling

It is not the intention here to discuss in any detail the management techniques available for planning and scheduling, these are discussed fully in Chapter 1, Business Management Techniques. However, in order to see how the computer may help us to plan and schedule a design project there is set out below, a brief reminder of one or two important *time planning* methods which you need to be acquainted with in order to *plan the design process.*

As you may be aware, the simplest planning aid is the Gantt chart, which is a form of bar chart where each bar represents an individual activity necessary to complete a project. The bar shows the scheduled start and finish time for each activity and the vertical arrows indicated their current status (Figure 2.3.17).

To produce a bar chart for a particular design project you would need to:

- Analyse the project and break it down in terms of activities.
- Estimate the time to perform each activity.
- Place the activities onto the chart in strict time sequence, having determined which activities must be performed sequentially and which may be performed sequentially.
- Ensure that the total time for the activities is planned to meet the specified completion date, if not adjust accordingly.

The primary advantage of the bar chart is that the plan, schedule, and progress of the project are all represented together on the one diagram. In spite of this important advantage, bar charts have not been too successful with complex engineering projects. The simplicity of the bar chart, precludes sufficient detail to determine early slippage of scheduled activities. Also the bar chart is essentially a manual–graphical procedure, which is difficult to modify quickly, as project changes occur.

As a result of the disadvantages associated with bar charts, a *network* based methodology had to be developed. The network model, like the bar chart, is a graphical representation of the planning process necessary for the project. It is based on the logic of *precedence planning*, which simply means that all activities which precede a given activity need to be completed, before the given activity may commence.

Figure 2.3.18 shows a typical *project network*. The *bracketed numbers* give an estimate of time for each activity (the units of time must be stated, hours, days, weeks, etc.). The *solid lines* denote activities which usually require resources to complete – time, manpower, equipment, the *dashed line* indicates a *dummy activity*, that is one which shows precedence only. Order of precedence is also shown for example, activities A_3 and B_2 must be completed before activity E is started. Whereas activities A_1, B_1 and D_1 may be commenced simultaneously. Each activity starts and finishes in a unique pair of *nodes* called *events*. The events denote a point in time, their occurrence signifies the completion of all activities terminating with the event in question. For

Figure 2.3.18 *Note: (i) Critical path (B₁—▷C—▷A₃—▷E) total 18 days. (ii) Dummy activity (5—▷6) precedence only .*

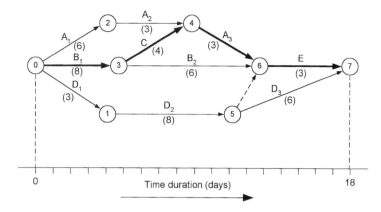

example, the occurrence of event 3 signals the completion of activities 1—▷3 and 2—▷3. Time flows from the tail to the head of the arrow.

The essential ingredient in all networks is the *critical path* which by definition is a sequence of activities which determines the duration of the whole project. If we again consider Figure 2.3.18, the activities B_1, C, A_3 and E total 18 days, which is the same as the time duration for the whole project (events 0 to 6), so this pathway has no *float*, for this reason it is referred to as the *critical path*. You should refer to Chapter 1 for a full explanation of this process, (see also Lockyer, 1991).

Summary of planning and scheduling process

Many computer programs exist for analysing critical path networks, and some of these can schedule resources. These programs are particularly useful for assisting with the planning of the design process. Below is summarised the planning and scheduling process for a design project, which encompasses the use of computers.

(1) Start by planning the work sequence *manually*, and producing a project network. Your sequence will include all the major steps given in Figure 2.2.1. Each of these major steps will be broken down into sub-activities. So for example, when considering *concept design*, this may include, information searching, lateral thinking exercises, group activities, the development and presentation of alternative design solutions, the evaluations of design solutions and selection of the optimum solution, and so on.

(2) Estimate the time duration and resource requirements for each activity. Remember that some of the activities will be able to run concurrently, while others will need to be completed prior to the start of subsequent activities. Resource requirements should include, costs, materials, tooling, jigs and fixtures, machinery, production requirements and human resources. Time scales will depend on the complexity and number of networks required for a particular design.

(3) Load the network into a suitable computer system, with appropriate software installed. Provide the system with supplementary data such as; time units, required start and finish dates, and the milestones for the design project, if applicable.

(4) Load the computer with information on total resource capacities, cost rates, calendar information and all other relevant data. Safety margins with respect to time, physical and human resources, should be considered. So for example, you may not wish to declare all human resources that are available,

but hold some back, so that in the event of the project running behind time you have additional human resources in reserve, that you can use to bring the project back on schedule.

(5)   Start the time-analysis run and rectify input errors, should they occur. Make sure that the network can be time-run, without error.

(6)   Carry out the resource scheduling run and produce the network printouts, work schedules and resource lists for monitoring and review by, marketing, administration, accounting, design, production and management staff, as required.

(7)   Continually update the computer, with information parameters, as they occur, repeating, as necessary, steps 1–6.

This concludes this short section on the use of computers in engineering design. Many aspects of the use of computers have been omitted, in the interests of space. No mention has been made of computer programming itself, software design quality control, or dynamic system modelling such as vibration analysis or control system design. Spreadsheets and databases have not been mentioned in any detail, these packages are particularly useful in fulfilling the numerous administrative and management accounting functions that surround all engineering design projects. Nevertheless it is hoped that some indication of the importance of the computer to the design engineer, and the design process, has been emphasised.

We leave this section with a few examples of the variety and detail of engineering drawings and diagrams, which may be achieved using commercially available drafting packages. Figures 2.3.22, 2.3.24 and 2.3.25 illustrate the use of specialist software for the production of engineering system diagrams. The graphical output from a standard mathematical modelling package is shown in Figure 2.3.23.

Figure 2.3.19

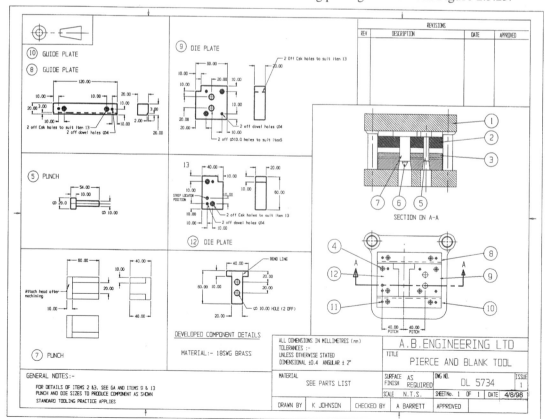

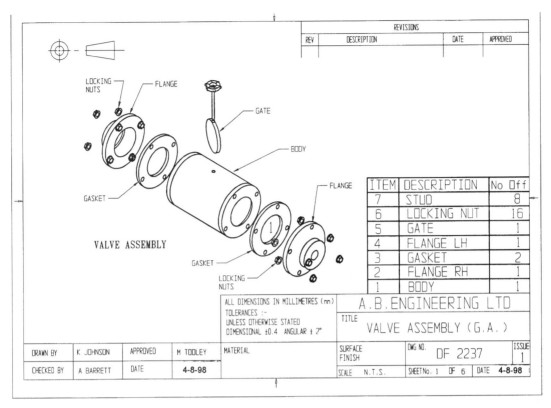

Figure 2.3.20

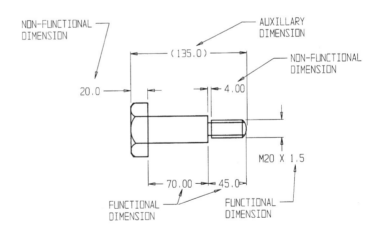

Figure 2.3.21

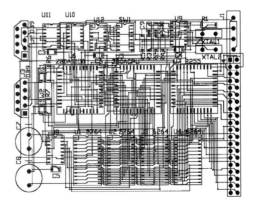

Figure 2.3.22

E:\traxmaker\Demosmd.pcb Check Print

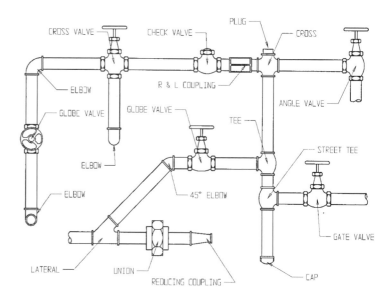

Figure 2.3.24

Create a surface plot of a unit sphere:

$N := 25$

$i := 0..N$ $\qquad \phi_i := i \cdot \dfrac{\pi}{N}$ $\qquad j := 0..N$ $\qquad \theta_j := j \cdot 2 \cdot \dfrac{\pi}{N}$

$X_{i,j} := \sin(\phi_i) \cdot \cos(\theta_j)$ $\qquad Y_{i,j} := \sin(\phi_i) \cdot \sin(\theta_j)$

$$Z_{i,j} := \cos(\phi_i)$$

X, Y, Z

Figure 2.3.23

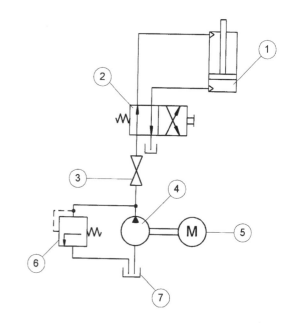

Figure 2.3.25
KEY: 1, Double-acting actuator; 2, 4/2-way directional control valve; 3, Shut-off valve; 4, Hydraulic pump; 5, Motor; 6, Pressure relief valve; 7, Tank (reservoir).

Questions 2.3.1

(1) State the advantages of 3D drafting packages over their 2D counterpart.

(2) Explain the operation of a cathode ray tube (CRT) *and* explain the differences in operating principle between the: DVST; vector refresh; and raster refresh systems.

(3) Describe the likely computer storage systems required with:

    (i)   a stand alone PC workstation; and
    (ii)   a networked CAD system, with local workstations.

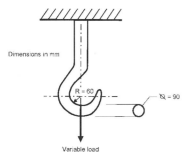
Dimensions in mm

R = 60

⌀ = 90

Variable load

Figure 2.3.26

(4) You have been tasked with the re-design of the front wing fairings of a new model of motor car, which encompasses new front light clusters. State, giving explicit reasons, which type of computer modelling package you would choose to assist you.

(5) Investigate an FEA package that will assist in the design and development of the load-carrying gantry hook, shown in Figure 2.3.26. Consider the analysis in terms of the ability of the hook to withstand the loading conditions, from the point of view of strength and rigidity.

(6) Using the design specification you produced for the 'hydraulic hose connector' in Question 2.1.2.

    (i) *Draw-up a network* for the design process from final specification to the production of the agreed final solution.

    (ii) With help from your tutor, load the network into a computer equipped with suitable software, carry out steps 1–5, of the summary of the planning and scheduling process given above.

Summary of the design process

We conclude this short chapter with a summary which highlights the philosophy and thinking necessary during the design process in order to; produce an engineering product or system, *on-time, at the right price and of the right quality.*

The first step in this process was to *analyse the design problem*, this included considerations such as the; likely market, need and economic viability of the design. The production of a suitable *design specification*, to suit the needs of the customer, whether internal or external, was emphasised as being of paramount importance.

In producing the design specification all the design parameters would need to be considered. These included; performance, ergonomics, manufacture and materials, maintenance, safety, installing and commissioning, as well as organisational factors such as; legal implications, costs, transportation, quality, and so on.

In order to produce a number of possible *design alternatives*, the use of lateral thinking techniques was discussed, these include methods such as; brain-storming, systematic searching, literature searching and the gallery method to name but a few. Next, evaluation techniques might need to be used to select possible design alternatives. Finally, matrix methods, with appropriate ranking methods could be used to assist in selecting the *optimum design solution*. It is at this stage that the *design report* is written, and design ideas communicated to those in a position to make management decisions.

Use of the above techniques could be repeated for all phases of the design, assisting with the problems that might be encountered during the *layout*, *detail*, after sales and service phases of the project.

During the layout and detail phases, design engineers will be required to draw upon their knowledge of; engineering principles and applications, materials, mathematics, machine component operation and selection, drafting techniques and the nature of numerous technological advances, in order to produce the final design solution.

Layout and detail design must also take into consideration the issues surrounding packaging and transportation, such as the need for modular construction, if appropriate.

Throughout the *whole design process* a constant check must be maintained to ensure that the design solution is produced within budget, and progresses on schedule. This may be achieved through the use of appropriate *computer software packages*, as discussed in section three of this chapter.

Legal and safety aspects must be continually monitored to ensure that the design complies with all relevant design standards and, where appropriate, meets the requirements of European and other International legislation.

Finally, an efficient *after sales service* must be set up, and appropriate help given with installing and commissioning the product, plant or system.

In order to draw together some of the ideas expressed in this chapter we finish with one or two design questions, which will help you put into practice some of the concepts needed to ensure a systematic approach to design. To assist you with these exercises you should refer to the references listed at the end of this book.

Problems

(1)   Your company intends to produce a steam iron, which is able to use ordinary tap water to fill the reservoir, without incurring fouling, scaling or discolouration. The power rating for the iron should not exceed 1.8 kW and, the iron should be light and, capable of easy handling by both right- and left-handed users.

Using the information provided in this chapter and any other information from the reference sources:

(i)   Produce a design specification which includes at least 15 requirements and constraints.

(ii)   Produce at least four preliminary design solutions, in the form of a concept sketch with accompanying explanation, for each possible solution.

(2)   A device to test the mechanical properties of toothpaste tubes is required by a manufacturer:

(i)   Identify at least four possible sources of energy that may be used for the testing device.

(ii)   Produce as many possible design alternatives as you can using different power sources, giving details of the design principles and proposed operation for each design solution.

(iii)   Using a suitable method, rank your possible design solutions.

(iv)   Select your optimum solution and produce a general arrangement drawing (layout drawing), using a computer aided drafting package.

(3)   There are currently a large number of microcomputers on the market, which utilise the *Pentium* processor. All these computers aim to fulfil the same function, but they differ in price and detail design features. After studying consumer magazines and/or other appropriate literature:

(i)   Prepare an evaluation matrix for six different models with no more than 10 selection criteria.

(ii)   Select the best product and explain in detail why, in your opinion, it is superior to the others.

3 Engineering science

Summary

This unit aims to provide you with an understanding of the scientific principles that underpin the design and operation of modern engineering systems. Unlike other core units, this unit covers both mechanical and electrical principles. It thus provides a valuable introduction to engineering science for anyone who has not studied engineering before.

It is important to realise that this unit contains the basic underpinning knowledge required for further study of the two *principles* core units: Electrical and Electronic Principles (Unit 6) and Mechanical Principles (Unit 7). The unit also provides a grounding for several of the specialist Option Units. It is divided into four sections dealing with: static and dynamic engineering systems, energy transfer in thermal and fluid systems, single phase AC theory and information and energy control systems.

3.1 Investigate static and dynamic engineering systems

The study of static structures such as bridges, buildings and frameworks, together with dynamic systems that include engines, power plant, electrical machines, chemical plant, fluid power, air, sea and rail transport, is of the utmost importance to all engineers. The general nature of static and dynamic engineering systems makes their study essential as a foundation for all engineers, irrespective of their specialisation.

This section introduces the reader to some fundamental statics and dynamics. The subject matter covered includes: thermal stressing, torsion in solid and hollow circular shafts, uniform acceleration and the implications of Newton's laws of motion and mechanical oscillations including simple harmonic motion.

Thermal stress and strain

Thermal stress and strain are considered important by engineers for a variety of reasons and need to be taken into account in many areas of engineering practice. For example, engineers often support the ends of bridges on rollers or use hydraulic dampers, to allow them to expand

and contract freely. The expansion of long runs of steel pipe when installing plant and equipment needs to be taken into account, if excessive thermal stresses due to expansion and possible damage are to be avoided. Engineers designing pressure vessels which contain liquids need to take into account the expansion and contraction of the liquid and the resulting thermal stress and strain, if a successful and safe design is to be achieved. For these and many other reasons, a study of temperature-dependent stress and strain in materials is considered essential for those who wish to practise as engineers.

Thermal strain

Changes in the temperature of a material give rise to thermal strain. You may remember from your previous studies that the linear strain of different materials in contact is dependent upon the type of material, length of material and the change in temperature the material is subjected to. These relationships may be represented mathematically by the formula:

$$x = \alpha \Delta TL \tag{3.1.1}$$

where x = change in dimension in metres, α = coefficient of linear thermal expansion of the material, ΔT is the change in temperature, and L is the length of material in metres.

There is no stress associated with this strain unless the material, for example in the form of a bar, is prevented from extending. In this case the load produced in the bar would be the same as the load required to compress the bar the distance equal to the free expansion of the bar Figure 3.1.1 illustrates this point.

For convenience I have listed in Table 3.1.1, typical values of coefficients of linear expansion for some of the more common materials.

Table 3.1.1   *Coefficients of linear thermal expansion*

Some common metals	α $(10^{-6}/^{\circ}C)$
Aluminium alloy	23
Brass	22
Copper	17
Grey cast iron	12
Magnesium alloy	28
Nickel alloy steel	13
Plain carbon steel	12
Titanium alloy	9.5
Zinc alloy	28

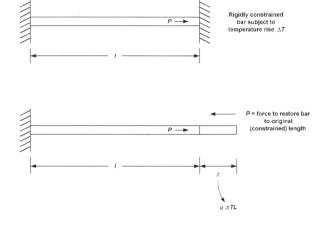

Figure 3.1.1   *Rigidly constrained bar subject to temperature rise*

Some polymers	$\alpha \ (10^{-5}/°C)$
Acetal	8
PTFE	7
Nylon 6/12	9
Polycarbonate	7
Polyethylene	11
Polystyrene	6.5

Note: the coefficient of volumetric or cubic expansion $\gamma = 3\alpha$.

The example below shows the importance of considering the stresses set up due to thermal expansion.

Example 3.1.1

An aluminium alloy expands 23 μm per metre length per °C rise in temperature. How does the magnitude of the thermal strain compare with the elastic strain? Well, for example, the elastic strain at a tensile stress of 70 MN/m^2 for our aluminium alloy having an elastic modulus E of 70 GN/m^2 is:

$$\epsilon = \frac{\sigma}{E} = \frac{70 \times 10^6}{70 \times 10^9} = 0.001.$$

A temperature rise of 43.48°C in the same aluminium alloy gives a thermal strain:

$$\epsilon = \alpha\Delta T \text{ (since unit length)} = 23 \times 10^{-6} \times 43.48 = 0.001.$$

If the extension of the aluminium bar due to this rise in temperature was completely prevented, a compressive stress of approximately 70 MN/m^2 would be set up in the bar (since it is this stress that is required to return the bar to its original length).

If our bar was also subject to external load, then both thermal strain and elastic strain could occur simultaneously. Thus the total strain ϵ is the sum of the two and our formula for extension may be represented algebraically as follows:

$$\epsilon = \alpha\Delta T + \frac{\sigma}{E}$$

and   extension = ϵL

extension = $\left(\alpha\Delta T + \frac{\sigma}{E}\right) L$ (3.1.2)

Let us consider one further example which illustrates the *effects* of thermal strain.

Effects of thermal strain

Example 3.1.2

Figure 3.1.2 shows a bar of copper alloy 400 mm long. For three-quarters of its length the diameter of the bar is 25 mm, for the remaining quarter of its length the diameter of the bar has been reduced to 15 mm. Assume that the bar is rigidly clamped at its ends, preventing movement, as shown, and is then subjected to a temperature rise of 40°C. Find the maximum

cont.

stress in the bar. Take the elastic modulus of the copper alloy as $E = 120$ GPa.

Use our value of α for copper given in Table 3.1.1. Then if the bar were allowed to expand freely without constraint, the expansion of the whole bar is given by:

$$\alpha \Delta TL = 17 \times 10^{-6} \times 40 \times 0.4$$
$$= 272 \times 10^{-6} \text{ m}.$$

Expansion is prevented by a compressive force exerted by the clamps, as shown in Figure 3.1.2.

Now each part of the copper bar carries the load, but we note that the extensions of each section of the bar are different because the lengths and cross-sectional areas supporting the load are different. Since the load is the same throughout, then the maximum stress will be felt in the smaller diameter section of the bar.

We know that load $P = \sigma A$ and that the cross-sectional area is given by the formula $\pi d^2/4$, since $\pi/4$ is a constant for both sections of the bar then the proportion of the stress taken by each section of the bar can be found from:

$$P = \sigma_1 A_1 = \sigma_2 A_2$$

(*note: suffix 1 used for larger diameter section*)
i.e. the proportion of the stress taken by each section of the bar is

$$\sigma_1 = \frac{A_2 \sigma_2}{A_1}$$

$$= \frac{15^2}{25^2} \sigma_2$$

$$= 0.36\sigma_2 \qquad (3.1.3)$$

Now remembering that the total length of the bar remains unchanged, then:

total contraction due to load = extension due to temperature change

$$x_1 + x_2 = 272 \times 10^{-6} \text{ m}$$

so: $\dfrac{\sigma_1 L_1}{E} + \dfrac{\sigma_2 L_2}{E} = 272 \times 10^{-6}$ m (from $x = \epsilon L$ and $\epsilon = \sigma/E$)

therefore, $\sigma_1 \times 0.75 \div \sigma_2 \times 0.25 = 272 \times 10^{-6} \times 120 \times 10^9$

$$3\sigma_1 + \sigma_2 = 130.56 \times 10^6 \text{ N/m}^2 \qquad (3.1.4)$$

then from equations (3.1.3) and (3.1.4)

$$\sigma_2 = 62.77 \times 10^6 \text{ N/m}^2$$

Hence maximum stress in the bar $= 62.77$ MN/m^2

It is important that you are able to follow the argument and calculations associated with Example 3.1.2, since it illustrates the effects of thermal strain on materials and, it leads us into the study of compound bars which follow.

To make sure you have grasped the principles of thermal strain and its effects try the following questions.

Figure 3.1.2 *Example 3.1.2 copper bar*

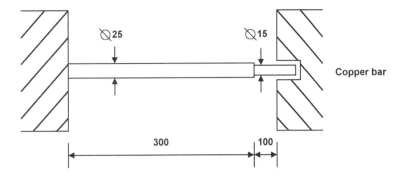

Problems 3.1.1

(1) A plain carbon steel support bar 300 mm long is firmly fixed at its ends so that it cannot expand when heated. If its temperature is now raised through 60°C, find the stress set up due to heating. Take $E_s = 210$ GN/m^2 and use the value for the linear coefficient of plain carbon steel given in Table 3.1.1.

(2) A brittle steel rod 400 mm long is subject to a temperature rise of 180°C. It is then clamped instantaneously to prevent contraction on cooling. Then it is allowed to cool and fractures at a temperature of 100°C. Calculate the fracture stress of the steel. Take $E = 210$ GPa and $\alpha = 12 \times 10^{-6}$/°C.

Compound bars and the effects of thermal strain

A compound bar may be defined as two or more parallel materials rigidly fixed together so that there is no relative movement between the ends. We will deal only with compound bars having a symmetrical cross-section. Non-symmetrical sections involve the analysis of complex bending forces, in addition to those created axially.

We have already established several mathematical relationships in our initial study of thermal strain, which will be useful when we analyse the stresses and strains set up in compound bars, resulting from external loads and temperature change. Before considering the effects of temperature on compound bars, let us first remind ourselves of the elastic stress/strain relationships that exist when we load a compound bar uni-axially. This is best achieved by use of an example.

Example 3.1.3

A compound bar consists of a piece of brass reinforced by two nickel alloy steel plates as shown in Figure 3.1.3.

Find the stresses in the brass and the nickel steel for an axial compressive load of 60 kN. Take the modulus of elasticity for brass as

$E_b = 110$ GPa, and for nickel steel as $E_s = 218$ GPa.

The end load of 60 kN is carried by both the brass and the steel. We have already established, that for elastic strain:

$$\epsilon_b = \frac{x_b}{E_b} \text{ (brass) and } \epsilon_s = \frac{x_s}{E_s} \text{ (steel).}$$

We know that the original lengths are the same, therefore $L_b = L_s$ and the compressions are the same, therefore $x_b = x_s$.

cont.

So $\epsilon_b = \epsilon_s$ (3.1.5)

Remember from our previous study that:

$$\epsilon_b = \frac{\sigma_b}{E_b}$$

and $\epsilon_s = \dfrac{\sigma_s}{E_s}$

So substituting these relationships into Equation (3.1.5) gives:

$$\frac{\sigma_b}{E_b} = \frac{\sigma_s}{E_s}$$

and since we have the elastic moduli, we are able to form an equation relating the stress in the brass with the stress in the steel,

i.e. $\dfrac{\sigma_b}{110 \times 10^9} = \dfrac{\sigma_s}{218 \times 10^9}$

$$\sigma_s = \frac{218 \times 10^9}{110 \times 10^9} \sigma_b$$

$$\sigma_s = 1.98\sigma_b \qquad\qquad (3.1.6)$$

Now, we also know that the load on the brass plus the load on the steel is equal to the total load.

i.e. $P_b + P_s = 60 \times 10^3$ N   (3.1.7)

By calculating the cross-sectional area, A, of both the brass and the steel (in mm²) and by using some algebra, we will be able to find the individual stresses imposed on the brass and steel.

So: load on brass $P_b = A_b \times \sigma_b = (30 \times 10^3)\,\sigma_b$
        load on steel $P_s\ = A_s \times \sigma_s = (15 \times 10^3)\,\sigma_s$.

If we now substitute the above expressions for P_b and P_s into Equation (3.1.7) then we have:

$$(30 \times 10^3)\,\sigma_b + (15 \times 10^3)\,\sigma_s = 60 \times 10^3$$

and remembering that Equation (3.1.6) gives an expression for σ_s in terms of σ_b, that is:

$$\sigma_s = 1.98\sigma_b \text{ then on substitution into Equation (3.7) we get:}$$

cont.

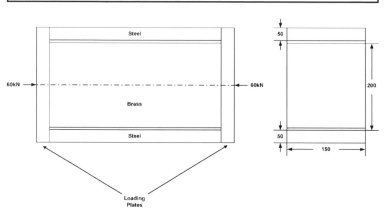

Figure 3.1.3 *Example 3.1.3 compound bar*

$(30 \times 10^3)\sigma_b + (15 \times 10^3)(1.98\sigma_b) = 60 \times 10^3$

from which $\sigma_b = 1.005$ N/mm²

that is the compressive stress in brass $= 1.005$ MN/m²
and by substituting this value for σ_b into Equation (3.1.6) we
have:

$\sigma_s = 1.98 \times 1.005$ MN/m²

thus compressive stress in steel bars $= 1.99$ MN/m²

The compressive nature of the stresses can be determined
by inspection, since the loading plates are being squeezed
together.

I hope the similarity of this technique, when compared with our solution for Example 3.1.2, has not gone unnoticed. Problems are frequently encountered with the manipulation of the algebra but, providing your solution to all compound bar problems is laid out in a logical manner and the equations needed for the solution are clearly identified for later use, mistakes should not occur!

We are now ready to consider compound bars loaded uniaxially being simultaneously subjected to temperature change. This situation can occur in many real engineering situations, for example the bonding together of dissimilar metals subject to temperature change. Bushes and dissimilar metal liners that are clamped or bolted in position and subject to temperature change, concrete reinforced by axial steel bars may also be modelled as a composite or compound bar, finally bolted and bushed inspection covers for pressure vessels, is yet another example.

In order to illustrate uniaxial and thermal strain in a compound bar, we will consider the following example, which brings together the techniques and formulae we have studied so far.

Example 3.1.4

A plain carbon steel rod with a diameter of 24 mm, threaded at both ends, is located in an aluminium tube 300 mm long with an external diameter of 60 mm and an internal diameter of 30 mm. The rod and tube are attached at their ends by rigid loading plates and bolted into position, as shown in Figure 3.1.4. The threads have a pitch of 1 mm and each end nut is tightened by one half turn, while at the same time, the assembly is subject to a 40°C temperature rise. Find the total stresses in the compound bar.

Take $E_s = 210$ GPa and $E_a = 75$ GPa. Use the values of the linear coefficients of expansion, given in Table 3.1.1, as required.

We know from our study of thermal strain that the total direct strain in any one member of a compound bar is equal to the

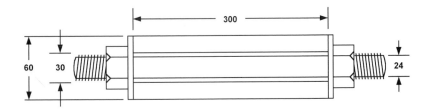

Figure 3.1.4 *Example 3.1.4 carbon steel rod in aluminium tube*

direct strain due to the uniaxial load, plus the direct strain due to temperature change.

Since elastic strain is proportional to stress it follows that; the direct stress in any one member of a compound bar is equal to the direct stress due to the uniaxial load, plus the direct stress due to temperature change.

So in our case we need to establish the following for both the aluminium tube and the steel rod:

(i)   stress due to tightening each nut by half a turn;
(ii) stress due to the temperature rise.

(i) Stresses due to tightening each nut by one half turn
In this case the extension of the steel rod, plus the compression of the aluminium tube equal the axial movement of each nut being tightened by half a turn, therefore:

$x_s + x_a = 1.0$ mm, and remembering that $x = \epsilon L$ then, $\epsilon_s L_s + \epsilon_a L_a = 1.0$.

Now making the assumption that, the original effective lengths of the steel rod and aluminium tube are equal, then:

$\epsilon_s(300) + \epsilon_a(300) = 1.0$
so $\epsilon_s + \epsilon_a = 0.0033$ (3.1.8)

and substituting into Equation (3.1.8) the relationship for strain in terms of the elastic modulus and stress we have:

$$\frac{\sigma_s}{210} + \frac{\sigma_a}{75} = 3.3 \qquad\qquad (3.1.9)$$

The above relationship maintains consistency of units. Make sure that you can verify this formula!

We also know that in the final position the *load* on the steel rod is equal to the *load* on the aluminium tube.

$$P_s = P_a \qquad\qquad (3.1.10)$$

so for the steel rod $P_s = A_s \times \sigma_s$ or $P_s = 452\sigma_s$

similarly for the aluminium tube $P_a = A_a \times \sigma_a$ or $P_a = 2120\sigma_a$

substituting the above values into Equation (3.1.10) gives:

$$\sigma_s = 4.69\sigma_a \qquad\qquad (3.1.11)$$

Now substituting Equation (3.1.11) into Equation (3.1.9) and simplifying gives:

$0.0223\sigma_a + 0.0133\sigma_a = 3.3$

from which $\sigma_a = 92.6$ N/mm²

therefore the stress in the aluminium tube = 92.6 MN/m² (compressive).

If we now substitute our value for σ_a into Equation (3.1.11) then:

$\sigma_s = 4.69 \times 92.6$

therefore the stress in the steel rod is = 434.3 MN/m² (tensile).

cont.

(ii) Stresses due to temperature rise

What do we know? If we consider the way the loads act in the compound bar assembly then we know that these loads cause the following strains:

(a) an extension of the steel bolt;
(b) a compression of the aluminium tube.

Now that heat is involved this needs to be *counter-balanced* by:

(c) the expansion of the aluminium tube;
(d) the expansion of the steel rod.

The above expansions are relative, so to find the strain between the rod and the tube, knowing that the tube will expand more for a given temperature we must subtract the expansions of the rod from that of the tube, i.e.

$$\alpha_a L \Delta T - \alpha_s L \Delta T$$

So since elastic strains and thermal strains are equal then:

$$x_s + x_a = \alpha_a L \Delta T - \alpha_s L \Delta T \qquad (3.1.12)$$

Again, if we assume that $L = L_a = L_s$ then we may substitute $x_s = \epsilon_s L$ and $x_a = \epsilon_a L$, then;

$$\epsilon_s L + \epsilon_a L = L \Delta T(\alpha_a - \alpha_s)$$

and after cancellation of the common factor L and using values of linear expansion coefficients from Table 3.1.1 we have:

$$\epsilon_s + \epsilon_a = 40(23 \times 10^{-6} - 12 \times 10^{-6})$$
$$\epsilon_s + \epsilon_a = 440 \times 10^{-6} \qquad (3.1.13)$$

Now using the relationships from part (i) given by equations (3.1.9 and 3.1.11). Equation (3.1.13) may be expressed as:

$$0.0223\sigma_a + 0.0133\sigma_a = 440 \times 10^{-3} \qquad (3.1.14)$$

from which $\sigma_a = 12.36$ N/mm$_2$

So *stress* due to temperature change in *aluminium tube* = 12.36 MN/m^2.

Substituting our value of σ_a into Equation (3.1.11) then:

$$\sigma_s = 4.69 \times 12.36.$$

So *stress due to temperature rise in steel rod = 57.97 MN/m²*.

Check that you can reproduce Equation (3.1.14) by following the work we did in part (i) of this example.

We are now in a position to find the *total* stresses in the compound bar assembly. The total stress in the aluminium tube and steel rod is caused by the stress due to tightening and the stress due to rise in temperature that is:

total stress in aluminium tube is equal to:
92.6 MN/m² + 12.36 MN/m² = 104.96 MN/m² (compressive).

Also:

total stress in steel rod is equal to:
434.3 MN/m² + 57.97 MN/m² = 492.27 MN/m² (tensile).

Example 3.1.4 concludes our study on compound bars. I hope you noticed the similarity in the way we dealt with both elastic and thermal strain. Before leaving this subject completely it will be worth our while to look at two closely related characteristics of materials very briefly. These are *Poisson's ratio* and *bulk modulus*.

Poisson's ratio

We have so far only been concerned with uniaxial strains. However, when we stretch a solid such as a bar or rod, the axial extension is accompanied by a lateral reduction (see Figure 3.1.5). If the solid is subjected to loads in the elastic range it obeys Hooke's law, so the axial strain at a point will be equal to the lateral strain at that point.

The ratio of the lateral strain to the axial strain is known as *Poisson's ratio*.

$$\text{Poisson's ratio} = -\frac{\text{lateral strain}}{\text{axial strain}}$$

The minus sign results from the convention that compressive strains are considered negative and tensile strains positive. Since Poisson's ratio always produces a tensile strain accompanied by a compressive strain, the laws of arithmetic always produce a minus sign. Typical values for engineering metals are centred around the value 0.3.

Bulk modulus

When dealing with liquids subjected to changes in pressure due to either temperature change or forces exerted within closed containers, axial strain has no relevance. Changes in dimension occur throughout the whole volume of the liquid. Therefore, we need to look at bulk properties of the liquid that may be subjected to pressure, p, which causes a reduction in volume V, when for example that liquid is contained within a pressure vessel.

The term *volumetric strain* is used for the fractional change in volume, i.e. $\delta V/V$ of a liquid when subjected to a change in pressure. The term *bulk modulus* (symbol K), is used for the pressure divided by the volumetric strain. Thus

$$K = -\frac{p}{\delta V/V} \qquad (3.1.15)$$

The minus sign occurs because an increase in pressure is accompanied by a corresponding reduction in volume. This concludes our brief introduction to thermal stress and strain. Now try the problems.

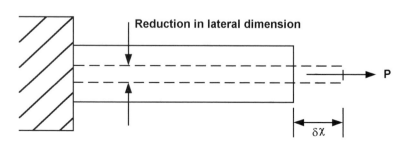

Figure 3.1.5 *Illustration of axial and lateral strain for a solid*

Problems 3.1.2

(1) Find the change in length of the aluminium tube in Example 3.1.4, due to the 40°C rise in temperature.

(2) Two walls 7 m apart are supported by a 25 mm diameter plain carbon steel tie rod as shown in Figure 3.1.6. The rod is heated to 150°C and the nuts are screwed up finger tight. If when the rods are cooled to 50°C, the walls are pulled in, so that the distance between them is reduced by 5 mm find the tension in the rod. Take $E_s = 210$ GPa and $\alpha_s = 12 \times 10^{-6}/°C$.

(3) A stainless steel rod is placed inside, and concentric with, a mild steel tube. There is a large radial clearance between the rod and tube but both are welded at each end to form a compound bar of length 500 mm. If the compound bar is subject to a temperature rise of 30°C and the extension is found to be 0.2 mm, find the stress in the rod given that E for the rod is 170 kN/mm².

Torsion

Drive shafts for pumps and motors, propeller shafts for motor vehicles and aircraft as well as pulley assemblies and drive couplings for machinery, are all subject to *torsional* or *twisting* loads. At the same time shear stresses are set up within these shafts resulting from these torsional loads. Engineers need to be aware of the nature and magnitude of these torsional loads and the subsequent shear stresses in order to design against premature failure and to ensure safe and reliable operation during service.

Review of shear stress and strain

In order to study the theory of torsion it is necessary to review the fundamental properties of shear stress and strain. For some of you this may be a new topic, for others the coverage will act as revision however, no matter what your background, it is a convenient place to start.

If two equal and opposite parallel forces, not acting in the same straight line, act on a body, then that body is said to be loaded in *shear*, Figure 3.1.7. Shear stress is defined in a similar way to tensile stress, except that the area in shear acts parallel to the load, so:

$$\text{Shear stress } (\tau) = \frac{\text{load causing shear } (F)}{\text{area in shear } (A)}$$

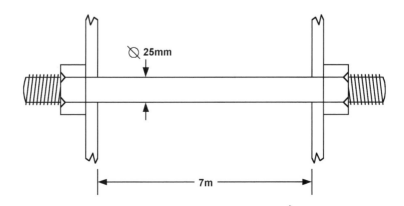

Figure 3.1.6 *Problem 3.1.2 tie rod*

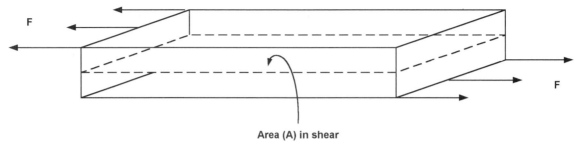

Figure 3.1.7   *Solid subject to shear stress*

Figure 3.1.8 illustrates the phenomenon known as shear strain. With reference to Figure 3.1.8, *shear strain* may be defined in two ways as:

(i)   the relative distance between the two surfaces on which the shear load acts divided by the distance between the two surfaces; or

(ii)   the angle of deformation θ (in radians).

The relationship between shear stress and shear strain is again similar to that of tensile stress and strain. In the case of shear the modulus is known as the modulus of rigidity or shear modulus (*G*), and is defined as:

$$G = \frac{\text{shear stress } (\tau)}{\text{shear strain } (\gamma)}.$$

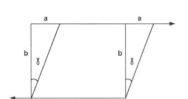

Figure 3.1.8   *Shear strain and the angle of deformation*

The modulus of rigidity (*G*), or shear modulus, measures the stiffness of the body, in a similar way to the elastic modulus, the units are the same, *G* being measured in N/m².

Riveted joints are often used where high shear loads are encountered, bolted joints being used where the primarily loads are tensile.

Example 3.1.5

Figure 3.1.9 shows a bolted coupling in which the two 12 mm securing bolts act at 45° to the axis of the load. If the pull on the coupling is 80 kN, calculate the direct and shear stresses in each bolt.

We know that the axis of the bolts is at 45° to the line of action of the load (*P*) = 80 kN. We will assume that the bolts are ductile and so the load is shared equally between the bolts.

Shear area of each bolt is $= \dfrac{\pi (12)^2}{4} = 113.1 \text{ mm}^2$

cont.

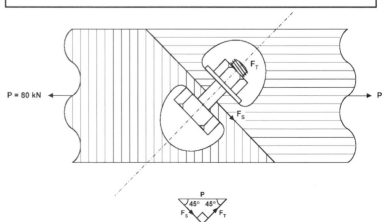

Figure 3.1.9   *Example 3.1.5 bolted coupling subject to tensile load*

Shear force (F_s)
from diagram = Psin 45 = $80 \times 10^3 \times 0.7071$ = 56568.5 N
This shear force is shared between the two bolts.
Thus shear force on each bolt = $\dfrac{56568.5}{2 \times 113.1}$ = 250.08 N/mm²

Similarly:
the direct tensile force = Pcos 45 = $80 \times 10^3 \times 0.7071$ = 56568.5 N
and the tensile force on each bolt = $\dfrac{56568.5}{2 \times 113.1}$ = 250.08 N/mm²

*So direct and shear stresses are, respectively, 250 MN/m² and
250 MN/m² .*

Engineers' theory of torsion

Shafts are the engineering components which are used to transmit
torsional loads and twisting moments or *torque*. They may be of any
cross-section but are often circular, since this cross-section is particularly
suited to transmitting torque from pumps, motors and other power
supplies used in engineering systems.

We have been reviewing shear stress because if a uniform circular
shaft is subject to a torque (twisting moment) then it can be shown that
every section of the shaft is subject to a state of pure shear. In order to
help us derive the engineers' theory of torsion which relates torque,
the angle of twist and shear stress, we must first make the following
fundamental assumptions.

(i)   The shaft material has uniform properties throughout.
(ii)   The shaft is not stressed beyond the elastic limit, in other words
       it obeys Hooke's law.
(iii) Each diameter of the shaft carries shear forces which are
       independent of and, do not interfere with their neighbours.
(iv)   Every cross-sectional diameter rotates through the same angle.
(v)   Circular sections which are radial before twisting are assumed to
       remain radial after twisting.

Let us first consider torsion from the point of view of the *angle of twist*.

Figure 3.1.10 shows a circular shaft of radius R which is firmly fixed
at one end and at the other is subject to a torque (twisting moment) T.
Imagine that on our shaft we have marked a radial line of length L,
which when subjected to torque T, twists from position p to position q.
Then the angle θ is the same angle of distortion we identified in Figure
3.1.8 and is therefore the *shear strain* of our shaft.

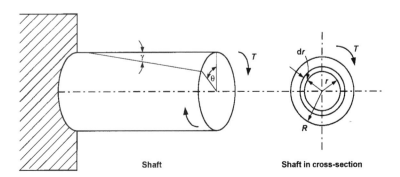

Figure 3.1.10 *Circular shaft
subject to torque*

You will remember that the angle of distortion is measured in radians. When considering torsion, shear strain or the angle of twist is measured in radians per unit length, that is in *rad/m*.

Thus from basic trigonometry:

arc length pq = the radius R multiplied by θ the angle of distortion seen in cross-section. It also follows that the arc length pq = the length of the radial line L multiplied by the angle of distortion γ so:

arc length $pq = R\theta = L\gamma$ from which

$$\gamma = \frac{R\theta}{L} \qquad (3.1.16)$$

Now from our previous study of the modulus of rigidity we know that:

$$G = \frac{\tau}{\gamma} \quad \text{or} \quad \gamma = \frac{\tau}{G} \qquad (3.1.17)$$

Then combining equations (3.1.16) and (3.1.17) and re-arranging gives the relationship:

$$\frac{\tau}{R} = \frac{G\theta}{L} \qquad (3.1.18)$$

The above relationship is independent of the value of the radius, R, so that any intermediate radius, emanating from the centre of the cross-section of the shaft can be considered. The related shear strain can then be determined for that radius.

With a little more algebraic manipulation we can also find expressions which relate the *shear stresses*, developed in a shaft subject to pure torsion. Their values are given by equations (3.1.17) and (3.1.18), if these two equations are combined and rearranged we have:

$$\tau = \gamma G = \frac{G\theta R}{L} \qquad (3.1.19)$$

Equation (3.1.19) is useful because it relates the shear stress and shear strain with the angle of twist per unit length. It can also be seen from this relationship that the shear stress is directly proportional to the radius, with a maximum value of shear stress occurring at the outside of the shaft at radius R. Obviously the shear stress at all other values of the radius will be less, since R is less. Other values of the radius apart from the maximum are conventionally represented by lower case r, see Figure 3.1.10.

Figure 3.1.11 shows the cross-section of a shaft subject to torsional loads, with the corresponding shear stress distribution, which increases as the radius increases.

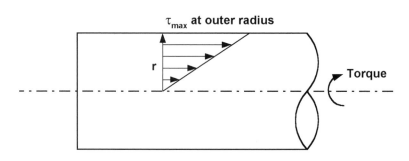

Figure 3.1.11 *Shear stress distribution for shaft subject to a torque*

Although not part of our study at this time, it should be noted that the shear stresses shown in Figure 3.1.11 have complimentary shears of equal value running normal to them, i.e. parallel with the longitudinal axis of the shaft.

We will now derive the relationship between the *stresses* and the *torque* that is imposed on a circular shaft, this will involve the use of the integral calculus. Figure 3.1.10 shows the cross-section of our shaft, which may be considered as being divided into minute parts or *elements*. These elements having radius r and thickness dr. From elementary theory we know that:

force set up on each element = stress $\times$ area.

The area of each of these cross-sectional elements of thickness dr at radius (approximately) equal to r is:

$$= 2\pi r \mathrm{d}r$$

now if τ is the shear stress at radius r then the shear force on the element is: $= 2\pi r \mathrm{d}r\tau$
so the torque carried by the element is:

$$= 2\pi r^2 \tau \mathrm{d}r \qquad (3.1.20)$$

also from Equation (3.1.19) $\tau = \dfrac{G\theta r}{L}$ and so substituting Equation (3.1.19)

into Equation (3.1.20) for the shear stress t then:

torque carried by element $= 2\pi r^2 \times \dfrac{G\theta}{L} \times \mathrm{d}r$

$$= 2\pi \dfrac{G\theta}{L} r^3 \mathrm{d}r.$$

Now the total torque carried by the solid shaft is the sum of all the element torques, from the centre of the cross-section to the extremities of the shaft where $r = R$. The integral calculus may now be used to sum all the elements.

Then total torque $T = 2\pi \dfrac{G\theta}{L} r^3 \mathrm{d}r$

Since all circular sections (elements) remain radial before and after torque is applied then θ remains the same for all the elements making up the solid shaft. Also because G and L are constants then:

$T = \dfrac{G\theta}{L} 2\pi r^3 \mathrm{d}r$ where $J = 2\pi r^3 \mathrm{d}r$ = polar second moment of area

$T = \dfrac{G\theta}{L} J$

or $\dfrac{T}{J} = \dfrac{G\theta}{L} \qquad (3.1.21)$

Now combining equations (3.1.18) and (3.1.21) gives the relationship known as the *engineers' theory of torsion*.

$$\dfrac{T}{J} = \dfrac{\tau}{R} = \dfrac{G\theta}{L} \qquad (3.1.22)$$

In practice, it has been shown that the engineers' theory of torsion, based on the assumptions given earlier, shows excellent correlation with experimental results. So although the derivation of Equation (3.1.22) has been rather arduous, you will find it very useful when dealing with problems related to torsion!

We will now consider in a little more detail the polar second moment of area J, identified above, i.e.

$$J = 2\pi r^3 dr.$$

One solution of the integral between the limits the polar second moment of area for a solid shaft is:

$$J = 2\pi[r^4/4] \quad \text{and so } J = \frac{2\pi R^4}{4} \text{ or } J = \frac{\pi D^4}{32} \tag{3.1.23}$$

The polar second moment of area defined above is a measure of the resistance to bending of a shaft, more will be said on this subject in Chapter 5.

The polar second moment for hollow shafts is analogous to that for solid shafts, except the area is treated as an annulus.

Therefore the polar second moment of area for a *hollow shaft* is:

$$J = \frac{\pi}{2}(R^4 - r^4) \text{ or } J = \frac{\pi}{32}(D^4 - d^4) \tag{3.1.24}$$

When the difference between the diameters is very small as in a very thin walled hollow shaft, the errors encountered on subtraction of two very large numbers close together, prohibits the use of Equation (3.1.23). We then have to use an alternative relationship, which measures the polar second moment of area for an individual element. For very thin walled hollow shafts this is a much better approximation to the real case.

That is, the polar second moment of area for a thin walled hollow cylinder is:

$$J = 2\pi r^3 t \text{ (approximately, where } t = dr) \tag{3.1.25}$$

Power transmitted by shafts

The power transmitted by shafts is the final topic we need to consider, in order to be equipped to put theory into practice. The most useful definition of power for a rotating shaft carrying a torque, relates this torque with the angular velocity of the shaft.

Then the power transmitted by a shaft in watts is given by:

Power (watts) = torque × angular velocity

where the torque is measured in newton metres (Nm) and the angular velocity is measured in radians/second (rad/s). Thus:

$$\text{Power} = T\omega \text{ watts} \tag{3.1.26}$$

We are now ready to look at one or two applications of our theory.

Example 3.1.6

A solid circular shaft 40 mm in diameter is subjected to a torque of 800 Nm.

(i)   Find the maximum stress due to torsion
(ii) Find the angle of twist over a 2 m length of shaft given that

cont.

the modulus of rigidity of the shaft is 60 GN/m²

(i) The maximum stress due to torsion occurs when $r = R$, i.e. at the outside radius of the shaft. So in this case $R = 20$ mm. Using the standard relationship

$$\frac{T}{J} = \frac{\tau}{R}$$

We have the values of R and T, so we only need to find the value of J for our solid shaft and then we will be able to find the maximum value of the shear stress τ_{max}.

Then for a solid circular shaft $J = \dfrac{\pi D^4}{32}$

so $J = \dfrac{\pi (40)^4}{32} = 0.251 \times 10^6$ mm⁴

and on substitution into the standard relationship given above we have:

$$\tau = \frac{(25)(800 \times 10^3)}{0.251 \times 10^6} \quad \frac{(mm)(N\ mm)}{mm^4}$$

giving $\tau_{max} = 79.7$ N/mm².

This value is the maximum value of the shear stress, which occurs at the outer surface of the shaft.
Notice the manipulation of the units, care must always be taken to ensure consistency of units, especially where powers are concerned!
(ii) To find θ we again use the engineers' theory of torsion, which after rearrangement gives:

$$\theta = \frac{LT}{GJ}$$

and substituting our known values for L, T, J and G we have:

$$\theta = \frac{(2000)(800 \times 10^3)}{(60 \times 10^3)(0.251 \times 10^6)} \quad \frac{(mm)(N\ mm)}{(N\ mm^{-2})(mm^4)}$$

$= 0.106$ radians
So angle of twist $= 6.07$ degrees.
Note once again the careful manipulation of units!

Example 3.1.7

Calculate the power which can be transmitted by a hollow circular propshaft, if the maximum permissible shear stress is 60 MN/m² and it is rotating at 100 rev/min. The propshaft has an external diameter of 120 mm and internal diameter of 60 mm.

Again we use the engineers' theory of torsion and note that the maximum shear stress (60 MN/m²) will be experienced on the outside surface of the propshaft, where $R = 60$ mm.

Then using: $T = \dfrac{\tau J}{R}$ where $J = \dfrac{\pi}{32}(D^4 - d^4)$

cont.

Then $J = 19.09 \times 10^6$ mm^4 (you should check this result)

So torque $T = \dfrac{(60)(19.09 \times 106)}{60} = 19.09 \times 106$ N mm

$= 19.09 \times 103$ N m.

Now the angular velocity in rad/s is $= \dfrac{2\pi 100}{60} = 10.47$ rad/s

and we know that power is $= T\omega$

So maximum power transmitted by propshaft:

$= (19.06 \times 103)(10.47) = 199.6$ kW.

In this example not all working has been shown, you are strongly advised to check all results and ensure again that the units correspond.

Problems 3.1.3

(1) A solid shaft rotating at 140 rev/min has a diameter of 80 mm and, transmits a torque of 5 kN m. If $G = 80$ GN/m^2, determine the value of τ_{max} and the angle of twist per metre length of the shaft.

(2) A hollow shaft has an external diameter of 100 mm and internal diameter of 70 mm. It transmits 750 kW of power at 1200 rev/min. Find the maximum and minimum shear stress and determine the angle of twist over a 2 m length, given $G = 75$ GN/m^2.

Uniformly accelerated motion

This section is concerned with the study of linear and angular motion, and the application of Newton's laws to moving bodies. We will look at linear and angular acceleration, angular kinetic energy and, linear and angular motion using energy methods.

Linear equations of motion

The equations of linear and angular motion may be familiar to you, their derivation and use are included here as useful revision. The important point to realise with respect to the equations of linear motion is that they are concerned with uniform or constant acceleration. If we also consider equations of angular motion with constant acceleration, they may be derived simply from the equation of linear motion, as you will see later. In real life constant acceleration rarely occurs, but linear and angular motion can be modelled very well, if we make this assumption and formulate relevant equations which simplify our analysis.

We can quite simply derive the linear equations of motion (for constant acceleration) by considering the velocity–time graph shown in Figure 3.1.12.

You should note that the gradient of the graph gives the acceleration a and the area under the graph gives the distance travelled s. Also note that the graph shows the initial velocity u, final velocity v and the time t.

So from Figure 3.1.12 we have:

Figure 3.1.12   *Velocity–time graph*

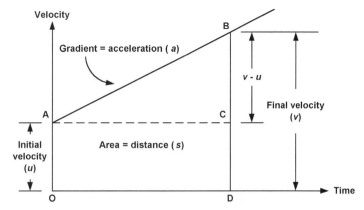

$$\text{gradient} = BC/AC \quad \text{or} \quad a = \frac{v-u}{t} \tag{3.1.27}$$

$$\text{from which} \quad v = u + at \tag{3.1.28}$$

now area $OABD$ = area $OACD$ + area ABC

$$\text{that is } s = ut + \tfrac{1}{2}(v-u)t \tag{3.1.29}$$

and on rearrangement and division by the common factor t, we get

$$\frac{s}{t} = \tfrac{1}{2}(v+u) \tag{3.1.30}$$

Equation 3.1.30 enables us to find the average velocity s/t.

Make sure you can obtain Equation 3.1.30 from Equation 3.1.29 by transposition and separating out the common factor t!

We can also find the distance in terms of final velocity and acceleration by considering Equation 3.1.27. where:

$$at = v - u$$

and substituting at for $v-u$ in Equation 3.1.29 we get:

$$s = ut + \tfrac{1}{2}at^2 \tag{3.1.31}$$

Now by rearranging Equation 3.1.27 we get,

$$t = \frac{v-u}{a}$$

and also rearranging Equation 3.1.30 we get,

$$s = \tfrac{1}{2}(v+u)t \tag{3.1.32}$$

and then substituting $\dfrac{v-u}{a}$ for t into Equation 3.1.32 gives:

$$s = \tfrac{1}{2}(v+u)\frac{(v-u)}{a} \quad \text{and on rearrangement we get:}$$

$$v^2 - u^2 = 2as \tag{3.1.33}$$

Mathematics in action

Note that we can *apply the calculus* in order to find some of the equations identified above. We know that the differential calculus may be used to find rates of change. For example, the rate of change of distance with respect to time is velocity, this may be represented mathematically as ds/dt. So at an instant in time we are able to determine the rate of

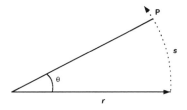

Angular distance, $s = r\theta$ (θ in radians)

Figure 3.1.13 *Illustration of radian measure*

change of distance or instantaneous velocity.

Consider Equation 3.1.31 where, $s = ut + \frac{1}{2}at^2$, now if we differentiate this equation with respect to time t, we have,

$$\frac{ds}{dt} = u + at \quad \text{where} \quad \frac{ds}{dt} = v = u + at$$

we obtain $v = u + at$ which is Equation 3.1.28!

I am sure you also know that, the rate of change of velocity with respect to time dv/dt gives us acceleration. Also, that the angular distance s, through which the point P travels (Figure 3.1.13) is given by the angle θ in radians subtended at the centre O, multiplied by the radius r. Then:

$$s = r\theta \tag{3.1.34}$$

Now differentiating distance s with respect to time t gives a linear tangential velocity of P and differentiating θ (angular distance in radians) with respect to time t, gives the angular velocity ω (in rad/s) of the rotating arm OP, so we have,

$$\frac{ds}{dt} = \frac{rd\theta}{dt} \tag{3.1.35}$$

where

$\frac{ds}{dt} = v$ (tangential linear velocity) and $\frac{d\theta}{dt} = \omega$ (angular velocity of arm OP)

then from Equation 3.1.35 we have,

$$v = r\theta \tag{3.1.36}$$

Similarly, again differentiating both sides of Equation (3.1.36) we obtain

$$\frac{dv}{dt} = \frac{r\,d\theta}{dt}$$

and knowing that dv/dt gives linear acceleration a and $d\theta/dt$ gives angular acceleration (rad/s²) α, we have,

$$a = r\alpha \tag{3.1.37}$$

You will find the above derived relationships very useful.

Angular equations of motion

Equations 3.1.34, 3.1.36 and 3.1.37 provide a relationship between linear and angular distance, velocity and acceleration, respectively. They enable us to use the linear equations of motion for constant acceleration and *transpose* them to find their equivalents for angular motion with constant acceleration. The linear equations of motion are now summarised, together with Equations 3.1.34, 3.1.36, and 3.1.37 which I will refer to as the transposition equations of motion.

Linear equations of motion:

$$v = u + at \tag{3.1.28}$$
$$s = ut + \frac{1}{2}at^2 \tag{3.1.31}$$
$$s = \frac{1}{2}(u + v)t \tag{3.1.32}$$
$$v^2 - u^2 = 2as \tag{3.1.33}$$

Transposition equations

$$s = r\theta \qquad\qquad (3.1.34)$$
$$v = r\omega \qquad\qquad (3.1.36)$$
$$a = r\alpha \qquad\qquad (3.1.37)$$

Now by substituting the *angular relationships* for distance (s), velocity (v, u) and acceleration (a), into the linear equations of motion we get their angular equivalent.

Angular equations of motion

$$\omega_f = \omega_i + \alpha t \qquad\qquad (3.1.38)$$
$$\theta = \tfrac{1}{2}(\omega_i + \omega_f)t \qquad\qquad (3.1.39)$$
$$\theta = \omega_i t + \tfrac{1}{2}\alpha t^2 \qquad\qquad (3.1.40)$$
$$\omega_f^2 - \omega_i^2 = 2\alpha\theta \qquad\qquad (3.1.41)$$

Where the units in the SI system are:

θ = angular distance in radians (rads);

t = time in seconds (s);

ω_f, ω_i = final and initial angular velocities in radians per second (rad/s);

α = angular acceleration in radians per second per second (rad/s²).

Example 3.1.8

The armature of an electric motor rotating at 1500 rev/min accelerates uniformly until it reaches a speed of 2500 rev/min. During the accelerating period the armature makes 300 complete revolutions. Determine the angular acceleration and the time taken.

Solving this type of problem is best achieved by first writing down the *knowns* in the correct SI units.

So in this case we have,

initial angular velocity of armature = 1500 rev/min or

$$\omega_i = \frac{1500 \times 2\pi}{60} = 157 \text{ rad/s}$$

Make sure you can convert rev/min into rad/sec and vice-versa!

Similarly, $\quad \omega_f = \dfrac{2500 \times 2\pi}{60} = 262 \text{ rad/s}$

$\theta = 300 \times 2\pi = 1885$ rads.

Using our angular motion equations, we must select the equation which allows us to use the most information and enables us to find one or more of the unknowns.

Then using the equation $\omega_f^2 - \omega_i^2 = 2\alpha s$ to find α rad/s²

$$262^2 - 157^2 = 2\alpha(1885)$$

from which $\alpha = 11.7$ rad/s²

and using the equation $\omega_f = \omega_i + \alpha t$ to find the time in seconds

then $262 = 157 + 11.7t$ which gives

$t = 9$ s.

Example 3.1.9

An aircraft sits on the runway ready for take-off. It has 1.4 m diameter wheels and accelerates uniformly from rest to 225 km/h (take-off speed) in 40 seconds.
  Determine:

(i)   the angular acceleration of the undercarriage wheels;
(ii) the number of revolutions made by each wheel during the take-off run.

Apart from identifying all the knowns in the correct SI units, in this example it will also be necessary to consider a combination of linear and angular motion. If we study Figure 3.1.14 we note that, in general, the angular rotation of the wheel causes linear motion along the ground, provided there are frictional forces sufficient to convert the rotating (torque) at the wheel into linear motion.

We have, $v_i = 0$ and $v_f = \dfrac{225 \times 1000}{60 \times 60}$ m/s = 62.5 m/s.

Then using the transformation equation for velocity we have,

$v = r\omega$ and so $\omega_i = 0$ and $\omega_f = \dfrac{62.5}{0.7} = 89.29$ rad/s

and the angular acceleration may be found using the equation:

$\omega_f = \omega_i + \alpha t$ i.e. $\dfrac{89.29}{40} = \alpha = 2.23$ rad/s²

and the number of radians turned through by each wheel may be found using the equation:

$\theta = \frac{1}{2}(\omega_i + \omega_f)t$ then $\theta = \frac{1}{2}(0 + 89.29)40 = 1785.8$ rads.
So the number of revolutions turned through by each wheel is $\dfrac{1785.8}{2\pi} = 284.2$ revs.

Before we take a look at moment of inertia and its relationship with torque. It will be useful to consider Newton's laws in a little detail, since these laws are fundamental to a study of motion.

Newton's laws of motion

Newton studied and developed Galileo's ideas about motion and subsequently stated the three laws which bear his name. His laws are a set of statements that we believe to be true in most circumstances, since they are in very exact agreement with the results obtained from experimentation.

Newton's first law states that: every body continues in its state of rest or of uniform motion in a straight line, unless acted upon by external forces to change that state.

This law really defines a force as something which changes the state of rest or uniform motion of a body. Note that this force may make contact with the body, such as a push with the hands or, it may have no direct contact as in the case of gravitational, electrical and magnetic forces.
  Newton's first law also suggests that matter has a built-in resistance to change its state from rest or uniform motion. This reluctance to change, possessed by all bodies is known as *inertia*. The mass of a body is a measure of its inertia, a small mass, such as a table tennis ball requires

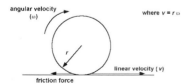

Figure 3.1.14 *Relationship between linear and angular velocity*

little force to change its state of motion, whereas a 60 ton train requires a very large force to change its state, in other words to accelerate or retard a body.

Newton's second law is concerned with how forces can be measured. *It states that the rate of change of momentum is proportional to the impressed force, and takes place in the direction of the straight line in which the force acts.*

The *momentum* of a body of constant *mass m* moving with *velocity u* is, by definition, equal to the product of the mass and velocity, that is

momentum = mass (m) × velocity (u).

If a force acts on a body for a period of time t and changes its velocity from u to v, then the body is subject to a change in momentum:

change in momentum = $mv - mu$

and, the rate of change of momentum = $\dfrac{mv - mu}{t}$.

Then by the second law: Force $(F) = km\dfrac{(v-u)}{t}$ where k is the constant of proportionality (which can be shown to equal unity).

Now we know that acceleration of a body $a = \dfrac{v-u}{t}$

therefore, since from Newton's second law $F = \dfrac{km(v-u)}{t}$ we have

$F = kma$

and with $k = 1$ then; $F = ma$ (3.1.42)

Equation 3.1.42 is one form of Newton's second law and it enables us to measure a force by finding the acceleration it produces on a known mass. An alternative form of Equation 3.1.42 may be used when considering the special acceleration due to gravity and weight force. We know that the weight W of a body is the produce of the mass of the body and the force due to gravity that is acting on this mass, then

$W = mg$ (3.1.43)

where, at sea-level (on earth) the value of the acceleration due to gravity which produces the weight force is taken as 9.81 m/s^2.

Newton's third law states that: *to every action there is an equal and opposite reaction.*

This law is stating that forces never occur singularly, but always in pairs. This is very important when we try to determine all the forces that may act on a body. Consider the forces acting in the pulley assembly shown in Figure 3.1.15. The weight produces *force pairs* in the restraining cables, as shown. Thus the weight exerts a force on the cable below the pulley, which is counter-balanced by the force exerted by the cable on the weight. Note the direction arrows for these forces.

Moment of inertia and angular motion

We now know from Newton's second law that:

Force (F) = mass (m) × acceleration (a).

There is a similar expression for angular motion which relates torque (turning moment), angular acceleration and the mass moment of inertia. It is given by the formula:

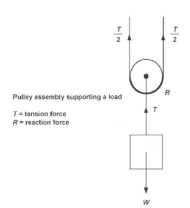

Figure 3.1.15 *Forces acting on a pulley assembly*

$$T = I\alpha \tag{3.1.44}$$

where T = torque (Nm)
α = angular acceleration (rad/s²)
$I = mr^2$

I is known as the *moment of inertia of the body about the axis of rotation*. The units of I are (kg m²), this is because the inertia of a body, from Newton's first law, is proportional to its mass in kg and, the moment of inertia is the mass multiplied by the distance squared. A mathematical derivation of the moment of inertia will be found in Chapter 6, Section 3. It should be remembered that the moment of inertia of a rotating body is equivalent to the mass of a body in linear motion.

The axis of rotation, if not stated, is normally obvious. For example a flywheel or electric motor rotates about its centre, which we refer to as its *polar axis*. When giving values of I they should always be stated with the reference axis.

Radius of gyration (k)

In order to use the above definition for the moment of inertia I, we needed to be able to determine the radii at which the mass or masses were situated from the centre of rotation of the body, see Figure 3.1.16.

For most engineering components the mass is *distributed* and *not* concentrated at any particular radius, so we need some way of finding an equivalent radius about which the whole mass of the rotating body is deemed to act.

The radius of gyration k is the radius at which a concentrated mass M (equal to the whole mass of the body) would have to be situated so that its moment of inertia is equal to that of the body.

So the moment of the body $I = Mk^2$.

All of this might, at first, appear a little confusing! In practice tables of the values of k for common engineering shapes may be used. To enable you to tackle engineering problems involving the inertia of rotating bodies, the values of k for some commonly occurring situations are given in Figure 3.1.17.

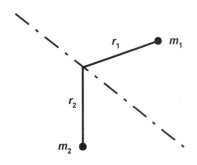

Figure 3.1.16 *Rotating masses concentrated at a point*

Example 3.1.10

A solid cylinder having a total mass of 140 kg and a diameter of 0.4 m, is free to rotate about its polar axis. It accelerates from 750 to 1500 rev/min in 15 s. There is a resistance to motion set up by a frictional torque of 1.1 Nm as a result of worn bearings. Find the torque which must be applied to the cylinder to produce the motion.

We need to find the torque T from $T = I\alpha$

where for a solid disc $I = \frac{1}{2}Mr^2$ (Figure 3.1.16).

The angular acceleration can be found by using Equation (3.1.38).

So from $\omega_f = \omega_i - \alpha t$ where $\omega_i = \dfrac{750 \times 2\pi}{60} = 78.5$ rad/s

$\omega_f = \dfrac{1500 \times 2\pi}{60} = 157$ rad/s

and $t = 15$ s then $157 = 78.5 + \alpha 15$ giving $\alpha = 5.23$ rad/s².

cont.

Figure 3.1.17 *Definition of the moment of inertia for some engineering components*

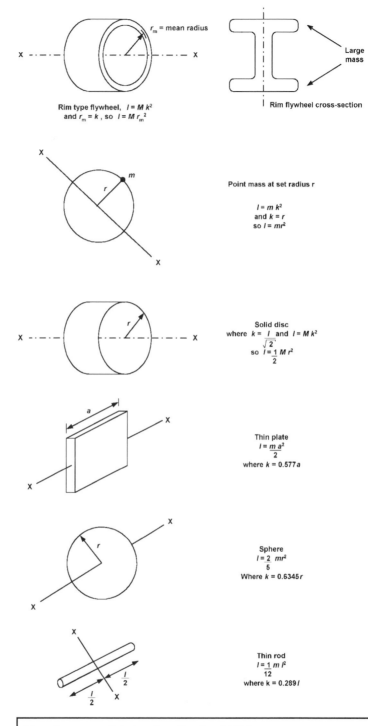

Rim type flywheel, $I = M k^2$ and $r_m = k$, so $I = M r_m^2$

Rim flywheel cross-section

r_m = mean radius

Large mass

Point mass at set radius r

$I = m k^2$
and $k = r$
so $I = mr^2$

Solid disc
where $k = \dfrac{I}{\sqrt{2}}$ and $I = M k^2$
so $I = \dfrac{1}{2} M r^2$

Thin plate
$I = \dfrac{m\,a^2}{2}$
where $k = 0.577 a$

Sphere
$I = \dfrac{2}{5} mr^2$
Where $k = 0.6345 r$

Thin rod
$I = \dfrac{1}{12} m\,l^2$
where $k = 0.289\,l$

Now for our solid circular disc $I = \tfrac{1}{2}Mr^2$

$I = \tfrac{1}{2}(140)(0.2)^2 = 2.8$ kg m^2

and from $T = I\alpha$ we have:

net accelerating torque $T = (2.8)(5.23) = 14.64$ Nm

then, since net accelerating torque = applied torque − friction torque
we have 14.64 = applied torque +1.1
giving: applied torque = 15.74 Nm.

Linear and angular kinetic energy

Kinetic energy is the energy possessed by a body due to its movement. This movement or motion can either be linear or angular (rotational).

Linear kinetic energy is given by the formula:

$$K.E. = \tfrac{1}{2}mv^2 \tag{3.1.45}$$

where m = mass (kg), v = velocity of body (m/s) and $K.E.$ = kinetic energy in Joules (J).

Note, that since *energy may be defined as the ability to do work*, the units of energy are identical to the units for work, i.e. 1 Joule = 1 Nm. Equation 3.1.45 can be derived from Newton's second law and the linear equations of motion, see Problem 3.4.1.

For the angular kinetic energy of a mass, a similar relationship exists to that given by Equation 3.1.45 for linear kinetic energy, which is derived from $T = I\alpha$ and the angular equations of motion.

Angular kinetic energy is given by the formula:

$$\text{angular } K.E. = \tfrac{1}{2}I\omega^2 \tag{3.1.46}$$

where I = mass moment of inertia, as before (kg m^2) and ω = angular velocity (rad/s). The angular kinetic energy is again measured in Joules (J).

Example 3.1.11

Two small masses both rotate at 1500 rev/min about a fixed central axis. The masses and their radii of rotation are; 6 *g* at 25 cm and 24 *g* at 21 cm. Find the:

(i)   mass moment of inertia of the system;
(ii) equivalent radius of gyration;
(iii) kinetic energy of the system.

In order to solve this problem it must be assumed that the size of the rotating masses are negligible when compared with their radius of rotation, so that the radius of rotation of each mass is the same as its radius of gyration.

Then:

(i) $\Sigma m^2 = (6)(0.25)^2$ g m^2 + $(24)(0.21)^2$ g m^2 = 0.375 g m^2 + 1.0584 g m^2

      So the moment of inertia of the system = 1.433 g m^2

(ii) Now the moment of inertia of the system also = Mk^2 where k is the radius of gyration and M is the total mass of the system.
      So from $(30)(k^2) = 1.433$ g m^2 we get k = 0.218 m and so the radius of gyration is 21.8 cm.

(iii) The kinetic energy of the system may be found using Equation 3.1.46

      where $\omega = \dfrac{1500 \times 2\pi}{60}$

      and $K.E. = \tfrac{1}{2}I\omega^2 = (\tfrac{1}{2})(1.433 \times 10^{-3})(157)^2 = 17.66$ J.

Problems 3.1.4

(1) A turbine and shaft assembly has a mass moment of inertia of 15 kg m². The assembly is accelerated from rest by the application of a torque of 40 N m. Determine the speed of the shaft after 20 seconds.

(2) A rim type flywheel gains 2.0 kJ of energy when its rotational speed is raised from 250 to 270 rev/min. Find the required inertia of the flywheel.

(3) The armature of an electric motor has a mass of 50 kg and a radius of gyration of 150 mm. It is retarded uniformly by the application of a brake, from 2000 rev/min to 1350 rev/min, during which time the armature makes 850 complete revolutions. Find the retardation and the braking torque.

(4) Using Newton's second law of motion $F = ma$ and the equation of motion $v_f^2 = v_i^2 + 2as$. Show that linear kinetic energy is given by $K.E. = \frac{1}{2}mv^2$.

Mechanical oscillations

In this short section we will look at simple harmonic motion and see how this motion is applied to linear and transverse systems. The effects of forcing and damping on oscillating systems will also be discussed.

Simple harmonic motion

When a body oscillates backwards and forwards so that every part of its motion recurs regularly, we say that it has *periodic motion*. For example a piston attached to a connecting rod and crankshaft, moving up and down inside the cylinder, has periodic motion.

If we study the motion of the piston P carefully (Figure 3.1.18), we note that when the piston moves towards A, its velocity v is from right to left. At A it comes instantaneously to rest and reverses direction. So before reaching A the piston must slow down, in other words the acceleration a, must act in the opposite direction to the velocity. This is also true (in the opposite sense) when the piston reverses its direction and reaches B at the other end of its stroke. At the times in between these two extremities the piston is being accelerated in the same direction as its velocity.

Neither the velocity nor the acceleration acting on the piston is uniform and so they cannot be modelled using the methods for uniform acceleration. However all is not lost, because although the periodic movement of the piston is complicated, its motion may be modelled using *simple harmonic motion*.

We may define *simple harmonic motion* (*s.h.m.*) as a special periodic motion in which:

(i) the acceleration of the body is always directed towards a fixed point in its path;

(ii) the magnitude of the acceleration of the body is proportional to its distance from the fixed point.

Thus the motion is similar to the motion of the piston described, except that the acceleration has been defined in a particular way.

Figure 3.1.18 *Reciprocal motion of spring–mass system*

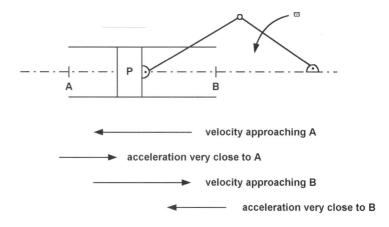

velocity approaching A

acceleration very close to A

velocity approaching B

acceleration very close to B

Note: Acceleration opposes motion as piston decelerates close to ends A & B, but is in the same direction as velocity during mid-travel.

Mathematics in action

In order to analyse s.h.m. we need to make use of a technique that involves trigonometrical identities.

Consider the point P moving round in a circle of radius r, with angular velocity ω (Figure 3.1.19a). We know from our study of the equations of motion that its tangential speed $v = \omega r$. Imagine that after leaving position A, the motion is frozen at a moment in time t, when the point is at position P as shown. Also note the point Q which is at the base of the perpendicular from the point P on the circle diameter AOB. We will now show that this point Q moves with s.h.m. about the centre O.

The point Q moves as the point P moves, therefore the *acceleration of Q* is the component of the acceleration of P parallel to OB. Now the acceleration of P along PO is given by $\omega^2 r$. Then using trigonometry (Figure 3.1.19b), the component of this acceleration parallel to AB is $\omega^2 r \cos \theta$. So we may write that the acceleration a of Q is

$$a = -\omega^2 r \cos \theta \qquad (3.1.47)$$

The negative sign results from our definition of s.h.m. given above. Where we state that the acceleration is always directed towards a fixed point. In our case towards the point O, which is represented mathematically by use of the negative sign.

Now we also know that $x = r \cos \theta$

therefore, $a = -\omega^2 x \qquad (3.1.48)$

It is important to note how the acceleration of Q varies with different values of x. For example, the acceleration of Q will be zero, when x is at the fixed point of rotation O. Also the acceleration will be at a maximum, when x is at the limits of the oscillation, that is at points A and B.

The time required for the point Q to make one complete oscillation from A to B and back, is known as the *period T*. The time period can be determined using:

$$T = \frac{\text{circumference of described circle}}{\text{speed of Q}}$$

Figure 3.1.19 *Analysis of simple harmonic motion*

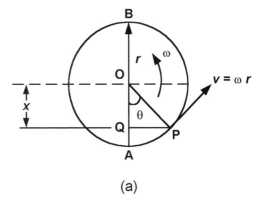

(a)

(b)   (c)

then $T = \dfrac{2\pi r}{v} = \dfrac{2\pi}{\omega}$ (3.1.49)

(since, from our transformation equations $v = \omega r$).

Also note that the frequency f is the number of complete oscillations (cycles), back and forth, made in unit time. The frequency is therefore the reciprocal of the time period T. The unit of frequency is the hertz (Hz) which is one cycle per second. Thus:

$f = \dfrac{1}{T}$ cycles per second $= \dfrac{\omega}{2\pi}$ Hz.   (3.1.50)

The *velocity* of Q is the component of P's velocity parallel to AB (Figure 3.1.19c), which is:

velocity of Q $= -v \sin \theta$

also velocity of Q $= -\omega r \sin \theta$ (from $v = \omega r$)   (3.1.51)

Note that $\sin \theta$ may be positive or negative dependent on the value of θ. A negative sign is added to the above equations so that, by convention, when the velocity acts upwards it is negative and when acting downwards it is positive.

The variation of the velocity of Q with time t assuming we start at time zero, is given by:

$= -\omega r \sin \omega t$ (since $\theta = \omega t$).

The variation of the velocity of Q with displacement x

$= -\omega r \sin \theta$

$= \pm \omega r \sqrt{(1 - \cos^2 \theta)}$ (since $\sin^2 \theta + \cos^2 \theta = 1$)

$= \pm \omega r \sqrt{(1 - (x/r)^2)}$

$= \pm \omega \sqrt{(r^2 - x^2)}$ (3.1.52)

In order to complete our analysis of the s.h.m. of Q we need to consider its *displacement*, which is given by

$$x = r \cos \theta$$

$$= r \cos \omega t \qquad (3.1.53)$$

The displacement, like the velocity and acceleration of Q is sinusoidal (Figure 3.1.20). Note from the diagram that when the velocity is zero the acceleration is a maximum.

If you found the mathematical analysis of s.h.m. difficult then you should refer to Chapter 6.

Example 3.1.12

A piston performs reciprocal motion in a straight line which can be modelled as s.h.m. Its velocity is 15 m/s when the displacement is 80 mm from the mid-position, and 3 m/s when the displacement is 160 mm from the mid-position. Determine:

(i)   the frequency and amplitude of the motion;
(ii)   the acceleration when the displacement is 120 mm from the mid-position.

We first need to determine the distance the piston moves on either side of the fixed point, in our case the mid-point. This is known as the *amplitude*.

We use the equation $v = \omega\sqrt{(r^2 - x^2)}$ and substitute the two sets of values for the velocity and displacement given in the question, then solve for r, using simultaneous equations.

From given data we have, $x = 0.08$ m when $v = 15$ m/s and $x = 0.16$ m when $v = 3$ m/s.

So substituting into above equation gives:

$$15 = \omega\sqrt{(r^2 - 0.0064)} \qquad (3.1.54)$$
$$3 = \omega\sqrt{(r^2 - 0.0256)} \qquad (3.1.55)$$

Dividing the top equation by the bottom eliminates ω and squaring both sides gives:

$$25 = \frac{r^2 - 0.0064}{r^2 - 0.0256}$$

and so $r = 162.5$ mm.

In order to determine the frequency we first need to find the angular velocity from Equation 3.1.54,

then $15 = \omega\sqrt{(0.0264 - 0.0064)}$

so $\omega = 106$ rad/s

and using equation $f = \dfrac{\omega}{2\pi} = \dfrac{106}{2\pi}$ then frequency = 16.9 Hz.

To find the acceleration when $x = 0.12$ m we use Equation 3.1.48, then $a = \omega^2 x$

$$a = (106)^2(0.12)$$

$$a = 1348.3 \text{ m/s}^2$$

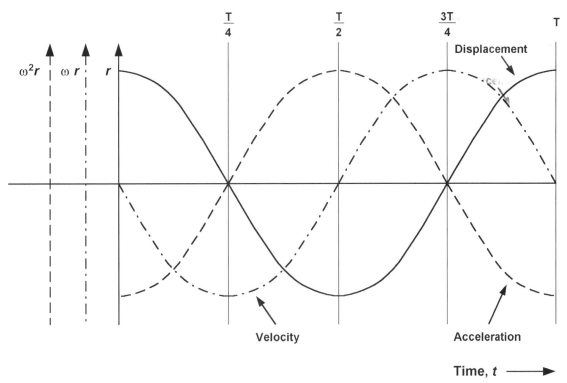

Figure 3.1.20 *Displacement, velocity and acceleration of a point subject to s.h.m.*

Motion of a spring

You will be aware that for a spring which obeys Hooke's law, the extension of the spring is directly proportional to the force applied to it. So it follows that the extension of the spring is directly proportional to the tension in the spring resulting from the applied force.

Consider the spring with the mass (m) attached to it hanging freely as shown in Figure 3.1.21. The mass exerts a downward tension (mg) on the spring, which stretches the spring by an amount (l), until the internal resistance of the spring (kl), balances the tension created by the mass. Then:

$$mg = kl \tag{3.1.56}$$

where: k = the spring constant or force required to stretch the spring by unit length (N/m). The spring constant is an inherent property of the spring and depends on its material properties and dimensions. l = the length the spring stretches in reaching equilibrium, measured in metres (m), m = the mass (kg) and g = acceleration due to gravity (m/s²). Note that both sides of Equation 3.1.56 have units of force.

Suppose the mass is now pulled down a further distance x below its equilibrium position, the stretching tension acting downwards is now $k(l + x)$ which is also the tension in the spring acting upwards. So we have

$$mg = k(l + x)$$

and so the resultant restoring force upwards acting on the mass will be

$$= k(l + x) - mg$$
$$= kl + kx - kl \quad (kl = mg \text{ from } 3.1.56)$$
$$= kx.$$

When we release the mass it will oscillate up and down. If it has acceleration (a) at extension (x) then by Newton's second law

$$-kx = ma \qquad\qquad (3.1.57)$$

The negative sign results from the assumption that at the instant shown the acceleration acts upwards, or in a negative direction (using our sign convention). At the same time the displacement (x) acts in the opposite direction, that is downwards and positive.

Now transposing Equation 3.1.57 for a gives:

$$a = -\frac{kx}{m}$$

also from Equation 3.1.48 $a = -\omega^2 x$ and so $k/m = \omega^2$ and since k and m are fixed for any system k/m is always positive. Thus the motion of the spring is a simple harmonic about the equilibrium point, providing the spring system obeys Hooke's law.

Now since the motion is simple harmonic, the time period for the spring system is given by:

$$T = \frac{2\pi}{\omega} \qquad \text{and from above} \quad \sqrt{\left(\frac{k}{m}\right)} = \omega$$

$$\text{so } T = 2\pi\sqrt{\left(\frac{m}{k}\right)}. \qquad\qquad (3.1.58)$$

Example 3.1.13

A spiral spring is loaded with a mass of 5 kg which extends it by 100 mm. Calculate the period of vertical oscillations.

The time period of the oscillations is given by Equation 3.1.58.

$$T = 2\pi \sqrt{\left(\frac{m}{k}\right)}.$$

Now k = force per unit displacement and is given from the information in the question as:

$$k = \frac{5 \times 9.81 \text{ N}}{0.1 \text{ m}} = 490.5 \text{ N/m}$$

cont.

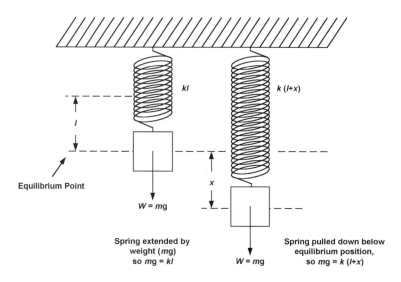

Figure 3.1.21 *Reciprocal motion of spring–mass system*

then $T = (2\pi)\sqrt{\left(\dfrac{5}{490.5}\right)} = 0.63$ s.

Motion of a simple pendulum

The simple pendulum consists of a small bob of mass m (which is assumed to act as a particle). A light inextensible cord of length l to which the bob is attached, is suspended from a fixed point O. Figure 3.1.22 illustrates the situation when the pendulum is drawn aside and oscillates freely in the vertical plane along the arc of a circle, shown by the dotted line.

It can be shown that the pendulum for small angles θ (rads) describes s.h.m. Figure 3.1.22 shows the forces resulting from the weight of the bob, together with its radial and tangential components. Note that the tangential component $mg \sin \theta$ is the unbalanced restoring force acting towards the centre, while the radial component $mg \cos \theta$ balances the force in the cord P. Then if a is the acceleration of the bob along the arc at A due to the unbalanced restoring force, the equation of motion of the bob is given by:

$$-mg \sin \theta = ma \qquad (3.1.59)$$

Again using our sign convention the restoring force is towards Q and is, therefore, considered to be negative, while the displacement x is measured from point Q and is, therefore, positive.

Now when θ is small $\sin \theta = \theta$ (in radians) using this fact it can be show that:

$$a = \frac{g}{l}x = -\omega^2 x \text{ (where } \omega^2 = g/l) \qquad (3.1.60)$$

which is *simple harmonic motion* and so the time period T for the motion is given by:

$$T = \frac{2\pi}{\omega} = 2\pi \sqrt{\left(\frac{l}{g}\right)} \qquad (3.1.61)$$

Example 3.1.14

A simple pendulum has a bob attached to an inextensible cord 50 cm long. Determine its frequency.

Knowing that the time period is equal to the reciprocal of the frequency, all that is needed is to find the time period T and then the frequency.

cont.

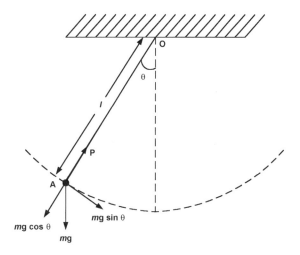

Figure 3.1.22 *Motion of a simple pendulum*

Then $T = 2\pi\sqrt{\left(\dfrac{l}{g}\right)}$

$= 2\pi\sqrt{\left(\dfrac{0.5}{9.81}\right)}$

$= 1.42$ s.

So the frequency in Hz is $= \dfrac{1}{1.42}$

$=\quad 0.704$ Hz.

Damping, forcing and resonance

If we observe a simple pendulum which is allowed to swing freely, we note that after a time the amplitude of the oscillations of the bob decrease to zero. Its motion, is therefore, not perfect s.h.m., it has been acted upon by the air which offers a resistance to its motion, we say that the pendulum has been *damped* by air resistance.

The rate at which our pendulum, or any body subject to oscillatory motion is brought to rest depends on the *degree of damping*. For example, in the case of our pendulum the air provides *light damping*, because the number of oscillations that occur before the displacement of the motion is reduced to zero, is large (Figure 3.1.23a). Similarly, a body subject to *heavy damping* has its displacement reduced quickly (Figure 3.2.23b). *Critical damping* occurs when the time taken for the displacement to become zero is a minimum (Figure 3.1.23cd).

Engineering examples of damped oscillatory motion are numerous. For example, on traditional motor vehicle suspension systems the motion of the springs are damped using oil shock absorbers. These prevent the onset of vibrations which are likely to make the control and handling of the vehicle difficult.

The design engineer needs to control the vibration when it is undesirable and to enhance the vibration when it is useful. So large machines may be placed on anti-vibration mountings to reduce unwanted vibration which, if left unchecked, would cause the loosening of parts leading to possible malfunction or failure. Delicate instrument systems in aircraft are insulated from the vibrations set up by the aircraft engines and atmospheric conditions, again by placing anti-vibration mountings between the instrument assembly and the aircraft structure. Occasionally vibration is considered useful, for example, shakers in foundries and vibrators for testing machines.

If we again return to the example of our simple pendulum the frequency of vibration is dependent on the length of the cord, so that each pendulum of given length, will vibrate at its *natural frequency*, when allowed to swing *freely*. If a body is subject to an external vibration force then it will oscillate at the frequency of this external vibration and we

Figure 3.1.23 *Effects of damping on a body subject to oscillatory motion*

(a) light damping

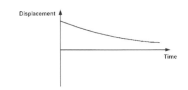

(b) heavy damping

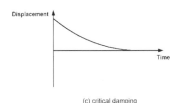

(c) critical damping

call this *forced vibration*, our foundry shaker subjects the melt to forced vibration to assist mixing and settlement.

When the frequency of the external driving force is equal to that of the natural frequency of the body or system, then we say *resonance* occurs. This is where the energy from the external driving force is most easily transferred to the body or system. Since no energy is required to maintain the vibrations of an undamped system at its natural frequency, then all the energy transferred from the driving force will be used to build up the amplitude of vibration which, at resonance, will increase without limit Hence the need to ensure such a system is adequately damped, Figure 3.1.24 illustrates the relationship between amplitude, frequency and damping at resonance.

Resonance is generally considered troublesome, especially in mechanical systems. The classic example used to illustrate the effects of unwanted resonance is the Tacoma Bridge disaster in America in 1940. The prevailing wind caused an oscillating force in resonance with the natural frequency of part of the bridge. Oscillations built up which were so large that they destroyed the bridge.

The rudder of an aircraft is subject to forced vibrations that result from the aircraft engines, flight turbulence and aerodynamic loads imposed on the aircraft. In order to prevent damage from possible resonant frequencies, the rudder often has some form of hydraulic or viscous damping fitted. The loss of a flying control resulting from resonant vibration could result in an aircraft accident, with subsequent loss of life.

So when might resonant frequencies be useful? Electrical resonance occurs when a radio circuit is tuned by making its natural frequency of oscillations equal to that of the incoming radio systems.

Problems 3.1.5

(1)   A particle moves with simple harmonic motion. Given that its acceleration is 12 m/s² when 8 cm from the mid-position. Find the time period for the motion. Also, if the amplitude of the motion is 11 cm find the velocity when the particle is 8 cm from the mid-position.

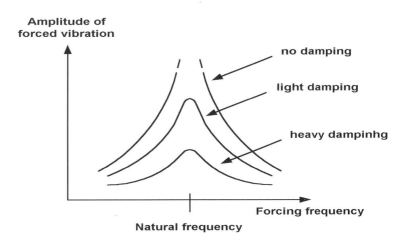

Figure 3.1.24 *Relationship between amplitude, frequency and damping at resonance*

(2)   Write a short account of simple harmonic motion, explaining the terms amplitude, time period and frequency.

(3)   A simple pendulum is oscillating with amplitude of 30 mm. If the pendulum is 80 cm long, determine the velocity of the bob as it passes through the mid-point of its oscillation.

(4)   A body of mass 2 kg hangs from a spiral spring. When the mass is pulled down 10 cm and released, it vibrates with s.h.m. If the time period for this motion is 0.4 s. Find the velocity when it passes through the mid-position of its oscillation and the acceleration when it is 4 cm from the same mid-position.

(5)   An engine consists of a simple crank and connecting rod mechanism, in which the crank is 7.5 cm long and the connecting rod is 45 cm long. When the crank is 40 degrees from top dead centre. Determine the velocity and acceleration of the piston when the crank rotates at 360 r.p.m.

3.2 Energy transfer in thermal and fluid systems

The fundamental concept of energy is crucial to our understanding of all engineering systems. Energy is often defined as, that which enables a body to do work. I am sure that you are familiar with terms such as kinetic energy (the energy due to motion) and potential energy (the energy due to the relative position of a body). There are many other examples, including electrical, chemical, strain, nuclear and thermal. All these forms of energy are present in all branches of engineering and so a study of their nature, and application, is essential for all those aspiring to become engineers.

In this section we are concerned with the study of energy in thermal and fluid systems. The section on heat transfer that follows will introduce you to the modes of transfer, conduction, convection and radiation. A number of important engineering applications of heat transfer are also covered including heat flow through materials and heat losses in fluid transport systems.

After heat transfer we look at the way energy is transferred in fluid systems. The fluid may be a liquid or a gas such as oil or air, energy losses occur as fluids travel over solid bodies. For example hydraulic oil experiences a resistance to flow from the sides of the pipes and components that contain it, similarly air flowing over the surface of an aircraft wing is subject to retarding forces near the wing surface, where it loses energy. The study of fluid energy transfer is a complicated branch of engineering science, only a brief introduction to the subject is given here.

Heat Transfer

Introduction

Engineers are concerned with designing, installing and maintaining plants and equipment based on thermodynamic principles such as boilers, refrigeration plants, heat exchangers, heaters, turbines and the associated systems that use one or more of these components. In order to understand the principles associated with these systems a study of the branch of science known as *heat transfer* is necessary.

Whenever a temperature difference exists within a system, or whenever two systems at different temperatures are brought into contact, energy is transferred. The process by which energy transport takes place is known as *heat transfer*. The energy in transit, called heat, cannot be measured or observed directly but we are able to measure and observe the effects it produces. Heat flow, like performing work, is a process where the internal energy of a system is changed.

The branch of science concerned with the relationship between heat and other forms of energy is known as *thermodynamics*. Engineering thermodynamics may be studied as a separate optional unit within this Higher National programme. We are only concerned here with the relationship between thermodynamics and heat transfer.

The principles of thermodynamics are based on observations and have, over a long period of time, been generalised into *laws* that are believed to hold true for all thermodynamic processes. The first of these principles, called *the first law of thermodynamics,* may be stated as: *energy can be neither created nor destroyed but only changed from one form to another.* You probably recognise this well known law, which governs all energy transformations and enables us to use calculations to determine results quantitatively. It has also been proven experimentally that no process is possible where the result is the net transfer of heat from a low temperature region to a high temperature region. In other words, *heat energy can only flow 'downhill' from a region of high temperature to one of low temperature.* This statement of experimental truth is known as *the second law of thermodynamics.*

The reason for the introduction of the laws of thermodynamics is that all heat transfer processes involving the transfer and change of energy must obey both of the basic laws of thermodynamics, defined previously. So we might therefore conclude that the principles of heat transfer can be determined from these laws of thermodynamics. Unfortunately this is not true, the full explanation is rather more complicated.

Classical thermodynamics is restricted to the study of systems which are in a state of *equilibrium*. So the study of thermodynamics by itself is of very little use when determining heat transformations that occur from non-equilibrium engineering processes. In real life situations equilibrium is seldom, if ever, attained! Therefore, since *heat flow* only occurs as a result of a temperature non-equilibrium, its quantitative treatment must be based on an area of science other than thermodynamics.

Engineering heat transfer

From a thermodynamic point of view the amount of heat transferred in a system, as the result of a process, is simply equal to the difference between the energy change of the system and the work done on or by the system.

When we consider *engineering heat transfer* we are primarily concerned with the determination of *the rate of heat transfer* for a specified difference in temperature. The dimensions of heaters, boilers, heat exchangers and refrigerators depend not only on the amount of heat to be transferred, but also on the rate of transfer under given external constraints. Engineering components such as transformers, electrical and mechanical machines, turbines and bearing assemblies require us to analyse heat transfer rates, in order to avoid conditions that will cause overheating and damage the equipment. These examples emphasise the importance of the solution of heat transfer problems in engineering, rather than just the use of thermodynamic reasoning alone.

Heat transfer may be defined as: *the transmission of energy from one region to another, as a result of a temperature difference between them.* In heat transfer, as in other branches of engineering, the successful solution of problems requires us to make assumptions and simplifications. This enables us to formulate mathematical models whose solution is possible either manually, or with the use of a computer. It is important to remember that all idealised models of real engineering problems should be treated with caution, since any simplified analysis may, in certain cases, severely limit the accuracy of the results.

Modes of heat transfer

Literature on heat transfer generally recognises three distinct modes of heat transmission, the names of which will be familiar to the reader, i.e. *conduction*, *convection* and *radiation*. Technically only conduction and radiation are true heat transfer processes, because both of these transmission methods depend totally and utterly on a temperature difference being present. Convection also depends on the transportation of a mechanical mass. Nevertheless, since convection also accomplishes transmission of energy from high to low temperature regions, it is conventionally regarded as a heat transfer mechanism.

Let us now consider these mechanisms of heat transfer in detail, together with examples of their use and the importance of heat transfer rates. It is important to remember that heat transfer can only take place when there is a temperature gradient caused by a difference in temperature between a system and its surroundings.

Conduction

Thermal conduction in solids and liquids seems to involve two processes. The first is concerned with atoms and molecules, the second with 'free' electrons.

Atoms at high temperatures vibrate more vigorously about their equilibrium positions in the lattice than their cooler neighbours. Since atoms and molecules are bonded to one another, they pass on some of their vibrational energy. This energy transfer occurs from atoms of high vibrational energy to those of lower vibrational energy, without appreciable displacement. This energy transfer has a knock-on effect, since high vibrational energy atoms increase the energy in adjacent low vibrational energy atoms, which in turn causes them to vibrate more energetically, causing thermal conduction to occur.

The second process involves material with a ready supply of free electrons. Since electrons are considerably lighter than atoms, then any gain in energy by electrons results in an increase in the electron's velocity and it is able to pass this energy on quickly to cooler parts of the material.

This phenomenon is one of the reasons why electrical conductors which have many free electrons are also good thermal conductors. Do remember that metals are not the only good thermal conductors, the first mechanism described above which does not rely on free electrons is a very effective method of thermal conduction, especially at low temperatures.

Examples of heat transfer by conduction are numerous. The exposed end of a metal spoon which is immersed in hot tea will eventually feel warm due to the conduction of energy through the spoon. On a cold winter's day there is likely to be a significant amount of energy lost through the walls of a heated room to the outside, primarily due to conduction. Yet another example of energy transfer by conduction is seen in the classic physics experiment where a copper rod is heated by a bunsen burner and the rod becomes warm remote from the heat source.

Mathematics in action

Fourier's law

As I have already mentioned, whenever a temperature gradient exists in a solid material, heat will flow from the high temperature to the low temperature region. The rate at which heat is transferred by conduction, Q, is proportional to the temperature gradient dT/dx multiplied by the area A through which heat is being transferred:

$$Q \propto \frac{A\,dT}{dx}$$

Note that the temperature gradient dT/dx may be expressed as the change in temperature with respect to the length of the path of the heat flow.

In the above relationship T is the local temperature and x is the distance in the direction of the heat flow. The actual rate of heat flow also depends on the thermal conductivity k, which is an inherent physical property of the material. So for conduction through a homogenous (i.e. having the same properties throughout) material, the rate of heat transfer is given by:

$$Q = -kA\frac{dT}{dx} \tag{3.2.1}$$

The minus sign results from the second law of thermodynamics, where you will remember, heat must flow from higher to lower temperature.

The above mathematical relationship was found by the French mathematician Fourier and, since this relationship has never been disproved, it is known as *Fourier's Law*. A full mathematical treatment of the Fourier Equation will be found in Chapter 6.

Fourier's law (Equation 3.2.1) gives us the rate of heat flow in the direction x per unit area perpendicular to the direction of transfer (also see Figure 3.2.1). The thermal conductivity k is a material property that indicates the amount of heat flow per unit time across a unit area when the temperature gradient is unity (horizontal $\times$ direction, Figure 3.2.1). So in the SI system, the area (A) is given in square metres (m^2) the temperature in Kelvin (K) and the rate of heat flow in watts (W). The thermal conductivity therefore has the units of watts per metre per kelvin (W/mK).

Under the steady state conditions shown in Figure 3.2.1, where the temperature distribution is linear (straight line between T_1 and T_2), the temperature gradient may be expressed as:

$$\frac{dT}{dx} = \frac{T_2 - T_1}{L}$$

and the heat flux is then

$$q = -k\frac{(T_1 - T_2)}{L} \tag{3.2.2}$$

Please remember that Equation (3.2.2) provides the *heat flux*, that

Figure 3.2.1   *One-dimensional heat transfer by conduction*

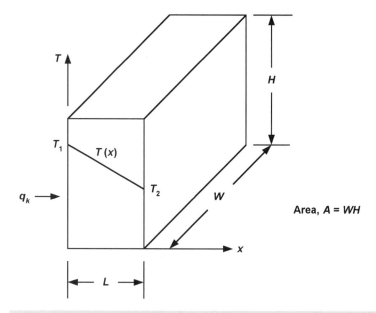

is, the rate of heat transfer per unit area. In order to find the *heat rate* by conduction, Q (W) through a plane wall of area A, we must find the product of the flux and this area, $Q = q.A$ (note the dot notation which is often used for multiplication).

So we may express Equation 3.2.2 in terms of Q, i.e:

$$Q = kA\frac{(T_1 - T_2)}{L} \qquad (3.2.3)$$

Example 3.2.1

Calculate the rate of heat loss through a pane of glass ($k = 0.78$ W/m K) 1.0 m high, 0.5 m wide and 0.75 cm thick, if the outer surface temperature is 15°C and the inner-surface temperature is 20° C. Assume that steady state conditions prevail.

Figure 3.2.2 illustrates the situation. We make the assumptions that:

(i)   steady state conditions prevail;
(ii) we are only considering conduction in the x direction through the glass (one dimensional conduction);
(iii) the glass is homogenous (has constant properties throughout).

Then we determine heat transfer through the glass using Equation 3.2.2

$$q = -k\frac{(T_2 - T_1)}{L} = -0.78\frac{(15.0 - 25.0)}{0.0075} = 1040 \text{ W/m}^2.$$

Now the heat flux represents the rate of heat transfer (heat loss per unit area in this case). Then the heat loss through the glass pane is:

$$Q = (1.0 \text{ m} \times 0.5)\ 1040 \text{ W/m}^2 = 520 \text{ W}.$$

Note that a temperature difference of 1°C is equal to 1 K,

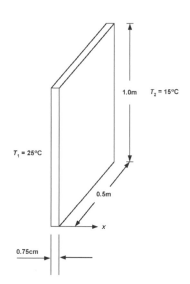

Figure 3.2.2   *Example 3.2.1 Heat conduction through pane of glass*

cont.

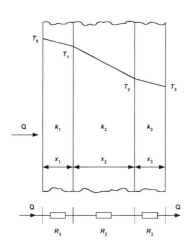

Figure 3.2.3   *Heat transfer throughout a composite wall*

therefore °C and degrees K may be used interchangeably when considering temperature difference.

Heat transfer by conduction through solid composite walls

So far we have only considered plane walls which consist of one homogenous material. There are however, many cases in practice when different materials are constructed in layers to form a *composite* wall. For example, furnace walls have an inner lining of refractory brick, followed by an insulating material and another layer of fire brick. There may also be an external layer of plaster or similar finishing material to complete the composite wall.

The flow of heat through a composite wall and the resistance to this heat flow by each of the materials is similar to the resistance to current flow, set up in an electric circuit. You will find this electrical analogy very useful when solving problems concerned with heat transfer.

Consider the diagram of a composite wall as shown in Figure 3.2.3. There are three layers of different materials of thickness x_1, x_2, x_3, with corresponding thermal conductivities k_1, k_2, k_3. The internal wall temperature is T_0 and the material interface temperatures are T_1, T_2 and T_3, as shown.

If we think of the flow of heat as being analogous to the flow of an electric current, then we know that the heat flow is caused by a temperature difference while the current flow is caused by a difference in potential. So it is possible to think of thermal resistance in a similar manner to the way we think of electrical resistance. Then from Ohm's law:

$$V = IR \text{ or } I = V/R \qquad (3.2.4)$$

(where V is potential difference, I is current and R is resistance).

Now if we compare Equation 3.2.4 with Equation 3.2.3 where

$$Q = kA\,\frac{(T_1 - T_2)}{L}$$

then,

Thermal resistance, $R = \dfrac{L}{k\,A}$ \qquad (3.2.5)

where Q is the equivalent of I, and $(T_1 - T_2)$ is the equivalent of V.

The solid composite wall is equivalent to a series of electrical resistances, as shown in Figure 3.2.3. These resistances can be added to find the total resistance of the solid composite wall to heat flow.

$$\text{Then } R_T = R_1 + R_2 + R_3 \ldots\ldots = \frac{x}{k_1 A} + \frac{x_2}{k_2 A} + \frac{x_3}{k_3 A} \qquad (3.2.6)$$

(where x_1, x_2, x_3 are the distance L through which heat flows)

It can be seen from Equation 3.2.6 that in this case the area, A remains constant throughout the wall, so it is usual to calculate the total thermal resistance for unit surface area in such problems. Again if we continue with the electrical analogy we can find an expression for the overall heat transfer through the wall.

$$\text{So we have: } Q = \frac{T_0 - T_3}{R_T} \qquad (3.2.7)$$

(from the electrical equivalent $I = V/R$).

Convection

In order to consider heat transfer from fluids to solids and vice versa, rather than just between solids we need to understand how heat is transferred by *convection*.

Heat transfer by convection consists of two mechanisms. In addition to energy transfer by *random molecular motion* (diffusion), there is also energy being transferred by the *bulk motion* of the fluid. So in the presence of a temperature gradient large numbers of molecules are moving together in bulk, at the same time as the random motion of the individual molecules takes place. The cumulative effect of both of these energy transfer methods is referred to as *heat transfer by convection*.

We are especially interested as engineers in heat transfer by convection which occurs between a fluid in motion and a solid boundary surface. At the fluid–surface interface there is a region where the velocity of the fluid varies from zero at the solid surface to some finite value, u associated with the flow velocity, this region of the fluid is known as the *velocity boundary layer*. Now if the surface and fluid flow temperatures differ there is also a region of the fluid through which the temperature varies from T_s at the solid surface to T_f in the flow, as the heat energy is transferred. This region is called the *thermal boundary layer*, Figure 3.2.4 illustrates boundary layer phenomenon.

The important point to note is that at the surface where the bulk velocity is zero, all heat energy is transferred by random molecular motion, that is by conduction. As we move further away from the surface the boundary layer grows and heat transfer is more and more dependent on the contribution made by the bulk fluid motion.

Convection heat transfer may be classified according to the nature of the flow. When the fluid flow is caused by external means, we refer to heat transfer as *forced convection*. As an example consider the use of a fan to provide forced convection air cooling of the electrical components within a personal computer. *Natural convection* is induced by forces which result from density differences caused by temperature variations within a fluid. So if we again consider electrical circuit boards as an example. Air makes contact with the hot electrical components and experiences

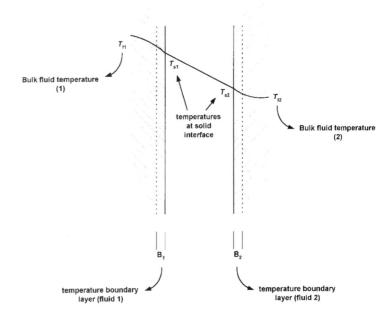

Figure 3.2.4 *Thermal boundary layer at fluid/solid interface*

an increase in temperature, which causes a reduction in density. Now in this lighter air, buoyancy forces cause vertical motion and the warm air rises. As it does so it is replaced by an inflow of cooler ambient air, these *convection currents* now act as the heat transfer mechanism.

Regardless of the heat transfer process the rate of heat transfer, which is of particular interest to engineers, may be determined by using *Newton's law of cooling*. The appropriate rate equation is of the form:

$$Q = hA(T_s - T_f) \tag{3.2.8}$$

where h is called the *heat transfer coefficient*. The units of h are W/m^2 K.

We are now ready to return to engineering problems involving composite walls, where heat transfer results from convection as well as conduction.

Heat transfer with convection through composite walls

If the materials of the composite wall involve heat transfer from fluids to solids and vice versa, rather than just between solids, then there is a similar electrical analogy to that for heat transfer between solids where, for fluids, as already mentioned, *the heat transfer coefficient 'h'* (units W/m^2 K) depends on the fluid and the fluid velocity.

Then in a similar way to the formulation of Equation 3.2.5 using our electrical analogy, (see Newton's law of cooling in mathematics, page n.n.) the thermal resistance of a fluid at the interface with a solid can be shown to be:

$$\text{Thermal resistance of fluid} = \frac{1}{hA} \tag{3.2.9}$$

(where Q is again the equivalent of I and the temperature difference is equivalent to V).

As well as discussing conduction and convection in some detail, I have also spent some time discussing thermal resistance using an electrical analogy. I would now like to show you how useful this technique can be when solving heat flow problems that involve composite materials.

Example 3.2.2

The wall of a house consists of an inner thermal block 125 mm wide and a 125 mm outer brick separated by an air gap. The inner surface of the wall is at a temperature of 25°C and the outside temperature is at 5°C. For simplicity no additional insulating material has been included in the construction.

Calculate the rate at which heat is lost per m^2 of the wall surface (see Figure 3.2.5).
Assume:

(i)   the heat transfer coefficient from the outside wall surface to the outside air is 5 W/m^2 K;
(ii) resistance to heat flow of the air gap is 0.15 K/W;
(iii) thermal conductivities of thermal block and outer brick are 0.2 and 0.5 W/m K, respectively.

Let us consider 1 m^2 of the surface area of the wall. Then using Equation 3.2.6 for any resistance $R = x/kA$, we have

Resistance of thermal block = $125 \times 10^{-3}/0.2 = 0.625$ K/W.

cont.

Resistance of outer brick = $125 \times 10^{-3}/0.5 = 0.25$ K/W.

Also using Equation 3.2.9 for the fluid, $R = 1/hA$, we get: resistance of the outside wall surface to the outside air = 1/5 K/W. Since we are given the thermal resistance of the air gap as 0.15 K/W then:

Total resistance R_T 0.625 + 0.25 + 0.2 + 0.15 = 1.225 K/W.

Now using Equation 3.2.7 we can find the rate of heat loss per square metre.

Then: $$Q = \frac{T_0 - T_A}{R_T} = \frac{25 - 5}{1.225} = 16.3 \text{ W.}$$

Suppose we also wanted to find the interface temperature T_1 (see Figure 3.2.5). We can apply the electrical analogy to each layer by again using Equation 3.2.7

So to find T_1 we have: $$Q = 16.3 = \frac{25 - T_1}{0.625}$$

from which $T_1 = 14.8$°C.

T_2 and T_3 can be found in a similar way.

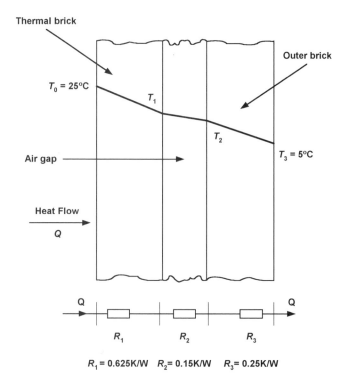

Figure 3.2.5 *Example 3.2.2 wall of house*

Heat transfer through a cylinder wall

Before we leave heat transfer by conduction and convection I would like to consider one more very common engineering situation, that of heat transfer through the wall of a cylinder. We again use our electrical analogy to find an expression for thermal resistance, but first it will be necessary to solve a simple differential equation in order to find an expression for the heat flow rate Q.

For a plane wall the area perpendicular to the heat flow is constant, this is not the case for a cylinder (Figure 3.2.6).

At any radius within the cylinder, assuming negligible axial and circumferential heat transfer, we have from the diagram:

$$Q = -k2\pi r(dt/dr) \qquad \text{per unit length of cylinder.}$$

Then separating the variables we get: $Qdr/r = -k2\pi dt$
now if we integrate the above equation, we can determine an expression for Q

then: $$Q = \frac{2k(t_1 - t_2)}{\ln(r_2/r_1)} \qquad\qquad (3.2.10)$$

The heat transfer per unit length Q has units of W/m.

Please note that in the mathematical treatment given above I have used lower case letters for thermodynamic temperature, this is a common approach and will be found in many text books. However, in science, thermodynamic temperature in degrees Kelvin is normally represented by an upper case T.

If you are unable to follow my abbreviated mathematical argument above, please refer to Chapter 6, page n.n. for a more detailed explanation.

If we use our electrical analogy again, then the temperature difference $(t_1 - t_2)$ represents the potential difference V and so comparing $I = V/R$ with Equation 3.2.9, then for a cylinder

$$R = \frac{\ln (r_2/r_1)}{2\pi k} \quad \text{for unit length} \qquad\qquad (3.2.11)$$

So Equation 3.2.11 enables us to find the thermal resistance for a cylinder wall per unit length.

In Table 3.2.1 I have, for convenience, listed the thermal conductivity of some common engineering materials. The thermal conductivities are approximate and reference should always be made to manufacturers' data for accurate values of individual materials.

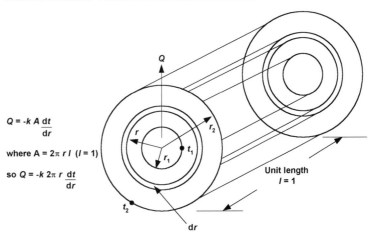

Figure 3.2.6 *Heat transfer through cylinder wall*

Table 3.2.1 *Thermal conductivity of some common engineering materials*

Material	Thermal conductivity (W/m K)
Air	0.026
Brick (common building)	0.3–0.7
Copper	390
Aluminium	235
Duralumin	166
Carbon steel (1%)	43
Cork	0.05
Cast iron	52
Concrete	0.8–1.3
Engine oil	0.15
Freon	0.07
Glass	0.8
Hydrogen	0.18
Lead	34.5
Plastics	0.2–0.4
Rubber	0.15
Water	0.6

Example 3.2.3

A steel pipe carrying steam at 250°C has an internal diameter of 100 mm and an external diameter of 120 mm. The pipe is insulated using an inner layer of specially treated plastic based composite 40 mm thick, with an outer layer of felt 50 mm thick. The ambient temperature surrounding the pipe is at 20°C. Calculate the overall thermal resistance of the assembly and so find the heat loss per unit length of the pipe. Also calculate the temperature on the outer surface of the pipe assembly. Assume that the heat transfer rates for the internal and external surfaces of the pipe assembly are 500 and 15 W/m² and the thermal conductivities of the steel, plastic composite and felt are 45, 0.2 and 0.05 W/m K, respectively.

Figure 3.2.7 illustrates the situation with the appropriate dimensions shown.

To find total resistance (R_T) we consider unit length of pipe (1 m) and using Equation 3.2.9 we have:

Resistance of steam film $= 1/hA$

$$= \frac{1}{500 \times 2\pi \times 50 \times 1 \times 10^{-3}} = 0.00637 \text{ K/W}.$$

Also from Equation 3.2.11 we have for the steel pipe:

$$\text{Resistance of pipe} = \frac{\ln (r_2/r_1)}{2\pi k} = \frac{\ln (60/50)}{2\pi \times 45} = 0.000645 \text{ K/W}.$$

In a similar manner using Equation 3.2.11 we obtain the resistances of the insulating materials.

$$\text{Resistance of plastic composite} = \frac{\ln (100/60)}{2\pi \times 0.2} = 0.406 \text{ K/W}.$$

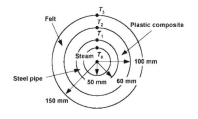

Figure 3.2.7 *Example 3.2.3 cross-section of insulated pipe assemb ly*

Resistance of felt $= \dfrac{\ln{(150/100)}}{2\pi \times 0.05} = 1.29$ K/W.

Finally, again using Equation 3.2.9 we obtain the resistance of the air film.

Resistance of air film $= \dfrac{1}{15 \times 2\pi \times 150 \times 1 \times 10^{-3}} = 0.071$ K/W.

Then total resistance (since by our electrical analogy the resistances are in series) is given by the addition of the resistances, i.e. $R_T = 1.77$ K/W.

To find the heat loss per unit length we continue with the electrical analogy, by using Equation 3.2.7, so in this case

$$Q = \frac{T_0 - T_4}{R_T} = \frac{250 - 20}{1.77} = 130 \text{ W}.$$

Finally, the temperature T_3 is given by again using Equation 3.2.7, i.e.

$$Q = 130 = \frac{T_3 - 20}{0.071} \text{ then } T_3 = 29.2^{\circ}\text{C}.$$

So, temperature of outside surface of pipe assembly $= 29.2^{\circ}$C.

In order to complete our introduction to heat transfer I will now look at heat transfer by radiation.

Radiation

Thermal radiation is energy emitted by matter that is at some finite temperature. Although we are primarily interested in radiation from solid substances, radiation can also be emitted by liquids and gases. Thermal radiation is attributed to the electron energy changes within atoms or molecules. As electron energy levels change energy is released which travels in the form of electromagnetic waves of varying wavelength.

When striking a body the emitted radiation is either absorbed by, reflected by, or transmitted through the body. If we assume that the quantity of radiation striking a body is unity, then we state that:

$$\alpha + \gamma + \tau = 1 \tag{3.2.12}$$

where α is the fraction of incident radiation energy absorbed, called the *absorptivity*, γ is the fraction of energy reflected, called the *reflectivity* and, τ is the fraction of energy transmitted, called the *transmissivity*.

Black and grey body radiation

It is useful to define an ideal body which absorbs all the incident radiation which falls on it. This body is known as a *black body*. From Equation 3.2.12 it follows that a black body is one where $\alpha = 1$.

So we say that a black body is a perfect absorber, because it absorbs all the radiation incident upon it, irrespective of wavelength. We would also expect a black body to be the best possible *emitter* at any given temperature, otherwise its temperature would rise above that of its surroundings, which is not the case.

In practice an almost perfect black body consists of an enclosure, such as a cylinder, with dull black interior and a small hole (Figure

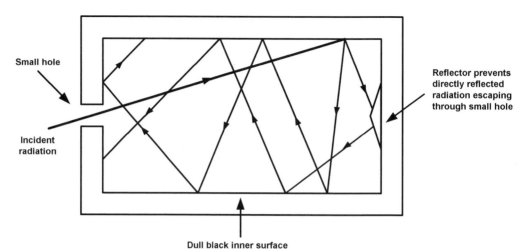

Figure 3.2.8 *Black body design*

3.2.8). Radiation entering the hole has little chance of escaping because any energy not absorbed when the wall of the cylinder is first struck, will continue to strike other surfaces gradually depleting its energy until none is left.

When the cylinder is heated by an external source, radiation appears from the hole (this radiation may be any colour according to its wavelength and temperature). Thus a whole spectrum of energy may be produced with the form shown in Figure 3.2.9.

You should note the following from the curves:

(i) as the temperature rises the energy in each wave band increases, the body becomes brighter;
(ii) at a temperature, T the energy radiated is a maximum for a certain wavelength, which decreases with rising temperature;
(iii) even at high temperatures (around 1000 K) only a small fraction of the radiation appears as visible light, temperatures of around 4000 K, are required to produce a maximum in the visible region.

So in a black body we have the equivalent of a perfect emitter and absorber. We can use this model to compare the radiation emitted by other bodies without these properties. The *emissivity*, ϵ, of a body is the ratio of the energy emitted by the body to that emitted by a black body at the same temperature and wavelength, such matter is known as a *grey body*. A comparison of the spectral emissive energy of a tungsten filament lamp (grey body) when compared with that of the corresponding black body is shown in Figure 3.2.10.

The Stefan–Boltzmann law

It was found experimentally by Stefan, and subsequently proved theoretically by Boltzmann, that the total energy radiated by all wavelengths per unit area per unit time by a black body is directly proportional to the fourth power of the absolute temperature of the body.

i.e. $E = \sigma T^4$ \hfill (3.2.13)

where $\sigma = 5.672 \, 10^{-8}$ W/m^2 K^4, is the Stefan–Boltzmann constant of proportionality, found by experimentation.

Note also from our discussion on emissivity above, that the energy emitted by a non-black body is:

$E = \epsilon \sigma T^4$ \hfill (3.2.14)

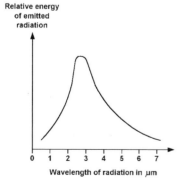

Figure 3.2.9 *Radiated energy spectrum*

Figure 3.2.10 *Grey body radiation*

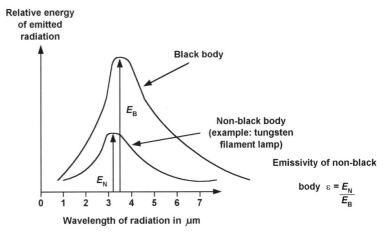

When a body is very small when compared to its large surroundings, a negligible amount of radiation is reflected from the surroundings onto the body, so in effect the surroundings are black. Under these circumstances it can be shown that, the rate of heat transfer from the body to its surroundings is,

$$Q = \epsilon\sigma A(T_s^4 - T_{sur}^4) \tag{3.2.15}$$

Now apart from radiation there are many instances where the surface of a body may transfer heat by convection from its surface to a surrounding gas (see Example 3.2.4). Then, the total rate of heat transfer from the surface is the sum of the heat rates due to these two modes. The heat transfer rate for convection is given by Equation 3.2.8, and for radiation by Equation 3.2.15. Then

$$Q = hA(T_s - T_f) + \epsilon\sigma A(T_s^4 - T_{sur}^4) \tag{3.2.16}$$

Finally there are many occasions where it is often necessary to express the net radiation exchange in the form

$$Q = h_r A (T_s - T_{sur}) \tag{3.2.17}$$

where from Equation 3.2.15 the *radiation heat transfer coefficient* h_r is

$$h_r = \epsilon\sigma(T_s + T_{sur})(T_s^2 + T_{sur}^2) \tag{3.2.18}$$

Equations 3.2.17 and 3.2.18 enable the radiation heat transfer coefficient to be determined in a similar way to equations 3.2.3 and 3.2.8 for conduction and convection.

Example 3.2.4

A steam pipe without insulation passes through a large area of a factory, in which the air and factory walls are at 20°C. The external diameter of the pipe is 80 mm, its surface temperature is 220°C, and it has an emissivity of 0.75. If the coefficient associated with free convection heat transfer from the surface to the air is 20 W/m² K, what is the rate of heat loss from the surface per unit length of pipe?

From the above we make the assumption that radiation exchange between the pipe and the factory is between a small surface and a very much larger surface, so Equation 3.2.15 derived from the Stefan–Boltzmann law applies. Also heat loss

cont.

from the pipe, to the factory air is by convection and by radiation exchange with the factory walls.

Then the heat loss from the pipe is given by:

$$Q = h\pi DL\,(T_s - T_f) + \epsilon\sigma\pi DL(T^4 - T^4)$$

where πDL is the surface area of pipe, and so heat loss per unit length of the pipe ($L = 1$) is then:

$$\frac{Q}{L} = q = (16)(\pi)(80 \times 10^{-3})(493 - 293) + (0.75)(5.67 \times 10^{-8})(\pi)(80 \times 10^{-3})$$
$$(493^4 - 293^4).$$

Then heat loss per unit length of pipe $q = 1357$ W/m.

Problems 3.2.1

(1) Explain the following terms:

    (i)   heat transfer;
    (ii)   heat transfer rate;
    (iii) conduction, convection and radiation.

(2) The inner surface of a brick wall is at 30°C and the outer surface is at 5°C. Calculate the rate of heat transfer per unit area of the wall surface. Given that the wall is 30 cm thick and the thermal conductivity of the brick is 0.45 W/m K.

(3) Calculate the rate of heat transfer by natural convection between a shed roof of area 10 m × 15 m and ambient air, if the roof surface temperature is 25°C, the air temperature is 8°C, and the average convection heat transfer coefficient is 10 W/m² K.

(4) A large pipe, 50 cm in diameter ($\epsilon = 0.85$) carrying steam has a surface temperature of 250°C. The pipe is located in a room at 15°C, and the convection heat transfer coefficient between the pipe surface and the air in the room is 20 W/m² K. Determine:

    (i)   the radiation heat transfer coefficient;
    (ii)   the rate of heat loss per metre of pipe length.

Fluid flow systems, viscosity and energy loss

Introduction

The term fluid mechanics refers to the study of fluid behaviour, either at rest or in motion. Fluids can be liquids such as water, oil and petroleum spirit or they can be gases such as air, nitrogen or oxygen. The use of fluids in engineering systems is large and varied, requiring engineers working in this field to be conversant with the concepts and applications of fluid mechanics.

For example the performance of an automated manufacturing machine which is controlled by fluid power, is dependent on the flow of hydraulic oil and the pressures exerted within the system. Heating, ventilating and pressurisation systems deliver air at relatively low pressures to living and working spaces to improve the comfort of the occupiers.

The internal combustion engine relies on the flow of fuel and oil to power and cool it. Aircraft rely on the flow of air over their wings and control surfaces to gain height and change direction (this special application of fluid mechanics being known as aerodynamics).

These examples are just a few of the many applications of the study of fluid systems. In this section we concentrate on *fluids in motion*, looking particularly at the way in which the viscosity (resistance to flow) of the fluid determines the losses that occur within a system and the effect of such losses.

Viscosity

The ease with which a fluid flows is an indication of its viscosity. Cold heavy oils such as those used to lubricate large gearboxes, have a high viscosity and flow very slowly, whereas petroleum spirit (an oil derivative) is extremely light and volatile and flows very easily and so has low viscosity. We thus define *viscosity* as the property of a fluid that offers resistance to the relative motion of the fluid molecules. The energy losses due to friction in a fluid are dependent on the fluid viscosity. We will look at friction losses in fluid systems later in this section.

As a fluid moves there is developed in it a shear stress, the magnitude of which depends on the viscosity of the fluid. You have already met the concept of shear stress and should remember that it can be defined as the force required to slide one unit area of a substance over the other. It thus has units of N/m² and is denoted by the Greek letter tau (τ).

Figure 3.2.11 illustrates the concept of velocity change in a fluid by showing a thin layer of fluid (*boundary layer*) sandwiched between a fixed and moving boundary. Now a fundamental condition exists between a fluid and a boundary, where the velocity of the fluid at the boundary surface is identical to that of the boundary. So in our case the velocity of the fluid next to the moving boundary (velocity *v*) also has velocity *v* and the fluid at the fixed or stationary boundary has velocity zero.

If the distance between the two surfaces is small the *rate of change* of velocity *v* with respect to distance *y* is linear. The velocity gradient is a measure of how the velocity changes and it is defined as $\Delta v/\Delta y$ that is :

velocity gradient or shear rate $= \Delta v/\Delta y$.

Now from our definition of shear stress we know that shear stress is directly proportional to the velocity gradient and using a constant of proportionality (μ) we have

$$\tau = \mu(\Delta v/\Delta y) \tag{3.2.19}$$

The constant of proportionality μ is known as the *dynamic viscosity*. What are the SI units of μ? The units of dynamic viscosity are easily

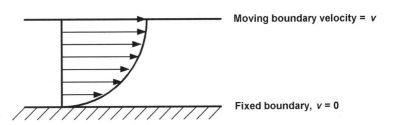

Figure 3.2.11   *Velocity change at boundary layer*

found by transposing Equation 3.2.19 for μ and looking at the units of the individual elements. Then

$$\mu = \frac{\tau \Delta y}{\Delta v} \tag{3.2.20}$$

so substituting the units for stress, distance and velocity into Equation 3.2.20 gives:

$$\mu = \frac{N}{m^2} \cdot \frac{m}{m/s} = \frac{N \cdot s}{m^2} \tag{3.2.21}$$

Note the dot notation for multiplication.

Alternative units expressing μ in terms of kg can be obtained by substituting for the unit of force, the Newton, from the defining equation of force $F = ma$.

$$1\,N = 1\,kg \cdot \frac{m}{s^2}$$

then from the units given for μ in Equation 3.2.21 we get

$$\mu = \frac{(kg \cdot m)}{s^2} \times \frac{s}{m^2} = \frac{kg}{m \cdot s} \tag{3.2.22}$$

You will still find many text books and manufacturers' literature in which the obsolete cgs units for viscosity are quoted, in the interests of completeness they are also given here.

The common units in this system are known as the Poise and centiPoise where

$$1\,P = 0.1\,Ns/m^2 \quad \text{and } 1cP = 10^{-3}\,Ns/m^2.$$

Many problems in fluid mechanics involve the use of the ratio of the dynamic velocity divided by the fluid density, this ratio defines the *kinematic viscosity*, which is more often quoted in literature and data sheets.

So kinematic viscosity $(v) = \dfrac{\mu}{\rho}$ $\qquad(3.2.23)$

The SI units for kinematic viscosity can be found simply by substituting the previously found units for dynamic viscosity and the units for density into Equation 3.2.23. Then:

$$v = \frac{kg}{m.s} \times \frac{m^3}{kg}$$

$$v = m^2/s.$$

The cgs units for kinematic viscosity are the Stoke and centiStoke where

$$1\,St = 10^{-4}\,m^2/s \qquad 1\,cSt = 10^{-6}\,m^2/s.$$

The units for dynamic and kinematic viscosity can at first seem confusing, you should take extreme care when using them. Remember that both the SI and the older cgs systems are used to accommodate the varying needs of manufacturers and users.

Example 3.2.5

The dynamic viscosity of pure water at 20°C is given as $1.0 \times 10^{-3}\,N.s/m^2$. Determine the kinematic viscosity of the water

cont.

in SI and cgs units, given that its density is 10^3 kg/m^3.

From Equation 3.2.23 we know that the kinematic viscosity $\nu = \dfrac{\mu}{\rho}$.

The units for dynamic viscosity and density, given above, are standard SI units and therefore may be substituted directly into Equation 3.2.23 to obtain the kinematic viscosity in m^2/s.

So we have: kinematic viscosity $= \dfrac{1 \times 10^{-3}}{10^3} = 1 \times 10^{-6}$ m^2/s

Now in cgs units we have that 1 cS $= 1 \times 10^{-6}$ m^2/s so kinematic viscosity of water in centiStokes is 1.0.

In Table 3.2.2 are listed some *typical values* for the dynamic and kinematic viscosity of a few common fluids, at given temperature, standard atmospheric pressure and selected densities.

Table 3.2.2 *Typical values for the dynamic and kinematic viscosity of a few common fluids*

Fluid	Temperature ($^\circ$C)	Dynamic viscosity (N.s/m^2)	Kinematic viscosity (m^2/s)
Air	0	1.72×10^{-5}	1.33×10^{-5}
Air	20	1.81×10^{-5}	1.51×10^{-5}
Water	20	1×10^{-3}	1.0×10^{-6}
Petroleum	25	2.87×10^{-4}	4.2×10^{-7}
Engine oil	20	3.5×10^{-1}	4.375×10^{-4}
Engine oil	80	1.9×10^{-1}	2.375×10^{-4}

Variation of viscosity with temperature

You will be aware from what was said earlier that cold gearbox oil has a high viscosity and is thus difficult to pour. In fact in very cold climates motorists use some form of heater underneath the engine sump to keep the oil warm so that they are able to turn the engine over on the initial start of the day. From this example and many others you are no doubt familiar with the fact that, fluid viscosity varies with temperature. In my example, as the temperature of the oil (*liquid*) is *increased*, the viscosity *decreases* noticeably.

It is important to realise that gases behave differently to liquids, with respect to temperature change and viscosity. For *gases* the viscosity *increases* with *increase* in temperature although the changes are generally smaller than those for liquids. Note the behaviour of air in Table 3.2.2, with change in temperature.

A measure of how greatly the viscosity of a fluid changes with temperature is given by its *viscosity index*, sometimes referred to as VI. Equipment, machinery and systems, which are lubricated or operated by oils and hydraulics fluids may need to operate at extremes of temperature. In these circumstances the VI becomes especially important in determining fluid behaviour.

The viscosity index system uses a method whereby the fluid (particularly oils) is compared with two representative families of fluids both at 40°C and 100°C, one showing a small change of viscosity and the other a large change. For oils, these families are chosen so that they have an index number of 100 and 0, respectively.

In general, *the higher the VI the better the temperature viscosity behaviour*, in other words the higher the viscosity index, the smaller the change in viscosity with temperature.

It is worth noting that changes in viscosity also occur with changes in *pressure*. In most *liquids* the viscosity *increases* with *increase* in pressure, this becomes significant at the pressures operating within hydraulic systems. There are thus certain similarities between the pressure sensitivity of a liquid and its temperature sensitivity, from which it might be deduced that liquids showing maximum change due to temperature would also be most affected by pressure. This is indeed the case, for example, oils of low VI increase in viscosity with pressure far more than those with a high VI.

Laminar and turbulent fluid flow

When analysing fluid flow it is often necessary to determine the characteristics of such flow. Consider the flow of air over the aircraft wing section shown in Figure 3.2.12a. It is represented by *stream lines*, which show the air flowing in layers in a smooth and regular manner, this type of flow is referred to as *laminar flow*. Another example of laminar flow occurs when the flow through the nozzle of a water hose is adjusted from the tap, until a smooth flow of water is expelled from the nozzle. If you continue to open up the tap eventually the water flows in a chaotic way, tumbling and turning over itself, this is an example of *turbulent flow* (Figure 3.2.12b).

Thus, more formally, in *laminar flow* the particles of fluid move smoothly in straight lines, although the velocity of the particles moving along one line may not necessarily be the same as that along another line. The viscosity of the fluid is of major importance in this type of flow.

In turbulent flow there is irregular and chaotic motion of the fluid particles such that a thorough mixing of the fluid takes place. In this type of flow fluid inertia effects are dominant over viscous effects.

Reynolds number

The behaviour of a fluid, particularly with regard to energy losses, is dependent on whether the flow is laminar or turbulent, for this reason it is important to have a way of predicting the type of flow, other than by direct observation which is often impossible and impractical in closed systems.

In the late nineteenth century Osborne Reynolds discovered that the flow condition in closed pipe systems depended on, fluid density, fluid viscosity, pipe diameter and the average velocity of flow. His experiments showed that these parameters, in the form of a dimensionless number, could be used to decide whether the flow was laminar or turbulent.

This dimensionless number known as Reynolds Number

$$(\text{Re}) = \frac{Vd}{\nu} \qquad (3.2.24)$$

Where: V = mean velocity of the fluid; d = the pipe diameter; ν = kinematic viscosity of the fluid.

We can show that the Reynolds number is dimensionless by substituting the standard SI units into Equation 3.2.24.

$$\text{Re} = \frac{m}{s} \cdot m \cdot \frac{s}{m^2} = \quad \text{dimensionless number.}$$

Figure 3.2.12 *Illustration of laminar and turbulent flow*

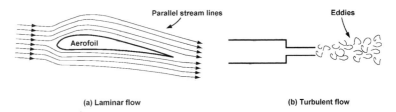

(a) Laminar flow (b) Turbulent flow

If a Reynolds number below about 2000 is obtained, the flow is assumed to be laminar and a Reynolds number of about 2500 the flow is turbulent. Between the values of 2000 and 2500 the flow is defined as *critical* and may be laminar or turbulent depending on external circumstances.

Example 3.2.6

Heavy oil flows in a pipe at an average velocity of 4 m/s. If the internal diameter of the pipe is 10 cm, determine whether the flow is laminar or turbulent. Take the density of the oil as 930 kg/m^2 and its dynamic viscosity as 1.08×10^{-1} N.s/m^2.

To answer this question we need to determine the Reynolds number, which requires the use of Equation 3.2.24.

The kinematic viscosity is given by

$$\mu/\rho = \frac{1.08 \times 10^{-1}}{930} = 1.16 \times 10^{-4} \text{ m}^2/\text{s}$$

Then $\text{Re} = \dfrac{Vd}{\nu} = \dfrac{4 \times 0.1}{1.16 \times 0^{-4}} = 3448$

Since Reynolds number is > 2500, then flow is turbulent.

Energy losses in fluid systems

Whenever a fluid flows there are energy losses which are used to overcome viscous drag forces within the system, friction effects, and heat transfer into and out of the fluid. For example, a loss of energy due to viscous effects within an enclosed hydraulic system manifests itself as a pressure loss in the direction of the fluid flow.

We are concerned in this section, mainly with the energy losses resulting from the flow of fluids in closed circular pipes and the devices used to control the flow. These include fluid power systems and fluid distribution systems and their associated fixtures and fittings.

Most of the problems associated with the flow of fluid in pipes involves the prediction of the conditions at one section of the system when the conditions at another section are known. To help us make these predictions we need to consider the *continuity* of the energy due to fluid flow, within the system. Fluid flow may be determined by considering the volume, weight or mass of fluid flowing in a system per unit of time. We refer to these as the *volume flow rate, weight flow rate* and the *mass flow rate*, respectively. The most commonly used of these flow rates is the volume flow rate, which is defined as:

$$Q = Av \tag{3.2.25}$$

where in the SI system: A = the area of section (m^2); v = the mean velocity of the fluid (m/s) ; Q = the volume flow rate (m^3/s).

Figure 3.2.13 *Equation of continuity*

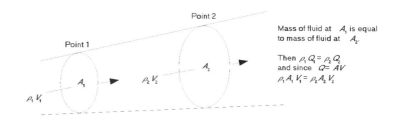

Volume flow rates are also commonly expressed in litres per min (l/min) or litres per second (l/s), where:

1 litre = 1000 cm³ or 1000 litres = 1 m³

Also the mass flow rate M may be expressed in terms of the volume flow rate Q by remembering that the volume of a substance is equal to its density multiplied by its mass, then:

mass flow rate $M = \rho Q$ (3.2.26)

So the units for mass flow rate are, from the right hand side of Equation 3.2.26, equal to:

kg/m³ × m³/s = kg/s.

Now one way in which we can determine the velocity of flow in an enclosed pipe system, is to use the principle of continuity, as mentioned before. Consider the circular pipe section shown in Figure 3.2.13. We assume that a fluid is flowing in the pipe at a steady rate, i.e. the amount of fluid flowing past any point at any time is constant. So in our case it is assumed that the amount (measured as the mass) of fluid flowing past point A_1 is equal to the amount flowing past A_2, at any moment in time.

Then since the mass flow rates are equal at both points in the pipe section we have:

$M_1 = M_2$ and so $\rho_1 Q_1 = \rho_2 Q_2$

then since $Q = Av$ the *equation of continuity* becomes:

$\rho_1 A_1 v_1 = \rho_2 A_2 v_2$ (3.2.27)

If the fluid in our pipe is a liquid it can be treated as virtually incompressible, therefore the density of the fluid will remain the same as it passes through both points. If the fluid is a gas such as air this condition is not valid. When studying airflow over aircraft lift surfaces and air for pitot static instruments, compressibility effects have to be taken into account. In our study of the continuity, we will only be concerned with *incompressible flow*.

So we may assume (for a liquid) equal densities at points 1 and 2 then:

$A_1 v_1 = A_2 v_2$ (3.2.28)

Note that Equation 3.2.28 tells us that, at any point, the flow velocity is reduced if the area of section is increased.

Example 3.2.7

If the internal diameters of the pipe at points 1 and 2 (Figure 3.2.13) are 100 mm and 150 mm, respectively, and water is flowing past point 1 with constant velocity of 6 m/s. Determine :

(i)   the velocity at point 2;
(ii) the volume flow rate;

(iii)   the mass flow rate;
(iv)   the weight flow rate.

(i)   The velocity of the water at point 2 is determined from the equation:

$A_1v_1 = A_2v_2$ where in our case $v_1 = 6$ m/s; $A_1 = 7854$ mm^2; $A_2 = 17672$ mm^2

$$v_2 = \frac{A_1v_1}{A_2} = \frac{7854 \times 6}{17672} = 2.67 \text{ m/s}.$$

Now because of the principle of continuity we may use the conditions at either section 1 or section 2, choosing section 1 we have:

(ii)   volume flow rate $Q = A_1v_1 = 7854 \times 10^{-6}$ m$^2 \times 6$ m/s
       $= 0.047$ m^3/s;
(iii)   from which the mass flow rate (assuming density of water is 1000 kg/m^3)

$M_1 = \rho Q_1 = 1000 \times 0.047 = 47$ kg/s

(iv)   the *weight flow rate* (*W*) is found by multiplying the volume flow rate Q (m^3/s) by the weight density (N/m^3). The weight density is referred to as the *specific weight* symbol (γ). Now for the water flowing in our pipe the specific weight is 9810 N/m^3.

Then from $W = \gamma Q$ we have $W = 9810$ N/m$^3 \times 0.047$ m^3/s $= 461$ N/s.

Bernoulli equation for incompressible flow

There are three forms of energy which are always considered when analysing fluid flow through a pipe, *potential energy* due to the height of the pipe, *kinetic energy* due to the fluid velocity and *flow energy or work done* to move the fluid against the forces opposing its motion.

You are already familiar with kinetic energy (KE) which we defined earlier as:

$$\text{KE} = \tfrac{1}{2}mv^2 \tag{3.2.29}$$

Potential energy (PE) is energy due to the position of a body. It is defined as:

$$\text{PE} = mgh \tag{3.2.30}$$

where m = mass (kg), g = acceleration due to gravity (m/s^2) and h = height (m) above some datum. Note that from the definition of the Newton, the units of the right hand side of Equation 3.2.30 gives PE in Nm or the Joule, as expected.

To establish an expression for the *work done* by the pressure forces in the system, consider Figure 3.2.14 which again shows a section of pipe, subject to fluid flow.

We treat the fluid as incompressible therefore the density of the fluid will be the same at stations 1 and 2 then by continuity:

$$A_1v_1 = A_2v_2 = Av \text{ (in general).}$$

Figure 3.2.14 *Determination of the Bernoulli equation*

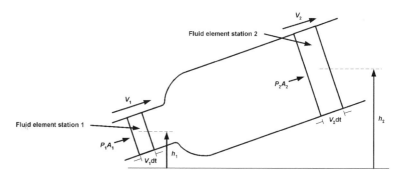

Now the work done on the fluid by the pressure force (*fluid energy*) at station 1 can be expressed as:

$$\text{fluid energy, FE} = p_1 A_1 L_1 \tag{3.2.31}$$

where p_1 = pressure (N/m²) acting at station 1; A = the area of section (m²), as before; L = the distance travelled (m) by the fluid a very short time later. It is derived from the velocity of the fluid multiplied by the time ($v.dt$).

Note that the units of the right hand side of the above equation for FE are Nm, as expected. Checking units to establish the validity of an equation is good practice, make sure you are able to use *dimensional analysis* when needed!

Also from Equation 3.2.31, $\text{FE} = p_1 V_1$ $\tag{3.2.32}$

where V_1 is the volume of the fluid subject to the pressure at station 1, and the volume (V) of the fluid is equal to the weight (w) divided by the weight density or specific weight (γ), or in symbols:

$$V_1 = \frac{w}{\gamma} \frac{\text{N}}{\text{N}} \text{ m}^3.$$

Note that since the mass of fluid is constant at both stations 1 and 2 (from the conservation of mass and equation of continuity) the weight is also constant, so the subscript 1 can be dropped.

Then $\text{FE} = \dfrac{p_1 w}{\gamma}$ $\tag{3.2.33}$

Now the *total amount of energy* possessed by the element of fluid positioned at station 1 is given by:

$$E_T = \text{FE} + \text{KE} + \text{PE}$$

where $\text{KE} = \frac{1}{2}mv_1^2 = \dfrac{\frac{1}{2}wv_1^2}{g}$ and $\text{PE} = mgh_1 = wh_1$.

So $E_1 = \dfrac{p_1 w}{\gamma} + \dfrac{wv_1^2}{2g} + wh_1.$

If the element of fluid at station 1 moves to station 2 (Figure 3.2.12), the values of pressure, height and velocity will change to p_2, h_2, v_2, as indicated.

Then by analogy, the total energy at station 2 is given by:

$$E_2 = \dfrac{p_2 w}{\gamma} + \dfrac{wv_2^2}{2g} + wh_2.$$

Now if no energy is added to the fluid or lost by the fluid as it travels between stations 1 and 2, then from the law of *the conservation of energy*, we know that:

$$E_1 = E_2$$
$$\text{or } \frac{p_1 w}{\gamma} + \frac{w v_1^2}{2g} + w h_1 = \frac{p_2 w}{\gamma} + \frac{w v_2^2}{2g} + w h_2.$$

You will note that by careful manipulation of the analysis, we have been able to make the weight (w) common to every term in the above equation, therefore, if we divide every term by the weight we get:

$$\frac{p_1}{\gamma} + \frac{v_1^2}{2g} + h_1 = p_2 + \frac{v_2^2}{\gamma} + \frac{h_2}{2g} \qquad (3.2.34)$$

This is known as *Bernoulli's equation*, and is fundamental to the study of fluid flow problems.

By carefully studying the units of the Bernoulli equation and remembering the analysis, we can interpret the true meaning of this formula. You will remember that Equation 3.2.34 was obtained by dividing each term which represented *energy*, by the *weight* of the element of the fluid. We can therefore refer to the resulting terms on each side of Equation 3.2.34 as *the energy possessed by the fluid per unit weight of fluid flow* (N.m/N). Now since we have divided the unit of energy (N.m) by the unit of weight (N). Each term in the Bernoulli equation represents the height (or *head*) of the fluid concerned with the fluid pressure energy, kinetic energy and potential energy, respectively.

So the term: (p/γ) is known as the *pressure head*; ($v^2/2\gamma$) as the *velocity head*; (h) as the *elevation head*.

Example 3.2.8

In a pipe system (Figure 3.2.15) water at 10°C is flowing from station 1 to station 2. The velocity of flow at station 1 is 3.5 m/s, the pressure is 350 kPa (remember that 1Pa = 1 N/m²) and the diameter is 50 mm. Station 2 is elevated 2.0 m above station 1 and has a diameter of 75 mm. Assuming there are no external energy losses or gains into or from the pipe system. Determine:

(i)   the velocity of the fluid at station 2;
(ii) the pressure of the fluid at station 2.

When tackling these problems, like so many others, it is best to write down the known values using the correct symbol.

cont.

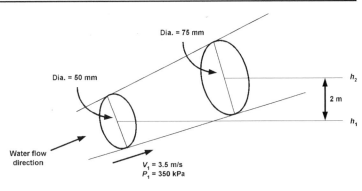

Figure 3.2.15   *Example 3.2.8*
pipe system

We have: $h_2 - h_1 = 2.0$ m, $v_1 = 3.5$ m/s, $d_1 = 50$ mm, $d_2 = 75$ mm, $p_1 = 350$ kPa (this will be the *gauge* pressure unless otherwise stated).

Having been given the diameters of the pipe at the two stations, we are required to find the corresponding areas of section, but why? Well in nearly all problems that require the use of the Bernoulli equation, we will first need to determine one unknown using the equation of continuity. You will remember that for *steady flow conditions* for a liquid, we use Equation 3.2.28.

$$A_1 v_1 = A_2 v_2.$$

We have been asked to determine v_2, then from the given diameters our areas of section are:

$$A_1 = 1963 \text{ mm}^2 \quad A_2 = 4418 \text{ mm}^2$$

and so $v_2 = \dfrac{A_1 v_1}{A_2} = \dfrac{1963 \cdot 3.5}{4418} \dfrac{\text{mm}^2}{\text{mm}^2} \cdot$ m/s $= 1.55$ m/s

Now if we use the Bernoulli equation, correctly laid out to indicate *the direction of flow*, in this case from station 1 to station 2, we have:

$$\frac{p_1}{\gamma} + \frac{v_1^2}{2g} + h_1 = \frac{p_2}{\gamma} + \frac{v_2^2}{2g} + h_2$$

Note that in many texts the symbol (z) is used to indicate the elevation from a fixed datum, rather than (h). We will continue to use (h) since it is easily remembered as height and each term in the whole equation is concerned with *head of fluid per unit weight*.

Now transposing the Bernoulli equation in terms of p_2 gives:

$$p_2 = p_1 + \gamma \left(h_1 - h_2 + \frac{v_1^2 - v_2^2}{2g} \right)$$

and knowing that the specific weight of water at 10°C is $= 9810$ N/m³ and the acceleration due to gravity $= 9.81$ m/s², we can substitute numerical values for all the variables into the above equation.

Then:

$$p_2 = 350 \text{ kPa} + 9.81 \frac{\text{kN}}{\text{m}^3} \left(-2.0 \text{ m} + \frac{(12.25 - 2.4025)\text{m}^2/\text{s}^2}{2 \times 9.81 \text{ m/s}^2} \right)$$

$$= 350 \text{ kPa} + 9.81 \frac{\text{kN}}{\text{m}^3} (-2 \text{ m} + 0.502 \text{ m})$$

$$= 350 \text{ kPa} - 14.7 \text{ kN/m}^2$$

So $p_2 = 335.3$ kPa.

Note the careful manipulation of units in this example. If you are in any doubt always revert to standard units when dealing with variables, beware of short cuts!

Now we are not quite there with respect to the energy balance within a fluid flow system. We have, in our derivation of the Bernoulli equation, made the assumption that there is no energy transferred to or removed from the fluid when it travels from station 1 to station 2 (Figure 3.2.14) in the system. In reality there are often energy losses resulting from pipe friction and losses through fixtures and fittings, as the fluid flows. Energy may be added and removed from the system as part of its function for example, a pump within the system will add energy whereas a motor will remove energy. These further energy losses and gains may be represented as *heads* in a similar manner to those already given in Bernoulli's equation. So we can represent these energies symbolically as follows:

energy added to the system $= h_A$;
energy removed from the system $= -h_R$;
energy losses $= -h_L$.

The sign convention used is *positive for energy added* to the system and *negative for energy lost or removed* from the system.

Now we can modify the Bernoulli equation to take account of these additional energies as follows.

$$\frac{p_1}{\gamma} + \frac{v_1^2}{2g} + h_1 + h_A - h_R - h_L = \frac{p_2}{\gamma} + \frac{v_2^2}{2g} + h_2 \qquad (3.2.35)$$

The above equation is known as the *general energy equation* for fluid flow.

Note that the *additional gains and losses of energy must occur in the direction of flow of the fluid*, in the above form, Equation 3.2.35 suggests flow from station 1 to station 2, read from left to right.

The Darcy equation and friction losses in fluid systems

In the general energy equation the term (h_L) defines the energy loss from a fluid system. One component of this energy loss is due to friction in the flowing fluid, in particular *losses in straight pipes* may be estimated from the *Darcy formula*, which can be written in terms of the head loss h_L as:

$$h_L = 4f.\frac{lv^2}{d2g} \qquad (3.2.36)$$

or in terms of the pressure loss as:

$$\Delta p = 4f.\frac{l\rho v^2}{d2g} \qquad (3.2.37)$$

where Δp = the pressure loss; h_L = the head loss per unit weight; f = the pipe friction factor; l = the pipe length; d = the pipe diameter; r = the fluid density; v = the mean velocity of the fluid; g = the acceleration due to gravity.

The Darcy Equation (3.2.36 and 3.2.37) may be determined from the general energy equation by considering the average shear stress at the pipe wall, created by the fluid flow.

Friction factor

The term (f) or *friction factor* in the Darcy equation, *depends on the nature of the flow in the pipe*. For *laminar flow only* the value of the friction factor depends on the *Reynolds number* and can easily be calculated from

Figure 3.2.16 *Moody diagram*

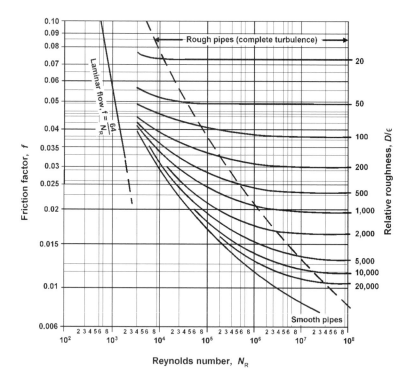

$$f = \frac{16}{\text{Re}}.$$

If the flow in the pipe is *turbulent*, then, the friction factor can only be determined from experimental data. The results of such data have been conveniently displayed on a *Moody diagram*, which enables us to estimate friction factors, where *turbulent flow* is present. It can also be seen from the Moody diagram (Figure 3.2.16), that the friction factor not only depends on the Reynolds number but also varies with *surface roughness*.

In practice when dealing with Moody diagrams there is one important difference in terminology between the UK and USA versions of the diagram. In the USA system the quantity 4*f* quoted in the Darcy equation is simply termed *f*. This means that when using the USA version of the Moody diagram the friction factor is given by

$$f = \frac{64}{N_R} \text{ where } N_R \text{ is used instead of Re for Reynolds number.}$$

Figure 3.2.16 shows the USA version of the Moody diagram (reproduced from; Pao, R. H. F. 1961. *Fluid Mechanics*. John Wiley & Sons, page 284).

Example 3.2.9

Heavy oil with a kinematic viscosity of 1.2×10^{-4} m²/s flows 50 m through a 10 cm diameter pipe, at a velocity of 1.5 m/s. Determine the energy loss resulting from the flow.

We first need to establish whether the flow is laminar or turbulent, by calculating the Reynold's number.

Then $\qquad \text{Re} = \dfrac{vd}{\nu} = \dfrac{1.5 \times 0.1}{1.2 \times 10^{-4}} \qquad = 1250.$

Re <2000 (laminar) therefore we may calculate (*f*) from the

cont.

Reynolds number.

Using the *head loss form* of the Darcy equation:

$$h_L = 4f \cdot \frac{l v^2}{d \, 2g} \quad \text{where} \quad f = \frac{16}{Re} = \frac{16}{1250} = 0.0128$$

we have $h_L = 4 \times 0.0128 \, \dfrac{(50 \times 1.5^2)}{0.1 \times 2 \times 9.81}$

and $h_L = 2.94$ m.

What this answer tells us, is that we lose 2.94 N.m of energy for each Newton of oil that flows along the 50 m of pipe.

Example 3.2.10

Water at 20°C flows at 3 m/s in a 30 m plastic pipe with an internal diameter of 25 mm. Determine the friction factor (*f*) and the energy loss due to the flow.

In order to determine whether or not the flow is laminar or turbulent, we must again find the Reynold's number.

We are not given the kinematic viscosity, but it can be found by consulting Table 3.2.2, where we find v for water at 20°C = 1×10^{-6} m²/s.

Then $Re = \dfrac{vd}{v} = \dfrac{3 \times 25 \times 10^{-3}}{1 \times 10^{-6}} = 75\,000$

which is > 2000 hence flow is turbulent.

Now since the flow is turbulent we cannot use the simple formula given in the previous example, to find the friction factor, we need to use the *Moody diagram*. So as well as the Reynold's number we need also to find the *relative roughness* (D/ϵ) where D is the pipe diameter and ϵ is the average pipe wall roughness. Note that in the Moody diagram (Figure 3.2.16) the relative roughness is shown as a set of parametric curves. The wall roughness (ϵ) is often quoted in tables for different materials.

Plastic pipes can be treated as smooth, so in this case to read the value of the friction factor from the diagram we move along the horizontal axis (note that the scales on both axes are logarithmic) until we find 75 000 which lies between 10^4 and 10^5 at the point three-quarters of the way between 6 and 8, now follow this line up until you meet the roughness curve for plastic (smooth pipes) and read across. You should find a friction factor which approximately equals 0.019.

Remember that since we have the USA version of the friction factor, this value 0.019 is the equivalent to 4*f* in our version of the Darcy equation. So having found the friction factor we can now use:

$$h_L = \frac{4f \, l v^2}{d \, 2g} \quad \text{where} \quad 4f = 0.019 \text{ and so}$$

$$h_L = \frac{0.019 \times 30 \times 3^2}{25 \times 10^{-3} \times 2 \times 9.81} = 10.46 \text{ m.}$$

So head loss due to flow is 10.46 m.

Note

In this book we are only interested in energy losses in circular cross-section pipes, *if the cross-section is non-circular* then instead of using the diameter to find an appropriate Reynolds number we use a characteristic dimension called the *hydraulic radius* where

$$\text{hydraulic radius } (R) = \frac{A}{L} = \frac{\text{cross-sectional area}}{\text{wetted perimeter}}.$$

This dimension is particularly useful for open section channels, sewerage systems and other mass fluid transport applications.

Finally, before we leave our very brief introduction to fluid mechanics, we will look at one more application concerned with fluid losses in bearings.

Energy loss in plain bearings

In this very short section we will look at the application of fluid mechanics to estimating power losses within *plain bearing* assemblies. We first need to establish a little terminology and one or two basic concepts.

Plain bearings carry the load directly on supports using a sliding motion, as opposed to roller bearings, where balls or rollers are interposed between the sliding surfaces.

Plain bearings are of two types: *journal or sleeve* bearings, which are cylindrical and support radial loads (those perpendicular to the shaft axis); and *thrust* bearings, which are generally flat and, in the case of a rotating shaft, support loads in the direction of the shaft axis (axial direction).

Figure (3.2.17) shows a crankshaft supported by two main bearings attached to the connecting rod by the connecting rod bearing. All three are journal bearings. Flanges on the main bearings act as thrust bearings, which restrain the axial motion of the shaft.

Lubrication

There are generally considered to be three different types of lubrication. *Hydrodynamic lubrication* where the surfaces are completely separated by a film of lubricant and, the loads generated between the surfaces are

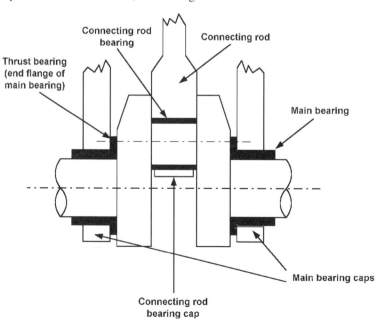

Figure 3.2.17 *Crankshaft journal and thrust bearing*

supported entirely by fluid pressure, generated by the motion of the surfaces. *Mixed film lubrication* where the surface peaks of the rubbing surfaces are intermittently in contact, and there is partial hydrodynamic support. Finally, there is *boundary lubrication* where surface contact is continuous, the lubricant being continually smeared over the surface providing a renewable surface film which reduces friction and wear.

In hydrodynamic lubrication, the film thickness varies typically between about 0.008 mm at its thinnest point and 0.02 mm at its thickest point. Figure 3.2.18a shows a loaded bearing journal, subject to hydrodynamic lubrication, at rest. The bearing clearance space is filled with oil, which has been squeezed out at the bottom by the load (W). Slow clockwise rotation of the shaft to the right (3.2.18b) causes it to roll to the right. However it stays in this position unable to climb to the right, because of the lack of friction due to the lubricant, thus lubrication at the boundaries takes place.

If the rotational speed of the shaft is progressively increased more and more fluid is forced forward and adheres to the journal surface, causing sufficient pressure to build-up to raise or float the shaft away from the boundary (Figure 3.2.18c). Hydrodynamic lubrication, therefore, depends on the rotational speed of the shaft (n), the dynamic viscosity of the oil μ (the higher the viscosity of the oil the lower the rotational speed required to float the shaft), and the bearing unit load P which is the load (W) divided by the bearing area (the diameter D multiplied by the length L of the bearing).

We can form a relationship between μ, n and P which optimises the efficiency of the hydrodynamic lubrication of the journal bearing, where: for hydrodynamic lubrication optimise the relationship: $\mu n/P$ for given conditions.

So, for example, very smooth surfaces may be lubricated using low viscosity oils, this reduces the effects of viscous friction, under these circumstances a low value of $\mu n/P$ is appropriate. In order to reduce power losses in bearings we should always endeavour to reduce the coefficient of friction between the surfaces. In hydrodynamic lubrication typical values for the coefficient of friction are 0.002 to 0.01, which are very low, this is why this form of lubrication is so effective!

Note

You will remember from your previous studies that the friction force between two rubbing surfaces is given by $F = \mu R$, where μ is the coefficient of friction between the surfaces and R is the normal reaction force created by the load applied to the surfaces. It is unfortunate that the symbol (μ) for the coefficient of friction is also the same as that

Figure 3.2.18 *Hydrodynamic lubrication for journal bearing*

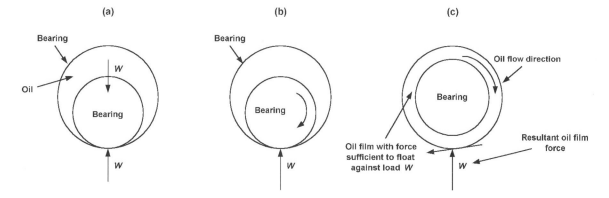

used for dynamic viscosity. To avoid confusion, we will for this particular subject, use the symbol (f) for the coefficient of friction and the normal symbol (μ) for dynamic viscosity.

You should also be aware of the relationship between the torque created in a rotating shaft and the power required to drive this torque. The power required to overcome the frictional torque (T_f) created in the bearings of a shaft is given by:

$$\text{Power} = 2\pi T_f n \qquad\qquad (3.2.38)$$

This formula will be used below, to estimate power loss in a hydrodynamic bearing.

Petroff method for estimating power loss in hydrodynamic bearings

The scientist Petroff in 1883 analysed viscous friction drag in what we now know as hydrodynamic bearings. He devised the Petroff equation, which provides a quick and simple method for obtaining reasonable estimates of coefficients of friction for lightly loaded bearings, enabling us to estimate power losses within such bearings. The derivation of the Petroff equation will not be given here, however its use is simple, and should not present you with too many problems.

Petroff equation $f = 2\pi^2 \dfrac{\mu n}{P} \dfrac{R}{c}$

Where: f = coefficient of viscous friction; μ = dynamic viscosity (N/m^2. s); n = rotational velocity (*revs per second*); R = shaft radius (*m*); c = radial clearance = (bearing diameter − shaft diameter)/2.

$$P = \frac{W}{DL} = \frac{W}{A} = \frac{\text{load}}{\text{bearing area}}.$$

Example 3.2.11

An 80 mm diameter shaft is supported by a bearing 60 mm in length with a radial clearance of 0.1 mm. It is lubricated by an oil having a dynamic viscosity of 60 mN/m².s (at the operating temperature).The shaft rotates at 480 rpm and carries a radial load of 4000 N. Determine an estimate for the bearing coefficient of friction and the power loss using the Petroff method.

Figure 3.2.19 illustrates the journal bearing given in the question. In order to apply the Petroff method we must make the assumption that for the loading conditions, there is no eccentricity

cont.

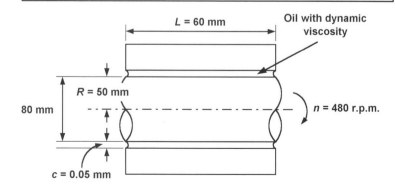

Figure 3.2.19   *Example 3.2.11 journal bearing*

between the bearing and the journal and, that no fluid flows in the axial direction. We must also assume that the simple friction condition $F_d = fR$ (where F_d = the friction drag force, and R = the radial load), applies.

Then using Petroff's equation we obtain the coefficient of viscous friction:

$$f = 2\pi^2 \frac{\mu n}{P} \frac{R}{c}$$

$$f = \frac{(2\pi^2)(60 \times 10^{-3})(\text{ 8 rps})}{(833\ 333)} \frac{(40\ \text{mm})}{(0.05\ \text{mm})}$$

$f = 0.0096$

where $P = \dfrac{\text{load}}{\text{bearing area}} = \dfrac{4000}{0.08 \times 0.06} = 833\ 333.$

So the frictional drag force $F_d = fR = (0.0096)(4000) = 36.38$ N. Then the torque resulting from this drag force $T_f = (36.38)(0.04) = 1.46$ N m. Remember the torque due to the friction, is equal to the frictional drag force multiplied by the shaft radius.

Then from Equation 3.2.38 Power $= 2\pi T_f n = (2\pi)(1.46)(6$ rps$) = 55$ watts.

Problems 3.2.2

(1) Explain the nature of viscosity. Your answer should differentiate between dynamic and kinematic viscosity and include the effects on viscosity of temperature change, for both liquids and gases.

(2) The dynamic viscosity of an engine oil is 0.25 N.s/m². Determine its kinematic viscosity, given that its density is 750 kg/m³.

(3) Water at 20°C flows in a pipe at an average velocity of 3.5 m/s. If the internal diameter of the pipe is 120 mm. Determine whether or not the flow is laminar or turbulent.

(4) If the internal diameters of the pipes at the points 1 and 2 (Figure 3.2.13) are 80 mm and 140 mm respectively, and oil with a density of 830 kg/m³ is flowing past *point 2 towards point 1* with a velocity of 5 m/s. Find:
     (i)   the velocity at point 1;
     (ii) the volume flow rate;
     (iii) the mass flow rate.

(5) Figure 3.2.20 shows a wind tunnel set-up in a laboratory.

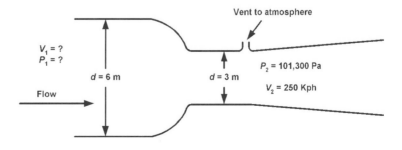

Figure 3.2.20 *Problem 3.2.2 wind tunnel set-up*

The wind tunnel is circular in cross-section with a diameter upstream of the test section of 6 m and a test section diameter of 3 m. The test section is vented to atmosphere. Air flows through the wind tunnel with a velocity, at the working section, of 250 kph. Given that atmospheric pressure is 101 300 Pa and the mass density of mercury is 13 600 kg/m³. Find:
(i)   the upstream section velocity;
(ii) the upstream pressure;
(iii) the height of the mercury column being used to regulate the wind tunnel speed.

Note that the mercury column height results from the measured difference between the upstream and working section pressures.

6. Find the friction factor, if water is flowing at 10 m/s in a cast iron pipe having an inside diameter 2.5 cm. The relative roughness of the pipe may be taken as approximately equal to 100.

7. Use the Darcy equation to determine the energy loss due to the flow of water in Question 6, if the water flows through 5m.

8. A 100 mm diameter shaft is supported by a bearing 80 mm in length, with a radial clearance of 0.08 mm. The dynamic viscosity of the lubricating oil may be taken as 50 mPa . s. If the shaft rotates at 900 rpm and carries a load of 6000 N. Determine estimates for the bearing coefficient of friction and the power loss.

3.3 Single-phase a.c. theory

If you have not studied a.c. theory before, this introductory section has been designed to quickly get you up to speed. If, on the other hand, you have previously studied Electrical and Electronic Principles at level N (or an equivalent GNVQ Advanced level unit) you should move on to 'Using complex notation' (see page 154).

Alternating voltage and current

Unlike direct currents which have steady values and always flow in the same direction, alternating currents flow alternately one way and then the other. The alternating potential difference (voltage) produced by an alternating current is thus partly positive and partly negative. An understanding of alternating currents and voltages is important in a number of applications including a.c. power distribution, amplifiers and filters.

A graph showing the variation of voltage or current present in a circuit is known as a *waveform*. Some common types of waveform are shown in Figure 3.3.1. Note that, since the waveforms of speech and music comprise many components at different frequencies and of different amplitudes, these waveforms are referred to as *complex*.

The equation for the sinusoidal voltage shown in Figure 3.3.2, at a time t, is:

$$v = V_{max} \sin(\omega t)$$

where v is the instantaneous voltage, V_{max} is the maximum value of voltage (also known as the *amplitude* or *peak value* of voltage), and ω is the angular velocity (in radians per second).

The frequency of a repetitive waveform is the number of cycles of the waveform which occur in unit time. Frequency is expressed in Hertz (Hz). A frequency of 1 Hz is equivalent to one cycle per second. Hence, if a voltage has a frequency of 400 Hz, four hundred cycles will occur in every second.

Figure 3.3.1 *Various waveforms; (a) sine; (b) square; (c) triangle; (d) pulse; (e) complex*

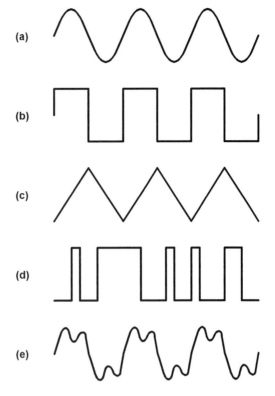

(a)

(b)

(c)

(d)

(e)

Since there are 2π radians in one complete revolution or *cycle*, a frequency of one cycle per second must be the same as 2π radians per second. Hence a frequency, f, is equivalent to:

$$f = \omega/(2\pi) \text{ Hz}$$

Alternatively, the angular velocity, ω, is given by:

$$\omega = 2\pi f \text{ rad/sec.}$$

We can thus express the instantaneous voltage in another way:

$$v = V_{max} \sin(2\pi f t).$$

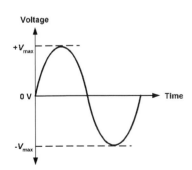

Voltage

$+V_{max}$

0 V

Time

$-V_{max}$

Figure 3.3.2   *One cycle of a sine wave*

Example 3.3.1

A sine wave voltage has a maximum value of 100 V and a frequency of 50 Hz. Determine the instantaneous voltage present (a) 2.5 ms and (b) 15 ms from the start of the cycle.

We can determine the voltage at any instant of time using:

$$v = V_{max} \sin(2\pi f t)$$

where $V_{max} = 100$ V and $f = 50$ Hz.

In (a), $t = 2.5$ ms hence:

$$v = 100 \sin(2\pi \times 50 \times 0.0025) = 100 \sin(0.785) = 100 \times 0.707 = 70.7 \text{ V.}$$

In (b), $t = 15$ ms hence:

$$v = 100 \sin(2\pi \times 50 \times 0.015) = 100 \sin(4.71) = 100 \times -1 = -100 \text{V.}$$

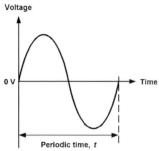

Figure 3.3.3 *Periodic time*

Periodic time

The periodic time of a waveform is the time taken for one complete cycle of the wave. The relationship between periodic time and frequency is thus:

$$t = 1/f \text{ or } f = 1/t$$

where t is the periodic time (in seconds) and f is the frequency (in Hz).

Example 3.3.2

A waveform has a frequency of 200 Hz. What is the periodic time of the waveform?

$t = 1/f = 1/200 = 0.005$ s (or 5 ms).

Example 3.3.3

A waveform has a periodic time of 2.5 ms. What is its frequency?

$f = 1/t = 1/(2.5 \times 10^{-3}) = 1 \times 10^{+3}/2.5 = 400$ Hz.

Average, peak, peak–peak, and r.m.s. values

The *average* value of an alternating current which swings symmetrically above and below zero will obviously be zero when measured over a long period of time. Hence average values of currents and voltages are invariably taken over one complete half-cycle (either positive or negative) rather than over one complete full-cycle (which would result in an average value of zero).

The *peak* value (or *amplitude*) of a waveform is a measure of the extent of its voltage or current excursion from the resting value (usually zero). The *peak-to-peak* value for a wave which is symmetrical about its resting value is twice its peak value.

The *r.m.s.* (or *effective*) value of an alternating voltage or current is the value which would produce the same heat energy in a resistor as a direct voltage or current of the same magnitude. Since the r.m.s. value of a waveform is very much dependent upon its shape, values are only meaningful when dealing with a waveform of known shape. Where the shape of a waveform is not specified, r.m.s. values are normally assumed to refer to sinusoidal conditions.

The following formulae apply to a sine wave:

$$V_{average} = 0.636 \times V_{peak}$$
$$V_{peak-peak} = 2 \times V_{peak}$$
$$V_{r.m.s.} = 0.707 \times V_{peak}.$$

Similar relationships apply to the corresponding alternating currents, thus:

$$I_{average} = 0.636 \times I_{peak}$$
$$I_{peak-peak} = 2 \times I_{peak}$$
$$I_{r.m.s.} = 0.707 \times I_{peak}.$$

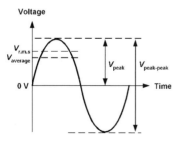

Figure 3.3.4 *Relationship between average, r.m.s., peak and peak–peak values*

Example 3.3.4

A sinusoidal voltage has an r.m.s. value of 220 V. What is the peak value of the voltage?

cont.

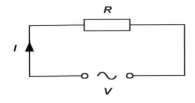

Figure 3.3.5 *Alternating current in a resistor*

Since $V_{r.m.s.} = 0.707\,V_{peak}$, $V_{peak} = V_{r.m.s.}/0.707 = 1.414 \times V_{r.m.s.}$

Thus $V_{peak} = 1.414 \times 220 = 311$ V.

Example 3.3.5

A sinusoidal alternating current has a peak–peak value of 4 mA. What is its r.m.s. value?

First we must convert the peak–peak current into peak current:

Since $I_{peak-peak} = 2 \times I_{peak}$, $I_{peak} = 0.5 \times I_{peak-peak}$

Thus $I_{peak} = 0.5 \times 4 = 2$ mA.

Now we can convert the peak current into r.m.s. current using:

$I_{r.m.s.} = 0.707 \times I_{peak}$.

Thus $I_{r.m.s.} = 0.707 \times 2 = 1.414$ mA.

Alternating current in a resistor

Ohm's Law is obeyed in an a.c. circuit just as it is in a d.c. circuit. Thus, when a sinusoidal voltage, V, is applied to a resistor, R, (as shown in Figure 3.3.5) the current flowing in the resistor will be given by:

$I = V/R.$

This relationship must also hold true for the instantaneous values of current, i, and voltage, v, thus:

$i = v/R$

and since $v = V_{max}\sin(\omega t)$

$i = V_{max}\sin(\omega t)/R.$

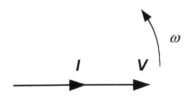

Figure 3.3.6 *Phasor diagram for the circuit in Figure 3.3.5*

The current and voltage will both have a sinusoidal shape and since they rise and fall together, they are said to be *in-phase* with one another. We can represent this relationship by means of the *phasor diagram* shown in Figure 3.3.6. This diagram shows two rotating phasors (of magnitude I and V) rotating at an angular velocity, ω. The applied voltage (V) is referred to as the *reference phasor* and this is aligned with the horizontal axis (i.e. it has a phase angle of 0°).

Phasor diagrams provide us with a quick way of illustrating the relationships that exist between sinusoidal voltages and currents in a.c. circuits without having to draw lots of time related waveforms. Figure 3.3.7 will help you to understand how the previous phasor diagram relates to time-related waveforms for the voltage and current in a resistor.

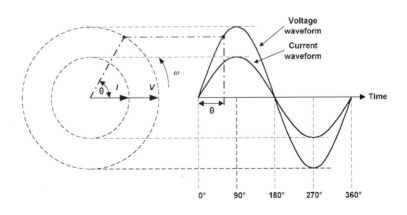

Figure 3.3.7 *Phasor diagram and time-related waveforms*

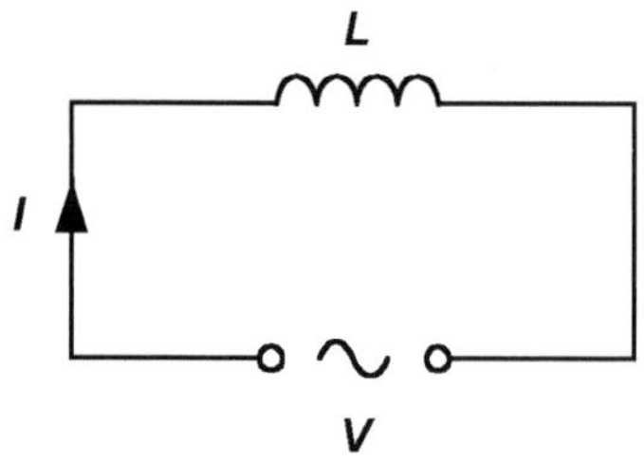

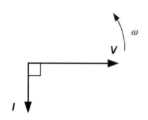

Figure 3.3.8 *Alternating current in an inductor*

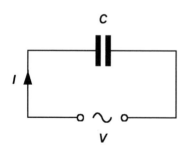

Figure 3.3.9 *Phasor diagram for the circuit in Figure 3.3.8*

Figure 3.3.10 *Alternating current in a capacitor*

Figure 3.3.11 *Phasor diagram for the circuit in Figure 3.3.10*

Example 3.3.6

A sinusoidal voltage 10 V$_{peak-peak}$ is applied to a resistor of 1 kΩ. What value of r.m.s. current will flow in the resistor?

   This problem must be solved in several stages. First we will determine the peak–peak current in the resistor and then we will convert this value into a corresponding r.m.s. quantity.

   Since: $I = V/R$, $I_{peak-peak} = V_{peak-peak}/R$.
   Thus: $I_{peak-peak} = 10V_{peak-peak}/1k\Omega = 10$ mA.
   Next: $I_{peak} = I_{peak-peak}/2 = 10/2 = 5$ mA.
   Finally: $I_{r.m.s.} = 0.707 \times I_{peak} = 0.707 \times 5$ mA $= 3.53$ mA.

Alternating current in an inductor

When a sinusoidal voltage, V, is applied to an inductor, L, (as shown in Figure 3.3.8) the current flowing in the inductor will be given by:

$$I = V/X$$

where X is the *reactance* of the inductor. Like resistance, reactance is measured in ohms (Ω).

   The current in an inductor lags behind the voltage by a *phase angle* of 90° ($\pi/2$ radians) and since $v = V_{max} \sin(\omega t)$

$$i = V_{max} \sin(\omega t - \pi/2) /X.$$

   Once again, the current and voltage will both have a sinusoidal shape but they are 90° apart and this relationship has been illustrated by means of the phasor diagram shown in Figure 3.3.9. The applied voltage (V) is the *reference phasor* (its phase angle is 0°) whilst the current flowing (I) has a *lagging phase angle* of 90°.

Alternating current in a capacitor

When a sinusoidal voltage, V, is applied to a capacitor, C, (as shown in Figure 3.3.10) the current flowing in the inductor will be given by:

$$I = V/X$$

where X is the *reactance* of the capacitor. Like resistance, reactance is measured in ohms (Ω).

   The current in a capacitor leads the applied voltage by a *phase angle* of 90° ($\pi/2$ radians) and since $v = V_{max} \sin(\omega t)$

$$i = V_{max} \sin(\omega t + \pi/2)/X.$$

   As before, the current and voltage both have a sinusoidal shape 90° apart and this relationship is illustrated in the phasor diagram shown in Figure 3.3.11. The applied voltage (V) is the *reference phasor* (its phase angle of 0°) whilst the current flowing (I) has a *leading phase angle* of 90°.

Reactance

When alternating voltages are applied to capacitors or inductors the amount of current flowing will depend upon the value of capacitance or inductance and on the frequency of the voltage. In effect, capacitors and inductors oppose the flow of current in much the same way as a resistor. The important difference being that the effective resistance (or *reactance*) of the component varies with frequency (unlike the case of a pure resistance where the magnitude of the current *does not* change with frequency).

Reactance, like resistance, is simply the ratio of applied voltage to the current flowing. Thus:

$$X = V/I$$

where X is the reactance in ohms (Ω), V is the alternating potential difference in volts (V) and I is the alternating current in amps (A).

In the case of *capacitive reactance* (i.e. the reactance of a capacitor) we use the suffix, C, so that the reactance equation becomes:

$$X_C = V_C/I_C.$$

Similarly, in the case of *inductive reactance* (i.e. the reactance of an inductor) we use the suffix, L, so that the reactance equation becomes:

$$X_L = V_L/I_L.$$

Inductive reactance

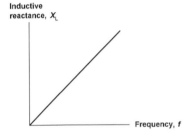

Inductive reactance is directly proportional to the frequency of the applied alternating current and can be determined from the following formula:

$$X_L = 2\pi f L$$

where X_L is the reactance in Ω, f is the frequency in Hz, and L is the inductance in H.

Since inductive reactance is directly proportional to frequency ($X_L \propto f$), the graph of inductive reactance plotted against frequency takes the form of a straight line (see Figure 3.3.12).

Figure 3.3.12   *Variation of inductive reactance with frequency*

Example 3.3.7

Determine the reactance of a 10 mH inductor at (a) 100 Hz and (b) at 10 kHz.

(a)   At 100 Hz, $X_L = 2\pi \times 100 \times 10 \times 10^{-3} = 6.28\ \Omega$.
(b)   At 10 kHz, $X_L = 2\pi \times 10\,000 \times 10 \times 10^{-3} = 628\ \Omega$.

Capacitive reactance

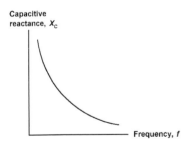

Capacitive reactance is inversely proportional to the frequency of the applied alternating current and can be determined from the following formula:

$$X_C = 1/(2\pi f C)$$

where X_C is the reactance in Ω, f is the frequency in Hz, and C is the capacitance in F.

Since capacitive reactance is inversely proportional to frequency ($X_C \propto 1/f$), the graph of inductive reactance plotted against frequency takes the form of a rectangular hyperbola (see Figure 3.3.13).

Figure 3.3.13   *Variation of inductive reactance with frequency*

Example 3.3.8

Determine the reactance of a 1μF capacitor at (a) 100 Hz and (b)   10 kHz.

cont.

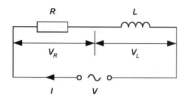

Figure 3.3.14   *Resistance and inductance in series*

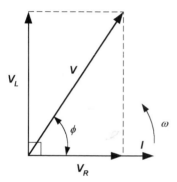

Figure 3.3.15   *Phasor diagram for the circuit in Figure 3.3.14*

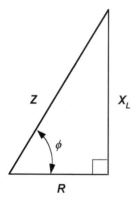

Figure 3.3.16   *Impedance triangle for the circuit in Figure 3.3.14*

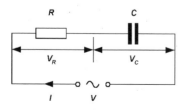

Figure 3.3.17   *Resistance and capacitance in series*

Resistance and inductance in series

When a sinusoidal voltage, V, is applied to a series circuit comprising resistance, R, and inductance, L, (as shown in Figure 3.3.14) the current flowing in the circuit will produce separate voltage drops across the resistor and inductor (V_R and V_L respectively). These two voltage drops will be 90° apart – with V_L leading V_R. We can illustrate this relationship using the phasor diagram shown in Figure 3.3.15. Note that we have used current as the reference phasor in this series circuit for the simple reason that the same current flows through each component (recall that earlier we used the applied voltage as the reference).

From Figure 3.3.15 you should note that the supply voltage (V) is simply the result of adding the two voltage phasors, V_R and V_L. Furthermore, the angle between the supply voltage (V) and supply current (I), ϕ, is known as the *phase angle*.

Now $\sin \phi = V_L/V$, $\cos \phi = V_R/V_L$, and $\tan \phi = V_L/V_R$.

Since $X_L = V_L/I$, $R = V_R/I$ and $Z = V/I$ (where Z is the *impedance* of the circuit), we can illustrate the relationship between X_L, R and Z using the *impedance triangle* shown in Figure 3.3.16.

Note that $Z = \sqrt{(R^2 + X_L^2)}$ and $\phi = \arctan(X_L/R)$.

Example 3.3.9

An inductor of 80 mH is connected in series with a 100 Ω resistor. If a sinusoidal current of 20 mA at 50 Hz flows in the circuit, determine:

(a)   the voltage dropped across the inductor;
(b)   the voltage dropped across the resistor;
(c)   the impedance of the circuit;
(d)   the supply voltage;
(e)   the phase angle.

(a) $V_L = I\,X_L = I \times 2\pi f\,L = 0.02 \times 25.12 = 0.5$ V.
(b) $V_R = I\,R = 0.02 \times 100 = 2$ V.
(c) $Z = \sqrt{(R^2 + X_L^2)} = \sqrt{(100^2 + 25.12^2)} = \sqrt{10\,631} = 103.1$ Ω.
(d) $V = I \times Z = 0.02 \times 103.1 = 2.06$ V.
(e) $\phi = \arctan(X_L/R) = \arctan(25.12/100) = \arctan(0.2512) = 14.1°$.

Resistance and capacitance in series

When a sinusoidal voltage, V, is applied to a series circuit comprising resistance, R, and inductance, L, (as shown in Figure 3.3.17) the current flowing in the circuit will produce separate voltage drops across the resistor and capacitor (V_R and V_C, respectively). These two voltage drops will be 90° apart – with V_C lagging V_R. We can illustrate this relationship using the phasor diagram shown in Figure 3.3.18. Note that once again

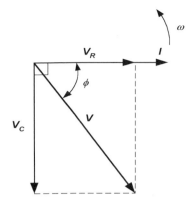

Figure 3.3.18 *Phasor diagram for the circuit in Figure 3.3.17*

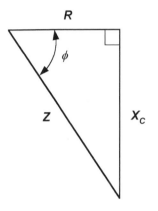

Figure 3.3.19 *Impedance triangle for the circuit in Figure 3.3.17*

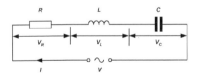

Figure 3.3.20 *Resistance, inductance and capacitance in series*

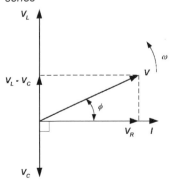

Figure 3.3.21 *Phasor diagram for the circuit in Figure 3.3.20 when $X_L > X_C$*

we have used current as the reference phasor in this series circuit.

From Figure 3.3.18 you should note that the supply voltage (V) is simply the result of adding the two voltage phasors, V_R and V_C. Furthermore, the angle between the supply voltage (V) and supply current (I), ϕ, is known as the *phase angle*.

Now $\sin \phi = V_C / V$, $\cos \phi = V_R / V_L$, and $\tan \phi = V_C / V_R$.

Since $X_L = V_C/I$, $R = V_R/I$ and $Z = V/I$ (where Z is the *impedance* of the circuit), we can illustrate the relationship between X_C, R and Z using the *impedance triangle* shown in Figure 3.3.19.

Note that $Z = \sqrt{(R^2 + X_C^2)}$ and $\phi = \arctan(X_C/R)$.

Example 3.3.10

A capacitor of 22 μF is connected in series with a 470 Ω resistor. If a sinusoidal current of 10 mA at 50 Hz flows in the circuit, determine:

(a) the voltage dropped across the capacitor;
(b) the voltage dropped across the resistor;
(c) the impedance of the circuit;
(d) the supply voltage;
(e) the phase angle.

(a) $V_C = I X_C = I \times 1/(2\pi f C) = 0.01 \times 144.5 = 1.4$ V.
(b) $V_R = I R = 0.01 \times 470 = 4.7$ V.
(c) $Z = \sqrt{(R^2 + X_C^2)} = \sqrt{(470^2 + 144.5^2)} = \sqrt{241\,780} = 491.7$ Ω.
(d) $V = I \times Z = 0.01 \times 491.7 = 4.91$ V.
(e) $\phi = \arctan(X_C/R) = \arctan(144.5/470) = \arctan(0.3074) = 17.1°$.

Resistance, inductance and capacitance in series

When a sinusoidal voltage, V, is applied to a series circuit comprising resistance, R, inductance, L, and capacitance, C, (as shown in Figure 3.3.20) the current flowing in the circuit will produce separate voltage drops across the resistor, inductor and capacitor (V_R, V_L and V_C, respectively). The voltage drop across the inductor will lead the applied current (and voltage dropped across V_R) by 90° whilst the voltage drop across the capacitor will lag the applied current (and voltage dropped across V_R) by 90°.

When the inductive reactance (X_L) is greater than the capacitive reactance (X_C), V_L will be greater than V_C and the resulting phasor diagram is shown in Figure 3.3.21.

Conversely, when the capacitive reactance (X_C) is greater than the inductive reactance (X_L), V_C will be greater than V_L and the resulting phasor diagram is shown in Figure 3.3.22.

Note that once again we have used current as the *reference phasor* in this series circuit.

From Figures 3.3.21 and 3.3.22, you should note that the supply voltage (V) is simply the result of adding the three voltage phasors, V_L, V_C and V_R and that the first stage in simplifying the diagram is that of resolving V_L and V_C into a single voltage ($V_L - V_C$ or $V_C - V_L$ depending upon whichever is the greater). Once again, the phase angle, ϕ, is the angle between the supply voltage and current.

Figures 3.3.23 and 3.3.24 show the impedance triangle for the circuit, for the cases when $X_L > X_C$ and $X_C > X_L$, respectively.

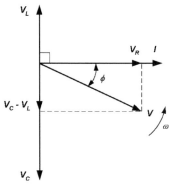

Figure 3.3.22 *Phasor diagram for the circuit in Figure 3.3.20 when $X_L < X_C$*

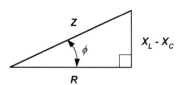

Figure 3.3.23 *Impedance triangle for the circuit in Figure 3.3.20 when $X_L > X_C$*

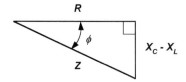

Figure 3.3.24 *Impedance triangle for the circuit in Figure 3.3.20 when $X_L < X_C$*

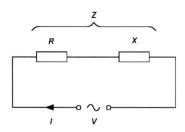

Figure 3.3.25 *A series circuit containing reactance and resistance*

Note that, when $X_L > X_C$, $Z = \sqrt{(R^2 + (X_L - X_C)^2)}$ and $\phi = \arctan(X_L - X_C)/R$; similarly, when $X_C > X_L$, $Z = \sqrt{(R^2 + (X_C - X_L)^2)}$ and $\phi = \arctan(X_C - X_L)/R$.

Example 3.3.11

A series circuit comprises an inductor of 80 mH, a resistor of 200 Ω and a capacitor of 22 μF. If a sinusoidal current of 40 mA at 50 Hz flows in this circuit, determine:

(a) the voltage developed across the inductor;
(b) the voltage dropped across the capacitor;
(c) the voltage dropped across the resistor;
(d) the impedance of the circuit;
(e) the supply voltage;
(f) the phase angle.

(a) $V_L = I X_L = I \times 2\pi f L = 0.04 \times 25.12 = 1$ V.
(b) $V_C = I X_C = I \times 1/(2\pi f C) = 0.04 \times 144.5 = 5.8$ V.
(c) $V_R = I R = 0.04 \times 200 = 8$ V.
(d) $Z = \sqrt{(R^2 + (X_C - X_L)^2)} = \sqrt{(200^2 + (144.5 - 25.12)^2)}$
    $= \sqrt{54\ 252}$
    $= 232.9$ Ω.
(e) $V = I \times Z = 0.04 \times 232.9 = 9.32$ V.
(f) $\phi = \arctan(X_C - X_L)/R = \arctan(119.38/200) = \arctan(0.597)$
    $= 30.8°$.

Impedance

Circuits that contain a mixture of both resistance and reactance (of either or both types) are said to exhibit *impedance*. Impedance, like resistance and reactance, is simply the ratio of applied voltage to the current flowing. Thus:

$$Z = V/I$$

where Z is the impedance in ohms (Ω), V is the alternating potential difference in volts (V) and I is the alternating current in amps (A).

The impedance of a series circuit (R in series with X – see Figure 3.3.25) is given by:

$$Z = \sqrt{(R^2 + X^2)}$$

where Z is the impedance (in Ω), X is the reactance, either capacitive or inductive (expressed in Ω), and R is the resistance (in Ω).

Example 3.3.12

A resistor of 30 Ω is connected in series with a capacitive reactance of 40 Ω. Determine the impedance of the circuit and the current flowing when the circuit is connected to a 115 V supply.

First we must find the impedance of the C–R series circuit:

$$Z = \sqrt{(R^2 + X_C^2)} = \sqrt{(30^2 + 40^2)} = \sqrt{2500} = 50.$$

The current taken from the supply can now be found:

$$I = V/Z = 115/50 = 2.3 \text{ A}.$$

cont.

Example 3.3.13

A coil is connected to a 50 V a.c. supply at 400 Hz. If the current supplied to the coil is 200 mA and the coil has a resistance of 60 Ω, determine the value of inductance.

We can find the impedance of the coil from:

$Z = V/I = 50/0.2 = 250\ \Omega$

Since $Z = \sqrt{(R^2 + X_L^2)}$, $Z^2 = (R^2 + X_L^2)$ and $X_L^2 = Z^2 - R^2$

$X_L^2 = Z^2 - R^2 = 250^2 - 60^2 = 62\,500 - 3600 = 58\,900.$

Thus $X_L = \sqrt{58\,900} = 243\ \Omega.$

Since $X_L = 2\pi f L$, $L = X_L/(2\pi f) = 243/(6.28 \times 400) = 0.097$ H.

Hence $L = 97$ mH.

Problems 3.3.1

(1)   A sinusoidal alternating current is specified by the equation, $i = 250 \sin(628\,t)$ mA.
      Determine:
      (a)   the peak value of the current;
      (b)   the r.m.s. value of the current;
      (c)   the frequency of the current;
      (d)   the periodic time of the current;
      (e)   the instantaneous value of the current at $t = 2$ ms.

(2)   A sinusoidal alternating voltage has a peak value of 160 V and a frequency of 60 Hz. Write down an expression for the instantaneous voltage and use it to determine the value of voltage at 3 ms from the start of a cycle.

(3)   A resistance of 45 Ω is connected to a 150 V 400 Hz a.c. supply. Determine:
      (a)   an expression for the instantaneous current flowing in the resistor;
      (b)   the r.m.s. value of the current flowing in the resistor.

(4)   Determine the reactance of a 680 nF capacitor at (a) 400 Hz and (b) 20 kHz.

(5)   Determine the reactance of a 60 mH inductor at (a) 20 Hz and (b) at 4 kHz.

(6)   A resistor of 120 Ω is connected in series with a capacitive reactance of 200 Ω. Determine the impedance of the circuit and the current flowing when the circuit is connected to a 200 V a.c. supply.

(7)   A coil has an inductance of 200 mH and a resistance of 40 Ω. Determine:
      (a)   the impedance of the coil at a frequency of 60 Hz;
      (b)   the current that will flow in the coil when it is connected to a 110 V 60 Hz supply.

(8)   A capacitor of 2 μF is connected in series with a 100 Ω resistor across a 24 V 400 Hz a.c. supply. Determine the current that will be supplied to the circuit and the voltage that will be dropped across each component.

(9)   A capacitor of 100 nF is used in a power line filter. Determine the frequency at which the reactance of the capacitor is equal to 10 Ω.

(10) An inductor is used in an aerial tuning unit. Determine the value

of inductance required if the inductor is to have a reactance of 300 Ω at a frequency of 3.5 MHz.

Question 3.3.1

A voltage, $v = 17 \sin(314\,t)$ volts, is applied to a pure resistive load of 68 Ω.

(a) Sketch a fully labelled graph showing how the current flowing in the resistor varies over the period from 0 to 30 ms.

(b) Mark the following on your graph and values for each:
(i) the peak value of current;
(ii) the periodic time.

(c) Determine the instantaneous values of voltage and current at $t = 3.5$ ms.

Question 3.3.2

A series circuit comprises an inductor of 60 mH, a resistor of 33 Ω and a capacitor of 47 μF. If a sinusoidal current of 50 mA at 50 Hz flows in this circuit, determine:

(a) the voltage developed across the inductor;
(b) the voltage dropped across the capacitor;
(c) the voltage dropped across the resistor;
(d) the impedance of the circuit;
(e) the supply voltage;
(f) the phase angle.

Sketch a phasor diagram showing the voltages and current present in the circuit. Label your drawing clearly and indicate values.

Using complex notation

Complex notation provides us with a simple yet powerful method of solving even the most complex of a.c. circuits. Complex notation allows us to represent electrical quantities that have both *magnitude* and *direction* (you will already know that in other contexts we call these *vectors*). The *magnitude* is simply the amount of resistance, reactance, voltage or current, etc. In order to specify the *direction* of the quantity, we use an *operator* to denote the phase shift relative to the *reference* quantity (this is usually current for a series circuit and voltage for a parallel circuit). We call this operator 'j'.

Another view

From studying complex numbers in mathematics, you will recall that every complex number consists of a *real part* and an *imaginary part*. In an electrical context, the *real part* is that part of the complex quantity that is in-phase with the reference quantity (current for series circuits and voltage for parallel circuits). The *imaginary part*, on the other hand, is that part of the complex quantity that is at 90° to the reference.

Mathematics in action
The j-operator

You can think of the j-operator as a device that allows us to indicate a phase shift of 90°. A phase shift of +90° is represented by +j whilst a phase shift of −90° is represented by −j.

A *phasor* is simply an electrical vector. A *vector*, as you will doubtless recall, has magnitude (size) and direction (angle relative to some reference direction). The j-operator can be used to rotate a phasor. Each successive multiplication by j has the effect of rotating the phasor through a further 90°.

The j-operator has a value which is equal to $\sqrt{-1}$. Thus we can conclude that:

$j = \sqrt{-1}$

multiplying by j gives, $j^2 = \sqrt{-1} \times \sqrt{-1} = -1$

multiplying again by j gives, $j^3 = \sqrt{-1} \times \sqrt{-1} \times \sqrt{-1} = -1 \times j = -j$

multiplying again by j gives,
$j^4 = \sqrt{-1} \times \sqrt{-1} \times \sqrt{-1} \times \sqrt{-1} = -1 \times -1 = +1$

multiplying again by j gives,
$j^5 = \sqrt{-1} \times \sqrt{-1} \times \sqrt{-1} \times \sqrt{-1} \times \sqrt{-1} = -1 \times -1 \times j = +j$

and so on.

Mathematics in action
The Argand diagram

The Argand diagram provides a useful method of illustrating complex quantities and allowing us to solve problems graphically. In common with any ordinary 'x–y' graph, the Argand diagram has two sets of axes as right angles. The horizontal axis is known as the *real axis* whilst the vertical axis is known as the *imaginary axis* (don't panic – the imaginary axis isn't really imaginary we simply use the term to indicate that we are using this axis to plot values that are multiples of the j-operator).

Complex impedances

The j-operator and the Argand diagram provide us with a useful way of representing impedances. Any complex impedance can be represented by the relationship:

$$Z = (R \pm j\,X)$$

where Z represents impedance, R represents resistance and X represents reactance.

All three quantities are, of course, measured in ohms (Ω).

The $\pm$ j term simply allows us to indicate whether the reactance is due to inductance, in which case the j term is positive (i.e. $+ j$) or whether it is due to capacitance, in which case the j term is negative (i.e. $- j$).

Consider, for example, the following impedances:

1. $Z_1 = 20 + j\,10$ this impedance comprises a resistance of 20 Ω connected in series with an inductive reactance (note the positive sign before the j-term) of 10 Ω.

2. $Z_2 = 15 - j\,25$ this impedance comprises a resistance of 15 Ω connected in series with a capacitive reactance (note the negative sign before the j-term) of 25 Ω.

3. $Z_3 = 30 + j\,0$ this impedance comprises a pure resistance of 30 Ω (there is no reactive component).

Figure 3.3.26 *Argand diagram showing complex impedances*

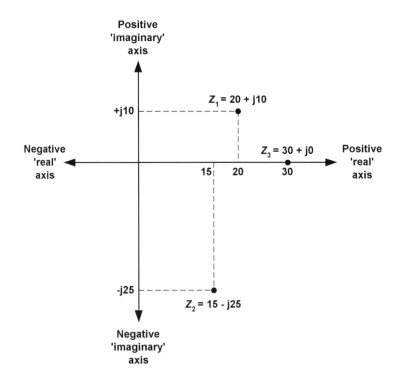

These three impedances are shown plotted on an Argand diagram in Figure 3.3.26.

Voltages and currents can also take complex values. Consider the following:

1. $I_1 = 2 + j\,0.5$ this current is the result of an in-phase component of 2 A and a reactive component (at $+90°$) of 0.5 A.

2. $I_2 = 1 - j\,1.5$ this current is the result of an in-phase component of 1 A and a reactive component (at $-90°$) of 1.5 A.

3. $I_3 = 3 + j\,0$ this current in-phase and has a value of 3 A.

These three currents are shown plotted on an Argand diagram in Figure 3.3.27.

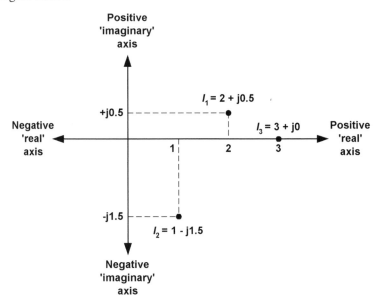

Figure 3.3.27 *Argand diagram showing complex currents*

Example 3.3.14

A current of 2 A flows in an impedance of $(100 + j\,120)\ \Omega$. Derive an expression, in complex form, for the voltage that will appear across the impedance.

Since $V = I \times Z$

$V = 2 \times (100 + j\,120) = (200 + j\,240)\ V.$

Note that, in this example we have assumed that the supply current is the *reference*. In other words, it could be expressed in complex form as $(2 + j\,0)$ A.

Example 3.3.15

An impedance of $(200 + j\,100)\ \Omega$ is connected to a 100 V a.c. supply. Determine the current flowing and express your answer in complex form.

Since $I = V/Z$

$I = 100 / (200 + j\,100) = 100 \times (200 - j\,100)/(200 + j\,100) \times (200 - j\,100)$

(here we have multiplied the top and bottom by the *complex conjugate*)

$I = 100 \times (200 - j\,100)/(200^2 + 100^2) = (20\,000 - j\,10\,000)/(40\,000 + 10\,000).$

Thus $I = (2 - j\,)/5 = (0.4 - j\,0.2)$ A.

Note that, in this example we have assumed that the supply voltage is the *reference*. In other words, it could be expressed in complex form as $(100 + j\,0)$ V.

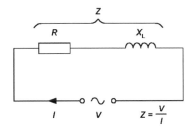

Figure 3.3.28 *Inductance and resistance in series*

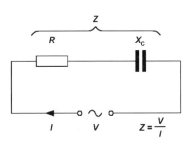

Figure 3.3.29 *Capacitance and resistance in series*

Inductance and resistance in series

A series circuit comprising resistance and inductance (see Figure 3.3.28) can be represented by:

$$Z = R + jX_L \text{ or } Z = R + j\omega L$$

where $\omega = 2\pi f$.

Capacitance and resistance in series

A series circuit comprising resistance and capacitance (see Figure 3.3.29) can be represented by:

$$Z = R - jX_C \text{ or } Z = R - j/\omega C$$

where $\omega = 2\pi f$.

Inductance, resistance and capacitance in series

A series circuit comprising inductance, resistance and capacitance in series (see Figure 3.3.30) can be represented by:

$$Z = R + jX_L - jX_C = R + j(X_L - X_C)$$

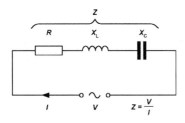

Figure 3.3.30 *Inductance, resistance and capacitance in series*

or $Z = R + j\omega L - j/\omega C = R + j(\omega L - 1/\omega C)$

where $\omega = 2\pi f$.

Example 3.3.16

Write down the impedance (in complex form) of each of the circuits shown in Figure 3.3.31.

(a) $Z_a = 45 + j\,30$.
(b) $Z_b = 5 - j\,4$.
(c) $Z_c = 15 + j\,9 - j\,12 = 15 - j\,3$.

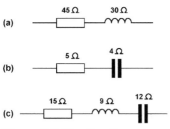

Figure 3.3.31 *See Example 3.3.16*

Complex admittances

When dealing with parallel circuits it is easier to work in terms of *admittance* (Y) rather than *impedance* (Z). Note that:

$$Y = 1/Z .$$

The impedance of a circuit comprising resistance connected in parallel with reactance is given by:

$$Y = (G \pm j\,B)$$

where Y represents *admittance*, G represents *conductance*, and B represents *susceptance*.

Similarly $G = 1/R$ and $B = 1/X$.

All three quantities are measured in Siemens (S).

The $\pm j$ term simply allows us to indicate whether the susceptance is due to capacitance (in which case the j term is positive, i.e. $+ j$) or whether it is due to inductance (in which case the j term is negative, i.e. $- j$).

Consider, for example, the following admittances:

1. $Y_1 = 0.05 - j\,0.1$ this admittance comprises a conductance of 0.05 S connected in parallel with a negative susceptance of 0.1 S. The value of resistance can be found from $R = 1/G = 1/0.05 = 20\,\Omega$ whilst the value of inductive reactance (note the minus sign before the j-term) can be found from $X = 1/B = 1/0.1 = 10\,\Omega$.

2. $Y_2 = 0.2 + j\,0.05$ this admittance comprises a conductance of 0.2 S connected in parallel with a positive susceptance of 0.05 S. The value of resistance can be found from $R = 1/G = 1/0.2 = 5\,\Omega$ whilst the value of capacitive reactance (note the plus sign before the j-term) can be found from $X = 1/B = 1/0.05 = 20\,\Omega$.

Example 3.3.17

A voltage of 20 V appears across an admittance of $(0.1 + j\,0.25)\,\Omega$. Determine the current flowing and express your answer in complex form.

Since $I = V/Z$ and $Z = 1/Y$

cont.

$I = V \times Y = 20 \times (0.1 + j\,0.25) = 2 + j\,5\,\text{A}.$

Note that, in this example we have assumed that the supply voltage is the *reference*. In other words, it could be expressed in complex form as $(20 + j\,0)$ A.

Inductance and resistance in parallel

A parallel circuit comprising resistance and inductance (see Figure 3.3.32) can be represented by:

$$Y = G - j(1/X_L) \text{ or } Y = G - j/\omega L$$

where $\omega = 2\pi f$.

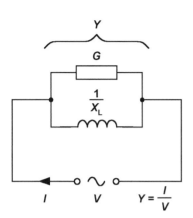

Figure 3.3.32 *Inductance and resistance in parallel*

Capacitance and resistance in parallel

A series circuit comprising resistance and capacitance (see Figure 3.3.33) can be represented by:

$$Y = G + j\,(1/X_C) \text{ or } Y = G + j\,\omega C$$

where $\omega = 2\pi f$.

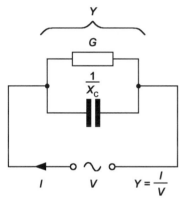

Figure 3.3.33 *Capacitance and resistance in parallel*

Inductance, resistance and capacitance in parallel

A series circuit comprising capacitance, inductance and resistance in parallel (see Figure 3.3.34) can be represented by:

$$Y = G + j\,(1/X_C) - j(1/X_L) = G + j\,(1/X_C - 1/X_L)$$

$$Y = G + j\,(\omega C - 1/\omega L)$$

where $\omega = 2\pi f$.

Example 3.3.18

Write down the admittance (in complex form) of each of the circuits shown in Figure 3.3.35.

(a) $Y_a = 0.5 + j\,0.2.$
(b) $Y_b = 0.25 - j\,0.125.$
(c) $Y_c = 0.1 - j\,0.04 + j\,0.05 = 0.1 + j\,0.01.$

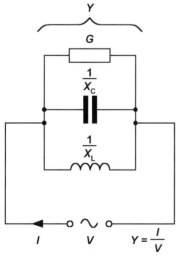

Figure 3.3.34 *Inductance, resistance and capacitance in parallel*

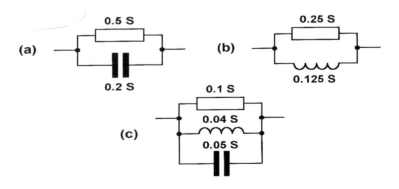

Figure 3.3.35 *See Example 3.3.18*

Resonant circuits

Thus far, we have considered circuits that contain either a combination of resistance and capacitance or a combination of resistance and inductance. These circuits can be described as 'non-resonant' since the voltage and current will not be in-phase at any frequency. More complex circuits, containing all three types of component are described as 'resonant' since there will be one frequency at which the two reactive components will be equal but opposite. At this particular frequency (known as the *resonant frequency*) the effective reactance in the circuit will be zero and the voltage and current will be in-phase. This can be a fairly difficult concept to grasp at first sight so we will explain it in terms of the way in which a resonant circuit behaves at different frequencies.

Series resonance

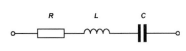

Figure 3.3.36 *Series resonant circuit*

A series resonant circuit comprises inductance, resistance and capacitance connected in series, as shown in Figure 3.3.36.

At the resonant frequency, the inductive reactance, X_L, will be equal to the capacitive reactance, X_C. In other words, $X_L = X_C$. In this condition, the supply voltage will be in-phase with the supply current. Furthermore, since $X_L = X_C$, the reactive components will cancel (recall that they are 180° out of phase with one another) the impedance of the circuit will take a minimum value, equal to the resistance, R.

At a frequency which is less than the resonant frequency (i.e. below resonance) the inductive reactance, X_L, will be smaller than the capacitive reactance, X_C. In other words, $X_L < X_C$. In this condition, the supply voltage will lag the supply current by 90°.

At a frequency which is greater than the resonant frequency (i.e. above resonance) the capacitive reactance, X_C, will be smaller than the inductive reactance, X_L. In other words, $X_C < X_L$. In this condition, the supply voltage will lag the supply current by 90°.

Resonant frequency

The frequency of resonance can be determined by equating the two reactive components, as follows:

$$X_L = X_C$$

thus $2\pi f_o L = 1/(2\pi f_o C)$.

We need to make f_o the subject of this equation (where f_o is the frequency of resonance):

Now $f_o^2 = 1/(4\pi^2 L C)$

or $f_o = \dfrac{1}{2\pi \sqrt{(LC)}}$

Another view

Using j notation, the impedance of the circuit shown in Figure 3.3.36 will be given by $Z = R + jX_L - j X_C = R + j(X_L - X_C)$. At resonance, $X_L = X_C$, thus $Z = R + j(X_L - X_C) = R + j\,0 = R$. In this condition, the circuit behaves like a pure resistor and thus the supply current and voltage will be in-phase.

Example 3.3.19

A series circuit consists of $L = 60$ mH, $R = 15\,\Omega$ and $C = 15$ nF. Determine the frequency of resonance and the voltage dropped across each component at resonance if the circuit is connected to a 300 mV a.c. supply.

$f_o = 1/2\pi \sqrt{(L C)} = 1/2\pi \sqrt{(60 \times 10^{-3} \times 15 \times 10^{-9})}$ Hz

cont.

$f_\text{o} = 1/2\pi \sqrt{(900 \times 10^{-12})} = 1/(2\pi \times 30 \times 10^{-6})$ Hz

$f_\text{o} = 1 \times 10^6/(2\pi \times 30) = 5.3 \times 10^3$ Hz = 5.3 kHz.

At resonance, the reactive components (X_L and X_C) will be equal but of opposite sign. The impedance at resonance will thus be R alone. We can determine the supply current from:

$I = V/Z = V/R = 300$ mV/15 Ω = 20 mA.

The reactance of the inductor is equal to:

$X_\text{L} = 2\pi f_\text{o} L = 2\pi \times 5.3 \times 10^3 \times 60 \times 10^{-3}\,\Omega = 2\pi \times 5.3 \times 60\ \Omega$
$= 2$ kΩ.

The voltage developed across the inductor will be given by:

$V_\text{L} = I \times X_\text{L} = 20$ mA $\times 2$ kΩ = 40 V.

Note that this voltage leads the supply current by 90°.
Since the reactance of the capacitor will be the same as that of the inductor, the voltage developed across the capacitor will be identical (but lagging the supply current by 90°). Thus:

$V_\text{C} = V_\text{L} = 40$ V.

Parallel resonance

A parallel resonant circuit comprises inductance, resistance and capacitance connected in parallel, as shown in Figure 3.3.37. A variation on this is when a capacitor is connected in parallel with a series combination of inductance and resistance, as shown in Figure 3.3.38.

In the case of the circuit shown in Figure 3.3.37, the frequency of resonance can once again be determined by simply equating the two reactive components, as follows:

$$X_\text{L} = X_\text{C}$$

thus $2\pi f_\text{o} L = 1/(2\pi f_\text{o} C)$

or $f_\text{o} = \dfrac{1}{2\pi \sqrt{(L\, C)}}.$

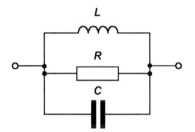

Figure 3.3.37 *Parallel resonant circuit*

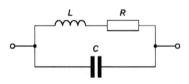

Figure 3.3.38 *Alternative form of parallel resonant circuit*

Example 3.3.20

A parallel circuit consists of L = 40 mH, R = 1 kΩ and C = 10 nF. Determine the frequency of resonance and the current in each component at resonance if the circuit is connected to a 2 V a.c. supply.

$f_\text{o} = 1/2\pi \sqrt{(L\, C)} = 1/2\pi \sqrt{(40 \times 10^{-3} \times 10 \times 10^{-9})}$ Hz

$f_\text{o} = 1/2\pi \sqrt{(400 \times 10^{-12})} = 1/(2\pi \times 20 \times 10^{-6})$ Hz

$f_\text{o} = 1 \times 10^6/(2\pi \times 20) = 7.96 \times 10^3$ Hz = 7.96 kHz.

At resonance, the reactive components (X_L and X_C) will be equal but of opposite sign. The impedance at resonance will thus be R alone. We can determine the supply current from:

$I = V/Z = V/R = 2$ V/1 kΩ = 2 mA.

The reactance of the inductor is equal to:

$X_\text{L} = 2\pi f_\text{o} L = 2\pi \times 7.96 \times 10^3 \times 40 \times 10^{-3}\ \Omega = 2\pi \times 7.96 \times 40\ \Omega = 2$ k Ω.

cont.

The current in the inductor will be given by:

$I = V_L/X_L = 2\text{ V}/2\text{ k}\Omega = 1\text{ mA}.$

Note that this current lags the supply voltage by 90°.

Since the reactance of the capacitor will be the same as that of the inductor, the current in the capacitor will be identical (but leading the supply current by 90°). Thus:

$I_C = I_L = 2\text{ mA}.$

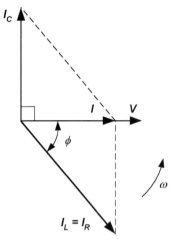

Figure 3.3.39 *Phasor diagram for the circuit of Figure 3.3.20*

In the case of the circuit shown in Figure 3.3.38, the frequency of resonance is once again the frequency at which the supply voltage and current are in phase. Note that the current in the inductor, I_L, lags the supply voltage by an angle, ϕ (see Figure 3.3.39).

Now $I_L = V/Z_L = V/\sqrt{(R^2 + X_L^2)}$

and $I_C = V/X_C.$

At resonance, $I_C = I_L \sin \phi$ and $\sin \phi = X_L/Z_L.$

Thus $V/X_C = V/\sqrt{(R^2 + X_L^2)} \times X_L/Z_L$

or $V\omega C = V/\sqrt{(R^2 + (\omega L)^2)} \times \omega L/\sqrt{(R^2 + X_L^2)}$

$C = L/(R^2 + (\omega L)^2)$

$L/C = (R^2 + (\omega L)^2$

$(\omega L)^2 = (L/C) - R^2$

$\omega^2 = 1/LC - R^2/L^2$

$\omega = \sqrt{(1/LC - R^2/L^2)}$

but, $\omega = 2\pi f_o$

thus $f_o = \dfrac{1}{2\pi} \sqrt{\left(\dfrac{1}{LC} - \dfrac{R^2}{L^2}\right)}.$

At resonance, the impedance of the circuit shown in Figure 3.3.38 is called its *dynamic impedance*. This impedance is given by:

$Z_d = V/I.$

From the phasor diagram of Figure 3.3.39:

$I = I_C/\tan \phi.$

Thus, $Z_d = V \times \tan \phi/I_C = (V/I_C) \times \tan \phi = X_C \times \tan \phi.$

Since $\tan \phi = X_L/R$

$Z_d = X_C \times X_L/R = 1/(2\pi f_o C) \times (2\pi f_o L/R) = \dfrac{L}{CR}.$

This value of impedance is known as the *dynamic impedance* of the circuit. You should note that the dynamic impedance increases as the ratio of L/C increases.

Example 3.3.21

A coil having an inductance of 1 mH and a resistance of 100 Ω is connected in parallel with a capacitor of 10 nF. Determine the frequency of resonance and the dynamic impedance of the circuit.

cont.

$$f_o = \frac{1}{2\pi} \sqrt{\left(\frac{1}{LC} - \frac{R^2}{L^2}\right)}$$

$f_o = 1/2\pi \sqrt{(1/(1 \times 10^{-3} \times 10 \times 10^{-9}) - (100)^2/(1 \times 10^{-3})^2)}$

$f_o = 1/2\pi \sqrt{(1/(1 \times 10^{-11}) - (10^4/10^{-6}))}$

$f_o = 1/2\pi \sqrt{(10^{11} - 10^{10})}$

$f_o = 1/2\pi \sqrt{(9 \times 10^{10})} = (3 \times 10^5)/2\pi = 47.7$ kHz.

The dynamic impedance is given by:

$Z_d = L/(CR) = 1 \times 10^{-3}/(10 \times 10^{-9} \times 100) = 1 \times 10^3 = 1$ kΩ.

Q-factor

The Q-factor (or *quality factor*) is a measure of the 'goodness' of a tuned circuit. Q-factor is sometimes also referred to as 'voltage magnification factor'. Consider the circuit shown in Figure 3.3.40. The Q-factor of the circuit tells you how many times greater the inductor or capacitor voltage is than the supply voltage. The better the circuit, the higher the 'voltage magnification' and the greater the Q-factor.

In Figure 3.3.40, $Q = V_L/V = V_C/V$.

Since $V_L = I \times X_L = I \times \omega L$ and $V = I \times Z = I \times R$ (at resonance)

$$Q = V_L/V = I \times \omega L/I\,R = \omega L/I \times R = \frac{2\pi f_o L}{R}.$$

Similarly, since $V_C = I \times X_C = I/\omega C$ and $V = I \times Z = I \times R$ (at resonance)

$$Q = V_C/V = I/\omega\,C \times I\,R = 1/\omega\,C\,R = \frac{1}{2\pi f_o CR}.$$

Bandwidth

As the Q-factor of a tuned circuit increases, the frequency response curve becomes sharper. As a consequence, the bandwidth of the tuned circuit is reduced. This relationship is illustrated in Figure 3.3.41.

We normally specify the range of frequencies that will be accepted by a tuned circuit by referring to the two *half-power frequencies*. These are the frequencies at which the power in a tuned circuit falls to 50% of its maximum value. At these frequencies:

(1)   The tuned circuit's resistive and reactive components (R and X) will be equal.
(2)   The phase angle (between current and voltage) will be $\pi/4$ (or 45°).
(3)   Both current and voltage will have fallen to $1/\sqrt{2}$ or 0.707 of their maximum value.

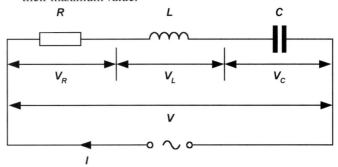

Figure 3.3.40 *Voltages in a resonant circuit*

Figure 3.3.41 *Relationship between Q-factor and bandwidth*

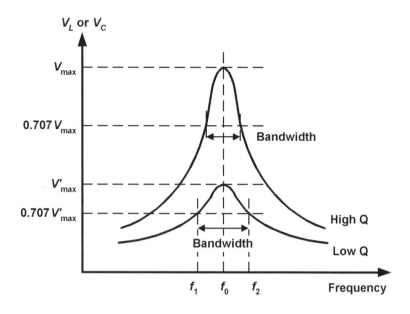

At the upper of the two cut-off frequencies, f_2

$R = X_{\text{effective}}$ (from 1 on page 163)

where $X_{\text{effective}} = X_L - X_C$ (since $X_L > X_C$ above f_o)

thus: $R = X_L - X_C = 2\pi f_2 L - 1/2\pi f_2 C$

rearranging gives:

$R - 1/2\pi f_2 L = -1/2\pi f_2 C$

thus $2\pi f_2 R - (2\pi f_2)^2 L = -1/C$

or $1/C = -2\pi f_2 R + (2\pi f_2)^2 L \dots$ (i).

Whereas, at the lower of the two cut-off frequencies, f_1 :

$R = X_{\text{effective}}$ (from 1 on page 163)

where $X_{\text{effective}} = X_C - X_L$ (since $X_C > X_L$ below f_o)

thus: $R = X_C - X_L = 1/2\pi f_1 C - 2\pi f_1 L$

rearranging gives:

$R + 2\pi f_1 L = 1/2\pi f_1 C$

thus $2\pi f_1 R + (2\pi f_1)^2 L = 1/C$

or $1/C = 2\pi f_1 R + (2\pi f_1)^2 L \dots$ (ii).

Equating (i) and (ii) gives:

$-2\pi f_2 R + (2\pi f_2)^2 L = 2\pi f_1 R + (2\pi f_1)^2 L$

$(2\pi f_2)^2 L - (2\pi f_1)^2 L = 2\pi f_1 R + 2\pi f_2 R$

$(2\pi f_2)^2 - (2\pi f_1)^2 = R/L \times (2\pi f_1 + 2\pi f_2)$

$(2\pi f_1 + 2\pi f_2)(2\pi f_1 - 2\pi f_2) = R/L \times (2\pi f_1 + 2\pi f_2)$

$(2\pi f_1 - 2\pi f_2) = R/L$

$f_1 - f_2 = R/2\pi L.$

But bandwidth, $f_w = f_1 - f_2$

thus $f_w = R/2\pi L$.

Dividing both sides by f_o gives:

$f_w/f_o = R/2\pi L f_o$.

Earlier, we defined Q-factor as :

$Q = 2\pi L f_o/R$.

Thus $f_w/f_o = 1/Q$

or $Q = f_o/f_w$.

Example 3.3.22

A tuned circuit comprises a 400 µH inductor connected in series with a 100 pF capacitor and a resistor of 10 Ω. Determine the resonant frequency of the tuned circuit, its Q-factor and bandwidth.

$f_o = 1/2\pi \sqrt{(LC)} = 1/2\pi \sqrt{(400 \times 10^{-6} \times 100 \times 10^{-12})}$

$f_o = 1/2\pi \sqrt{(4 \times 10^{-14})} = 1/2\pi \times 2 \times 10^{-7}$

$f_o = 0.159/2 \times 10^{-7} = 0.0795 \times 10^7$

$f_o = 0.795 \times 10^6$ Hz or 795 kHz.

The Q-factor is determined from:

$Q = 2\pi f_o L/R = 6.28 \times 795 \times 10^3 \times 400 \times 10^{-6}$

$Q = 199.7 \times 10^4 \times 10^3 \times 10^{-6} = 19.97$.

The bandwidth is determined from:

$f_w = f_o/Q = 795 \times 10^3/19.97 = 39.8 \times 10^3$ Hz.

Thus $f_w = 39.8$ kHz.

Example 3.3.23

A series tuned circuit filter is to be centred on a frequency of 73 kHz. If the filter is to use a capacitor of 1nF, determine the value of inductance required and the maximum value of series loss resistance that will produce a bandwidth of less than 1kHz.

Now $f_o = 1/2\pi \sqrt{(LC)}$.

Thus $L = 1/(4\pi^2 C f_o^2)$

$L = 1/(4\pi^2 \times 1 \times 10^{-9} \times (73 \times 10^3)^2)$

$L = 1/(4\pi^2 \times 1 \times 10^{-9} \times 5329 \times 10^6)$

$L = 10^3/(4\pi^2 \times 5329) = 10^3/210435 = 0.00475$ H

or $L = 4.75$ mH.

Now $f_w = f_o/Q$ and $Q = 2\pi f_o L/R$

Thus $f_w = f_o/(2\pi f_o L/R)$ or $f_w = f_o \times R/(2\pi f_o L)$.

Hence $R = f_w \times (2\pi f_o L)/f_o = f_w \times 2\pi L$

thus $R = 1 \times 10^3 \times 2\pi \times 4.75 \times 10^{-3} = 29.8$ Ω.

Thus, to achieve a bandwidth of 1 kHz or less the total series loss resistance must not be greater than 29.8 Ω.

Problems 3.3.2

(1)   A sinusoidal current of 2 A flows in an impedance of (100 + j 120) Ω. Derive an expression, in complex form, for the voltage that will appear across the impedance.

(2)   A sinusoidal voltage of 20 V appears across an admittance of (0.1 + j 0.25) Ω. Determine the current flowing and express your answer in complex form.

(3)   A coil has an inductance of 60 mH and a resistance of 20 Ω. If a sinusoidal current of 20 mA at 100 Hz flows in the coil, derive an expression for the voltage that will appear across it. Express your answer in complex form.

(4)   A capacitor of 2 μF is connected in parallel with a resistor of 500 Ω. If the parallel circuit is connected to a 20 V 50 Hz sinusoidal supply, derive an expression for the current that will be supplied. Express your answer in complex form.

(5)   A series circuit consists of L = 60 mH, R = 15 Ω and C = 15 nF. Determine the frequency of resonance and the voltage dropped across each component at resonance if the circuit is connected to a 300 mV a.c. supply.

(6)   A coil having an inductance of 1 mH and a resistance of 100 Ω is connected in parallel with a capacitor of 10 nF. Determine the frequency of resonance and the dynamic impedance of the circuit.

(7)   The aerial tuned circuit of a long wave receiver is to tune from 150 to 300 kHz. Determine the required maximum and minimum values of tuning capacitor if the inductance of the aerial coil is 900 μH.

(8)   The parallel tuned circuit to be used in a variable frequency oscillator comprises a coil of inductance 500 μH and resistance 45 Ω and a variable tuning capacitor having maximum and minimum values of 1000 pF and 100 pF, respectively. Determine the tuning range for the oscillator and the Q-factor of the oscillator tuned circuit at each end of the tuning range.

Power in an a.c. circuit

We have already shown how power in an a.c. circuit is only dissipated in resistors and not in inductors or capacitors. This merits further consideration as we need to be aware of the implications, particularly in applications where appreciable amounts of power may be present.

Power in a pure resistance

The voltage dropped across a pure resistance rises and falls in sympathy with the current. The voltage and current are said to be *in-phase*. Since the power at any instant is equal to the product of the *instantaneous voltage* and the *instantaneous current*, we can determine how the power supplied varies over a complete cycle of the supply current. This relationship is illustrated in Figure 3.3.42. From this you should note that:

Figure 3.3.42 *Voltage, current and power in pure resistance*

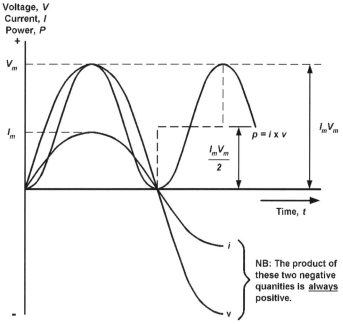

(a)   the power curve represents a cosine law at twice the frequency of the supply current, and

(b)   all points on the power curve are positive throughout a complete cycle of the supply current.

From the foregoing, the power at any instant of time (known as the *instantaneous power*) is given by:

$$p = i \times v$$

where i is the *instantaneous current*, and v is the *instantaneous voltage*.

since $i = I_{max} \sin(\omega t)$ and $v = V_{max} \sin(\omega t)$

$$p = i \times v = I_{max} \sin(\omega t) \times V_{max} \sin(\omega t) = I_{max} V_{max} \sin^2(\omega t)$$

Thus $p = \dfrac{I_{max} V_{max}}{2} \times (1 - \cos(2\omega t))$.

Mathematics in action

The double angle formula

The foregoing relationship is derived from the 'double angle formula', $\sin^2 \theta = \frac{1}{2}(1 - \cos(2\theta))$. There are two important results to note:

(1)   The power function is *always positive* (i.e. a graph of p plotted against t will always be above the x-axis – see Figure 3.3.42).

(2)   The power function has *twice the frequency* of the current or voltage.

(3)   The average value of power over a complete cycle is equal to $\dfrac{I_{max} V_{max}}{2}$.

We normally express power in terms of the r.m.s. values of current and voltage. In this event, the *average power* over one complete cycle of current is given by:

$$P = I \times (I \times R) = I^2 \times R \text{ or } P = V \times V/R = V^2/R$$

where I and V are r.m.s. values of voltage and current.

Question 3.3.3

Show that the instantaneous power in a resistive load connected to sinusoidal a.c. supply is given by:

$$p = \frac{I_{max}^2 R}{2} \times (1 - \cos(2\omega t)).$$

Using a graph of the power function, show that the average power over a complete cycle of the supply is given by:

$$P = I_{max}^2 R/2.$$

Power in a pure reactance

We have already shown how the voltage dropped across a pure inductive reactance leads the current flowing in the inductor by an angle of 90°. Since the power at any instant is equal to the product of the instantaneous voltage and current, we can once again determine how the power supplied varies over a complete cycle of the supply current, as shown in Figure 3.3.43.

From this you should note that:

(a) the power curve represents a sine law at twice the frequency of the supply current, and

(b) the power curve is partly positive and partly negative and the average value of power over a complete cycle of the supply current is zero.

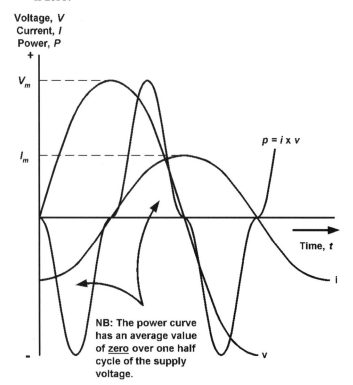

Figure 3.3.43 *Voltage, current and power in a pure inductive reactance*

Figure 3.3.44 *Voltage, current and power in a pure capacitive reactance*

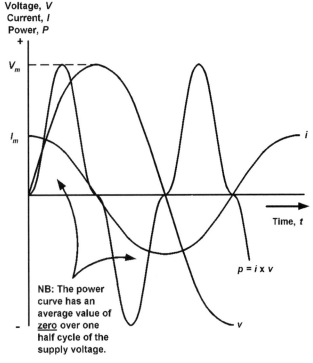

Voltage, V
Current, I
Power, P

V_m

I_m

Time, t

$p = i \times v$

v

NB: The power curve has an average value of <u>zero</u> over one half cycle of the supply voltage.

Similarly, we have already shown how the voltage dropped across a pure capacitive reactance lags the current flowing in the capacitor by an angle of 90°. Since the power at any instant is equal to the product of the instantaneous voltage and current, we can once again determine how the power supplied varies over a complete cycle of the supply current, as shown in Figure 3.3.44.

From this you should note that, once again:

(a) the power curve represents a sine law at twice the frequency of the supply current, and
(b) the power curve is partly positive and partly negative and the average value of power over a complete cycle of the supply current is zero.

Power factor

The *true power* in an a.c. circuit is simply the average power that it consumes. Thus:

True power $= I^2 \times R$.

True power is measured in watts (W).

The *apparent power* in an a.c. circuit is simply the product of the supply voltage and the supply current. Thus:

Apparent power $= V \times I$

Apparent power is measured in volt–amperes (VA).

The *power factor* of an a.c. circuit is simply the ratio of the *true power* to the *apparent power*. Thus:

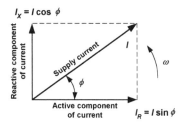

Figure 3.3.45 *Relationship between active and reactive components of supply current*

Power factor = True power/Apparent power

Power factor provides us with an indication of how much of the power supplied to an a.c. circuit is converted into useful energy. A high power factor (i.e. a value close to 1) in an L–R or C–R circuit indicates that most of the energy taken from the supply is dissipated as heat produced by the resistor. A low power factor (i.e. a value close to zero) indicates that, despite the fact that current and voltage is supplied to the circuit, most of the energy is returned to the supply and very little of it is dissipated as heat.

In terms of units, power factor = watts/volt–amperes.

The current supplied to a reactive load can be considered to have two components acting at right angles. One of these components is in-phase with the supply current and it is known as the *active component of current*. The other component is 90° out-of-phase with the supply current ($+90°$ in the case of a capacitive load and $-90°$ in the case of an inductive load) and this component is known as the *reactive component of current*. The relationship between the active and reactive components of the supply current is illustrated in Figure 3.3.45.

From Figure 3.3.45, you should note that the active component of current is given by ($I \times \cos \phi$) whilst the reactive component of current is given by ($I \times \sin \phi$).

The true power is given by the product of supply voltage and the active component of current. Thus:

True power = $V \times (I \times \cos \phi) = VI \times \cos \phi$ ... (i)

Similarly, the reactive power is given by the product of supply voltage and the reactive component of current. Thus:

Reactive power = $V \times (I \times \sin \phi) = VI \times \sin \phi$ ...(ii)

Also from Figure 3.3.45, the product of supply current and voltage (i.e. the *apparent power*) is given by:

Apparent power = $V \times I$ ... (iii)

Combining formula (i) and formula (iii) allows us to express power factor in a different way:

Power factor = True power/Apparent power = $(VI \cos \phi)/(VI) = \cos \phi$

Hence power factor = $\cos \phi$.

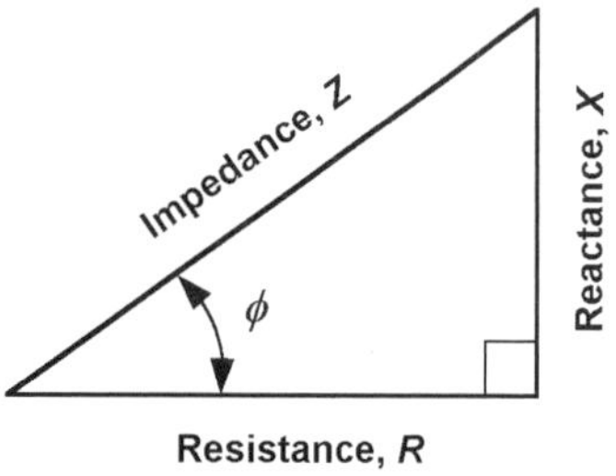

(Note that power factor can have values of between 0 and 1).

We can also express power factor in terms of the ratio of resistance, R, and reactance, X, present in the load. This is done by referring to the impedance triangle (Figure 3.3.46) where:

power factor = $\cos \phi = R/Z$.

The ratio of reactive power to true power can be determined by dividing equation (ii) by equation (i). Thus:

Reactive power/True power = $VI \sin \phi/VI \cos \phi = \tan \phi$.

Figure 3.3.46 *Impedance triangle*

Example 3.3.24

A power of 1 kW is supplied to an a.c. load which operates from a 220 V 50 Hz supply. If the supply current is 6 A, determine the power factor of the load.

Power factor = True power/Apparent power = 1 kW/(220 V × 6 A).

Thus power factor = 1 kW/1.32 kVA = 0.758.

Example 3.3.25

The phase angle between the voltage and current supplied to an a.c. load is 25°. If the supply current is 3 A and the supply voltage is 220 V, determine the true power and reactive power in the load.

Power factor = $\cos \phi = \cos 25° = 0.906$.

True power = $220 \times 3 \times \cos 25° = 220 \times 3 \times 0.906 = 598$ W.

Reactive power = $220 \times 3 \times \sin 25° = 220 \times 3 \times 0.423 = 279$ W.

Example 3.3.26

An a.c. load comprises an inductance of 0.5 H connected in series with a resistance of 85 Ω. If the load is to be used in conjunction with a 220 V 50 Hz a.c. supply, determine the power factor of the load and the supply current.

The reactance of the inductor, X_L, can be found from:

$X_L = 2\pi fL = 2\pi \times 50 \times 0.5 = 157$ Ω.

Power factor = $\cos \phi = R/Z = \dfrac{R}{\sqrt{R^2 + X^2_L}} = \dfrac{85}{\sqrt{85^2 + 157^2}} = 0.48$.

The supply current, I, can be calculated from:

$I = V/Z = V/\sqrt{(R^2 + X_L^2)} = 220/\sqrt{(85^2 + 157^2)}$.

$I = 220/\sqrt{(7225 + 24\,649)} = 220/\sqrt{(7225 + 24\,649)}$.

$I = 220/\sqrt{(31\,874)} = 220/178.5 = 1.23$ A.

Power factor correction

When considering the utilisation of a.c. power, it is very important to understand the relationship between power factor and supply current. We have already shown that:

Power factor = True power/Apparent power.

Thus:

Apparent power = True power/Power factor

or:

VI = True power/Power factor.

Since V is normally a constant (i.e. the supply voltage) we can infer that:

$I \propto$ True power/Power factor.

This relationship shows that the supply current is inversely proportional to the power factor. This has some important implications. For example, it explains why a large value of supply current will flow when a load has a low value of power factor. Conversely, the smallest value of supply current will occur when the power factor is 1 (the largest possible value for the power factor).

Another view

It might help to explain the relationship between power factor and supply current by taking a few illustrative values. Let's assume that a motor produces a power of 1 kW and it operates with an efficiency 60%. The input power (in other words, the product of the voltage and current taken from the supply) will be (1 kW/0.6) or 1.667 kW. The supply current that would be required at various different power factors (assuming a 220 V a.c. supply) is given below:

Power factor	Supply current
0.2	38 A
0.4	19 A
0.6	12.6 A
0.8	9.5 A
1.0	7.6 A

High values of power factor clearly result in the lowest values of supply current!

Example 3.3.27

A power of 400 W is supplied to an a.c. load when it is connected to a 110 V 60 Hz supply. If the supply current is 6 A, determine

cont.

the power factor of the load and the phase angle between the supply voltage and current.

True power = 400 W.

Apparent power = 110 V × 6 A = 660 VA.

Power factor = True power/Apparent power = 400 W/660 VA = 0.606.

Power factor = cos ϕ thus ϕ = cos^{-1}(0.606) = 52.7°.

Example 3.3.28

A motor produces an output of 750 W at an efficiency of 60% when operated from a 220 V a.c. mains supply. Determine the power factor of the motor if the supply current is 9.5 A. If the power factor is increased to 0.9 by means of a 'power factor correcting circuit', determine the new value of supply current.

True power = 750 W/0.6 = 1.25 kW.

Power factor = 1.25 kW/(220 V × 7.5 A) = 1.25 kW/1.65 kVA = 0.76.

Supply current (for a power factor of 0.9) = 1.25 kW/(220 V × 0.9) = 6.3 A.

It is frequently desirable to take steps to improve the power factor of an a.c. load. When the load is inductive, this can be achieved by connecting a capacitor in parallel with the load. Conversely, when the load is capacitive, the power factor can be improved by connecting an inductor in parallel with the load. In order to distinguish between the two types of load (inductive and capacitive) we often specify the power factor as either *lagging* (in the case of an inductor) or *leading* (in the case of a capacitor).

In either case, the object of introducing a component of *opposite* reactance is to reduce the overall phase angle (i.e. the angle between the supply current and the supply voltage). If, for example, a load has an (uncorrected) phase angle of 15° lagging, the phase angle could be reduced to 0° by introducing a component that would exhibit an equal, but opposite, phase angle of 15° leading. To put this into perspective, consider the following example:

Example 3.3.29

A 1.5 kW a.c. load is operated from a 220 V 50 Hz a.c. supply. If the load has a power factor of 0.6 lagging, determine the value of capacitance that must be connected in parallel with it in order to produce a unity power factor.

cos ϕ = 0.6 thus ϕ = cos^{-1}(0.6) = 53.1°.

Reactive power = True power × tan ϕ = 1.5 kW × tan (53.1°).

Thus reactive power = 1.5 × 1.33 kVA = 2 kVA.

In order to increase the power factor from 0.6 to 1, the capacitor connected in parallel with the load must consume an equal reactive power. We can use this to determine the current that must flow in the capacitor and hence its reactance.

cont.

Capacitor current, I_c = 2 kVA/220 V = 9.1 A.

Now X_c = V/I_c = 220 V/9.1 A = 24.12 Ω.

Since X_c = $1/(2\pi f C)$.

C = $1/(2\pi f X_c)$ = $1/(2\pi \times 50 \times 24.12)$ = 1/ 7578 F.

Thus C = 1.32×10^{-4} F = 132 μF.

Problems 3.3.3

(1)   An a.c. load draws a current of 2.5 A from a 220 V supply. If the load has a power factor of 0.7, determine the active and reactive components of the current flowing in the load.

(2)   An a.c. load has an effective resistance of 90 Ω connected in series with a capacitive reactance of 120 Ω. Determine the power factor of the load and the apparent power that will be supplied when the load is connected to a 415 V 50 Hz supply.

(3)   The phase angle between the voltage and current supplied to an a.c. load is 37°. If the supply current is 2 A and the supply voltage is 415 V, determine the true power and reactive power in the load.

(4)   A coil has an inductance of 0.35 H and a resistance of 65 Ω. If the coil is connected to a 220 V 50 Hz supply determine its power factor and the value of capacitance that must be connected in parallel with it in order to raise the power factor to unity.

(5)   An a.c. load comprises an inductance of 1.5 H connected in series with a resistance of 85 Ω. If the load is to be used in conjunction with a 220 V 50 Hz a.c. supply, determine the power factor of the load and the supply current.

(6)   A power of 400 W is supplied to an a.c. load when it is connected to a 110 V 60 Hz supply. If the supply current is 6 A, determine the power factor of the load and the phase angle between the supply voltage and current.

(7)   A motor produces an output of 750 W at an efficiency of 60% when operated from a 220 V a.c. mains supply. Determine the power factor of the motor if the supply current is 9.5 A. If the power factor is increased to 0.9 by means of a 'power factor correcting circuit', determine the new value of supply current.

(8)   An a.c. load consumes a power of 800 W from a 110 V 60 Hz supply. If the load has a lagging power factor of 0.6, determine the value of parallel connected capacitor required to produce a unity power factor.

Waveforms

The sinusoidal waveform

This is the most fundamental of all wave shapes and all other waveforms can be synthesised from sinusoidal components. To specify a sine wave, we need to consider just three things; amplitude, frequency and phase. Since no components are present at any other frequency (other than that of the fundamental sine wave) a sinusoidal wave is said to be a *pure tone*. All other waveforms can be reproduced by adding together sine waves of the correct amplitude, frequency, and phase. The study of these techniques is called Fourier Analysis (see Chapter 4).

Fundamental and harmonic components

An integer multiple of a *fundamental* frequency is known as a *harmonic*. In addition, we often specify the order of the harmonic (second, third, and so on). Thus the *second harmonic* has twice the frequency of the fundamental, the *third harmonic* has three times the frequency of the fundamental, and so on. Consider, for example, a fundamental signal at 1 kHz. The second harmonic would have a frequency of 2 kHz, the third harmonic a frequency of 3 kHz, and the fourth harmonic a frequency of 4 kHz.

Note that, in musical terms, the relationship between notes that are one *octave* apart is simply that the two frequencies have a ratio of 2:1 (in other words, the higher frequency is double the lower frequency).

Complex waveforms

Complex waveforms comprise a fundamental component together with a number of harmonic components, each having a specific amplitude and with a specific phase relative to the fundamental. The following example shows that this is not quite so complicated as it sounds!

Example 3.3.30

Consider a sinusoidal signal with an amplitude of 1 V at a frequency of 1 kHz. The waveform of this fundamental signal is shown in Figure 3.3.47.

Now consider the second harmonic of the first waveform. Let's suppose that this has an amplitude of 0.5 V and that it is in-phase with the fundamental. This 2 kHz signal is shown in Figure 3.3.48.

Finally, let's add the two waveforms together at each point in time. This produces the complex waveform shown in Figure 3.3.49.

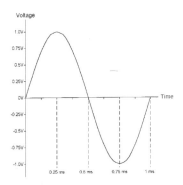

Figure 3.3.47 *Fundamental component*

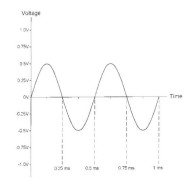

Figure 3.3.48 *Second harmonic component*

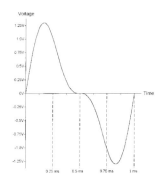

Figure 3.3.49 *Fundamental plus second harmonic components*

We can describe a complex wave using an equation of the form:

$$v = V_1 \sin(\omega t) + V_2 \sin(2\omega t \pm \phi_2) + V_3 \sin(3\omega t \pm \phi_3) + V_4 \sin(4\omega t \pm \phi_4) + \ldots$$

Where v is the instantaneous voltage of the complex waveform at time, t. V_1 is the amplitude of the fundamental, V_2 is the amplitude of the second harmonic, V_3 is the amplitude of the third harmonic, and so on. Similarly, ϕ_2 is the phase angle of the second harmonic (relative to the fundamental), ϕ_3 is the phase angle of the third harmonic (relative to the fundamental), and so on.

The important thing to note from the foregoing equation is that *all of the individual components that make up a complex waveform have a sine wave shape*.

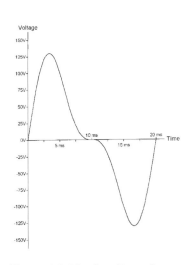

Figure 3.3.50 *See Example 3.3.31*

Example 3.3.31

The complex waveform shown in Figure 3.3.50 is given by:

$v = 100 \sin(100\,\pi t) + 50 \sin(200\,\pi t)$ V.

Determine:

(a) The amplitude of the fundamental.
(b) The frequency of the fundamental.
(c) The order of any harmonic components present.
(d) The amplitude of any harmonic components present.
(e) The phase angle of any harmonic components present (relative to the fundamental).

Comparing the foregoing equation with the general equation (see above) yields the following:

(a)   The first term is the fundamental and this has an amplitude (V_1) of 100 V.
(b)   The frequency of the fundamental can be determined from:

$\sin(\omega t) = \sin(100\,\pi t)$.

Thus $\omega = 100\pi$ but $\omega = 2\pi f$

thus $2\pi f = 100\pi$ and $f = 100\pi/2\pi = 50$ Hz.

(c)   The frequency of the second term can similarly be determined from:

$\sin(\omega t) = \sin(200\pi t)$.

Thus $\omega = 200\,\pi$ but $\omega = 2\pi f$

thus $2\pi f = 200\pi$ and $f = 200\pi/2\pi = 100$ Hz.

Thus the second term has twice the frequency of the fundamental, i.e. it is the second harmonic.

(d)   The second harmonic has an amplitude (V_2) of 50 V.
(e)   Finally, since there is no phase angle included within the expression for the second harmonic (i.e. there is no $\pm \phi_2$ term), the second harmonic component must be in phase with the fundamental.

Figure 3.3.51 *Fundamental and low-order harmonic components of a square wave*

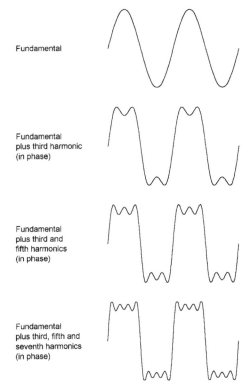

Fundamental

Fundamental plus third harmonic (in phase)

Fundamental plus third and fifth harmonics (in phase)

Fundamental plus third, fifth and seventh harmonics (in phase)

The square wave

The square wave can be created by adding a fundamental frequency to an infinite series of odd harmonics (i.e. the third, fifth, seventh, etc.). If the fundamental is at a frequency of f, then the third harmonic is at $3f$, the fifth is at $5f$, and so on.

The amplitude of the harmonics should decay in accordance with their harmonic order and they must all be in phase with the fundamental. Thus a square wave can be obtained from:

$$v = V \sin(\omega t) + V/3 \sin(3\omega t) + V/5 \sin(5\omega t) + V/7 \sin(7\omega t) + \ldots$$

where V is the amplitude of the fundamental and $\omega = 2\pi f$.

Figure 3.3.51 shows this relationship graphically.

The triangular wave

The triangular wave can similarly be created by adding a fundamental frequency to an infinite series of odd harmonics. However, in this case the harmonics should decay in accordance with the square of their harmonic order and they must alternate in phase so that the third, seventh, eleventh, etc. are in antiphase whilst the fifth, ninth, thirteenth, etc. are in phase with the fundamental. Thus a triangular wave can be obtained from:

$$v = V \sin(\omega t) + V/9 \sin(3\omega t - \pi) + V/25 \sin(5\omega t) + V/49 \sin(7\omega t - \pi) + \ldots$$

where V is the amplitude of the fundamental and $\omega = 2\pi f$.

Figure 3.3.52 shows this relationship graphically.

Question 3.3.4

A complex waveform is described by the equation:

Figure 3.3.52 *Fundamental and low-order harmonic components of a triangle wave*

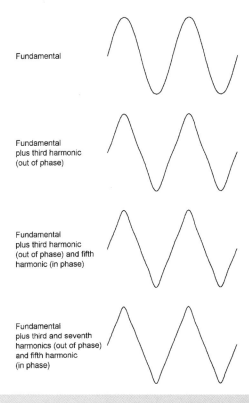

Fundamental

Fundamental plus third harmonic (out of phase)

Fundamental plus third harmonic (out of phase) and fifth harmonic (in phase)

Fundamental plus third and seventh harmonics (out of phase) and fifth harmonic (in phase)

$v = 100 \sin (100\pi t) + 50 \sin (200\pi t - \pi/2)$.

Determine, graphically, the shape of the waveform and estimate the peak–peak voltage.

Question 3.3.5

A complex waveform is described by the equation:

$v = 100 \sin (100\pi t) + 250 \sin (200\pi t - \pi/2) + 450 \sin (300\pi t)$.

Determine, graphically, the shape of the waveform and estimate the peak–peak voltage.

Effect of adding harmonics on overall waveshape

The results of adding harmonics to a fundamental waveform with different phase relationships are listed below:

(1) When odd harmonics are added to a fundamental waveform, regardless of their phase relative to the fundamental, the *positive and negative half-cycles will be similar* in shape and the resulting waveform will be symmetrical about the time axis – see Figure 3.3.53(a).

(2) When even harmonics are added to a fundamental waveform with a phase shift other than 180°, the *positive and half-cycles will be dissimilar* in shape – see Figure 3.3.53(b).

(3) When even harmonics are added to a fundamental waveform with a phase shift of 180°, the *positive half-cycles will be the mirror image of the negative half cycles* when reversed – see Figure 3.3.53(c).

Figure 3.3.53   *Effects of harmonic components on waveshape*

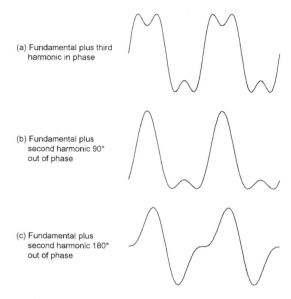

(a) Fundamental plus third harmonic in phase

(b) Fundamental plus second harmonic 90° out of phase

(c) Fundamental plus second harmonic 180° out of phase

Generation of harmonic components

Unwanted harmonic components may be produced in any system that incorporates non-linear circuit elements (such as diodes and transistors). Figure 3.3.54 shows the circuit diagram of a simple half-wave rectifier. The voltage and current waveforms for this circuit are shown in Figure 3.3.55. The important thing to note here is the shape of the current waveforms which comprises a series of relatively narrow pulses as the charge in the reservoir capacitor, C, is replenished at the crest of each positive-going half-cycle. These pulses of current are rich in unwanted harmonic content.

Harmonic components can also be generated in magnetic components (such as inductors and transformers) where the relationship between magnetic flux and applied current is non-linear (see Figure 3.3.56). In the case of a transformer, the voltage applied to the primary is sinusoidal

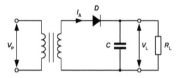

Figure 3.3.54   *A simple half-wave rectifier*

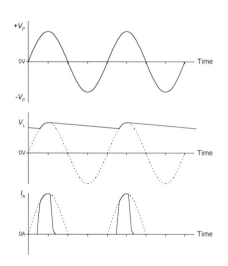

Figure 3.3.55   *Current and voltage waveforms for the simple half-wave rectifier*

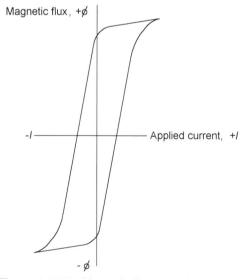

Figure 3.3.56   *Magnetic flux plotted against applied current for a transformer*

Figure 3.3.57 *Magnetic flux and applied current waveforms for a typical transformer*

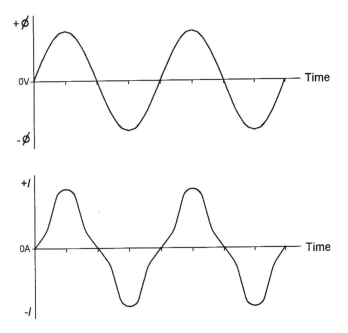

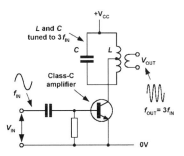

Figure 3.3.58 *A Class-C frequency multiplier stage*

and the flux (which lags the voltage by 90°) will also be sinusoidal. However, by virtue of the non-linearity of the magnetic flux/current characteristic, the current will not be purely sinusoidal but will contain a number of harmonic components, as shown in Figure 3.3.57.

Alternatively, there are a number of applications in which harmonic components are actually desirable. A typical application might be the frequency multiplier stage of a transmitter in which the non-linearity of an amplifier stage (operating in Class-C – see page 192) can be instrumental in producing an output which is rich in harmonic content. The desired harmonic can then be selected by means of a resonant circuit tuned to the desired multiple of the input frequency – see Figure 3.3.58.

Problems 3.3.4

(1)   A sinusoidal current is specified by the equation, $i = 0.1 \sin (200\pi t + \pi/2)$ A.
Determine:
(a) the peak value of the current;
(b) the frequency of the current;
(c) the phase angle of the current;
(d) the instantaneous value of the current at $t = 5$ ms.

(2)   A complex waveform comprises a fundamental (sinusoidal) voltage which has a peak value of 36 V and a frequency of 400 Hz together with a third harmonic component having a peak value of 12 V leading the fundamental by 90°. Write down an expression for the instantaneous value of the complex voltage, and use it to determine the value of voltage at 3 ms from the start of a cycle.

(3)   A complex waveform is given by the equation:
$i = 10 \sin (100\pi t) + 3 \sin (300\pi t - \pi/2) + 1 \sin (500\pi t + \pi/2)$ mA.
Determine:
(a) the amplitude of the fundamental;
(b) the frequency of the fundamental;

(c) the order of any harmonic components present;
(d) the amplitude of any harmonic components present;
(e) the phase angle of any harmonic components present (relative to the fundamental).

3.4 Information and energy control systems

Information systems

Much of electronics is concerned with the generation, transmission, reception, and storage of information using electronic circuits and devices. This information takes various forms including:

- speech or music (audio signals)
- visual images (video signals)
- alphanumeric characters used in computer systems
- numeric and other coded data in computer systems.

Electronic circuits provide different ways to process this information, including amplification of weak signals to usable levels, filtering to reduce noise and unwanted signal components, superimposition of a baseband signal onto a high-frequency carrier wave (*modulation*), recovery of a baseband signal from high-frequency carrier waves (*demodulation*), and logical operations such as those that take place in a computer.

The term *information system* applies to a system in which information is gathered, processed and either displayed or stored for later analysis. *Information engineering* is used to describe the 'engineering' of information systems, i.e. the specification, design, construction, implementation and maintenance of systems used for gathering, processing and using information. So far, so good – but what actually is 'information'?

To put it in its most simple terms, *information* is the way in which meaning is conveyed. Let's take an example to illustrate this. Imagine that you are in the driving seat of a car travelling on a motorway. You will be constantly receiving information. This information conveys meaning to you about the state of the car, of the motorway and of other vehicles around you. You receive this information in various ways using the majority of your senses; sight, hearing, touch and even smell. Signals are sent from sensory receptors via your nervous system to your brain which then uses the information presented to it in order to decide what to do next.

In the decision making process, you might also make use of information that you previously stored in your memory. For example, you might recall that a long hill is around the next bend and this prompts you to move out to overtake a slow-moving vehicle well in advance.

In an engineering context, information usually takes the forms of analogue or digital signals (which may either be *baseband* signals or signals *modulated* onto a high frequency carrier wave) transmitted by line, cable, optical fibre, point–point radio, microwave or satellite.

Information theory

Information theory was first developed in 1948 by the American electrical engineer Claude Shannon. Shannon was concerned with the measurement of information and, in particular the information content of a message.

The term *information content* relates to the probability that a given message will contain completely new and unpredictable information and that this information is different from all other possible messages received. Essentially, when several possible messages are present in a communications channel:

- the highest value for the information content is assigned to the message that is the least probable
- the least value for the information content is assigned to the message that is the most probable.

Shannon showed that the information content of a message is given by:

$$I = \log_2 (1/p) \text{ or } I = - \log_2 (p)$$

where I is the information content and p is the probability of a message being transmitted.

Mathematics in action
Manipulating logarithms

We can prove the relationship, $\log_2(1/p) = - \log_2 (p)$ as follows:

Let $y = \log_n(1/x)$ thus $y = \log_n(x^{-1})$.

Now if $y = \log_n(x^m)$ then $y = m \times \log_n (x)$.

Hence $y = \log_n(x^{-1}) = -1 \log_n(x)$.

Thus $y = \log_n(1/x) = -1 \log_n(x)$ and so $\log_2(1/p) = - \log_2 (p)$.

Example 3.4.1

A rotary position sensor has a resolution of 22.5° and provides its output in digital form using a 4-bit code. What is the information content of each of the following signals produced by this sensor?

(a) 0001
(b) x001
(c) xxx1

(where x = don't care, i.e. either 0 or 1).

(a)   There will only be one position that produces this code. Since there are 16 possible states (corresponding to codes 0000 to 1111) the probability of receiving this code (out of the 16 possible) is 1/16 or 0.0625. In this case:

Now $I = \log_2 (16) = 4$.

(b)   Two positions can produce this code, 0001 and 1001. The probability of receiving either of these two codes (out of the 16 possible) is 2/16 or 0.125. In this case:

Now $I = \log_2 (1/0.125) = \log_2 (8) = 3$.

(c)   Eight positions can produces this code; 0001, 0011, 0101, 0111, 1001, 1011, 1101, and 1111. The probability of receiving any one of these eight codes (out of the 16 possible) is 8/16 or 0.5. In this case:

Now $I = \log_2 (1/0.5) = \log_2 (2) = 1$.

ASCII code

The American Standard Code for Information Interchange (ASCII) is universally used in digital communications (see Table 3.4.1). The ASCII code is based on seven bits and these 128 different states are sufficient to convey upper and lower case letters, numbers and basic punctuation as well as 32 *control characters* needed to provide information about how the data is to be transferred and how the characters are to be displayed on the printed page or on a VDU screen. The control characters use the lowest five bits of the binary code and they are present when the two most-significant bits of the seven bit code are zero. Thus 0011011 is a control character (ESC) whereas 1001011 is a *printable character* (K).

Note that the decimal (Dec.) and hexadecimal (Hex.) values are often used to refer to ASCII characters rather than the binary codes that the digital circuits recognise directly.

ASCII characters are often transmitted along an eight bit data path where the most significant bit (MSB) is *redundant* (this simply means that, as far as the standard ASCII code is concerned, the MSB conveys no meaning). Many computer and IT manufacturers (e.g. IBM, Olivetti, etc.) have made use of the redundant MSB to extend the basic character set with foreign accents and block graphic characters. These codes are referred to as an *extended character set*.

Question 3.4.1

How many extra codes are available with an extended character set (i.e. a character set based on a full eight bits rather than a the standard ASCII seven bit code). What is the information content of these additional codes?

Elements of an information system

The basic elements of an information system are shown in Figure 3.4.1. In this simple model, the *input device* is responsible for gathering or extracting information from the outside world. Input elements can be a wide variety of sensing devices and transducers, such as temperature sensors, strain gauges, proximity sensors, keyboards, microphones, etc.

The *processing device* can typically perform a variety of tasks, including making decisions and storing the information for later analysis. Processing elements can be microprocessors, microcomputers, programmed logic arrays (PLA), programmable logic controllers (PLC), etc.

The *output device* is responsible for displaying or communicating the information (or the results of the analysis performed by the processing element). Output elements can be cathode ray tube (CRT) displays, alphanumeric displays, indicators, warning devices, etc.

Figure 3.4.2 shows an information system in which a number of input devices (*sensors* or *transducers*) produce electrical signals that are an analogy of the particular physical quantity present at the input. In order

Figure 3.4.1 *Basic elements of an information system*

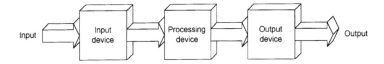

Table 3.4.1 *ASCII code*

Dec.	Binary	Hex.	ASCII
0	0000000	00	NUL
1	0000001	01	SOH
2	0000010	02	STX
3	0000011	03	ETX
4	0000100	04	EOT
5	0000101	05	ENQ
6	0000110	06	ACK
7	0000111	07	BEL
8	0001000	08	BS
9	0001001	09	HT
10	0001010	0A	LF
11	0001011	0B	VT
12	0001100	0C	FF
13	0001101	0D	CR
14	0001110	0E	SO
15	0001111	0F	SI
16	0010000	10	DLE
17	0010001	11	DC1
18	0010010	12	DC2
19	0010011	13	DC3
20	0010100	14	DC4
21	0010101	15	NAK
22	0010110	16	SYN
23	0010111	17	ETB
24	0011000	18	CAN
25	0011001	19	EM
26	0011010	1A	SUB
27	0011011	1B	ESC
28	0011100	1C	FS
29	0011101	1D	GS
30	0011110	1E	RS
31	0011111	1F	US
32	0100000	20	SP
33	0100001	21	!
34	0100010	22	"
35	0100011	23	#
36	0100100	24	$
37	0100101	25	%
38	0100110	26	&
39	0100111	27	'
40	0101000	28	(
41	0101001	29	)
42	0101010	2A	*
43	0101011	2B	+
44	0101100	2C	,
45	0101101	2D	-
46	0101110	2E	.
47	0101111	2F	/
48	0110000	30	0
49	0110001	31	1
50	0110010	32	2
51	0110011	33	3
52	0110100	34	4
53	0110101	35	5
54	0110110	36	6
55	0110111	37	7
56	0111000	38	8
57	0111001	39	9
58	0111010	3A	:
59	0111011	3B	;
60	0111100	3C	<
61	0111101	3D	=
62	0111110	3E	>
63	0111111	3F	?
64	1000000	40	@
65	1000001	41	A
66	1000010	42	B
67	1000011	43	C
68	1000100	44	D
69	1000101	45	E
70	1000110	46	F
71	1000111	47	G
72	1001000	48	H
73	1001001	49	I
74	1001010	4A	J
75	1001011	4B	K
76	1001100	4C	L
77	1001101	4D	M
78	1001110	4E	N
79	1001111	4F	O
80	1010000	50	P
81	1010001	51	Q
82	1010010	52	R
83	1010011	53	S
84	1010100	54	T
85	1010101	55	U
86	1010110	56	V
87	1010111	57	W
88	1011000	58	X
89	1011001	59	Y
90	1011010	5A	Z
91	1011011	5B	[
92	1011100	5C	\
93	1011101	5D	]
94	1011110	5E	^
95	1011111	5F	_
96	1100000	60	`
97	1100001	61	a
98	1100010	62	b
99	1100011	63	c
100	1100100	64	d
101	1100101	65	e
102	1100110	66	f
103	1100111	67	g
104	1101000	68	h
105	1101001	69	i
106	1101010	6A	j
107	1101011	6B	k
108	1101100	6C	l
109	1101101	6D	m
110	1101110	6E	n
111	1101111	6F	o
112	1110000	70	p
113	1110001	71	q
114	1110010	72	r
115	1110011	73	s
116	1110100	74	t
117	1110101	75	u
118	1110110	76	v
119	1110111	77	w
120	1111000	78	x
121	1111001	79	y
122	1111010	7A	z
123	1111011	7B	{
124	1111100	7C		
125	1111101	7D	}
126	1111110	7E	~
127	1111111	7F	DEL

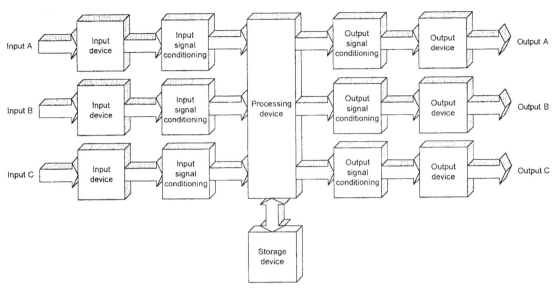

Figure 3.4.2 *A more complex information system*

to standardise the signal levels that are presented to the processing device, the electrical signals from each input device are subjected to *signal conditioning*. For example, some input transducers produce continuously variable analogue outputs which will require conversion to equivalent digital signals that can be applied to a *digital processing* device (e.g. a microprocessor or microcomputer). This conversion is carried out using a *digital-to-analogue converter* (*DAC*). Conversely, some output transducers produce discrete state digital outputs which will require conversion to equivalent analogue signals that can be applied to an analogue output device (e.g. a d.c. motor). This conversion is carried out using an *analogue-to-digital converter* (*ADC*). Other common examples of signal conditioning devices are level-shifters, amplifiers, filters, attenuators, modulators and demodulators.

Examples of information systems

Typical examples of information systems are plant monitoring and control systems, networked computer systems, and many other systems based on microcomputers and microprocessors in which information (or *data*) is exchanged between 'intelligent' devices. Note also that, whether they are based on analogue or digital technology, all types of communication system can be classified as 'information systems'.

Table 3.4.2 lists some examples of information systems.

Question 3.4.2

A local radio station provides entertainment and information for its listening audience.

Treat this as an information system and identify the following elements:

Input devices, input signal conditioning, processing, storage, output signal conditioning, output devices.

Table 3.4.2 *Some examples of information systems*

	Plant monitoring system	*Satellite global positioning system*	*Traffic information system*
Input devices	Temperature sensors, flow-rate sensors, etc.	UHF antenna, keypad	Magnetic sensors, TV cameras, keyboard
Input signal conditioning	Analogue-to-digital converters, amplifiers, filters	Frequency changing, amplification, demodulation	Optical fibre transmitters and receivers, micro-wave radio links
Processing	Industrial microcomputer	Dedicated microprocessor	Networked computer system
Storage	Hard disk drive with backup tape drive	Semiconductor memory	Magneto-optical disk
Output signal conditioning	CRT interface card, alphanumeric display driver, parallel output port	LCD display driver	Solid-state switching devices, serial interface to remote information displays
Output devices	CRT display, alpha-numeric display, piezoelectric warning device	LCD screen	Hazard warning indicators, traffic information displays

Signals

Within an information system, information is conveyed by means of *signals*. These may take a number of different forms:

- switching between two states (i.e. on/off, open/closed or high/low).
- a varying d.c. voltage or current (the instantaneous voltage or current represents the value of the signal at the time of sampling).
- a waveform in which one or more properties are varied (e.g. frequency or amplitude).

Signals can be sent using wires, cables, optical and radio links and they can be processed in various ways using amplifiers, modulators, filters, etc. Signals can also be classified as either *analogue* (continuously variable) or *digital* (based on discrete states).

Figure 3.4.3 shows various forms of signal, including baseband analogue and digital signals, amplitude modulated (AM) and frequency modulated (FM) signals, and a pulse width modulated (PWM) signal.

Noise

Signals in information systems are susceptible to corruption by stray unwanted signals (both *natural* and *man-made*) which we refer to as *noise*. Noise may occur regularly (in which case it is referred to as *impulse noise*) or may occur randomly (in which case it is called *random noise*).

Impulse noise

Noise generated by an electrical plant (such as transformers, rectifiers, and electrical machines) is, by its very nature, impulsive since it is likely

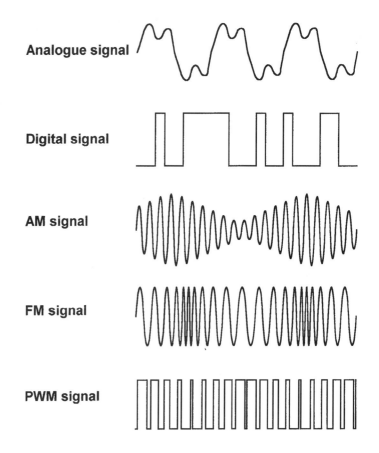

to repeat at the frequency of the supply. Impulse noise can also be generated by natural causes such as electrical discharges in the atmosphere.

Random noise

Random noise results from thermal agitation of electrons within electrical conductors and components such as resistors, transistors and integrated circuits. Random noise in electrical conductors and components is very much dependent on temperature.

Effective noise voltage

The effective noise voltage generated within an electrical conductor is given by:

$$V_n = \sqrt{(4\,k\,T\,R f_w)}$$

where V_n is the effective noise voltage (in V), k is Boltzmann's constant (1.38×10^{-23} J/K), T is the absolute temperature (in K), R is the resistance of the conductor (Ω) and f_w is the bandwidth in which the noise is measured.

The Thevenin equivalent circuit of a noise source is shown in Figure 3.4.4. Note that the voltage source (V_n) appears in series with its associated resistance (R).

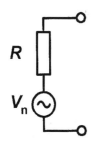

Figure 3.4.4 *Equivalent circuit of a noise source (Thevenin equivalent circuit)*

Example 3.4.2

Determine the effective noise voltage produced in a bandwidth of 10 kHz by a conductor having a resistance of 10 kΩ at a temperature of 17°C.

$V_n = \sqrt{(4kTRf_w)} = \sqrt{(4 \times 1.372 \times 10^{-23} \times (273 + 17) \times 10 \times 10^3 \times 10 \times 10^3)}$

$V_n = \sqrt{(4 \times 1.372 \times 10^{-15} \times 290)} = 2 \times 10^{-7} \times \sqrt{(1.372 \times 29)} = 1.26\ \mu V.$

Noise and its effect on small signals

In a communication system, thermally generated noise ultimately limits the ability of a system to respond to very small signals. What is of paramount importance here is the ratio of signal power to noise power. Let's illustrate this with an example; suppose the level of the signal applied to an information system is progressively reduced whilst the level of noise generated internally remains constant. A point will eventually be reached where the signal becomes lost in the noise. At this point, the signal will have become masked by the noise to the extent that it is no longer discernible. You might now be tempted to suggest that all we need to do is to increase the amount of amplification (gain) within the system to improve the level of the signal. Unfortunately, this won't work simply because the extra gain will increase the level of the noise as well as the signal. In fact, *no amount of amplification will result in an improvement* in the ratio of signal power to noise power.

Signal-to-noise ratio

The signal-to-noise ratio in a system is normally expressed in decibels (dB) and is defined as:

$S/N = 10 \log_{10}(P_{signal}/P_{noise})$ dB.

In practice, it is difficult to separate the signal present in a system from the noise. If, for example, you measure the output power produced by an amplifier you will actually be measuring the signal power *together with* any noise that may be present. Hence a more practical quantity to measure is the ratio of (signal-plus-noise)-to-noise. Furthermore, provided that the noise power is very much smaller than the signal power, there will not be very much difference between the signal-to-noise-ratio and the ratio of (signal-plus-noise)-to-noise.

Thus:

$(S+N)/N = 10 \log_{10}(P_{signal+noise}/P_{noise})$ dB

Example 3.4.3

In the absence of a signal, the noise power present at the output of an amplifier is 10 mW. When a signal is present, the power increases to 2 W. Determine the (signal-plus-noise)-to-noise ratio expressed in decibels.

cont.

Now $(S+N)/N = 10 \log_{10}(P_{\text{signal+noise}}/P_{\text{noise}})$ dB

thus $(S+N)/N = 10 \log_{10}(2 \text{ W}/10 \text{ mW}) = 10 \log_{10}(200) = 10 \times 2.3 = 23$ dB.

Note that (signal-plus-noise)-to-noise ratio may also be determined from the *voltages* present at a point in a circuit, in which case:

$$(S+N)/N = 20 \log_{10}(V_{\text{signal+noise}}/V_{\text{noise}}) \text{ dB.}$$

As a rough guide, Table 3.4.3 will give you some idea of the effect of different values of signal-to-noise ratio on the quality of audio signals when noise is present. Alternatively, Figure 3.4.5 shows the effects of different degrees of noise on a visual image.

Table 3.4.3 *Effect of signal-to-noise ratio on signal quality*

$(S+N)/N$ ratio	Effect
70 dB	Sound produced by a good quality compact disk audio system (signal appears to be completely free from noise)
50 dB	Sound produced by a good quality FM radio system when receiving a strong local signal (no noise detectable)
30 dB	Noise not noticeable – sound quality produced by an AM radio when receiving a strong local signal
20 dB	Noise becomes noticeable – sound quality not sufficient for the reproduction of music but acceptable for speech
10 dB	Signal noticeably degraded by noise – quality only sufficient for voice grade communication
6 dB	Very noisy voice communication channel – signal badly degraded by noise
0 dB	Noise and signal powers equal – voice signal severely degraded by noise and of unacceptable quality
-10 dB	Voice signal completely lost in noise and unusable

Problems 3.4.1

(1) A satellite receiver has an equivalent noise input resistance of 50 Ω. If it operates with an effective bandwidth of 50 kHz at a temperature of 20°C, determine the effective noise voltage at the input of the receiver.

(2) An instrumentation amplifier employs an input transducer which has an effective noise resistance of 100 Ω. If the transducer operates at a temperature of −10°C and it is connected to an amplifier with a gain of 2000 and a bandwidth of 2 kHz, determine the noise voltage produced at the output of the amplifier (you may assume that the amplifier itself is 'noise free').

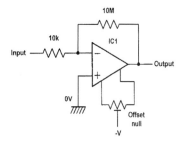 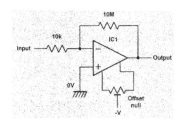 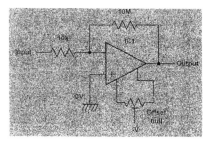

Figure 3.4.5 *Effect of various degrees of noise present with a visual image: (a) no noise present; (b) 20 dB signal-to-noise ratio; (c) 10 dB signal-to-noise ratio*

(3)   In the absence of a signal, the *Y*-amplifier of an oscilloscope produces an output of 50 mV at a temperature of 26°C. If the amplifier has a bandwidth of 10 MHz and a voltage gain of 1000, determine the effective noise resistance at the input of the *Y*-amplifier.

(4)   An amplifier produces an output of 4 W when a signal is applied. When the input signal is disconnected, the output power falls to 10 mW. Determine the (signal-plus-noise)-to-noise ratio expressed in decibels.

(5)   The output signal from an audio preamplifier has a (signal-plus-noise)-to-noise ratio of 40 dB. If the output power is 10 W when a signal is present, determine the output power when the signal is absent.

Amplifiers

Types of amplifier

Many different types of amplifier are found in electronic circuits. Before we explain the operation of transistor amplifiers in detail, it is worth describing some of the types of amplifier used in electronic circuits:

a.c. coupled amplifiers

In a.c. coupled amplifiers, stages are coupled together in such a way that d.c. levels are isolated and only the a.c. components of a signal are transferred from stage to stage.

d.c. coupled amplifiers

In d.c. (or direct) coupled, stages are coupled together in such a way that stages are not isolated to d.c. potentials. Both a.c. and d.c. signal components are transferred from stage to stage.

Large-signal amplifiers

Large-signal amplifiers are designed to cater for appreciable voltage and/or current levels (typically from 1 V to 100 V, or more).

Small-signal amplifiers

Small-signal amplifiers are designed to cater for low level signals (normally less than 1 V and often much smaller).

Audio frequency amplifiers

Audio frequency amplifiers operate in the band of frequencies that is normally associated with audio signals (e.g. 20 Hz to 20 kHz).

Wideband amplifiers

Wideband amplifiers are capable of amplifying a very wide range of frequencies, typically from a few tens of Hz to several MHz (see Figure 3.4.6).

Radio frequency amplifiers

Radio frequency amplifiers operate in the band of frequencies that is normally associated with radio signals (e.g. from 100 kHz to over

Figure 3.4.6 *Frequency response of various types of amplifier*

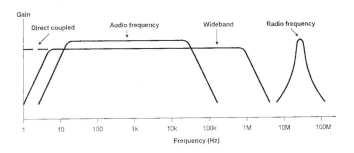

1 GHz). Note that it is desirable for amplifiers of this type to be *frequency selective* and thus their frequency response may be restricted to a relatively narrow band of frequencies (see Figure 3.4.6).

Low-noise amplifiers

Low-noise amplifiers are designed so that they contribute negligible noise (signal disturbance) to the signal being amplified. These amplifiers are usually designed for use with very small signal levels (usually less than 10 mV, or so).

Gain

One of the most important parameters of an amplifier is the amount of amplification or *gain* that it provides. Gain is simply the ratio of output voltage to input voltage, output current to input current, or output power to input power (see Figure 3.4.7). These three ratios give, respectively, the voltage gain, current gain, and power gain. Thus:

voltage gain, $A_v = V_{out}/V_{in}$

current gain, $A_i = I_{out}/I_{in}$

and power gain, $A_p = P_{out}/P_{in}$.

Note that, since power is the product of current and voltage $(P = IV)$, we can infer that:

$$A_p = P_{out}/P_{in} = (I_{out} V_{out})/(I_{in} V_{in}) = (I_{out}/I_{in}) \times (V_{out}/V_{in}).$$

Thus $A_p = A_i \times A_v$.

Example 3.4.4

An amplifier produces an output voltage of 2 V for an input of 50 mV. If the input and output currents in this condition are, respectively, 4 mA and 200 mA, determine:

(a)   the voltage gain;

cont.

Figure 3.4.7 *Definition of amplifier gain*

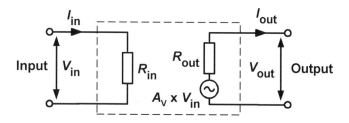

> (b)   the current gain;
> (c)   the power gain.
>
> (a)   The voltage gain is calculated from:
>
> $$A_v = V_{out}/V_{in} = 2\,V/50\,mV = 40.$$
>
> (b)   The current gain is calculated from:
>
> $$A_i = I_{out}/I_{in} = 200\,mA/4\,mA = 50.$$
>
> (c)   The power gain is calculated from:
>
> $$A_p = P_{out}/P_{in} = (2V \times 200\,mA)/(50\,mV \times 4\,mA) =$$
> $$0.4\,W/00\,\mu W = 2000.$$
>
> (Note that $A_p = A_i \times A_v = 50 \times 40 = 2000$.)

Gain can also be expressed in decibels (dB) rather than as a straightforward ratio. The decibel is one tenth of a Bel. This, in turn, is defined as the logarithm (to the base 10) of the ratio of the two powers levels.

Bels $= \log_{10}(P_{out}/P_{in})$ and decibels (dB) $= 10 \times \log_{10}(P_{out}/P_{in})$.

Thus, power gain (in dB), $A_{p(dB)} = 10\log_{10}(P_{out}/P_{in})$.

We use decibels (rather than Bels) simply because they are a more convenient unit to work with. Voltage and current gain can also be expressed in dB.

Voltage gain (in dB), $A_{V(dB)} = 20\log_{10}(V_{out}/V_{in})$

and, current gain (in dB), $A_{I(dB)} = 20\log_{10}(I_{out}/I_{in})$

> **Example 3.4.5**
>
> A telephone line amplifier provides a voltage gain of 18 dB. Determine the input signal voltage required to produce an output of 2 V.
>
> Now, $A_{V(dB)} = 20\log_{10}(V_{out}/V_{in})$.
>
> Thus $A_{V(dB)}/20 = \log_{10}(V_{out}/V_{in})$.
>
> $V_{out}/V_{in} = \text{antilog}_{10}(A_{V(dB)}/20)$
>
> or $V_{in} = V_{out}/\text{antilog}_{10}(A_{V(dB)}/20) = 2/\text{antilog}_{10}(18/20)$
>
> $V_{in} = 2/\text{antilog}_{10}(18/20) = 2/\text{antilog}_{10}(0.9) = 2/7.94 = 0.252\,V$
>
> $= 252\,mV.$

Class of operation

A requirement of most amplifiers is that the output signal should be a faithful copy of the input signal, albeit somewhat larger in amplitude. Other types of amplifier are 'non-linear', in which case their input and output waveforms will not necessarily be similar. In practice, the degree of linearity provided by an amplifier can be affected by a number of factors including the amount of bias applied and the amplitude of the input signal. It is also worth noting that a linear amplifier will become non-linear when the applied input signal exceeds a threshold value.

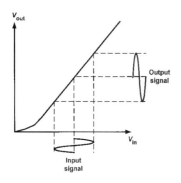

Figure 3.4.8 *Class A (linear operation)*

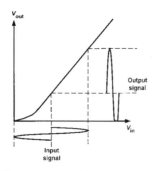

Figure 3.4.9 *Effect of reducing bias and increasing signal amplitude (the output waveform is no longer a faithful replica of the input)*

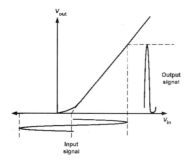

Figure 3.4.10 *Class AB operation (bias set at projected cut-off)*

Beyond this value the amplifier is said to be 'over-driven' and the output will become increasingly distorted if the input signal is further increased.

Amplifiers are usually designed to be operated with a particular value of *bias* supplied to the active devices (i.e. transistors). For linear operation, the active device(s) must be operated in the linear part of their transfer characteristic (V_{out} plotted against V_{in}).

Figure 3.4.8 shows the input and output signals for an amplifier operating in linear mode. This form of operation is known as *Class A* and the bias point is adjusted to the mid-point of the linear part of the transfer characteristic. Furthermore, current will flow in the active devices used in a Class A amplifier during a complete cycle of the signal waveform. At no time does the current fall to zero.

Figure 3.4.9 shows the effect of moving the bias point down the transfer characteristic and, at the same time, increasing the amplitude of the input signal. From this, you should notice that the extreme negative portion of the output signal has become distorted. This effect arises from the non-linearity of the transfer characteristic that occurs near the origin (i.e. the zero point). Despite the obvious non-linearity in the output waveform, the active device(s) will conduct current during a complete cycle of the signal waveform.

Now consider what will happen if the bias is even further reduced and the amplitude of the input signal is further increased. Figure 3.4.10 shows the bias point set at the *projected cut-off point*. The negative-going portion of the output signal becomes cut-off (or *clipped*) and the active device(s) will cease to conduct for this part of the cycle. This mode of operation is known as *Class AB*.

Now let's consider what will happen if no bias at all is applied to the amplifier (see Figure 3.4.11). The output signal will only comprise a series of positive-going half-cycles and the active device(s) will only be conducting during half-cycles of the waveform (i.e. they will only be operating 50% of the time). This mode of operation is known as *Class B* and is commonly used in push–pull power amplifiers where the two active devices in the output stage conduct on alternate half-cycles of the waveform.

Finally, there is one more class of operation to consider. The input and output waveforms for *Class C* operation are shown in Figure 3.4.12. Here the bias point is set at beyond the cut-off (zero) point and a very large input signal is applied. The output waveform will then comprise a series of quite sharp positive-going pulses. These pulses of current or voltage can be applied to a tuned circuit load in order to recreate a sinusoidal signal. In effect, the pulses will excite the tuned circuit and its inherent *flywheel action* will produce a sinusoidal output waveform. This

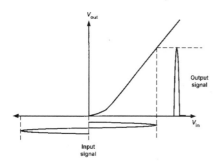

Figure 3.4.11 *Class B operation (no bias applied)*

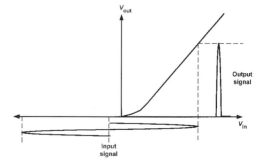

Figure 3.4.12 *Class B operation (bias is set beyond cut-off)*

mode of operation is only used in radio frequency power amplifiers which must operate at high levels of efficiency. Table 3.4.4 summarises the classes of operation used in amplifiers.

Table 3.4.4 *Classes of operation*

Class of operation	Bias point	Conduction angle (°)	Efficiency (typical) (%)	Application (typical)
A	Mid-point	360	5–25	Linear audio amplifiers
AB	Projected	210	25–45	Push–pull audio amplifiers
B	At cut-off	180	45–65	Push–pull audio amplifiers
C	Beyond cut-off	120	65–85	Radio frequency amplifiers

Input and output resistance

Input resistance is the ratio of input voltage to input current and it is expressed in ohms. The input of an amplifier is normally purely resistive (i.e. any reactive component is negligible) in the middle of its working frequency range (i.e. the *mid-band*). In some cases, the reactance of the input may become appreciable (e.g. if a large value of stray capacitance appears in parallel with the input resistance). In such cases we would refer to *input impedance* rather than input resistance.

Output resistance is the ratio of open-circuit output voltage to short-circuit output current and is measured in ohms. Note that this resistance is internal to the amplifier and should not be confused with the resistance of a load connected externally. As with input resistance, the output of an amplifier is normally purely resistive and we can safely ignore any reactive component. If this is not the case, we would once again refer to *output impedance* rather than output resistance.

The *equivalent circuit* in Figure 3.4.13 shows how the input and output resistances are 'seen', respectively looking into the input and output terminals.

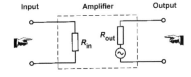

Figure 3.4.13 *Input and output resistance 'seen' looking into the input and output terminals respectively*

Frequency response

The frequency response of an amplifier is usually specified in terms of the upper and lower *cut-off frequencies* of the amplifier. These frequencies are those at which the output power has dropped to 50%

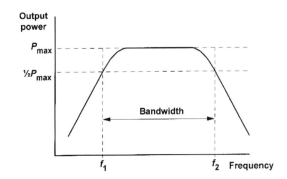

Figure 3.4.14 *Frequency response and bandwidth (output power plotted against frequency)*

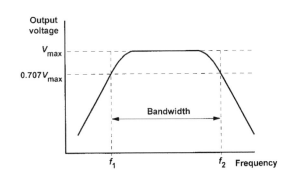

Figure 3.4.15 *Frequency response and bandwidth (output voltage plotted against frequency)*

Figure 3.4.16 *See example 3.4.6*

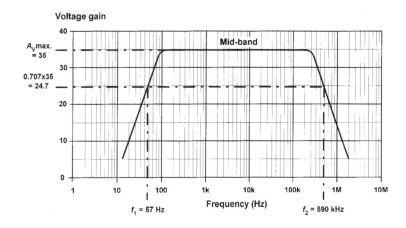

(otherwise known as the $-3\,dB$ points) or where the voltage gain has dropped to 70.7% of its mid-band value. Figures 3.4.14 and 3.4.15 respectively, show how the bandwidth can be expressed in terms of power or voltage. In either case, the cut-off frequencies (f_1 and f_2) and bandwidth are identical.

Note that, for response curves of this type, frequency is almost invariably plotted on a logarithmic scale.

Example 3.4.6

Determine the mid-band voltage gain and upper and lower cut-off frequencies for the amplifier whose frequency response is shown in Figure 3.4.16.

The mid-band voltage gain corresponds to the flat part of the frequency response characteristic. At the point the voltage gain reaches a maximum of 35 (see Figure 3.4.16).

The voltage gain at the cut-off frequencies can be calculated from:

A_V cut-off $= 0.707 \times A_V$ max. $= 0.707 \times 35 = 24.7$

This value of gain intercepts the frequency response at $f_1 = 57$ Hz and $f_2 = 590$ kHz (see Figure 3.4.16).

Bandwidth

The bandwidth of an amplifier is usually taken as the difference between the upper and lower cut-off frequencies (i.e. $f_2 - f_1$ in Figures 3.4.14 and 3.4.15). The bandwidth of an amplifier must be sufficient to accommodate the range of frequencies present within the signals that it is to be presented with. Many signals contain *harmonic* components (i.e. signals at $2f$, $3f$, $4f$, etc. where f is the frequency of the *fundamental* signal). To perfectly reproduce a square wave, for example, requires an amplifier with an infinite bandwidth (note that a square wave comprises an infinite series of harmonics). Clearly it is not possible to perfectly reproduce such a wave but it does explain why it can be desirable for an amplifier's bandwidth to greatly exceed the highest signal frequency that it is required to handle!

Phase shift

Phase shift is the phase angle between the input and output voltages measured in degrees. The measurement is usually carried out in the mid-band where, for most amplifiers, the phase shift remains relatively constant. Note also that conventional single-stage transistor amplifiers usually provide phase-shifts of either 180° or 360° (i.e. 0°).

Negative feedback

Many practical amplifiers use negative feedback in order to precisely control the gain, reduce distortion and improve bandwidth by feeding back a small proportion of the output. The amount of feedback determines the overall (or *closed-loop*) gain. Because this form of feedback has the effect of reducing the overall gain of the circuit, this form of feedback is known as *negative feedback*. An alternative form of feedback, where the output is fed back in such a way as to reinforce the input (rather than to subtract from it) is known as *positive feedback*. This form of feedback is used in oscillator circuits (see page 197).

Figure 3.4.17 shows the block diagram of an amplifier stage with negative feedback applied. In this circuit, the proportion of the output voltage fed-back to the input is given by β and the overall voltage gain will be given by:

Overall gain $= V_{out}/V_{in}$.

Now $V_{in}' = V_{in} - \beta V_{out}$ (by applying Kirchhoff's Voltage Law)

(note that the amplifier's input voltage has been *reduced* by applying negative feedback)

thus $V_{in} = V_{in}' + \beta V_{out}$

and $V_{out} = A_v \times V_{in}'$ (A_v is the *internal gain* of the amplifier).

Hence, overall gain $= V_{out}/V_{in} = (A_v \times V_{in}')/(V_{in}' + \beta V_{out}) = A_v/(1 + \beta A_v)$.

The overall gain with negative feedback applied, will thus be less than the gain without feedback. Furthermore, if A_v is very large (as is the case with an operational amplifier) the overall gain with negative feedback applied will be given by:

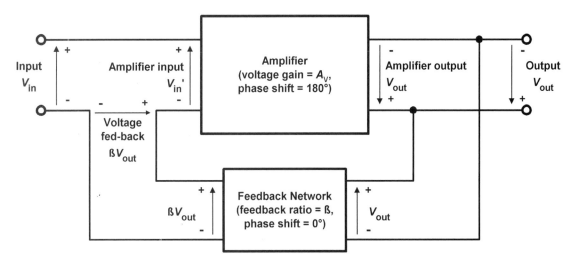

Figure 3.4.17 *Amplifier with negative feedback applied*

Overall gain (when A_v is very large) = $1/\beta$

Note, also, that the *loop gain* of a feedback amplifier is defined as the product of β and A_v.

Example 3.4.7

An amplifier with negative feedback applied has an open-loop voltage gain of 50 and one tenth of its output is fed back to the input (i.e. $\beta = 0.1$). Determine the overall voltage gain with negative feedback applied.

With negative feedback applied the overall voltage gain will be given by:

$A_v/(1 + \beta A_v) = 50/(1 + (0.1 \times 50)) = 50/(1 + 5) = 50/6 = 8.33$.

Example 3.4.8

If, in Example 3.4.7, the amplifier's open-loop voltage gain increases by 20%, determine the percentage increase in overall voltage gain.

The new value of open-loop gain will be given by:

$A_v = A_v + 0.2A_v = 1.2 \times 50 = 60$.

The overall voltage gain with negative feedback will then be:

$A_v/(1 + \beta A_v) = 60/(1 + (0.1 \times 60)) = 60/(1 + 6) = 60/7 = 8.57$.

The increase in overall voltage gain, expressed as a percentage will thus be:

$(8.57 - 8.33)/8.33 \times 100\% = 2.88\%$.

(Note that this example illustrates one of the important benefits of negative feedback in stabilising the overall gain of an amplifier stage.)

Example 3.4.9

An integrated circuit that produces an open-loop gain of 100 is to be used as the basis of an amplifier stage having a precise voltage gain of 20. Determine the amount of feedback required.

Rearranging the formula, $A_v/(1 + \beta A_v)$, to make β the subject gives:

$\beta = 1/A_v' - 1/A_v$

where A_v' is the overall voltage gain with feedback applied, and A_v is the open-loop voltage gain.

Thus $\beta = 1/20 - 1/100 = 0.05 - 0.01 = 0.04$.

Problems 3.4.2

(1)   An amplifier produces an output voltage of 6 V for an input of 150 mV. If the input and output currents in this condition are, respectively, 600 µA and 120 mA, determine:

(a) the voltage gain;
(b) the current gain;
(c) the power gain.

(2) An amplifier with negative feedback applied has an open-loop voltage gain of 40 and one fifth of its output is fed back to the input. Determine the overall voltage gain with negative feedback applied.

(3) A manufacturer produces integrated circuit amplifier modules that have voltage gains which vary from 12 000 to 24 000 due to production tolerances. If the modules are to be used with 5% negative feedback applied, determine the maximum and minimum values of voltage gain.

(4) An amplifier has an open-loop gain of 200. If this device is to be used in an amplifier stage having a precise voltage gain of 50, determine the amount of negative feedback that will be required.

Oscillators

This section describes some basic oscillator circuits that generate sine wave, square wave, and triangular waveforms. These circuits form the basis of clocks and timing arrangements as well as signal and function generators.

On page 195, we showed how negative feedback can be applied to an amplifier to form the basis of a stage which has a precisely controlled gain. An alternative form of feedback, where the output is fed back in such a way as to reinforce the input (rather than to subtract from it) is known as *positive feedback*. Figure 3.4.18 shows the block diagram of an amplifier stage with positive feedback applied. Note that the amplifier provides a phase shift of 180° and the feedback network provides a further 180°. Thus the overall phase shift is 0°. The overall voltage gain is given by:

Overall gain $= V_{out}/V_{in}$.

Now $V_{in}' = V_{in} + \beta V_{out}$ (by applying Kirchhoff's Voltage Law)

thus $V_{in} = V_{in}' - \beta V_{out}$

and $V_{out} = A_V \times V_{in}'$ (A_V is the *internal gain* of the amplifier).

Figure 3.4.18 *Amplifier with positive feedback applied*

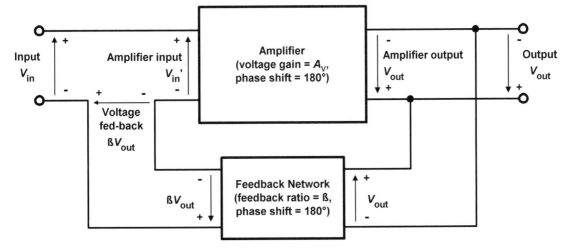

Hence, overall gain $= A_V \times V_{in}'/ V_{in}' - \beta V_{out} = A_V \times V_{in}'/ V_{in}' - \beta(A_V \times V_{in}')$

Thus overall gain $= A_V/(1 - \beta A_V)$.

Now consider what will happen when the *loop gain* (βA_V) approaches unity. The denominator $(1 - \beta A_V)$ will become close to zero. This will have the effect of increasing the overall gain, i.e. the overall gain with positive feedback applied will be *greater* than the gain without feedback.

It is worth illustrating this difficult concept using some practical figures. Assume that an amplifier has a gain of 9 and one tenth of its output is fed back to the input (i.e. $\beta = 0.1$). In this case the loop gain ($\beta \times A_V$) is 0.9.

With negative feedback applied the overall voltage gain will fall to:

$A_V/ (1 + \beta A_V) = 9/(1 + 0.1\,(9)) = 9/1.9 = 4.7$.

With positive feedback applied the overall voltage gain will be:

$A_V/ (1 - \beta A_V) = 9/(1 - 0.1 \times 9) = 9/(1 - 0.9) = 9/0.1 = 90$.

Now assume that the amplifier has a gain of 10 and, once again, one tenth of its output is fed back to the input (i.e. $\beta = 0.1$). In this example the loop gain ($\beta \times A_V$) is exactly 1.

With negative feedback applied, the overall voltage gain will fall to:

$A_V/(1 + \beta A_V) = 10/(1 + 0.1 \times 10) = 10/2 = 5$.

With positive feedback applied the overall voltage gain will be:

$A_V/(1 - \beta A_V) = 10/(1 - 0.1 \times 10) = 10/(1 - 1) = 10/0 = \infty$.

This simple example shows that a loop gain of unity (or larger) will result in infinite gain and an amplifier which is unstable. In fact, the amplifier will *oscillate* since any small disturbances will be amplified and result in an output. Clearly, as far as an amplifier is concerned, positive feedback may have an undesirable effect – instead of reducing the overall gain the effect is that of reinforcing any signal present and the output can build up into continuous oscillation if the loop gain is 1, or greater. To put this another way, an oscillator can simply be thought of as an amplifier that generates an output signal without the need for an input signal!

Conditions for oscillation

From the foregoing we can deduce that the conditions for oscillation are:

(a)   the feedback must be positive (i.e. the signal fed back must arrive back in phase with the signal at the input);

(b)   the overall loop voltage gain must be greater than 1 (i.e. the amplifier's gain must be sufficient to overcome the losses associated with any frequency selective feedback network).

Hence, to create an oscillator we simply need an amplifier with sufficient gain to overcome the losses of the network that provides positive feedback. Assuming that the amplifier provides 180° phase shift, the frequency of oscillation will be that at which there is 180° phase shift in the feedback network. A number of circuits can be used to provide 180° phase shift, one of the simplest being a three-stage C–R ladder network. Alternatively, if the amplifier produces 0° phase shift, the circuit will oscillate at the frequency at which the feedback network produces 0° phase shift. In both cases, the essential point is that the feedback should be positive so that the output signal arrives back at the input in such a sense as to reinforce the original signal.

Types of oscillator

Many different types of oscillator are used in electronic circuits. Oscillators tend to fall into two main categories:

- oscillators that produce sine wave outputs (i.e. *sinusoidal oscillators*)
- oscillators that produce square wave or rectangular pulse wave outputs (i.e. square wave oscillators). These types of oscillator are sometimes also known as *astable oscillators* (or *multivibrators*).

We shall briefly explore the operation of some of the most common types of sinusoidal and astable oscillator.

C–R ladder network oscillator

A simple phase-shift oscillator based on a three-stage C–R ladder network is shown in Figure 3.4.19. The total phase shift provided by the C–R ladder network (connected between collector and base) is 180° at the frequency of oscillation. The amplifier provides the other 180° phase shift in order to realise an overall phase shift of 360° or 0°.

The approximate frequency of oscillation of the circuit shown in Figure 3.4.19 will be given by:

$$f = 1/(2\pi\sqrt{6}CR).$$

However, since the transistor's input and output resistance appears in parallel with the ladder network (i.e. the transistor *loads* the network), the actual frequency of operation will be slightly different. At the operating frequency, the loss associated with the ladder network is 29, thus the transistor amplifier must provide a gain of at least 29 in order for the circuit to oscillate.

Example 3.4.10

Determine the approximate frequency of oscillation of a three stage ladder network oscillator where C = 10 nF and R = 10 kΩ.

Using $f = 1/(2\pi\sqrt{6}CR)$

cont.

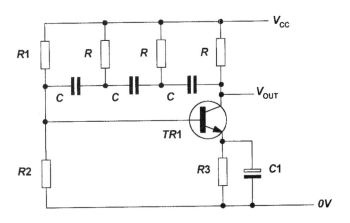

Figure 3.4.19 *Sine wave oscillator based on a three-stage C–R ladder network*

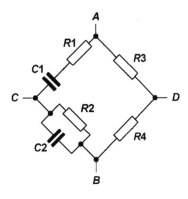

Figure 3.4.20 *A Wien bridge network*

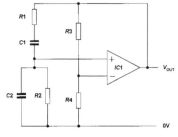

Figure 3.4.21 *Sine wave oscillator based on a Wien bridge (see Example 3.4.11)*

gives $f = 1/(6.28 \times 2.45 \times 10 \times 10^{-9} \times 10 \times 10^3)$ Hz

thus $f = 1/(6.28 \times 2.45 \times 10^{-4}) = 1 \times 10^4/15.386 = 647$ Hz.

Wien bridge oscillator

An alternative approach to providing the phase shift required is the use of a Wien bridge network (Figure 3.4.20). Like the C–R ladder, this network provides a phase shift which varies with frequency. The input signal is applied to A and B whilst the output is taken from C and D. At one particular frequency, the phase shift produced by the network will be exactly zero (i.e. the input and output signals will be in-phase). If we connect the network to an amplifier producing 0° phase shift which has sufficient gain to overcome the losses of the Wien bridge, oscillation will result.

The minimum amplifier gain required to sustain oscillation is given by:

$$A_V = 1 + C1/C2 + R2/R1$$

and the frequency at which the phase-shift will be zero is given by:

$$f = 1/(2\pi \sqrt{(C1 \times C2 \times R1 \times R2)}).$$

In practice, we normally make $R1 = R2$ and $C1 = C2$ hence:

$$A_V = 1 + C/C + R/R = 1 + 1 + 1 = 3$$

and the frequency at which the phase-shift will be zero is given by:

$$f = 1/(2\pi CR)$$

where $R = R1 = R2$ and $C = C1 = C2$.

Example 3.4.11

Figure 3.4.21 shows the circuit of a Wien bridge oscillator based on an operational amplifier. If $C1 = C2 = 100$ nF, determine the output frequencies produced by this arrangement (a) when $R1 = R2 = 1$ kΩ and (b) when $R1 = R2 = 6$ kΩ.

(a)   When $R1 = R2 = 1$ kΩ

$$f = 1/(2\pi CR)$$

where $R = R1 = R2$ and $C = C1 = C2$.

Thus $f = 1/(2\pi CR) = 1/(6.28 \times 100 \times 10^{-9} \times 1 \times 10^3) = 1/(6.28 \times 10^{-4})$

or $f = 10\,000/6.28 = 1592$ Hz $= 1.592$ kHz.

(b) When $R1 = R2 = 6$ kΩ

$$f = 1/(2\pi CR)$$

where $R = R1 = R2$ and $C = C1 = C2$.

Thus $f = 1/(2\pi CR) = 1/6.28 \times 100 \times 10^{-9} \times 6 \times 10^3 = 1/37.68 \times 10^{-4}$

or $f = 10\,000/37.68 = 265.4$ Hz.

Multivibrators

Multivibrators are a family of oscillator circuits that produce output waveforms consisting of one or more rectangular pulses. The term *multivibrator* simply originates from the fact that this type of waveform is rich in harmonics (i.e. 'multiple vibrations').

Multivibrators use regenerative (positive) feedback; the active devices present within the oscillator circuit being operated as switches, being alternately cut-off and driven into saturation.

The principal types of multivibrator are:

- *astable multivibrators* that provide a continuous train of pulses these are sometimes also referred to as *free-running multivibrators*
- *monostable multivibrators* that produce a single output pulse (they have one stable state and are thus sometimes also referred to as *one-shot* circuits)
- *bistable multivibrators* that have two stable states and require a trigger pulse or control signal to change from one state to another.

The astable multivibrator

Figure 3.4.22 shows a classic form of astable multivibrator based on two transistors. Figure 3.4.23 shows how this circuit can be re-drawn in an arrangement that more closely resembles a two-stage common emitter amplifier with its output connected back to its input.

In Figure 3.4.22 the values of the base resistors, $R3$ and $R4$ are such that the sufficient base current will be available to completely saturate the respective transistor. The values of the collector load resistors, $R1$ and $R2$, are very much smaller than $R3$ and $R4$. When power is first applied to the circuit, assume that TR2 saturates before TR1 (in practice one transistor would always saturate before the other due to variations in component tolerances and transistor parameters).

As TR2 saturates, its collector voltage will fall rapidly from $+V_{CC}$ to 0 V. This drop in voltage will be transferred to the base of TR1 via $C2$. This negative going voltage will ensure that TR1 is initially placed in the non-conducting state. As long as TR1 remains cut-off, TR2 will continue to be saturated. During this time, $C2$ will charge via R4 and TR1's base voltage will rise exponentially from $-V_{CC}$ towards $+V_{CC}$. However, TR1's base voltage will not rise much above 0V because, as soon as it reaches $+0.7$ V (sufficient to cause base current to flow) TR1 will begin to conduct. As TR1 begins to turn on, its collector voltage will fall rapidly from $+V_{CC}$ to 0 V. This fall in voltage is transferred to the base of TR2 via $C1$ and, as a consequence, TR2 will turn off. $C1$ will then charge via $R3$ and TR2's base voltage will rise exponentially from $-V_{CC}$ towards $+V_{CC}$. As before, TR2's base voltage will not rise much above 0 V because, as soon as it reaches $+0.7$ V (sufficient to cause base current to flow) TR2 will start to conduct. The cycle is then repeated indefinitely.

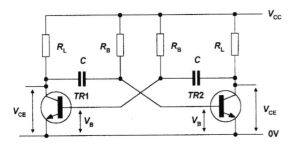

Figure 3.4.22 *Classic form of astable multivibrator using bipolar transistors*

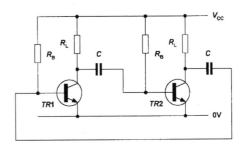

Figure 3.4.23 *Circuit of Figure 3.4.22 redrawn to show two common-emitter stages*

Figure 3.4.24 *Waveforms for the bipolar trnsistor astable multivibrator*

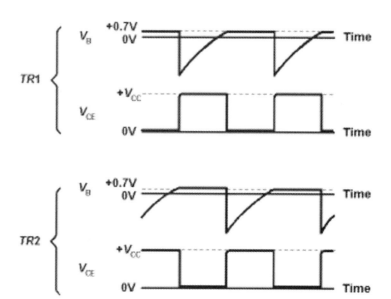

The time for which the collector voltage of TR2 is low and TR1 is high (T1) will be determined by the time constant, $R4 \times C2$. Similarly, the time for which the collector voltage of TR1 is low and TR2 is high (T2) will be determined by the time constant, $R3 \times C1$.

The following approximate relationships apply:

T1 = 0.7 C2 R4 and T2 = 0.7 C1 R3.

Since one complete cycle of the output occurs in a time, T = T1 + T2, the periodic time of the output is given by:

T = 0.7 (C2 R4 + C1 R3).

Finally, we often require a symmetrical 'square wave' output where T1 = T2. To obtain such an output, we should make $R3 = R4$ and $C1 = C2$, in which case the periodic time of the output will be given by:

T = 1.4 CR

where $C = C1 = C2$ and $R = R3 = R4$.

Waveforms for the astable oscillator are shown in Figure 3.4.24.

Example 3.4.12

The astable multivibrator in Figure 3.4.22 is required to produce a square wave output at 1 kHz. Determine suitable values for $R3$ and $R4$ if $C1$ and $C2$ are both 10 nF.

Since a square wave is required and $C1$ and $C2$ have identical values, $R3$ must be made equal to $R4$. Now:

$T = 1/f = 1/1 \times 10^3 = 1 \times 10^{-3}$ s.

Rearranging T = 1.4 CR to make R the subject gives:

$R = T/1.4\,C = 1 \times 10^{-3}/1.4 \times 10 \times 10^{-9} = 0.071 \times 10^6\ \Omega = 71$ kΩ.

Other forms of astable oscillator

Figure 3.4.25 shows the circuit diagram of an alternative form of astable oscillator which produces a triangular output waveform. Operational

Figure 3.4.25 *Astable oscillator using operational amplifiers*

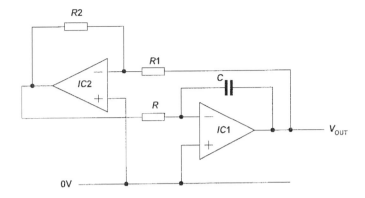

amplifier, IC1 forms an integrating stage whilst IC2 is connected with positive feedback to ensure that oscillation takes place.

Assume that the output from IC2 is initially at or near the supply voltage, $+V_{CC}$ and capacitor, C, is uncharged. The voltage at the output of IC2 will be passed, via R, to IC1. Capacitor, C, will start to charge and the output voltage of IC1 will begin to fall. Eventually, the output voltage will have fallen to a value that causes the polarity of the voltage at the non-inverting input of IC2 to change from positive to negative. At this point, the output of IC2 will rapidly fall to $-V_{CC}$. Again, this voltage will be passed, via R, to IC1. Capacitor, C, will then start to charge in the other direction and the output voltage of IC1 will begin to rise. Eventually, the output voltage will have risen to a value that causes the polarity of the non-inverting input of IC2 to revert to its original (positive) state and the cycle will continue indefinitely.

The *upper threshold voltage* (i.e. the maximum positive value for V_{out}) will be given by:

$$V_{UT} = V_{CC} \times (R1/R2).$$

The *lower threshold voltage* (i.e. the maximum negative value for V_{out}) will be given by:

$$V_{LT} = -V_{CC} \times (R1/R2).$$

Single-stage astable oscillator

A simple form of astable oscillator producing a square wave output can be built using just one operational amplifier, as shown in Figure 3.4.26. This circuit employs positive feedback with the output fed back to the non-inverting input via the potential divider formed by $R1$ and $R2$.

Assume that C is initially uncharged and the voltage at the inverting input is slightly less than the voltage at the non-inverting input. The output voltage will rise rapidly to $+V_{CC}$ and the voltage at the inverting input will begin to rise exponentially as capacitor C charges through R. Eventually, the voltage at the inverting input will have reached a value that causes the voltage at the inverting input to exceed that present at the non-inverting input. At this point, the output voltage will rapidly fall to $-V_{CC}$. Capacitor, C, will then start to charge in the other direction and the voltage at the inverting input will begin to fall exponentially. Eventually, the voltage at the inverting input will have reached a value that causes the voltage at the inverting input to be less than that present at the non-inverting input. At this point, the output voltage will rise rapidly to $+V_{CC}$ once again and the cycle will continue indefinitely.

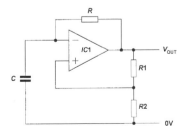

Figure 3.4.26 *Single-stage astable oscillator using an operational amplifier*

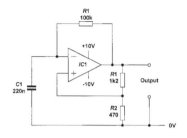

Figure 3.4.27 *See problem 3.4.3*

The *upper threshold voltage* (i.e. the maximum positive value for the voltage at the inverting input) will be given by:

$$V_{UT} = V_{CC} \times R2/(R1 + R2).$$

The *lower threshold voltage* (i.e. the maximum negative value for the voltage at the inverting input) will be given by:

$$V_{UT} = -V_{CC} \times R1/(R1 + R2).$$

Problems 3.4.3

(1) An amplifier with a gain of 8 has 10% of its output fed back to the input. Determine the gain of the stage (a) with negative feedback (b) with positive feedback.

(2) A phase shift oscillator is to operate with an output at 1 kHz. If the oscillator is based on a three-stage ladder network, determine the required values of resistance if three capacitors of 10 nF are to be used.

(3) A Wien bridge oscillator is based on the circuit shown in Figure 3.4.21 but $R1$ and $R2$ are replaced by a dual-gang potentiometer. If $C1 = C2 = 22n$ determine the values of $R1$ and $R2$ required to produce an output at exactly 400 Hz.

(4) Determine the peak–peak output voltage produced by the circuit shown in Figure 3.4.27.

(5) An astable multivibrator circuit is required to produce an asymmetrical rectangular output which has a period of 4 ms and is to be 'high' for 1 ms and 'low' for 3 ms. If the timing capacitors are both to be 10 nF, determine the values of the two timing resistors required.

Transducers

Transducers are devices that convert energy in the form of sound, light, heat, etc. into an equivalent electrical signal, or vice versa. Before we go further, let's consider a couple of examples that you will already be familiar with. A loudspeaker is a device that converts low-frequency electrical current into sound. A thermocouple, on the other hand, is a transducer that converts temperature into voltage.

Transducers may be used both as system inputs and system outputs. From the two previous examples, it should be obvious that a loudspeaker is an *output transducer* designed for use in conjunction with an audio system. Whereas, a thermocouple is an *input transducer* designed for use in conjunction with a temperature control system.

Tables 3.4.5 and 3.4.6 provide examples of transducers that can be used to input and output three important physical quantities; sound, temperature, and angular position:

Table 3.4.5 *Some examples of input transducers*

Physical quantity	Input transducer	Notes
Sound (pressure change)	Dynamic microphone	Diaphragm attached to a coil is suspended in a magnetic field. Movement of the diaphragm causes current to be induced in the coil.

Temperature	Thermocouple	Small e.m.f. generated at the junction between two dissimilar metals (e.g. copper and constantan). Requires reference junction and compensated cables for accurate measurement.
Angular position	Rotary potentiometer	Fine wire resistive element is wound around a circular former. Slider attached to the control shaft makes contact with the resistive element. A stable d.c. voltage source is connected across the ends of the potentiometer. Voltage appearing at the slider will then be proportional to angular position.

Table 3.4.6 *Some examples of output transducers*

Physical quantity	Output transducer	Notes
Sound (pressure change)	Loudspeaker	Diaphragm attached to a coil is suspended in a magnetic field. Current in the coil causes movement of the diaphragm which alternately compresses and rarefies the air mass in front of it.
Temperature	Resistive heating element	Metallic conductor is wound onto a ceramic or mica former. Current flowing in the conductor produces heat.
Angular position	Stepper motor	Multi-phase motor provides precise rotation in discrete steps of $15°$ (24 steps per revolution), $7.5°$ (48 steps per revolution) and $1.8°$ (200 steps per revolution).

Sensors

A *sensor* is a special kind of transducer that is used to generate an input signal to a measurement, instrumentation or control system. The signal produced by a sensor is an *electrical analogy* of a physical quantity, such as distance, velocity, acceleration, temperature, pressure, light level, etc. The signals returned from a sensor, together with control inputs from the operator (where appropriate) will subsequently be used to determine the output from the system. The choice of sensor is governed by a number of factors including accuracy, resolution, cost, and physical size.

Sensors can be categorised as either *active* or *passive*. An active sensor *generates* a current or voltage output. A passive transducer requires a source of current or voltage and it modifies this in some way (e.g. by virtue of a change in the sensor's resistance). The result may still be a voltage or current *but it is not generated by the sensor on its own*.

Sensors can also be classed as either *digital* or *analogue*. The output of a digital sensor can exist in only two discrete states, either 'on' or 'off', 'low' or 'high', 'logic 1' or 'logic 0', etc. The output of an analogue sensor can take any one of an infinite number of voltage or current levels. It is thus said to be *continuously variable*.

Table 3.4.7 shows some common types of sensor:

Table 3.4.7 *Some common types of sensor*

Physical parameter	Type of sensor	Notes
Angular position	Resistive rotary position sensor	Rotary track potentiometer with linear law produces analogue voltage proportional to angular position.

	Optical shaft encoder	Encoded disk interposed between optical transmitter and receiver (infra-red LED and photodiode or phototransistor).
Angular velocity	Tachogenerator	Small d.c. generator with linear output characteristic. Analogue output voltage proportional to shaft speed.
	Toothed rotor tachometer	Magnetic pick-up responds to the movement of a toothed ferrous disk. The pulse repetition frequency of the output is proportional to the angular velocity.
Flow	Rotating vane flow sensor	Turbine rotor driven by fluid. Turbine interrupts infra-red beam. Pulse repetition frequency of output is proportional to flow rate.
Linear position	Resistive linear position sensor	Linear track potentiometer with linear law produces analogue voltage proportional to linear position. Limited linear range.
	Linear variable differential transformer (LVDT)	Miniature transformer with split secondary windings and moving core attached to a plunger. Requires a.c. excitation and phase-sensitive detector.
	Magnetic linear position sensor	Magnetic pick-up responds to movement of a toothed ferrous track. Pulses are counted as the sensor moves along the track.
Light level	Photocell	Voltage-generating device. The analogue output voltage produced is proportional to light level.
	Light dependent resistor (LDR)	An analogue output voltage results from a change of resistance within a cadmium sulphide (CdS) sensing element. Usually connected as part of a potential divider or bridge.
	Photodiode	Two-terminal device connected as a current source. An analogue output voltage is developed across a series resistor of appropriate value.
	Phototransistor	Three-terminal device connected as a current source. An analogue output voltage is developed across a series resistor of appropriate value.
Liquid level	Float switch	Simple switch element which operates when a particular level is detected.
	Capacitive proximity switch	Switching device which operates when a particular level is detected. Ineffective with some liquids.
	Diffuse scan proximity switch	Switching device which operates when a particular level is detected. Ineffective with some liquids.
Pressure	Microswitch pressure sensor	Microswitch fitted with actuator mechanism and range setting springs. Suitable for high-pressure applications.
	Differential pressure vacuum switch	Microswitch with actuator driven by a diaphragm. May be used to sense differential pressure. Alternatively, one chamber may be evacuated and the sensed pressure applied to a second input.
	Piezo-resistive pressure sensor	Pressure exerted on diaphragm causes changes of resistance in attached piezo-resistive transducers. Transducers are usually arranged in the form of a four active element bridge which produces an analogue output voltage.
Proximity	Reed switch	Reed switch and permanent magnet actuator. Only effective over short distances.
	Inductive proximity switch	Target object modifies magnetic field generated by the sensor. Only suitable for metals (non-ferrous metals with reduced sensitivity).
	Capacitive proximity switch	Target object modifies electric field generated by the sensor. Suitable for metals, plastics, wood, and some liquids and powders.
	Optical proximity switch	Available in diffuse and through scan types. Diffuse scan types require reflective targets. Both types employ optical transmitters and receivers (usually infra-red emitting LEDs and photodiodes or photo-transistors). Digital input port required.

Strain	Resistive strain gauge	Foil type resistive element with polyester backing for attachment to body under stress. Normally connected in full bridge configuration with temperature-compensating gauges to provide an analogue output voltage.
	Semiconductor strain gauge	Piezo-resistive elements provide greater outputs than comparable resistive foil types. More prone to temperature changes and also inherently non-linear.
Temperature	Thermocouple	Small e.m.f. generated by a junction between two dissimilar metals. For accurate measurement, requires compensated connecting cables and specialised interface.
	Thermistor	Usually connected as part of a potential divider or bridge. An analogue output voltage results from resistance changes within the sensing element.
	Semiconductor temperature sensor	Two-terminal device connected as a current source. An analogue output voltage is developed across a series resistor of appropriate value.
Weight	Load cell	Usually comprises four strain gauges attached to a metal frame. This assembly is then loaded and the analogue output voltage produced is proportional to the weight of the load.
Vibration	Electromagnetic vibration sensor	Permanent magnet seismic mass suspended by springs within a cylindrical coil. The frequency and amplitude of the analogue output voltage are respectively proportional to the frequency and amplitude of vibration.

Energy control systems

Energy control systems are systems that control or regulate the supply of energy to a plant or process. Unlike information systems, energy control systems usually operate continuously and are invariably automatic, requiring little or no human intervention under normal operating conditions.

The basic elements of an energy control system are shown in Figure 3.4.28. In this simple model, the *input setting device* is responsible for setting the required level of energy output. Input setting devices can be switches, rotary or linear potentiometers, keypads or keyboards.

The *controlling element* or *controller* accepts the signal from the input device and, like the processing element of an information system, it can typically perform a variety of tasks, including making decisions and operating according to a stored algorithm (or *program*). Controllers can take the form of power amplifiers or power switching devices

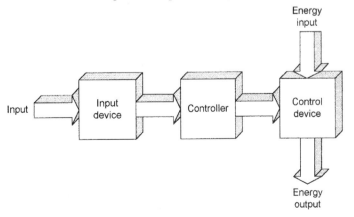

Figure 3.4.28 *Basic elements of an energy control system*

Figure 3.4.29 *An automatic energy control system*

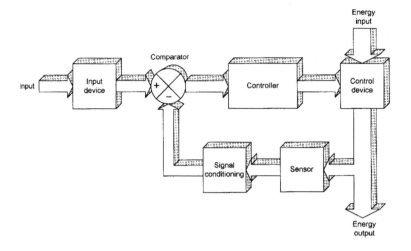

together with logic circuits, microprocessors, microcontrollers, or programmable logic controllers (PLC).

The *controlled device* is a specialised output device that is capable of regulating the energy input in order to provide the required energy output. Typical examples of control devices are motorised valves, pumps, heaters, etc. However, when an electrical supply is being controlled, the control device can be a relay, or a solid-state switching device such as a thyristor, triac, or a power transistor (see pages 212–214).

Automatic energy control systems

Figure 3.4.29 shows an energy control system which employs negative feedback in order to automatically regulate its output. This shows how the energy level at the output is determined by means of a *sensor* and, after signal conditioning, is compared with the desired signal level provided by the input device. The comparator produces an output which is the difference between the input signal and the signal that has been fed-back from the output. This output signal is often referred to as the *error signal* and, when it is present it indicates that there is a difference between the desired output and the actual output from the system. The error signal is fed to the controller which, in turn acts upon the control device. As with earlier systems, there may be a need for signal conditioning in order to regularise signal levels and to provide digital-to-analogue conversion or analogue-to-digital conversion, as appropriate.

Examples of energy control systems

Typical examples of energy control systems are shown in Table 3.4.8:

Table 3.4.8 *Some examples of energy control systems*

	Industrial heating system	*Hydraulic ram*	*Belt conveyor*
Input setting device	Variable resistor	Keypad	Switches
Comparator	Operational amplifier		Microcontroller Programmable logic controller
Controller	Power amplifier	Servo driver	Programmable logic controller
Controlled device	Motorised gas valve	Servo valve	Optically coupled triacs (three-phase)

Output sensing device	Thermistor	Resistive displacement transducer	Infra-red proximity sensors
Signal conditioning	None	ADC and DAC	Amplifier/level-shifter

Question 3.4.3

A gravel washing plant uses a constant head water supply which is pumped from a lake into a concrete reservoir. Treat this as an automatic energy control system and identify the following elements: input setting device, comparator, controller, controlled device, output sensing device, signal conditioning.

Power control

Many engineering processes involve the control of appreciable levels of electrical energy. The electronic circuits that provide this control are often referred to as *power controllers*. Power controllers are designed to switch high values of current and/or voltage. Since we are usually concerned with average power (rather than instantaneous values), the supply of energy is often controlled by rapidly interrupting an electrical supply (i.e. switching) rather than using a resistive control device (which would dissipate unwanted energy as heat). This process is best illustrated by taking an example.

Example 3.4.13

Two alternative forms of energy controller are shown in Figure 3.4.30(a) and (b). The circuit in Figure 3.4.30(a) makes use of a resistive control device whilst that in Figure 3.4.30(b) uses a switching device. For the purpose of this example we will assume that the switching device is perfect (i.e. that its 'on' resistance is zero, its 'off' resistance is infinite and that it changes from 'on' to 'off' in zero time).

For the resistive control device (Figure 3.4.30(a)):

Now $R_L = 1 \, \Omega$ thus the load voltage (V_L) and load current (I_L) for a load power (P_L) of 100 W will be given by:

$V_L = \sqrt{(P \times R_L)} = \sqrt{(100 \times 1)} = \sqrt{100} = 10$ V.

$I_L = P_L/V_L = 100/10 = 10$ A.

The power supplied to the circuit (P_S) will be given by:

$P_S = V_S \times I_L = 50 \times 10 = 500$ W.

The power dissipated in the resistive control device (P_R) will be:

$P_R = P_S - P_L = 500 - 100 = 400$ W.

The efficiency of the energy control system under these conditions will be:

$\eta = (P_L/P_S) \times 100\% = (100/500) \times 100\% = 20\%$.

For the switching control device (Figure 3.4.30(b):
When the switch is closed (see Figure 3.4.30(c)) the full supply will appear across the load. In this condition the peak power

cont.

supplied to the load will be:

$$P_{PK} = V_S{}^2/R_L = 50^2/1 = 2500 \text{ W}.$$

The waveform showing the instantaneous power delivered to the load is shown in Figure 3.4.30(c). From this you should note that, from the area under the graph, the peak power (P_{PK}) is related to the average power in the load ($P_{L(AV)}$) as follows:

$$P_{PK} \times t_{on} = P_{L(AV)} \times (t_{on} + t_{off})$$

thus $P_{L(AV)} = P_{PK} \times t_{on}/(t_{on} + t_{off}) = \tau \times P_{PK}$

where τ is the *duty cycle* of the control signal and $\tau = t_{on}/(t_{on} + t_{off})$.

Now $P_{L(AV)} = 100$ W and $P_{PK} = 2500$ W

thus $\tau = P_{L(AV)}/P_{PK} = 100/2500 = 0.04$.

Assuming that the switching frequency is 1 kHz (i.e. a complete on/off cycle takes 1 ms), the duty cycle for this control signal is thus 1 ms (i.e. $t_{on} + t_{off} = 1$ ms) hence:

$\tau = t_{on}/(t_{on} + t_{off})$ or $t_{on} = \tau \times (t_{on} + t_{off}) = 0.04 \times 1 \text{ ms} = 40 \text{ } \mu\text{s}$

and $t_{off} = 1 \text{ ms} - 40 \text{ } \mu\text{s} = 960 \text{ } \mu\text{s}$.

Thus, to produce an average load power of 100 W using a 1 kHz switching frequency, the 'on' time (t_{on}) must be 40 μs and the 'off' time (t_{off}) must be 960 μs.

Note that, since the switching control device has been assumed to be perfect, no power will be dissipated in it and thus the average power supplied to the circuit, P_S will be the same as that supplied to the load (i.e. 100 W). The circuit thus has an efficiency which approaches 100%.

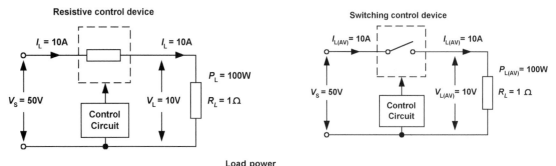

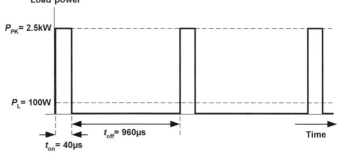

Figure 3.4.30 *Two alternative forms of electrical energy controller (a) (above left) using a resistive control device; (b) above right) using a switching control device; (c) (right) variation of load power with time*

Power switching devices

The switching device employed in a power control circuit is crucial to the successful operation of the circuit. The basic requirements for a power switching device are:

(a) The switching device should only require a very small input current in order to control a very much larger current (flowing through the load).

(b) The switching device should operate very rapidly (i.e. the time between 'on' and 'off' states should be negligible).

(c) During conduction, the switching device should be capable of continuously carrying the rated load current. It should also be capable of handling momentary surge currents (which may be an order of magnitude greater than the continuously rated load current).

(d) There should be minimal power dissipation within the switching device (it should exhibit a very low resistance in the on/conducting state and a very high resistance in the off/non-conducting state).

(e) In the non-conducting state, the switching device should be capable of continuously sustaining the peak value of rated supply voltage. It should also be capable of coping with momentary surge voltages (which may exceed the normal peak supply voltage by a factor of two, or more).

Relays

The traditional method of switching current through a load which requires isolation from the controlling circuit involves the use of an electromechanical relay. Relays offer many of the desirable characteristics of an 'ideal' switching device (notably a very low 'on' resistance and virtually infinite 'off' resistance coupled with a coil to contact breakdown voltage which is usually in excess of several kV). Unfortunately, relays also have several shortcomings which prevent their use in a number of applications. These shortcomings include:

- the *contact bounce* which occurs during the transitory state which exists when the contacts make contact;
- *arcing* (ionisation of the air between the contacts) which may occur when the contacts break and which can result in the generation of heat which can literally burn out the contact surfaces and which can produce a large amount of radio frequency interference;
- the need for regular inspection and routine maintenance with periodic replacement of relays when contacts wear out.

A typical electromechanical relay may be rated for around 1 000 000 operations, or more. To put this into context, if operated once every minute, the contact set on such a relay can be expected to give satisfactory operation for a period of about 2 years. It is important to note, however, that electromechanical relays are prone to *both* mechanical and electrical failure (the latter being more prevalent if the device is operated at, or near, its maximum rating).

Despite thus, in simple low-speed 'on/off' switching applications, the conventional electromechanical relay can still provide a cost-effective solution to controlling currents of up to about 10 A or more, at voltages of up to 250 V a.c. and 100 V d.c. Furthermore (unlike solid-state switching devices), relays are available with a variety of different contact

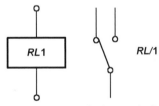

Figure 3.4.31 *Relay symbol*

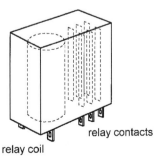

relay contacts

relay coil

Figure 3.4.32 *A typical relay package*

Figure 3.4.33   *On/off power switching arrangement based on a relay*

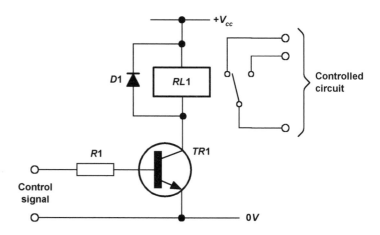

sets, including single-pole on/off switching, single-pole changeover (SPCO), double-pole changeover (DPCO), and four-pole changeover (4PCO). The coils which provide the necessary magnetic flux to operate a relay are available for operation on a variety of voltages between 5 V and 115 V d.c.

The symbol for a relay is shown in Figure 3.4.31 whilst Figure 3.4.32 shows the typical packaging used for such a device.

A typical on/off power switching arrangement based on a relay is shown in Figure 3.4.33. The diode connected across the relay coil is designed to absorb the *back e.m.f.* produced when the magnetic flux rapidly collapses at the point at which the current is removed from the coil.

Thyristors

Semiconductor power control devices provide fast, efficient and reliable methods of switching high currents and voltages. Thyristors (or 'silicon controlled rectifiers' as they are sometimes called) are three-terminal semiconductor devices which can switch very rapidly from a conducting to a non-conducting state. In the 'off' state, the thyristor exhibits negligible leakage current, whilst in the 'on' state the device exhibits very low resistance. This results in very little power loss within the thyristor even when appreciable power levels are being controlled.

Once switched into the conducting state, the thyristor will remain conducting (i.e. it is *latched* into the 'on' state) until the forward current is removed from the device. It is important to note that, in d.c. applications, this necessitates the interruption (or disconnection) of the supply before the device can be reset into its non-conducting state. However, where a thyristor is used with an alternating supply, the device will automatically become reset whenever the mains supply reverses. The device can then be triggered on the next half-cycle having correct polarity to permit conduction.

The symbol for a thyristor is shown in Figure 3.4.34 whilst Figure 3.4.35 shows the typical encapsulation used for such a device.

A typical power control arrangement based on a thyristor is shown in Figure 3.4.35. This circuit provides *half-wave control*. Assuming that the thyristor is perfect, the following formulae apply:

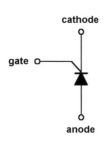

Figure 3.4.34   *Thyristor symbol*

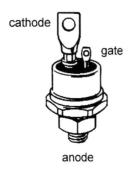

Figure 3.4.35   *A typical thyristor package*

Parameter	Minimum		Maximum
Load voltage (V_L)	0 V		$0.707\,V_L$
Load power (P_L)	0 W		$0.5\,V_S^2/R_L$
Power curve frequency		50 Hz	

Figure 3.4.36 *Half-wave power controller using a thyristor*

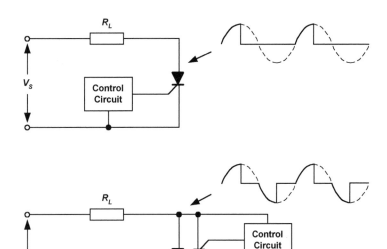

Figure 3.4.37 *Full-wave power controller using a thyristor*

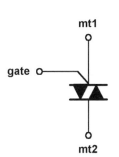

Figure 3.4.38 *Triac symbol*

Figure 3.4.39 *A typical triac package*

An improved power control arrangement based on two thyristors is shown in Figure 3.4.36. This circuit provides *full-wave control*. Assuming that each thyristor is perfect, the following formulae apply:

Parameter	Minimum	Maximum
Load voltage (V_L)	0 V	V_S
Load power (P_L)	0 W	V_S^2/R_L
Power curve frequency		100 Hz

Triacs

Triacs are a refinement of the thyristor which, when triggered, conduct on both positive and negative half-cycles of the applied voltage. Triacs have three terminals, known as main-terminal one (MT1), main-terminal two (MT2), and gate (G), as shown in Figure 3.4.38. Triacs can be triggered by both positive and negative voltages applied between G and MT1 with positive and negative voltages present at MT2, respectively. Triacs thus provide *full-wave control*.

In order to simplify the design of triggering circuits, triacs are often used in conjunction with diacs (equivalent to a bi-directional zener diode). A typical *diac* conducts heavily whenever the applied voltage exceeds approximately 30 V in either direction. Once triggered into the conducting state, the resistance of the diac falls to a very low value and thus a relatively large current will flow. The typical encapsulation used for a triac is shown in Figure 3.4.39.

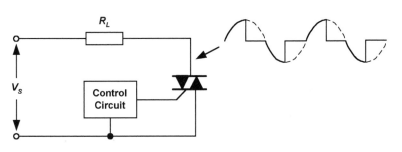

Figure 3.4.40 *Power controller using a triac*

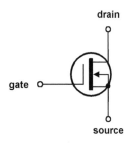

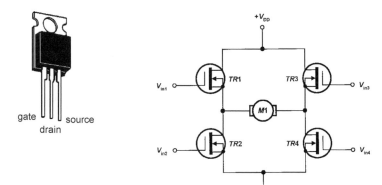

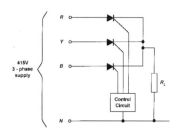

Figure 3.4.41 *(above) A typical power MOSFET*

Figure 3.4.42 *(above right) Power MOSFET symbol*

Figure 3.4.43 *(right) D.C. motor controller based on power MOSFETs*

Figure 3.4.44 *See Question 3.4.4*

A power control arrangement based on a triac is shown in Figure 3.4.40. This circuit provides *full-wave control*. Assuming that each triac is perfect, the following formulae apply:

Parameter	Minimum		Maximum
Load voltage (V_L)	0 V		V_S
Load power (P_L)	0 W		V_S^2/R_L
Power curve frequency		100 Hz	

Transistors

When direct current, rather than alternating current is to be controlled, transistors (both bipolar and field-effect types) can be used. Both types offer the ability to control high currents and high voltages when supplied with only a low-level signal. However, unlike thyristors and triacs, transistors do not remain *latched* in the conducting state. The symbol and packaging for a typical power MOSFET transistor is shown in Figures 3.4.41 and 3.4.42, respectively. Figure 3.4.43 shows a typical d.c. motor controller based on a bridge arrangement of four power MOSFET devices.

Question 3.4.4

A three-phase power controller based on thyristors is shown in Figure 3.4.44. Sketch time related waveforms showing the gate control signals and load voltage assuming that each thyristor is triggered for one third of a cycle of the supply current.

4 Electrical and electronic principles

Summary

This unit covers the electrical principles that students in many branches of electrical and electronic engineering need to understand. It builds on some of the basic theory introduced in Chapter 3 and provides the basis for further study in the more specialised electrical units.

4.1 Circuit theory

This section develops some of the basic electrical and electronic principles that we met in the previous chapter. It begins by explaining a number of useful circuit theorems including those developed by Thévenin and Norton. If you have previously studied Electrical and Electronic Principles at level N (or an equivalent GNVQ Advanced level unit) you should be quite comfortable with this material. If not, you should make sure that you *fully* understand Chapter 3.3 before going further!

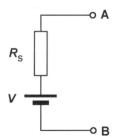

Figure 4.1.1 *Equivalent circuit of a voltage source*

Voltage and current sources

Let's start by considering the properties of a 'perfect' battery. No matter how much current is drawn from it, this device – if it were to exist – would produce a constant voltage between its positive and negative terminals. In practice, you will be well aware that the terminal voltage of a battery falls progressively as we draw more current from it. We account for this fact by referring to its *internal resistance*, R_s (see Figure 4.1.1).

Internal resistance

The notion of internal resistance is fundamental to understanding the behaviour of electrical and electronic circuits and it is worth exploring this idea a little further before we take a look at circuit theorems in some detail. In Figure 4.1.1, the internal resistance is shown as a single discrete resistance connected in series with a voltage source. The e.m.f. source has *exactly* the same voltage as that which would be measured between the battery's terminals when no current is being drawn from

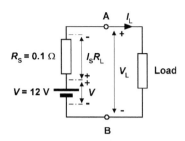

Figure 4.1.2 *See Example 4.1.1*

it. As more current is drawn from a battery, its terminal voltage will fall whereas the internal e.m.f. will remain constant. It might help to illustrate this with an example:

Example 4.1.1

A battery has an open-circuit (no-load) terminal voltage of 12 V and an internal resistance of 0.1 Ω. Determine the terminal voltage of the battery when it supplies currents of (a) 1 A, (b) 10 A and (c) 100 A.

The terminal voltage of the battery when 'on-load' will always be less than the terminal voltage when there is no load connected. The reduction in voltage (i.e. the voltage that is effectively 'lost' inside the battery) will be equal to the load current multiplied by the internal resistance. Applying Kirchhoff's voltage law to the circuit of Figure 4.1.2 gives:

$$V_L = V - I_L R_S$$

where V_L is the terminal voltage on-load, V is the terminal voltage with no load connected, I_L is the load current, and R_S is the internal resistance of the battery.

In case (a); I = 1 A, V = 12 V, and R_S = 0.1 Ω and thus:

$$V_L = V - I_L R_S = 12\,V - (1\,A \times 0.1\,\Omega) = 12 - 0.1 = 11.9\,V.$$

In case (b); I = 10 A, V = 12 V, and R_S = 0.1 Ω and thus:

$$V_L = V - I_L R_S = 12\,V - (10\,A \times 0.1\,\Omega) = 12 - 1 = 11\,V.$$

In case (a); I = 1 A, V = 12 V, and R_S = 0.1 Ω and thus:

$$V_L = V - I_L R_S = 12\,V - (100\,A \times 0.1\Omega) = 12 - 10 = 2\,V.$$

The previous example shows that, provided that the internal resistance of a battery is very much smaller than the resistance of any circuit connected to the battery's terminals, a battery will provide a reasonably constant source of voltage. In fact, when a battery becomes exhausted, its internal resistance begins to rise sharply. This, in turn, reduces the terminal voltage when we try to draw current from the battery as the following example shows:

Example 4.1.2

A battery has an open-circuit (no-load) terminal voltage of 9 V. If the battery is required to supply a load which has a resistance of 90 Ω, determine the terminal voltage of the battery when its internal resistance is (a) 1 Ω and (b)10 Ω.

As we showed before, the terminal voltage of the battery when 'on-load' will always be less than the terminal voltage when there is no load connected. Applying Kirchhoff's voltage law gives:

$$V_L = V - I_L R_S$$

where V_L is the terminal voltage on-load, V is the terminal voltage with no load connected, I_L is the load current, and R_S is the internal resistance of the battery.

Now $I_L = V_L/R_L$.

cont.

Thus $V_L = V - (V_L/R_L) \times R_S$

or $V_L + (V_L/R_L) \times R_S = V$

or $V_L (1 + (1/R_L) \times R_S) = V$

or $V_L (1 + R_S/R_L) = V.$

Hence $V_L = V/(1 + R_S/R_L).$

In case (a); $V = 12$ V, $R_S = 1\ \Omega$ and $R_L = 90\ \Omega.$

Thus $V_L = V/(1 + R_S/R_L) = 9/(1 + 1/90) = 9/(1 + 0.011)$

or $V_L = 9/1.011 = 8.902$ V.

In case (b); $V = 12$ V, $R_S = 10\ \Omega$ and $R_L = 90\ \Omega.$

Thus $V_L = V/(1 + R_S/R_L) = 9/(1 + 10/90) = 9/(1 + 0.111)$

or $V_L = 9/1.111 = 8.101$ V.

The constant voltage source

A battery is the most obvious example of a voltage source. However, as we shall see later, any linear circuit with two terminals, no matter how complex, can be represented by an equivalent circuit based on a voltage source which uses just two components; a source of e.m.f., V, connected in series with an internal resistance, R_S. Furthermore, the voltage source can either be a source of direct current (i.e. a battery), or a source of alternating current (i.e. an a.c. generator, a signal source, or the output of an oscillator).

An ideal voltage source – a *constant voltage source* – would have negligible internal resistance ($R_S = 0$) and its voltage/current characteristic would look like that shown in Figure 4.1.3.

In practice, $R_S \neq 0$ and the terminal voltage will fall as the load current, I_L, increases, as shown in Figure 4.1.4.

The internal resistance, R_S, can be found from the slope of the graph in Figure 4.1.4 as follows:

$$R_S = \left(\frac{\text{change in output voltage, } V_L}{\text{corresponding change in output current, } I_L} \right)$$

The symbols commonly used for constant voltage d.c. and a.c. sources are shown in Figure 4.1.5.

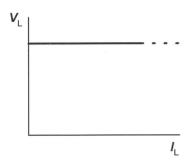

Figure 4.1.3 *V/I characteristic for a constant voltage source*

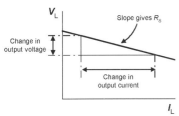

Figure 4.1.4 *V/I characteristic for a voltage source showing the effect of internal resistance*

Figure 4.1.5 *Symbols used for d.c. and a.c. constant voltage sources*

Example 4.1.3

A battery has a no-load terminal voltage of 12 V. If the terminal voltage falls to 10.65 V when a load current of 1.5 A is supplied, determine the internal resistance of the battery.
Now:

$$R_S = \left(\frac{\text{change in output voltage, } V_L}{\text{corresponding change in output current, } I_L} \right)$$

$$= \left(\frac{12 - 10.65}{1.5 - 0} \right) = \frac{1.35}{1.5} = 0.9\ \Omega.$$

cont.

Note that we could have approached this problem in a different way by considering the voltage 'lost' inside the battery (i.e. the voltage dropped across R_s). The lost voltage will be $(12 - 10.65) = 1.35$ V. In this condition, the current flowing is 1.5 A, hence applying Ohm's law gives:

$$R_s = V/I = 1.35/1.5 = 0.9 \; \Omega.$$

Thévenin's theorem

Thévenin's theorem states that:

Any two-terminal network can be replaced by an equivalent circuit consisting of a voltage source and a series resistance equal to the internal resistance seen looking into the two terminals.

In order to determine the Thévenin equivalent of a two-terminal network you need to:

(a)   Determine the voltage that will appear between the two terminals with no load connected to the terminals. This is the *open-circuit voltage*.

(b)   Replace any voltage sources with a short-circuit connection (or replace any current sources with an open circuit connection) and then determine the *internal resistance* of the circuit (i.e. the resistance that appears between the two terminals).

Example 4.1.4

Determine the Thévenin equivalent of the two-terminal network shown in Figure 4.1.6.

First we need to find the voltage that will appear between the two terminals, A and B, with no load connected. In this condition, no current will be drawn from the network and thus there will be no voltage dropped across R_3. The voltage that appears between A and B, V_{AB}, will then be identical to that which is dropped across R_2 (V_{CB}).

The voltage dropped across R_2 (with no load connected) can be found by applying the potential divider theorem, thus:

$$V_{CB} = V \times R_2/(R_1 + R_2) = 3 \times 6/(3 + 6) = 18/9 = 2 \text{ V}.$$

Thus the open-circuit output voltage will be 2 V.

cont.

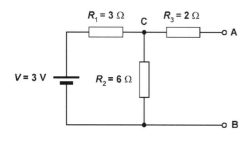

Figure 4.1.6   *See Example 4.1.4*

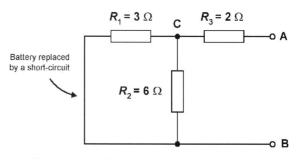

Figure 4.1.7   *See Example 4.1.4*

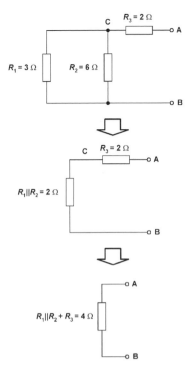

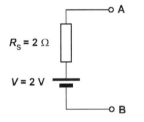

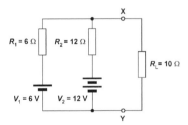

Figure 4.1.8 *See Example 4.1.4*

Figure 4.1.9 *See Example 4.1.4*

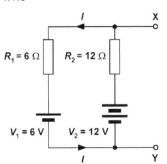

Figure 4.1.10 *See Example 4.1.5*

Figure 4.1.11 *See Example 4.1.5*

Next we need to find the internal resistance of the circuit with the battery replaced by a short-circuit (see Figure 4.1.7).

The circuit can be progressively reduced to a single resistor as shown in Figure 4.1.8.

Thus the internal resistance is 4 Ω.

The Thévenin equivalent of the two-terminal network is shown in Figure 4.1.9.

Example 4.1.5

Determine the current flowing in, and voltage dropped across, the 10 Ω resistor shown in Figure 4.1.10.

First we will determine the Thévenin equivalent circuit of the network.

With no load connected to the network (i.e. with R_L disconnected) the current supplied by the 12 V battery will flow in an anti-clockwise direction around the circuit as shown in Figure 4.1.11.

The total e.m.f. present in the circuit will be:

$V = 12 - 6 = 6$ V (note that the two batteries *oppose* one another).

The total resistance in the circuit is:

$R = R_1 + R_2 = 6 + 12 = 18$ Ω (the two resistors are in series with one another).

Hence, the current flowing in the circuit when R_L is disconnected will be given by:

$I = V/R = 6/18 = 0.333$ A.

The voltage developed between the output terminals (X and Y) will be equal to 12 V *minus* the voltage dropped across the 12 Ω resistor (or 6 V plus the voltage dropped across the 6 Ω resistor).

Hence $V_{XY} = 12 - (0.333 \times 12) = 12 - 4 = 8$ V

(or $V_{XY} = 6 + (0.333 \times 6) = 6 + 2 = 8$ V).

Thus the open-circuit output voltage will be 8 V.
The internal resistance with V_1 and V_2 replaced by short-circuit will be given by:

$R = 6 \times 12/(6 + 12) = 72/18 = 4$ Ω.

Thus the internal resistance will be 4 Ω.

Figure 4.1.12 shows the Thévenin equivalent of the circuit.

Now, with $R_L = 10$ Ω connected between X and Y the circuit looks like that shown in Figure 4.1.13.

cont.

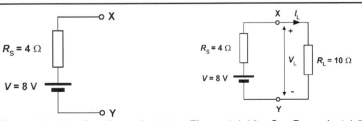

Figure 4.1.12 *See Example 4.1.5* **Figure 4.1.13** *See Example 4.1.5*

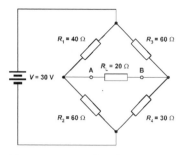

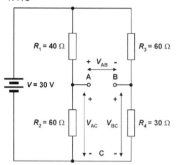

Figure 4.1.14 *See Example 4.1.6*

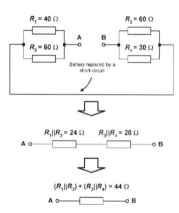

Figure 4.1.15 *See Example 4.1.6*

The total resistance present will be:

$$R = R_S + R_L = 4 + 10 = 14 \ \Omega.$$

The current drawn from the circuit in this condition will be given by:

$$I_L = V/R = 8/14 = 0.57 \text{ A}.$$

Finally, the voltage dropped across R_L will be given by:

$$V_L = I \times R_L = 0.57 \times 10 = 5.7 \text{ V}.$$

Example 4.1.6

Determine the current flowing in a 20 Ω resistor connected to the Wheatstone bridge shown in Figure 4.1.14.

Once again, we will begin by determining the Thévenin equivalent circuit of the network. To make things a little easier, we will redraw the circuit so that it looks a little more familiar (see Figure 4.1.15).

With A and B open-circuit, the circuit simply takes the form of two potential dividers (R_1 and R_2 forming one potential divider, whilst R_3 and R_4 form the other potential divider). It is thus a relatively simple matter to calculate the open-circuit voltage drop between A and B, since:

$$V_{AB} = V_{AC} - V_{BC} \text{ (from Kirchhoff's voltage law)}$$

where $V_{AC} = V \times R_2/(R_1 + R_2) = 30 \times 60/(40 + 60) = 18 \text{ V}$

and $V_{BC} = V \times R_4/(R_3 + R_4) = 30 \times 30/(30 + 60) = 10 \text{ V}.$

Thus $V_{AB} = 18 - 10 = 8 \text{ V}.$

Next, to determine the internal resistance of the network with the battery replaced by a short-circuit, we can again re-draw the circuit and progressively reduce it to one resistance. Figure 4.1.16 shows how this is done.

Thus the internal resistance is 44 Ω.

The Thévenin equivalent of the bridge network is shown in Figure 4.1.17.

Now, with $R_L = 20 \ \Omega$ connected between A and B the circuit looks like that shown in Figure 4.1.18.

The total resistance present will be:

$$R = R_S + R_L = 44 + 20 = 64 \ \Omega.$$

cont.

Figure 4.1.16 *See Example 4.1.6*

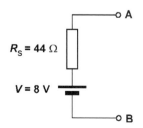

Figure 4.1.17 *See Example 4.1.6*

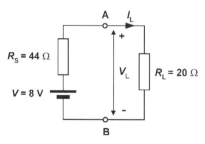

Figure 4.1.18 *See Example 4.1.6*

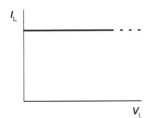

Figure 4.1.19 *I/V characteristic for a constant current source*

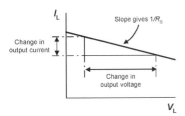

Figure 4.1.20 *I/V characteristic for a constant current source showing the effect of internal resistance*

Figure 4.1.21 *Symbols used for d.c. and a.c. constant current sources*

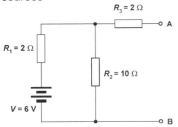

Figure 4.1.22 *See Example 4.1.7*

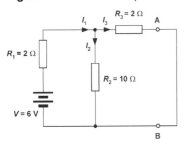

Figure 4.1.23 *See Example 4.1.7*

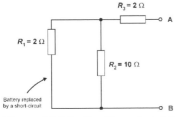

Figure 4.1.24 *See Example 4.1.7*

The current drawn from the circuit in this condition will be given by:

$$I_L = V/R = 8/64 = 0.125 \text{ A}.$$

The constant current source

By now, you should be familiar with the techniques for analysing simple electric circuits by considering the voltage sources that may be present and their resulting effect on the circuit in terms of the currents and voltage drops that they produce. However, when analysing some types of circuit it can often be more convenient to consider sources of current rather than voltage. In fact, any linear circuit with two terminals, no matter how complex, can be thought of as a current source; i.e. a supply of current, I, connected in parallel with an internal resistance, R_p. As with voltage sources, current sources can either be a source of direct or alternating current.

An ideal current source – a *constant current source* – would have infinite internal resistance ($R_p = \infty$) and its voltage/current characteristic would look like that shown in Figure 4.1.19.

In practice, $R_p \neq \infty$ and the current will fall as the load voltage, V_L, increases, as shown in Figure 4.1.20.

The internal resistance, R_p, can be found from the reciprocal of the slope of the graph in Figure 4.1.20 as follows:

$$R_P = \cfrac{1}{\left(\cfrac{\text{change in output current, } I_L}{\text{corresponding change in output voltage, } V_L}\right)}$$

$$= \left(\cfrac{\text{change in output voltage, } V_L}{\text{corresponding change in output current, } I_L}\right).$$

The symbols commonly used for constant current d.c. and a.c. sources are shown in Figure 4.1.21.

Norton's theorem

Norton's theorem states that:

Any two-terminal network can be replaced by an equivalent circuit consisting of a current source and a parallel resistance equal to the internal resistance seen looking into the two terminals.

In order to determine the Norton equivalent of a two-terminal network you need to:

(a) Determine the current that will appear between the two terminals when they are short-circuited together. This is the *short-circuit current.*

(b) Replace any current sources with an open-circuit connection (or replace any voltage sources with a short-circuit connection) and then determine the *internal resistance* of the circuit (i.e. the resistance that appears between the two terminals).

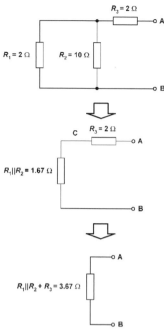

Figure 4.1.25 *See Example 4.1.7*

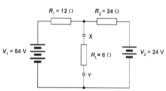

Figure 4.1.26 *See Example 4.1.7*

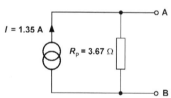

Figure 4.1.27 *See Example 4.1.8*

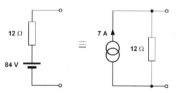

Figure 4.1.28 *See Example 4.1.8*

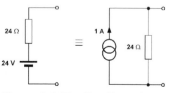

Figure 4.1.29 *See Example 4.1.8*

Example 4.1.7

Determine the Norton equivalent of the two-terminal network shown in Figure 4.1.22.

First, we need to find the current that would flow between terminals A and B with the two terminals linked together by a short-circuit (see Figure 4.1.23).

The total resistance that appears across the 6 V battery with the output terminals linked together is given by:

$$R_1 + (R_2 \times R_3)/(R_2 + R_3) = 2 + (2 \times 10)/(2 + 10) = 2 + 20/12 = 3.67\ \Omega.$$

The current supplied by the battery, I_1, will then be given by:

$$I_1 = V/R = 6/3.67 = 1.63\ \text{A}.$$

Using the current divider theorem, the current in R_3 (and that flowing between terminals A and B) will be given by:

$$I_3 = I_1 \times R_2/(R_2 + R_3) = 1.63 \times 10/(10 + 2) = 16.3/12 = 1.35\ \text{A}.$$

Thus the short-circuit current will be 1.35 A.

Second, we need to find the internal resistance of the circuit with the battery replaced by a short-circuit (see Figure 4.1.24).

The circuit can be progressively reduced to a single resistor as shown in Figure 4.1.25.

Thus the internal resistance is 3.67 Ω.

The Norton equivalent of the two-terminal network is shown in Figure 4.1.26.

Example 4.1.8

Determine the voltage dropped across a 6 Ω resistor connected between X and Y in the circuit shown in Figure 4.1.27.

First, we will determine the Norton equivalent circuit of the network. Since there are two voltage sources (V_1 and V_2) in this circuit, we can derive the Norton equivalent of each branch and then combine them together into one current source and one parallel resistance. This makes life a lot easier!

The left-hand branch of the circuit is equivalent to a current source of 7 A with an internal resistance of 12 Ω (see Figure 4.1.28).

The right-hand branch of the circuit is equivalent to a current source of 1 A with an internal resistance of 24 Ω (see Figure 4.1.29).

The combined effect of the two current sources is a single constant current generator of $(7+1) = 8$ A with a parallel resistance of $12 \times 24/(12 + 24) = 8\ \Omega$ (see Figure 4.1.30).

Now, with $R_L = 6\ \Omega$ connected between X and Y the circuit looks like that shown in Figure 4.1.31.

cont.

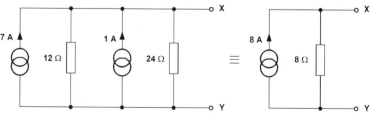

Figure 4.1.30 *See Example 4.1.8*

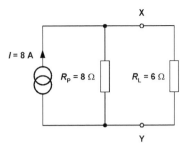

Figure 4.1.31   *See Example 4.1.8*

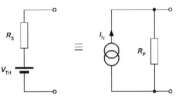

Figure 4.1.32   *Norton and Thévenin equivalent circuits*

The total resistance present will be:

$$R = R_\mathrm{P} \times R_\mathrm{L}/(R_\mathrm{P} \times R_\mathrm{L}) = 8 \times 6/(8 + 6) = 3.43\ \Omega.$$

When 8 A flows in this resistance, the voltage dropped across the parallel combination of resistors will be:

$$V_\mathrm{L} = 8/3.43 = 2.33\ \mathrm{V}.$$

Thévenin to Norton conversion

The previous example shows that it is relatively easy to convert from one equivalent circuit to the other (don't overlook the fact that, whichever equivalent circuit is used to solve a problem, the result will be identical!).

To convert from the Thévenin equivalent circuit to the Norton equivalent circuit:

$$R_\mathrm{P} = R_\mathrm{S}\ \text{and}\ V_\mathrm{TH} = I_\mathrm{N} \times R_\mathrm{P}.$$

To convert from Norton equivalent circuit to the Thévenin equivalent circuit:

$$R_\mathrm{S} = R_\mathrm{P}\ \text{and}\ I_\mathrm{N} = V_\mathrm{TH}/R_\mathrm{S}.$$

This equivalence is shown in Figure 4.1.32.

Superposition theorem

The superposition theorem is a simple yet powerful method that can be used to analyse networks containing a number of voltage sources and linear resistances.

The superposition theorem states that:

In any network containing more than one voltage source, the current in, or potential difference developed across, any branch can be found by considering the effects of each source separately and adding their effects. During this process, any temporarily omitted source must be replaced by its internal resistance (or a short-circuit if it is a perfect voltage source).

This might sound a little more complicated than it really is so we shall explain the use of the theorem with a simple example:

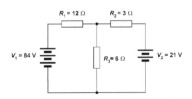

Figure 4.1.33   *See Example 4.1.9*

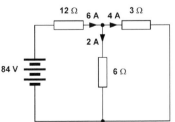

Figure 4.1.34   *See Example 4.1.9*

Example 4.1.9

Use the superposition theorem to determine the voltage dropped across the 6 Ω resistor in Figure 4.1.33.

First we shall consider the effects of the left-hand voltage source (V_1) when taken on its own. Figure 4.1.34 shows the voltages and currents (you can easily check these by calculation, if you wish!).

Next we shall consider the effects of the right-hand voltage source (V_2) when taken on its own. Figure 4.1.35 shows the voltages and currents (again, you can easily check these by calculation!).

Finally, we can combine these two sets of results, taking into

cont.

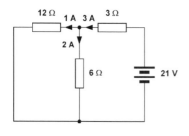

Figure 4.1.35 *See Example 4.1.9*

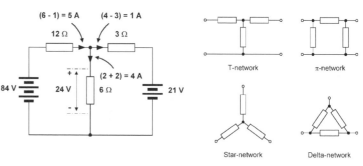

Figure 4.1.36 *See Example 4.1.9*

Figure 4.1.37 *Four basic electrical network configurations*

account the direction of current flow to arrive at the values of current in the complete circuit. Figure 4.1.36 shows how this is done. The current in the 6 Ω resistor is thus 4 A and the voltage dropped across it will be:

$$V = I \times R = 4 \times 6 = 24 \text{ V}.$$

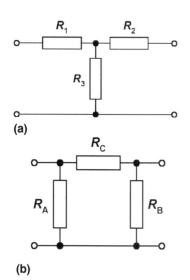

Figure 4.1.38 *T and π-networks*

T, π, star and delta networks

Figure 4.1.37 shows four basic electrical network configurations that you will encounter in various forms. In order to solve network problems it is often very helpful to convert from one type of circuit to another that has equivalent properties. We shall consider the T and π circuits first.

T–π and π–T transformation

T and π-networks are commonly encountered in attenuators and filters. Figure 4.1.38(a) shows a T-network (comprising R_1, R_2 and R_3) whilst Figure 4.1.38(b) shows a π-network (comprising R_A, R_B and R_C).

The T equivalent of the π-network can be derived as follows:
The resistance seen looking into each network at X–Z will be:

$$R_1 + R_3 = \frac{R_A \times (R_B + R_C)}{R_A + R_B + R_C}. \tag{i}$$

The resistance seen looking into each network at Y–Z will be:

$$R_2 + R_3 = \frac{R_B \times (R_A + R_C)}{R_A + R_B + R_C}. \tag{ii}$$

The resistance seen looking into each network at X–Y will be:

$$R_1 + R_2 = \frac{R_C \times (R_A + R_B)}{R_A + R_B + R_C} \tag{iii}$$

Subtracting (ii) from (i) gives:

$$R_1 - R_2 = \frac{R_A(R_B + R_C) - R_B(R_A + R_C)}{R_A + R_B + R_C} \tag{iv}$$

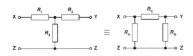

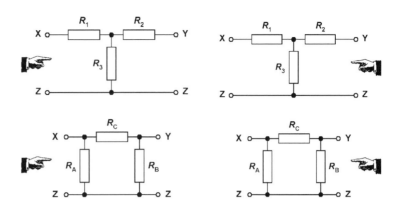

Figure 4.1.39 (above) *T-net-work equivalent of a π-network*
Figure 4.1.40 (right) *Comparing the resistance seen looking into the networks at X–Z*
Figure 4.1.41 (far right) *Comparing the resistance seen looking into the networks at Y–Z*

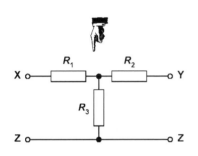

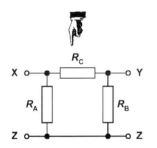

Figure 4.1.42 *Comparing the resistance seen looking into the networks at X–Y*

Adding (iii) and (iv) gives:

$$2R_1 = \frac{R_A(R_B + R_C) - R_B(R_A + R_C) + R_C(R_A + R_B)}{R_A + R_B + R_C}$$

thus $2R_1 = \dfrac{R_A R_B + R_A R_C - R_A R_B - R_B R_C + R_A R_C + R_B R_C}{R_A + R_B + R_C}$

$$= \frac{2R_A R_C}{R_A + R_B + R_C}$$

thus $R_1 = \dfrac{R_A R_C}{R_A + R_B + R_C}$. \hfill (v)

Similarly:

$$R_2 = \frac{R_B R_C}{R_A + R_B + R_C} \hfill \text{(vi)}$$

and

$$R_3 = \frac{R_A R_B}{R_A + R_B + R_C} \hfill \text{(vii)}$$

The π equivalent of the T-network is derived as follows:
Dividing (vi) by (vii) gives:

$$\frac{R_2}{R_1} = \frac{R_B R_C}{R_A R_C} = \frac{R_B}{R_A}$$

therefore, $R_B = \dfrac{R_2}{R_1} \times R_A$. \hfill (viii)

Similarly, dividing (vi) by (v) gives:

$$\frac{R_2}{R_3} = \frac{R_B R_C}{R_A R_B} = \frac{R_C}{R_A}.$$

therefore, $R_C = \dfrac{R_2}{R_3} \times R_A$. \hfill (ix)

Substituting (viii) and (ix) for R_B and R_C, respectively, in (v) gives:

$$R_1 = \frac{R_A \times \left(\frac{R_2}{R_3} \times R_A\right)}{R_A + \left(\frac{R_2}{R_1} \times R_A\right) + \left(\frac{R_2}{R_3} \times R_A\right)} = \frac{R_A{}^2 \times \frac{R_2}{R_3}}{R_A \times \left(1 + \frac{R_2}{R_1} + \frac{R_2}{R_3}\right)} = \frac{R_A \times \frac{R_2}{R_3}}{1 + \frac{R_2}{R_1} + \frac{R_2}{R_3}}$$

thus $R_A = \dfrac{R_1 R_3}{R_2} \times \left(1 + \dfrac{R_2}{R_1} + \dfrac{R_2}{R_3}\right) = \dfrac{R_1 R_3}{R_2} + \dfrac{R_1 R_2 R_3}{R_1 R_2} + \dfrac{R_1 R_2 R_3}{R_2 R_3}$

$$= \frac{R_1 R_3}{R_2} + \frac{R_2 R_3}{R_2} + \frac{R_1 R_2}{R_2}$$

hence $R_A = \dfrac{R_1 R_2 + R_1 R_3 + R_2 R_3}{R_2}$. $\qquad$ (x)

Similarly:

$$R_B = \frac{R_1 R_2 + R_1 R_3 + R_2 R_3}{R_1} \qquad \text{(xi)}$$

and

$$R_C = \frac{R_1 R_2 + R_1 R_3 + R_2 R_3}{R_3}. \qquad \text{(xii)}$$

The six transformation equations can be summarised as follows:

The equivalent π-network resistance between two terminals of a T-network is equal to the sum of the products of each pair of adjacent resistances divided by the opposite resistance – see equations (x)–(xii).

The equivalent T-network resistance between two adjacent terminals of a π-network is equal to the product of the two adjacent resistances divided by the sum of the three resistances – see equations (v)–(vii).

Example 4.1.10

Determine the π-network equivalent of the T-network shown in Figure 4.1.43.

From Figure 4.1.43, $R_1 = R_2 = 30\ \Omega$ and $R_3 = 120\ \Omega$.

Now $R_A = \dfrac{R_1 R_2 + R_1 R_3 + R_2 R_3}{R_2}$.

Now $R_1 R_2 + R_1 R_3 + R_2 R_3 = (30 \times 30) + (30 \times 120) + (30 \times 120)$
$= 8100$.

cont.

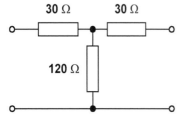

Figure 4.1.43 *See Example 4.1.10*

30 Ω   30 Ω

120 Ω

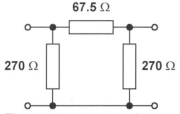

67.5 Ω

270 Ω   270 Ω

Figure 4.1.44 *π-network equivalent of the T-network shown in Figure 4.1.43*

$$\text{and } R_A = \frac{R_1 R_2 + R_1 R_3 + R_2 R_3}{R_2} = \frac{8100}{30} = 270\ \Omega$$

$$\text{similarly } R_B = \frac{R_1 R_2 + R_1 R_3 + R_2 R_3}{R_1} = \frac{8100}{30} = 270\ \Omega$$

$$\text{and } R_C = \frac{R_1 R_2 + R_1 R_3 + R_2 R_3}{R_3} = \frac{8100}{120} = 67.5\ \Omega.$$

Figure 4.1.44 shows the π-equivalent of the T-network shown in Figure 4.1.43.

Star-delta and delta-star transformation

Star and delta networks are commonly encountered in three-phase systems. Figure 4.1.45(a) shows a star network (comprising R_1, R_2 and R_3) whilst Figure 4.1.45(b) shows a delta network (comprising R_A, R_B and R_C).

The star equivalent of the delta network is derived as follows:
The resistance seen looking into each network at X–Z will be:

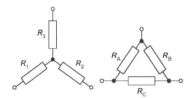

Figure 4.1.45 *Star and delta networks*

$$R_1 + R_3 = \frac{R_A \times (R_B + R_C)}{R_A + R_B + R_C}. \tag{i}$$

The resistance seen looking into each network at Y–Z will be:

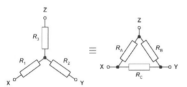

Figure 4.1.46 *Star equivalent of a delta network*

$$R_2 + R_3 = \frac{R_B \times (R_A + R_C)}{R_A + R_B + R_C}. \tag{ii}$$

The resistance seen looking into each network at X–Y will be:

$$R_1 + R_2 = \frac{R_C \times (R_A + R_B)}{R_A + R_B + R_C}. \tag{iii}$$

Subtracting (ii) from (i) gives:

$$R_1 - R_2 = \frac{R_A(R_B + R_C) - R_B(R_A + R_C)}{R_A + R_B + R_C}. \tag{iv}$$

Adding (iii) and (iv) gives:

$$2R_1 = \frac{R_A (R_B + R_C) - R_B(R_A + R_C) + R_C(R_A + R_B)}{R_A + R_B + R_C}$$

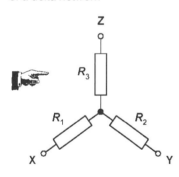

Figure 4.1.47 *Comparing the resistance seen looking into the networks at X–Z*

$$\text{thus } 2R_1 = \frac{R_A R_B + R_A R_C - R_A R_B - R_B R_C + R_A R_C + R_B R_C}{R_A + R_B + R_C}$$

$$= \frac{2R_A R_C}{R_A + R_B + R_C}$$

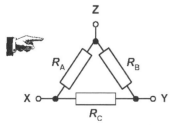

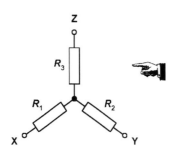

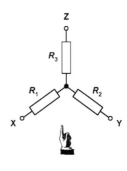

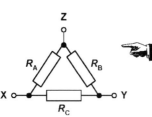

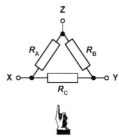

Figure 4.1.48 *Comparing the resistance seen looking into the networks at Y–Z*

Figure 4.1.49 *Comparing the resistance seen looking into the networks at X–Y*

thus $R_1 = \dfrac{R_A R_C}{R_A + R_B + R_C}$. (v)

Similarly:

$$R_2 = \dfrac{R_B R_C}{R_A + R_B + R_C} \quad\text{(vi)}$$

and

$$R_3 = \dfrac{R_A R_B}{R_A + R_B + R_C}. \quad\text{(vii)}$$

The delta equivalent of the star network is derived as follows:
Dividing (vi) by (vii) gives:

$$\frac{R_2}{R_1} = \frac{R_B R_C}{R_A R_C} = \frac{R_B}{R_A}$$

therefore, $R_B = \dfrac{R_2}{R_1} \times R_A$. (viii)

Similarly, dividing (vi) by (v) gives:

$$\frac{R_2}{R_3} = \frac{R_B R_C}{R_A R_B} = \frac{R_C}{R_A}$$

therefore, $R_C = \dfrac{R_2}{R_3} \times R_A$. (ix)

Substituting (viii) and (ix) for R_B and R_C, respectively, in (v) gives:

$$R_1 = \frac{R_A \times \left(\dfrac{R_2}{R_3} \times R_A\right)}{R_A + \left(\dfrac{R_2}{R_1} \times R_A\right) + \left(\dfrac{R_2}{R_3} \times R_A\right)} = \frac{R_A{}^2 \times \dfrac{R_2}{R_3}}{R_A \times \left(1 + \dfrac{R_2}{R_1} + \dfrac{R_2}{R_3}\right)} = \frac{R_A \times \dfrac{R_2}{R_3}}{1 + \dfrac{R_2}{R_1} + \dfrac{R_2}{R_3}}$$

thus $R_A = \dfrac{R_1 R_3}{R_2} \times \left(1 + \dfrac{R_2}{R_1} + \dfrac{R_2}{R_3}\right) = \dfrac{R_1 R_3}{R_2} + \dfrac{R_1 R_2 R_3}{R_1 R_2} + \dfrac{R_1 R_2 R_3}{R_2 R_3}$

$$= \dfrac{R_1 R_3}{R_2} + \dfrac{R_2 R_3}{R_2} + \dfrac{R_1 R_2}{R_2}$$

hence $R_A = \dfrac{R_1 R_2 + R_1 R_3 + R_2 R_3}{R_2}$. $\qquad$ (x)

Similarly:

$$R_B = \frac{R_1 R_2 + R_1 R_3 + R_2 R_3}{R_1} \qquad \text{(xi)}$$

and

$$R_C = \frac{R_1 R_2 + R_1 R_3 + R_2 R_3}{R_3} \qquad \text{(xii)}$$

The six transformation equations can be summarised as follows:

The equivalent delta resistance between two terminals of a star network is equal to the sum of the products of each pair of adjacent resistances divided by the opposite resistance – see equations (x)–(xii).

The equivalent star resistance between two adjacent terminals of a delta network is equal to the product of the two adjacent resistances divided by the sum of the three resistances – see equations (v)–(vii).

Example 4.1.11

Determine the star network equivalent of the delta network shown in Figure 4.1.50.

From Figure 4.1.50, $R_A = 4\ \Omega$, $R_B = 10\ \Omega$, and $R_C = 6\ \Omega$.

Now $R_1 = \dfrac{R_A R_C}{R_A + R_B + R_C}$.

Now $R_A + R_B + R_C = 4 + 10 + 6 = 20\ \Omega$

and $R_1 = \dfrac{R_A R_C}{R_A + R_B + R_C} = \dfrac{4 \times 6}{20} = \dfrac{24}{20} = 1.2\ \Omega$

cont.

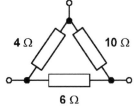

Figure 4.1.50 *See Example 4.1.11*

4 Ω   10 Ω

6 Ω

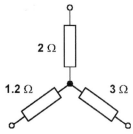

Figure 4.1.51 *Star network equivalent of the delta network shown in Figure 4.1.50*

similarly $R_2 = \dfrac{R_B R_C}{R_A + R_B + R_C} = \dfrac{10 \times 6}{20} = \dfrac{60}{20} = 3\ \Omega$

and $R_3 = \dfrac{R_A R_B}{R_A + R_B + R_C} = \dfrac{4 \times 10}{20} = \dfrac{40}{20} = 2\ \Omega$.

Figure 4.1.51 shows the star equivalent of the delta network shown in Figure 4.1.50.

Questions 4.1.1

(1)   The terminal voltage of a battery falls from 24 V with no load connected to 21.6 V when supplying a current of 6 A. Determine the internal resistance of the battery. Also determine the terminal voltage of the battery when it supplies a current of 10 A.

(2)   Determine the Thévenin equivalent of the circuit shown in Figure 4.1.52.

(3)   Determine the Norton equivalent of the circuit shown in Figure 4.1.53.

(4)   Use Thévenin's theorem to determine the current flowing in R_L in the circuit shown in Figure 4.1.54.

(5)   Use Norton's theorem to determine the voltage developed across R_L in the circuit shown in Figure 4.1.55.

(6)   Use the Superposition theorem to determine the voltage dropped across the 15 Ω resistor in Figure 4.1.56.

(7)   Determine the π-network equivalent of the T-network shown in Figure 4.1.57.

(8)   Determine the T-network equivalent of the π-network shown in Figure 4.1.58.

(9)   Determine the star network equivalent of the delta network shown in Figure 4.1.59.

(10) Determine the delta network equivalent of the star network shown in Figure 4.1.60.

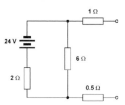

Figure 4.1.52 *See Questions 4.1.1*

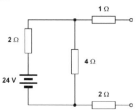

Figure 4.1.53 *See Questions 4.1.1*

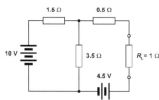

Figure 4.1.54 *See Questions 4.1.1*

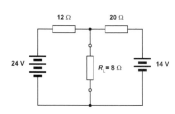

Figure 4.1.55 *See Questions 4.1.1*

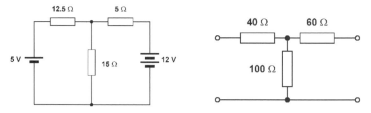

Figure 4.1.56 *See Questions 4.1.1*   **Figure 4.1.57** *See Questions 4.1.1*

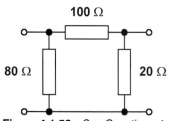

Figure 4.1.58 *See Questions 4.1.1*   **Figure 4.1.59** *See Questions 4.1.1*   **Figure 4.1.60** *See Questions 4.1.1*

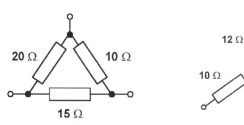

Networks of complex impedances

Until now, we have only considered networks of pure resistance and d.c. sources in our explanation of the basic circuit theorems. However, many real circuits use a.c. sources and contain a mixture of resistance and reactance in which case it is necessary to consider the effects of *impedance* rather than resistance. Before we begin, we shall introduce another technique for representing complex impedances which will save us a great deal of time and effort later on!

Another way of representing complex impedances

In Chapter 3 we showed how a complex impedance could be represented using the j-operator. In particular you should recall that any complex impedance can be represented by the relationship:

$$Z = (R \pm jX)$$

where Z represents impedance, R represents resistance, and X represents reactance.

You should also recall that a positive j-term $(+jX)$ indicates an inductive reactance whilst a negative j-term $(-jX)$ indicates a capacitive reactance. Since we use a set of rectangular axes (an Argand diagram) to represent $Z = (R \pm jX)$ graphically, this form of notation is sometimes referred to as *rectangular notation*.

An alternative way of representing a complex quantity is that of using a notation based on a polar, rather than a rectangular notation. Two quantities are required to specify an impedance using polar notation; Z and θ (where Z is the *modulus*, and θ is the angle, or *argument*).

Figure 4.1.61 shows the relationship between the rectangular and polar methods of representing a complex impedance.

From Figure 4.1.61, $Z = \sqrt{R^2 + X^2}$ and $\theta = \arctan (X/R)$.

Notice also that $R = Z \cos \theta$ and $X = Z \sin \theta$.

Using the foregoing relationships it is relatively simple to convert from rectangular form to polar form, and vice versa. Here are a few examples of some identical impedances:

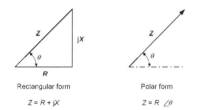

Rectangular form

$Z = R + jX$

Polar form

$Z = R \angle \theta$

Figure 4.1.61 *Relationship between rectangular and polar methods of representing complex impedances*

Rectangular form	Polar form
10 or (10 + j0)	$10\angle 0°$
+j10	$10\angle 90°$
−j10	$10\angle -90°$
10 + j10	$14.14\angle 45°$
10 − j10	$14.14\angle -45°$

Example 4.1.12

A circuit comprises a 30 Ω resistor connected in series with an inductor having an inductive reactance of 40 Ω. Determine the impedance of the circuit and express your answer in both rectangular and polar form.

Now $Z = R \pm jX$

where $R = 30 \; \Omega$ and $X = +j40 \; \Omega$.

Thus $Z = (30 + j40) \; \Omega$ (rectangular form).

Since $Z = \sqrt{R^2 + X^2}$ (we sometimes refer to this as *modulus-Z* or $|Z|$)

$Z = \sqrt{R^2 + X^2} = \sqrt{30^2 + 40^2} = \sqrt{250} = 50 \; \Omega$

and $\theta = \arctan(X/R) = \arctan(40/30) = 53.13°$.

Thus $Z = 50 \angle 53.13° \; \Omega$ (polar form)

(note that $30 + j40 \equiv 50\angle 53.13°$).

Multiplying and dividing impedances expressed in the rectangular $(R \pm jX)$ form can be quite tedious and error prone. To find the product of two impedances we have to multiply out the brackets, rearrange the terms and replace any j^2 by -1. Having done that we must 'tidy up' the result by grouping the terms together and writing the result in the $R \pm jX$ form. This is how it's done:

Let $Z_1 = R_1 + jX_1$ and $Z_2 = R_2 + jX_2$.

Then $Z_1 \times Z_2 = (R_1 + jX_1)(R_2 + jX_2) = R_1 R_2 + R_1 jX_2 + R_2 jX_1 + j^2 X_1 X_2$

or $Z_1 \times Z_2 = R_1 R_2 + j(R_1 + R_2) + (-1)X_1 X_2 = R_1 R_2 + j(R_1 + R_2) - X_1 X_2$

hence $Z_1 \times Z_2 = (R_1 R_2 - X_1 X_2) + j(R_1 + R_2)$.

The division of impedances expressed in complex form is even more complicated as it requires us to multiply the top and bottom of the expression by the *complex conjugate* of the denominator. This is how it's done:

Once again, let $Z_1 = R_1 + jX_1$ and $Z_2 = R_2 + jX_2$.

Then $\dfrac{Z_1}{Z_2} = \dfrac{(R_1 + jX_1)}{(R_2 + jX_2)} = \dfrac{(R_1 + jX_1)(R_2 - jX_2)}{(R_2 + jX_2)(R_2 - jX_2)} = \dfrac{(R_1 R_2 + X_1 X_2) + j(R_2 X_1 - R_1 X_2)}{(R_2^2 - R_2 jX_2 + R_2 jX_2 - j^2 X_2^2)}$

or $\dfrac{Z_1}{Z_2} = \dfrac{(R_1 R_2 + X_1 X_2) + j(R_2 X_1 - R_1 X_2)}{(R_2^2 - j^2 X_2^2)} = \dfrac{(R_1 R_2 + X_1 X_2) + j(R_2 X_1 - R_1 X_2)}{R_2^2 - (-1)X_2^2}$

or $\dfrac{Z_1}{Z_2} = \dfrac{(R_1 R_2 + X_1 X_2) + j(R_2 X_1 - R_1 X_2)}{R_2^2 + X_2^2} = \dfrac{R_1 R_2 + X_1 X_2}{R_2^2 + X_2^2} + \dfrac{j(R_2 X_1 - R_1 X_2)}{R_2^2 + X_2^2}$.

Multiplication and division of impedances is very much simpler using polar notation, since:

$$Z_1 \angle \theta_1 \times Z_2 \angle \theta_2 = Z_1 Z_2 \angle (\theta_1 + \theta_2) \text{ and } \dfrac{Z_1 \angle \theta_1}{Z_2 \angle \theta_2} = \dfrac{Z_1}{Z_2} \angle (\theta_1 - \theta_2)$$

where $Z_1\angle\theta_1$ is Z_1 expressed in polar form and $Z_2\angle\theta_2$ is Z_2 expressed in polar form.

Mathematics in action

Multiplication, division and powers of complex quantities expressed in polar form

The rules for multiplying and dividing complex quantities expressed in polar form are as follows:

Consider two complex numbers expressed in polar form, $Z_1\angle\theta_1$ and $Z_2\angle\theta_2$.

Multiplying these numbers gives:

$$Z_1\angle\theta_1 \times Z_2\angle\theta_2 = Z_1 Z_2 \angle(\theta_1 + \theta_2)$$

whilst dividing these numbers gives: $\dfrac{Z_1\angle\theta_1}{Z_2\angle\theta_2} = \dfrac{Z_1}{Z_2}\angle(\theta_1 - \theta_2)$.

Now $(Z_1\angle\theta_1)^2 = Z_1\angle\theta_1 \times Z_1\angle\theta_1 = Z_1 Z_1\angle(\theta_1 + \theta_1) = Z_1^2\angle(2\theta_1)$

similarly $(Z_2\angle\theta_2)^2 = Z_2\angle\theta_2 \times Z_2\angle\theta_2 = Z_2 Z_2\angle(\theta_2 + \theta_2) = Z_2^2\angle(2\theta_2)$.

Thus $(Z\angle\theta)^2 = Z^2\angle(2\theta)$

also $(Z\angle\theta)^3 = Z^3\angle(3\theta)$ whilst $(Z\angle\theta)^4 = Z^4\angle(4\theta)$.

Hence $(Z\angle\theta)^n = Z^n\angle(n\theta)$.

Now let's use polar notation to solve a problem and compare this method with rectangular notation:

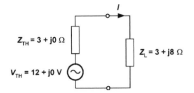

Figure 4.1.62   *See Example 4.1.13*

$Z_{TH} = 3 + j0\ \Omega$

$V_{TH} = 12 + j0\ V$

$Z_L = 3 + j8\ \Omega$

Example 4.1.13

Determine the current flowing in the load, Z_L, in the circuit shown in Figure 4.1.62.
  From Figure 4.1.62:

Thévenin equivalent circuit components are:

$V_{TH} = 12 + j0\ V$ and $Z_{TH} = 3 + j0\ \Omega$ whilst $Z_L = 3 + j8\ \Omega$.

Firstly, using the rectangular form of impedance:

The total impedance present, $Z_T = Z_{TH} + Z_L = 3 + j0 + 3 + j8 = 6 + j8$.
The current flowing can be determined from, $I = V_{TH}/Z_T$

thus $I = \dfrac{V_{TH}}{Z_T} = \dfrac{12 + j0}{6 + j8} = \dfrac{(12 + j0)(6 - j8)}{(6 + j8)(6 - j8)} = \dfrac{72 - j96}{36 + 64} = \dfrac{72 - j96}{100}$

$= 0.72 - j0.96\ A$.

cont.

Secondly, using the polar form of impedance (see Example 4.1.12):

$V_{TH} = 12\angle 0°$ V and $Z_{TH} = 3\angle 0°\ \Omega$ and $Z_L = 8.54\ \angle 69.4°\ \Omega$

and $Z_T = 6 + j8$ or $10\angle 53.1°\ \Omega$.

Once again, $I = \dfrac{V_{TH}}{Z_T} = \dfrac{12\angle 0°}{10\angle 53.1°} = \dfrac{12}{10}\angle (0° - 53.1°) = 1.2\angle -53.1°$ A

(note that $0.72 - j0.960 \equiv 1.2\angle -53.1°$).

Using the network theorems

The examples that follow illustrate how the network theorems can be applied to networks of complex impedances. All we need do is replace resistances by impedances – expressed in either rectangular or polar form, whichever is more convenient – and remember that voltages and currents will also take a complex form (which may also be expressed in either rectangular or polar format).

The first example can be solved quite easily using the rectangular form. The remainder are solved by converting from rectangular to polar form, and vice versa, as required. When deciding which form to use, just recall that, in general, it is easier to:

- Add impedances when they are represented in rectangular form.
- Multiply and divide impedances when they are represented in polar form.

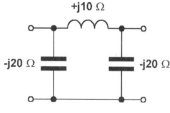

+j10 Ω

-j20 Ω   -j20 Ω

Figure 4.1.63 *See Example 4.1.14*

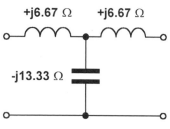

+j6.67 Ω   +j6.67 Ω

-j13.33 Ω

Figure 4.1.64 *T-network equivalent of the π-network shown in figure 4.1.63*

Example 4.1.14

Determine the T-network equivalent of the π-network shown in Figure 4.1.63.

The network components are: $Z_A = Z_B = -j20\ \Omega$ whilst $Z_C = +j10\ \Omega$.

Now $Z_1 = \dfrac{Z_A Z_C}{Z_A + Z_B + Z_C}$

hence $Z_1 = \dfrac{(-j20) \times (j10)}{-j20 - j20 + j10} = \dfrac{-j^2 200}{-j30} = \dfrac{j200}{30} = j6.67\ \Omega$

and since the network is symmetrical, $Z_2 = Z_1 = j6.67\ \Omega$.

Finally, $Z_3 = \dfrac{Z_A Z_B}{Z_A + Z_B + Z_C}$

thus $Z_3 = \dfrac{(-j20) \times (-j20)}{-j20 - j20 + j10} = \dfrac{-j^2 400}{-j30} = \dfrac{j400}{-30} = \dfrac{j400}{30}$

$= -j13.33\ \Omega$.

The T-equivalent of the π-network shown in Figure 4.1.63 is shown in Figure 4.1.64.

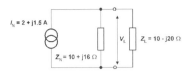

Figure 4.1.65   *See Example 4.1.5*

Example 4.1.15

Determine the voltage developed across the load, Z_L, in the circuit shown in Figure 4.1.65.

From Figure 4.1.62, the Norton equivalent circuit components are:

$I_N = 2 + j1.5$ A and $Z_N = 10 + j16\ \Omega$ whilst $Z_L = 10 - j20\ \Omega$.

Writing these in polar form (see Example 4.1.12) gives:

$I_N = 2.5\angle 36.9°$ A, $Z_N = 18.87\angle 60°\ \Omega$ and $Z_L = 22.36\angle -63.4\ \Omega$.

The total impedance, Z_N in parallel with Z_L, is given by:

$$Z_T = \frac{Z_N \times Z_L}{Z_N + Z_L}. \tag{i}$$

Note that it is worth evaluating $Z_N + Z_L$ in rectangular form and converting the result into polar form before entering values in equation (i).

Now $Z_N + Z_L = 10 + j16 + 10 - j20 = 20 - j4 = 20.4\angle 11.3°$.

$$\text{Thus } Z_T = \frac{Z_N \times Z_L}{Z_N + Z_L} = \frac{18.87\angle 60° \times 22.36\angle -63.4°}{20.4\angle 11.3°} = \frac{18.87 \times 22.36}{20.4}$$

$\angle(60 - 63.4 - 11.3°)$

or $Z_T = 20.68\angle -14.7°\ \Omega$.

Now $V_L = I \times Z_T = 2.5\ \angle 36.9° \times 20.68\angle -14.7° = 2.5 \times 20.68\angle$
$(36.9 - 14.7°)$

or $V_L = 51.7\angle 22.2°$ V.

Example 4.1.16

Determine the Thévenin equivalent of the circuit shown in Figure 4.1.66. Hence determine the current developed in a load impedance of $(50 - j25)\ \Omega$ connected between terminals A and B.

In order to determine the Thévenin equivalent of the circuit we need to find the open-circuit voltage developed between terminals A and B. This is the Thévenin equivalent voltage, V_{TH}.

$$V_{TH} = V \times \frac{Z_P}{Z_P + Z_S} = (100 + j80)\left(\frac{0 - j75}{50 - j75 + j75}\right) = \frac{(100 + j80)(50 - j75)}{50}$$

Thus $V_{TH} = 220 - j70$ V.

Converting this into polar form gives:

$V_{TH} = 230.87\angle -17.65°$ V.

Next we need to find the impedance seen looking into the network between A and B with the voltage source replaced by a short-circuit. This is the Thévenin equivalent impedance, Z_{TH}.

cont.

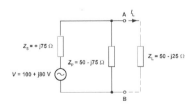

Figure 4.1.66   *See Example 4.1.16*

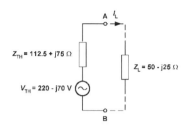

Figure 4.1.67 *Thévenin equivalent of Figure 4.1.66*

$$Z_{TH} = \frac{Z_s \times Z_P}{Z_P + Z_s} = \frac{(j75)(50-j75)}{50-j75+j75} = \frac{(j75)(50-j75)}{50}.$$

Thus $Z_{TH} = 112.5 + j75\ \Omega$.

The Thévenin equivalent circuit is shown in Figure 4.1.67. Now, to find the current flowing when terminals A and B are linked by a load impedance of $(50-j25)\ \Omega$ connected between terminals A and B, we need calculate the total impedance present, Z_T.

Now $Z_T = Z_{TH} + Z_L$

Thus $Z_T = 112.5 + j75 + 50 - j25 = 162.5 + j50\ \Omega$.

Converting the result into polar form gives:

$Z_T = 170.02 \angle 17.1°\ \Omega$.

Finally, $I = \dfrac{V_{TH}}{Z_T} = \dfrac{230.87\angle -17.65°}{170.02\angle 17°} = 1.36\angle -34.7°\,\text{A}$.

Converting this to rectangular form gives:

$I = 1.12 - j0.77\ \text{A}$.

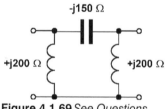

Figure 4.1.68 *See Questions 4.1.2*

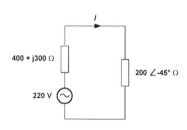

Figure 4.1.69 *See Questions 4.1.2*

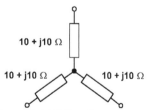

Figure 4.1.70 *See Questions 4.1.2*

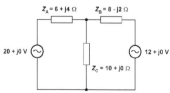

Figure 4.1.71 *See Questions 4.1.2*

Questions 4.1.2

(1) Express the following impedances in polar form:

   (a) $Z = 15 + j30\ \Omega$; (b) $Z = j12\ \Omega$; (c) $Z = 0.3 - j0.4\ \Omega$; (d) $Z = -j1000\ \Omega$.

(2) Express the following impedances in rectangular form:

   (a) $Z = 25\angle 60°\ \Omega$; (b) $Z = 125\angle -45°\ \Omega$; (c) $Z = 650\angle 85°\ \Omega$.

(3) Determine the current flowing in the circuit shown in Figure 4.1.68.

(4) An a.c. source of $(100 + j50)$ V and internal impedance $(20 + j0)\ \Omega$ is connected, in turn, to each of the following loads:

   (a) $Z = 80 + j0$; (b) $Z = 60 + j80$; (c) $Z = 20 - j100$.

   Determine the current that will flow in each case.

(5) Determine the T-network equivalent of the π-network shown in Figure 4.1.69.

(6) Determine the delta network equivalent of the star network shown in Figure 4.1.70.

(7) Use the superposition theorem to determine the current in Z_C in the figure shown in Figure 4.1.71.

Maximum power transfer theorem

The ability to transfer maximum power from a circuit to a load is crucial in a number of electrical and electronic applications. The relationship between the internal resistance of a source, the resistance of the load to which it is connected, and the power dissipated in that load can be easily illustrated by considering a simple example:

> ## Example 4.1.17
>
> A source consists of a 100 V battery having an internal resistance of 100 Ω. What value of load resistance connected to this source will receive maximum power?
>
> The equivalent circuit of this arrangement is shown in Figure 4.1.72.
>
> Now $P = V^2/R$ therefore we might consider what power would appear in resistor R_L for different values of R_L. The power in R_L can be determined from:
>
> $$P_L = V_L^2/R_L = (V \times R_L/(R_L + R_S))^2/R_L = V^2 \times R_L/(R_L + R_S)^2.$$
>
> Now, when $R_L = 10\ \Omega$:
>
> $$P_L = (10/(10 + 100) \times 100)^2/10 = (10/110 \times 100)^2/10 = 8.26\ \text{W}.$$
>
> This process can be repeated for other values of R_L (in, say, the range 0–200 Ω) and a graph can be plotted from which the maximum value can be located, as shown in Figure 4.1.73.
>
> Taking the maximum value for P_L from the graph, we can conclude that the value of R_L in which maximum power would be dissipated is 100 Ω.

From the foregoing, it should be noted that maximum power is dissipated in the load when its resistance is equal to that of the internal resistance of the source, i.e. when $R_L = R_S$. Note also that, in the extreme cases (i.e. when $R_L = 0$ and also when $R_L = \infty$) the power in the load, $P_L = 0$.

A more powerful approach would be to apply differential calculus to prove that the maximum value for P_L occurs when $R_L = R_S$. The method is as follows:

$$P_L = V^2 \times R_L/(R_L + R_S)^2 = \frac{V^2 R_L}{(R_L + R_S^2)}.$$

Applying the quotient rule gives:

$$\frac{dP_L}{dR_L} = \frac{d}{dR_L}\left(\frac{V^2 R_L}{(R_L + R_S^2)}\right) = \frac{V^2[(R_L + R_S)^2 - 2R_L(R_L + R_S)]}{(R_L + R_S)^4}.$$

Figures 4.1.72 *(right)* **and 4.1.73** *(far right) See Example 4.1.17*

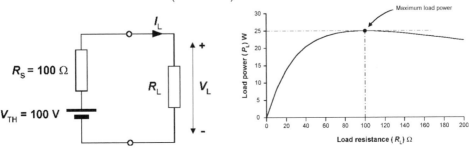

For maximum power in R_L the numerator must be zero. Hence:

$$(R_L + R_S)^2 - 2R_L(R_L + R_S) = 0$$

thus $(R_L + R_S)^2 = 2R_L(R_L + R_S)$

or $R_L + R_S = 2R_L$

or $R_L = R_S$.

The maximum power transfer theorem states that, in the case of d.c. circuits:

> *Maximum power will be dissipated in a load when the resistance of the load is equal to the internal resistance of the source (i.e. when $R_L = R_S$).*

In the case of a.c. circuits in which the source and load impedances are not purely resistive:

> *Maximum power will be dissipated in a load when the impedance of the load is the conjugate of the internal impedance of the source (i.e. when $Z_L = R + jX$ and $Z_S = R - jX$, and vice versa).*

Note that, in the case of an a.c. circuit in which the load impedance is purely resistive, maximum power will be dissipated in a load when $R_L = \sqrt{(R_S^2 + X_S^2)}$. Furthermore, when both the source and load impedances are purely resistive, maximum power will be dissipated in a load when $R_L = R_S$.

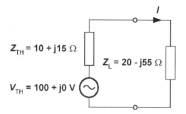

$Z_{TH} = 10 + j15\ \Omega$

$Z_L = 20 - j55\ \Omega$

$V_{TH} = 100 + j0\ V$

Figure 4.1.74 *See Example 4.1.18*

Example 4.1.18

Determine the power delivered to the load in the circuit shown in Figure 4.1.7.4. Also find the impedance that will receive maximum power.

The total impedance in the circuit, Z_T, will be given by:

$$Z_T = Z_{TH} + Z_L = 10 + j15 + 20 - j55 = 30 - j40\ \Omega$$

and the resistive part of this impedance, $R = 30\ \Omega$.

The current flowing will be given by:

$$I = \frac{V_{TH}}{Z_T} = \frac{100 + j0}{30 - j40} = 2\angle 53°\ A.$$

The power dissipated in the load will be given by $P_L = I^2R$.

Thus $P_L = 2^2 \times 30 = 120\ W$.

The impedance that will receive maximum power will simply be the conjugate of the impedance of the source. Since $Z_{TH} = 10 + j15\ \Omega$ the value of Z_L for maximum power transfer will be $Z_L = 10 - j15\ \Omega$. In this condition, the power dissipated in the load will be:

$$P_L = 5^2 \times 10 = 250\ W.$$

When maximum power is transferred between a source and the load to which it is connected, they are said to be *matched*. The process of *matching* a source to a load is important in applications such as measurement, instrumentation, telecommunications, and data transmission.

Matching can often be achieved by adding resistance or reactance to a circuit in order to satisfy the criteria of the maximum power transfer theorem. The following example shows how this is done:

Example 4.1.19

A transmitter having an output impedance of $(40 - j15)\ \Omega$ is to be matched to an aerial system having an impedance of $(50 + j35)\ \Omega$. Determine the impedance required in a matching network that will be connected between the transmitter and the aerial system.

In order to obtain a perfect match between the aerial system and the transmitter, the optimum value of Z_s would be $(50 - j35)\ \Omega$. Hence the impedance of the series-connected matching network should comprise of a resistance of $10\ \Omega$ together with a reactance of $-j20\ \Omega$.

The new source impedance, Z_s' then becomes:

$$Z_s' = (40 - j15) + (10 - j25) = 50 - j35\ \Omega$$

(note that Z_s' is now equal to the conjugate of the impedance of the aerial system).

Figure 4.1.75 illustrates this arrangement.

Another view

It is worth noting that, when maximum power is transferred from a source to a load, the efficiency is only 50% since equal amounts of power are dissipated internally (i.e. in R_s) and externally (i.e. in R_L). In many practical applications this condition may not be desirable and we often take steps to ensure that the impedance of a source (for example, a mains voltage outlet) is *very much less* than that of the load to which it is connected (for example, an electric heater).

Unfortunately, the problem with introducing additional impedances (sometimes referred to as *pads*) into a circuit in order to achieve a matched condition is that power will be lost in any resistive matching components. In some applications, this loss of power can be tolerated however, in many applications power loss can be undesirable and an alternative method of matching is required.

Transformer matching

Later, in this chapter we will explain the principle of the transformer but, for the moment, we will simply assume that a transformer is 'ideal' and can provide us with a virtually loss-free method of coupling an alternating current from one circuit to another. Transformers also provide us with a means of matching a source to a load. This is because an impedance connected to the secondary of a transformer becomes 'reflected' into its primary circuit. Consider the circuit shown in Figure 4.1.76.

The impedance seen 'looking into' the transformer's primary winding, Z_p, will be given by:

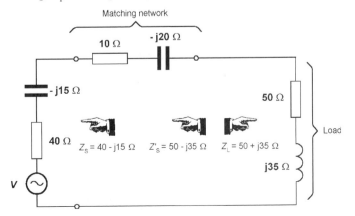

Figure 4.1.75 *See Example 4.1.19*

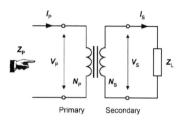

Figure 4.1.76 *Transformer matching*

$$Z_P = V_P/I_P \tag{i}$$

whereas the load impedance connected to the secondary winding of the transformer, Z_L, will be given by:

$$Z_L = V_S/I_S. \tag{ii}$$

Now the voltages, currents and 'turns ratio', N_P/N_S, are connected by the relationship:

$$\frac{V_P}{V_S} = \frac{I_S}{I_P} = \frac{N_P}{N_S} \tag{iii}$$

thus $V_P = \dfrac{N_P}{N_S} \times V_S.$

From (i) and (iii), $Z = \dfrac{N_P}{N_S} \times \dfrac{V_S}{I_P}.$

But $V_S = I_S \times Z_L$ from (ii)

thus $Z = \dfrac{N_P}{N_S} \times \dfrac{I_S Z_L}{I_P}.$

Now $\dfrac{I_S}{I_P} = \dfrac{N_P}{N_S}$ from (iii).

Thus $Z = \dfrac{N_P}{N_S} \times \dfrac{N_P}{N_S} \times Z_L = \left(\dfrac{N_P}{N_S}\right)^2 \times Z_L$

hence $Z = \left(\dfrac{N_P}{N_S}\right)^2 \times Z_L$ or $Z_L = \left(\dfrac{N_S}{N_P}\right)^2 Z.$

Example 4.1.20

A transformer having a turns ratio of 20:1 has its secondary connected to a load having an impedance of $(6 + j8)$ Ω. Determine the impedance seen looking into the primary winding.

Now $Z = \left(\dfrac{N_P}{N_S}\right)^2 \times Z_L$

where $N_P/N_S = 20$ and $Z_L = 6 + j8$ Ω

thus $Z = (20)^2(6 + j8) = 2400 + j3200 = 4000\angle 53°$ Ω.

Example 4.1.21

A transformer is required to match a load having an impedance of $(600 + j150)$ Ω to a source having an impedance of $(90 + j10)$ Ω. Determine the required turns ratio.

Now $Z = \left(\dfrac{N_P}{N_S}\right)^2 Z_L$ thus $\dfrac{N_P}{N_S} = \sqrt{\dfrac{Z}{Z_L}} = \sqrt{\dfrac{600 + j150}{90 + j10}} = \sqrt{\dfrac{618.5}{90.6}}$

$= 6.83.$

The required turns ratio is 6.83:1.

Questions 4.1.3

(1) A transformer is required to match a loudspeaker of impedance $(4 + j0)$ Ω to the output stage of an amplifier which has an impedance of 2 kΩ. Determine the required turns ratio of the transformer and the primary voltage required to produce an output power of 10 W in the loudspeaker.

(2) An aerial matching transformer having 20 primary turns and 4 secondary turns is connected between a transmitter and an aerial. If the impedance seen looking into the primary of the transformer is $(200 - j50)$ Ω determine the impedance of the aerial.

Mesh current analysis

In mesh current analysis, we identify a number of closed loops (or *meshes*) in which currents flow in the circuit and then use these to develop a set of equations based on Kirchhoff's voltage law. Consider the circuit with two voltage sources, V_1 and V_2, shown in Figure 4.1.77.

In mesh A: $V_1 = I_1(Z_1 + Z_2) - I_2 Z_2$.

In mesh B: $0 = I_2(Z_1 + Z_2 + Z_3) - I_1 Z_2 - I_3 Z_4$.

In mesh C: $V_2 = -I_3(Z_4 + Z_5) + I_2 Z_4$

where I_1, I_2 and I_3 are the *mesh currents*. Note that the true current in Z_1 is I_1, in Z_2 it is $(I_1 - I_2)$, in Z_3 it is I_2, in Z_4 it is $(I_2 - I_3)$, and in Z_5 it is I_3.

The use of mesh current analysis is best illustrated by an example:

Example 4.1.22

Determine the current in each component in the circuit of Figure 4.1.77 if the components have the following values:

$V_1 = 20 + j0$, $V_2 = 10 + j0$, $Z_1 = 10 + j0$, $Z_2 = 20 + j0$, $Z_3 = 5 + j10$, $Z_4 = 10 + j0$, and $Z_5 = 5 + j0$.

Forming the mesh current equations for loops A, B and C gives:

In mesh A: $V_1 = I_1(Z_1 + Z_2) - I_2 Z_2$
$\qquad 20 + j0 = I_1(10 + j0 + 20 + j0) - I_2(20 + j0)$
$\qquad 20 = 30I_1 - 20I_2.$ \hfill (i)

In mesh B: $0 = I_2(Z_1 + Z_2 + Z_3) - I_1 Z_2 - I_3 Z_4$
$\qquad 0 = I_2(10 + j0 + 20 + j0 + 5 + j10) - I_1(20 + j0) - I_3(10 + j0)$
$\qquad 0 = I_2(35 + j10) - I_1 20 - I_3 10$ \hfill (ii)

cont.

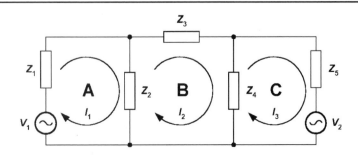

Figure 4.1.77 *Mesh current analysis*

In mesh C: $V_2 = -I_3(Z_4 + Z_5) + I_2Z_4$
$10 + j0 = -I_3(10 + j0 + 5 + j0) + I_2(10 + j0)$
$10 = -I_315 + I_210.$ (iii)

From (i) $I_1 = (20 + 20I_2)/30 = 0.667 + 0.667I_2.$

From (iii) $I_3 = 0.667I_2 - 0.667.$

Substituting for I_1 and I_3 in (ii) gives:

$0 = I_2(35 + j10) - 20 \times (0.667 + 0.667I_2) - 10 \times (0.667I_2 - 0.667)$

$0 = (35 + j10)I_2 - 13.34 - 13.33I_2 - 6.67I_2 + 6.67$

$0 = (35 + j10)I_2 - 13.34 - 13.33I_2 - 6.67I_2 + 6.67$

$13.33 - 6.67 = (35 - 13.33 - 6.67 + j10)I_2$

thus $6.67 = (15 + j10) I_2$

thus $I_2 = 6.67/(15 + j10) = 5.55 - j12.02.$

From (i) $I_1 = 0.667 + 0.667 (5.55 - j12.02)$

thus $I_1 = 0.667 + 3.7 - j8.02 = 4.37 - j8.02.$

From (iii) $I_3 = 0.667I_2 - 0.667$

thus $I_3 = 0.667(5.55 - j12.02) - 0.667 = 3.7 - j8.02 - 0.667 = 3.03 - j8.02.$

Current in Z_1 is $I_1 = 4.37 - j8.02$ A.

Current in Z_2 is $(I_1 - I_2) = 4.37 - j8.02 - 5.55 + j12.02 = 11.18 + j4$ A.

Current in Z_3 is $I_2 = 5.55 - j12.02$ A.

Current in Z_4 is $(I_2 - I_3) = 5.55 - j12.02 - 3.03 + j8.02 = 2.52 - j4$ A.

Current in Z_5 is $I_3 = 3.03 - j8.02$ A.

The simultaneous equations in mesh current analysis can often be most easily solved by using matrices, as the following example shows:

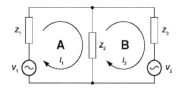

Figure 4.1.78 *See Example 4.1.23*

Example 4.1.23

Determine the current in each component, and the voltage dropped across Z_2, in the circuit shown in Figure 4.1.78.   The components have the following values:

$V_1 = 10 + j0, V_2 = 4 + j0, Z_1 = 6 + j0, Z_2 = 2 + j6, Z_3 = 2 + j0.$

Forming the mesh current equations for loops A and B gives:

In mesh A: $V_1 = I_1 Z_1 + (I_1 - I_2) Z_2$
$V_1 = I_1 Z_1 + I_1 Z_2 - I_2 Z_2$
or $V_1 = I_1 (Z_1 + Z_2) - I_2 Z_2.$ (i)

In mesh B: $V_2 = -I_2 Z_3 + (I_1 - I_2) Z_2$
$V_2 = -I_2 Z_3 + I_1 Z_2 - I_2 Z_2$
or $V_2 = I_1 Z_2 - I_2 (Z_2 + Z_3).$ (ii)

From (i) $10 + j0 = I_1(6 + j0 + 2 + j6) - I_2(2 + j6)$
$10 = I_1(8 + j6) - I_2(2 + j6).$ (iii)

cont.

From (ii) $4 = I_1(2 + j6) - I_2(2 + j6 + 2 + j0)$
$4 = I_1(2 + j6) - I_2(4 + j6).$ (iv)

From (iii) $10 = (8 + j6) I_1 - (2 + j6) I_2.$

From (iv) $4 = (2 + j6) I_1 - (4 + j6) I_2.$

Writing these simultaneous equations in matrix form gives:

$$\begin{bmatrix} (8 + j6) - (2 + j6) \\ (2 + j6) - (4 + j6) \end{bmatrix} \begin{bmatrix} I_1 \\ I_2 \end{bmatrix} = \begin{bmatrix} 10 \\ 4 \end{bmatrix}.$$

Hence:

$$\frac{I_1}{\begin{vmatrix} 10 & -(2 + j6) \\ 4 & -(4 + j6) \end{vmatrix}} = \frac{I_2}{\begin{vmatrix} (8 + j6) & 10 \\ (2 + j6) & 4 \end{vmatrix}} = \frac{1}{\begin{vmatrix} (8 + j6) & -(2 + j6) \\ (2 + j6) & -(4 + j6) \end{vmatrix}}$$

$$\frac{I_1}{-10(4 + j6) + 4(2 + j6)} = \frac{I_2}{4(8 + j6) - 10(2 + j6)}$$

$$= \frac{1}{-(4 + j6)(8 + j6) + (2 + j6)^2}$$

$$\frac{I_1}{-32 - j36} = \frac{I_2}{12 - j36} = \frac{I}{-28 - j48}.$$

Since:

$$\frac{I_1}{-32 - j36} = \frac{1}{-28 - j48}$$

$$I_1 = \frac{-32 - j36}{-28 - j48} = \frac{32 + j36}{28 + j48} = 0.85 - j0.17 = 0.87\angle - 11° \text{ A}.$$

Similarly:

$$\frac{I_2}{12 - j36} = \frac{1}{-28 - j48}$$

$$I_2 = \frac{12 - j36}{-28 - j48} = -\left(\frac{12 - j36}{28 + j48}\right) = -(-0.45 - j0.51) = 0.45 + j0.51$$

$$= 0.68 \angle 48.6° \text{ A}.$$

The current in Z_2 is $I_1 - I_2 = (0.85 - j0.17) - (0.45 + j0.51)$

or $I_1 - I_2 = 0.4 - j0.68 = 0.79 \angle -59.5° \text{ A}.$

The voltage dropped across Z_2 will be given by:

$(I_1 - I_2) \times Z_2 = (0.79\angle - 59.5°) \times (6.32\angle 71.6°) = 4.99\angle 12.1° = 4.88 + j1.05 \text{ V}.$

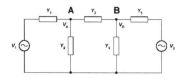

Figure 4.1.79 *Node voltage analysis*

Node voltage analysis

In node voltage analysis, we identify a number of junctions (or *nodes*) at which currents divide and then use these to develop a set of equations based on Kirchhoff's current law. Consider the circuit with two voltage sources, V_1 and V_2, shown in Figure 4.1.79.

At node A: $0 = (V_A - V_1)Y_1 + V_A Y_2 + (V_A - V_B)Y_3.$

At node B: $0 = (V_B - V_A)Y_3 + V_B Y_4 + (V_B - V_2) Y_5.$

The use of node voltage analysis is best illustrated by an example:

Example 4.1.24

Determine the voltage at each node in Figure 4.1.79 if the components have the following values:

$V_1 = 20 + j0, V_2 = 4 + j0, Z_1 = 10 + j0, Z_2 = 2 + j6, Z_3 = 2 - j2,$
$Z_4 = 4 + j3,$ and $Z_5 = 2 + j0.$

Applying Kirchhoff's current law at nodes A and B gives:

$$0 = (V_A - V_1)Y_1 + V_A Y_2 + (V_A - V_B)Y_3 \qquad\qquad \text{(i)}$$

and

$$0 = (V_B - V_A)Y_3 + V_B Y_4 + (V_B - V_2) Y_5. \qquad\qquad \text{(ii)}$$

Now:

$$Y_1 = \frac{1}{Z_1} = \frac{1}{10 + j0} = 0.1$$

$$Y_2 = \frac{1}{Z_2} = \frac{1}{2 + j6} = 0.05 - j0.15$$

$$Y_3 = \frac{1}{Z_3} = \frac{1}{2 - j2} = 0.25 + j0.25$$

$$Y_4 = \frac{1}{Z_4} = \frac{1}{4 + j3} = 0.16 - j0.12$$

$$Y_5 = \frac{1}{Z_5} = \frac{1}{2 + j0} = 0.5.$$

From (i):

$0 = 0.1(V_A - (20 + j0)) + V_A(0.05 - j0.15) + (V_A - V_B)(0.25 + j0.25)$

$0 = 0.1V_A - 2 + j0.05V_A - j0.15V_A + 0.25V_A + j0.25V_A - 0.25V_B - j0.25 V_B$

$$0 = 0.4V_A - j0.1V_A - 0.25V_B - j0.25V_B - 2. \qquad\qquad \text{(iii)}$$

From (ii):

$0 = (V_B - V_A)(0.25 + j0.25) + V_B(0.16 - j0.12) + 0.5 (V_B - (4 + j0))$

$0 = 0.25V_B + j0.25V_B - 0.25V_A - j0.25V_A + 0.16V_B - j0.12V_B + 0.5V_B - 2$

cont.

$$0 = -0.25V_A - j0.25V_A + 0.91V_B + j0.13V_B - 2. \qquad \text{(iv)}$$

From (iii):

$$2 = (0.4 - j0.1)V_A - (0.25V_B + j0.25V_B).$$

From (iv)

$$2 = -(0.25 + j0.25)V_A + (0.91 + j0.13)V_B.$$

Writing these two simultaneous equations in matrix form gives:

$$\begin{bmatrix} (0.4 - j0.1) & -(0.25 + j0.25) \\ -(0.25 + j0.25) & (0.91 + j0.13) \end{bmatrix} \begin{bmatrix} V_A \\ V_B \end{bmatrix} = \begin{bmatrix} 2 \\ 2 \end{bmatrix}$$

Hence:

$$\frac{V_A}{\begin{vmatrix} 2 & -(0.25 + j0.25) \\ 2 & (0.91 + j0.13) \end{vmatrix}} = \frac{V_B}{\begin{vmatrix} (0.4 - j0.1) & 2 \\ -(0.25 + j0.25) & 2 \end{vmatrix}} = \frac{1}{\begin{vmatrix} (0.4 - j0.1) & -(0.25 + j0.25) \\ -(0.25 + j0.25) & (0.91 + j0.13) \end{vmatrix}}$$

$$\frac{V_A}{2(0.91 + j0.13) + 2(0.25 + j0.25))} = \frac{V_B}{2(0.4 - j0.1) + 2(0.25 + j0.25)}$$

$$= \frac{1}{(0.4 - j0.1)(0.91 + j0.13) - (0.25 + j0.25)^2}$$

$$\frac{V_A}{2.32 + j0.76} = \frac{V_B}{1.3 + j0.3} = \frac{1}{0.377 - j0.747}$$

$$\frac{V_A}{2.32 + j0.76} = \frac{1}{0.377 - j0.747}$$

Thus $V_A = \dfrac{2.32 + j0.76}{0.377 - j0.747} = 0.438 + j2.88 = 2.92\underline{/81.36°}$

$$\frac{V_B}{1.3 + j0.3} = \frac{1}{0.377 - j0.747}.$$

Thus $V_B = \dfrac{1.3 + j0.3}{0.377 - j0.747} = 0.38 + j1.55 = 1.59\underline{/76.2°}.$

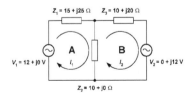

$Z_1 = 15 + j25\ \Omega$ $Z_3 = 10 + j20\ \Omega$

A   B

$V_1 = 12 + j0\ V$ I_1 I_2 $V_2 = 0 + j12\ V$

$Z_2 = 10 + j0\ \Omega$

Figure 4.1.80   *See Questions 4.1.4*

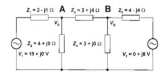

$Z_1 = 2 - j1\ \Omega$ A $Z_3 = 3 + j4\ \Omega$ B $Z_5 = 4 - j4\ \Omega$

V_A V_B

$Z_2 = 4 + j0\ \Omega$ $Z_4 = 3 + j0\ \Omega$

$V_1 = 10 + j0\ V$ $V_2 = 0 + j8\ V$

Figure 4.1.81   *See Questions 4.1.4*

Questions 4.1.4

(1) Determine the currents flowing in mesh A and B in Figure 4.1.80.

(2) Determine the voltage that appears at nodes A and B in Figure 4.1.81.

Coupled magnetic circuits

Any circuit in which a change of magnetic flux is produced by a change of current is said to possess *self-inductance*. In turn, the change of flux will produce an e.m.f. across the terminals of the circuit. We refer to this as an *induced e.m.f.*

When the flux produced by one coil links with another we say that the two coils exhibit *mutual inductance*, *M*. The more flux that links the two coils the greater the amount of mutual inductance that exists between them. This important principle underpins the theory of the transformer.

Coefficient of coupling

Let's assume that two coils, having self inductances, L_1 and L_2, are placed close together. The fraction of the total flux from one coil that links with the other coil is called the *coefficient of coupling, k*.

If all of the flux produced by L_1, links with, L_2, then the magnetic coupling between them is perfect and we can say that the *coefficient of coupling, k*, is unity (i.e. $k = 1$). Alternatively, if only half the flux produced by L_1 links with L_2, then $k = 0.5$.

Thus $k = \Phi_2/\Phi_1$

where Φ_1 is the flux in the first coil (L_1) and Φ_2 is the flux in the second coil (L_2).

Example 4.1.25

A coil, L_1, produces a magnetic flux of 80 mWb. If 20 mWb appears in a second coil, L_2, determine the coefficient of coupling, *k*.
   Now $k = \Phi_2/\Phi_1$
where $\Phi_1 = 80$ mWb and $\Phi_2 = 20$ mWb.
Thus $k = 20/80 = 0.25$.

Mutual inductance

In practice, inductive circuits are *coils*. By placing the coils close together or by winding them on the same closed magnetic core it is possible to obtain high values of mutual inductance. In this condition, we say that the two coils are *close coupled* or *tightly coupled*. Where the two coils are not in close proximity or are wound on separate magnetic cores, values of mutual inductance will be quite small. Under these circumstances we say that the two coils are *loose coupled*.

We said earlier that, when two inductive circuits are coupled together by magnetic flux, they are said to possess *mutual inductance*. The unit of mutual inductance is the same as that for self-inductance and two circuits are said to have a mutual inductance of one Henry (H) if an e.m.f. of one volt is produced in one circuit when the current in the other circuit varies at a uniform rate of one ampere per second. Hence, if two circuits have a mutual inductance, *M*, and the current in one circuit (the *primary*) varies at a rate di/dt (A/s) the e.m.f. induced in the second circuit (the *secondary*) will be given by:

$$e = -M \times (\text{rate of change of current}) \text{ or: } e = -M \times \frac{di}{dt} \text{ V.} \qquad (i)$$

Note the minus sign which indicates that the e.m.f. in the second circuit opposes the increase in flux produced by the first circuit.
The e.m.f. induced in an inductor is given by:

$$e = -N \times (\text{rate of change of flux}) \text{ or: } e = -N \times \frac{d\Phi}{dt} \text{ V}$$

where *N* is the number of turns and Φ is the flux (in Wb).

Hence the e.m.f. induced in the secondary coil of two magnetically coupled inductors will be given by:

$$e = -N_2 \times \frac{d\Phi}{dt} \quad V(ii)$$

where N_2 is the number of secondary turns.
Combining equations (i) and (ii) gives:

$$-M \times \frac{di}{dt} = -N_2 \times \frac{d\Phi}{dt}.$$

Thus: $M = N_2 \times \dfrac{d\Phi}{dt} \times \dfrac{dt}{di} = N_2 \times \dfrac{d\Phi}{di}.$

If the relative permeability of the magnetic circuit remains constant, the rate of change of flux with current will also be a constant. Thus we can conclude that:

$$M = N_2 \times \frac{d\Phi}{di} = N_2 \times \frac{\Phi_2}{I_1} \tag{iii}$$

where Φ_2 is the flux linked with the secondary circuit and I_1 is the primary current.

Expression (iii) provides us with an alternative definition of mutual inductance. In this case, M is expressed in terms of the secondary flux and the primary current.

Example 4.1.26

Two coils have a mutual inductance of 500 mH. Determine the e.m.f. produced in one coil when the current in the other coil is increased at a uniform rate from 0.2 A to 0.8 A in a time of 10 ms.

Now $e = -M \times \dfrac{di}{dt}.$

Since the current changes from 0.2 A to 0.8 A in 10 ms the rate of change of current with time, $\dfrac{di}{dt}$, is $\dfrac{(0.8 - 0.2)}{0.01} = 600$ A/s.

Thus $e = -0.5 \times 600 = -300$ V.

Example 4.1.27

An e.m.f. of 40 mV is induced in a coil when the current in a second coil is varied at a rate of 3.2 A/s. Determine the mutual inductance of the two coils.

Now $e = -M \times \dfrac{di}{dt}$ thus $M = -\dfrac{e}{\left(\dfrac{di}{dt}\right)} = \dfrac{0.04}{3.2} = 12.5$ mH.

Coupled circuits

Consider the case of two coils that are coupled together so that all of the flux produced by one coil links with the other. These two coils are, in effect, perfectly coupled. Since the product of inductance (L) and current (I) is equal to the product of the number of turns (N) and the flux (Φ), we can deduce that:

$$L_1 I_1 = N_1 \Phi \ \text{ or } L_1 = N_1 \Phi / I_1 \qquad \text{(i)}$$

similarly:

$$L_2 I_2 = N_2 \Phi \ \text{ or } L_2 = N_2 \Phi / I_2 \qquad \text{(ii)}$$

where L_1 and L_2 represent the inductance of the first and second coils, respectively, N_1 and N_2 represent the turns of the first and second coils, respectively, and ϕ is the flux shared by the two coils.

Since the product of reluctance (S) and flux (Φ) is equal to the product of the number of turns (N) and the current (I), we can deduce that:

$$S\Phi = N_1 I_1 = N_2 I_2 \ \text{ or } \Phi = N_1 I_1 / S = N_2 I_2 / S.$$

Hence, from (i): $L_1 = N_1 \Phi / I_1 = N_1 (N_1 I_1) / I_1 S = N_1^2 / S$

and from (ii): $L_2 = N_2 \Phi / I_2 = N_2 (N_2 I_2) / I_2 S = N_2^2 / S.$

Now $L_1 \times L_2 = (N_1^2 / S) \times (N_2^2 / S) = N_1^2 \, N_2^2 / S^2.$ $\qquad$ (iii)

Now $M = N_2 \times \dfrac{\Phi}{I_1}$ (see Equation (iii) on page 247).

Multiplying top and bottom by N_1 gives:

$$M = N_1 N_2 \Phi / (N_1 I_1) = N_1 N_2 / S. \qquad \text{(iv)}$$

Combining equations (iii) and (iv) gives:

$$M = \sqrt{(L_1 L_2)}.$$

In deriving the foregoing expression we have assumed that the flux is perfectly coupled between the two coils. In practice, there will always be some leakage of flux. Furthermore, there may be cases when we do not wish to couple two inductive circuits tightly together. The general equation is:

$$M = k\sqrt{(L_1 L_2)}$$

where k is the coefficient of coupling (note that k is always less than 1).

Example 4.1.28

Two coils have self-inductances of 10 mH and 40 mH. If the two coils exhibit a mutual inductance of 5 mH, determine the coefficient of coupling.

Now $M = k\sqrt{(L_1 L_2)}$ thus $k = \dfrac{M}{\sqrt{(L_1 L_2)}} = \dfrac{5}{\sqrt{400}} = 0.25.$

Series connection of coupled coils

There are two ways of series connecting inductively coupled coils, either in *series aiding* (so that the mutual inductance adds to the combined self-inductance of the two coils) or in *series opposition* (in which case the mutual inductance subtracts from the combined self-inductance of the two coils).

In the former, *series aiding* case, the effective inductance is given by:

$$L = L_1 + L_2 + 2M$$

whilst in the latter, *series opposing* case, the effective inductance is given by:

$$L = L_1 + L_2 - 2M$$

where L_1 and L_2 represent the inductance of the two coils and M is their mutual inductance.

Example 4.1.29

When two coils are connected in series, their effective inductance is found to be 1.2 H. When the connections to one of the two coils are reversed, the effective inductance is 0.8 H. If the coefficient of coupling is 0.5, find the inductance of each coil and also determine the mutual inductance.

In the first, series aiding case:

$$L = L_1 + L_2 + 2M \text{ or } 1.2 = L_1 + L_2 + 2M.$$

In the second, series opposing case:

$$L = L_1 + L_2 - 2M \text{ or } 0.8 = L_1 + L_2 - 2M.$$

Now $M = k\sqrt{(L_1 L_2)}$ thus $M = 0.5\sqrt{(L_1 L_2)}$.

Hence:

$$1.2 = L_1 + L_2 + 2\left(0.5\sqrt{(L_1 L_2)}\right) = L_1 + L_2 + \sqrt{(L_1 L_2)} \qquad \text{(i)}$$

and

$$0.8 = L_1 + L_2 - 2\left(0.5\sqrt{(L_1 L_2)}\right) = L_1 + L_2 - \sqrt{(L_1 L_2)}. \qquad \text{(ii)}$$

Adding (i) and (ii) gives:

$$1.2 + 0.8 = L_1 + L_2 + \sqrt{(L_1 L_2)} + L_1 + L_2 - \sqrt{(L_1 L_2)}.$$

Hence:

$$2 = 2\,(L_1 + L_2) \text{ or } 1 = L_1 + L_2 \text{ thus } L_1 = 1 - L_2.$$

Substituting for L_1 in equation (i) gives:

$$1.2 = (1 - L_2) + L_2 + \sqrt{((1 - L_2) \times L_2)}$$

$$1.2 = 1 + \sqrt{(L_2 - L_2^2)}$$

$$0.2 = \sqrt{(L_2 - L_2^2)}$$

$$0.04 = L_2 - L_2^2$$

$$0 = -0.04 + L_2 - L_2^2$$

$$0 = 0.04 - L_2 + L_2^2.$$

Solving this quadratic equation gives:

$$L_2 = 0.959 \text{ H or } 0.041 \text{ H}$$

thus

$$L_1 = 0.041 \text{ H or } 0.959 \text{ H}.$$

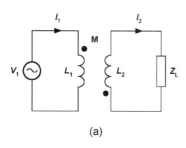

(a)

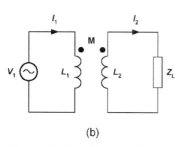

(b)

Figure 4.1.82 *Dot notation*

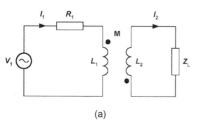

(a)

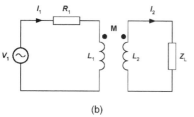

(b)

Figure 4.1.83 *Effect of primary winding resistance*

Dot notation

When two coupled circuits are drawn on a circuit diagram it is impossible to tell from the drawing the direction of the voltage induced in the second circuit (called the *secondary*) when a changing current is applied to the first circuit (known as the *primary*). When the direction of this induced voltage is important, we can mark the circuit with dots to indicate the direction of the currents and induced voltages, as shown in Figure 4.1.82.

In Figure 4.1.82(a), the total magneto-motive force produced will be the sum of the magneto-motive force produced by each individual coil. In this condition, the mutual inductance, M, will be *positive*.

In Figure 4.1.82(b), the total magneto-motive force produced will be the difference of the magneto-motive force produced by each individual coil. In this condition, the mutual inductance, M, will be *negative*.

Assuming that the primary coil in Figure 4.1.82(a) has negligible resistance, the primary voltage, V_1, will be given by:

$$V_1 = (I_1 \times j\omega L_1) + (I_2 \times j\omega M).$$

Conversely, in Figure 4.1.82(b), the primary voltage, V_1, will be given by:

$$V_1 = (I_1 \times j\omega L_1) - (I_2 \times j\omega M).$$

where I_1 and I_2 are the primary and secondary currents, respectively, L_1 and L_2 are the primary and secondary inductances, respectively, ω is the angular velocity of the current, and M is the mutual inductance of the two circuits.

If the primary coil has resistance, as shown in Figure 4.1.83(a), the primary voltage, V_1, will be given by:

$$V_1 = I_1 (R_1 + j\omega L_1) + (I_2 \times j\omega M).$$

Conversely, in Figure 4.1.83(b), the primary voltage, V_1, will be given by:

$$V_1 = I_1 (R_1 + j\omega L_1) - (I_2 \times j\omega M).$$

Transformers

E.m.f. equation for a transformer

If a sine wave current is applied to the primary winding of a transformer, the flux will change from its negative maximum value, $-\Phi_M$, to its positive maximum value, $+\Phi_M$, in a time equal to half the periodic time of the current, $t/2$. The total change in flux over this time will be $2\Phi_M$.

The average rate of change of flux in a transformer is thus given by:

(total flux change)/(time) $= 2\Phi_M/(t/2) = 4\Phi_M/t.$

Now $f = 1/t$ thus:

Average rate of change of flux $= 4f\Phi_M.$

The average e.m.f. induced per turn will be given by:

$E/N = 4f\Phi_M$

Hence for a primary winding with N_1 turns, the average induced primary voltage will be given by:

$E_1 = 4N_1 f \Phi_M$

Similarly, for a secondary winding with N_2 turns, the average induced secondary voltage will be given by:

$E_2 = 4N_2 f \Phi_M$

(Note that we have assumed that there is perfect flux linkage between the primary and secondary windings, i.e. $k = 1$).

For a sinusoidal voltage, the effective or r.m.s. value is 1.11 times the average value. Thus:

$$E_1 = 4.44\, N_1 f \Phi_M \tag{i}$$

and

$$E_2 = 4.44\, N_2 f \Phi_M. \tag{ii}$$

From equations (i) and (ii):

$E_1/E_2 = N_1/N_2.$

Assuming that the transformer is 'loss free', the power delivered to the secondary circuit will be the same as that in the primary circuit. Thus:

$P_1 = P_2$ or $E_1 \times I_1 = E_2 \times I_2.$

Thus:

$E_1/E_2 = I_2/I_1 = N_1/N_2.$

Example 4.1.30

A transformer operates with a 220 V 50 Hz supply and has 800 primary turns. Determine the maximum value of flux present.

Now $E_1 = 220$ V, $N_1 = 800$ and $f = 50$ Hz.

Since $E_1 = 4.44\, N_1 f \Phi_M$ V

$\Phi_M = E_1/4.44 N_1 f = 220/(4.44 \times 800 \times 50) = 220/177\,600$

$= 1.24$ mWb.

Example 4.1.31

A transformer has 455 primary turns and 66 secondary turns. If the primary winding is connected to a 110 V a.c. supply and the secondary is connected to an 8 Ω load, determine:

(a) the current that will flow in the secondary;
(b) the supply current.

You may assume that the transformer is 'loss-free'.

Now $E_1 = 110$V, $N_1 = 480$ and $N_2 = 60$

$E_1/E_2 = I_2/I_1 = N_1/N_2$ thus $E_2 = E_1 \times (N_2/N_1) = 110 \times 66/455 = 16$ V.

The current delivered to an 8 Ω load will thus be:

cont.

$$I_2 = E_2/12 = 16/8 = 2 \text{ A.}$$

$$I_1 = I_2 \times (N_2/N_1) = 2 \times 66/455 = 0.29 \text{ A.}$$

Equivalent circuit of a transformer

An *ideal transformer* would have coil windings that have negligible resistance and a magnetic core that is perfect. Furthermore, all the flux produced by the primary winding would be coupled into the secondary winding and no flux would be lost in the space surrounding the transformer.

In practice, a real transformer suffers from a number of imperfections. These are summarised below:

Leakage flux

Not all of the magnetic flux produced by the primary winding of a transformer is coupled into its secondary winding. This is because some flux is lost into the space surrounding the transformer (even though this has a very much higher reluctance than that of the magnetic core).

The leakage flux increases with the primary and secondary current and its effect is the same as that produced by an inductive reactance connected in series with the primary and secondary windings. We can take this into account by including this additional inductance to the simplified equivalent circuit of a transformer shown in Figure 4.1.84.

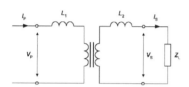

Figure 4.1.84 *Transformer equivalent circuit showing the effect of leakage flux*

Winding resistance

The windings of a real transformer exhibit both inductance *and* resistance. The primary resistance, R_1, effectively appears in series with the primary inductance, L_1, whilst the secondary resistance, R_2, effectively appears in series with the secondary inductance, L_2 (see Figure 4.1.85).

It is sometimes convenient to combine the primary and secondary resistances and inductances into a single pair of components connected in series with the primary circuit. We can do this by *referring* the secondary resistance and inductance to the primary circuit.

The amount of secondary resistance referred into the primary, R_2', is given by:

$$R_2' = R_2 \times \left(\frac{V_1}{V_2}\right)^2$$

thus the effective total primary resistance is given by:

$$R_e = R_1 + R_2 \left(\frac{V_1}{V_2}\right)^2.$$

Similarly, the amount of secondary inductance referred into the primary, L_2', is given by:

$$L_2' = L_2 \times \left(\frac{V_1}{V_2}\right)^2$$

thus the effective total primary inductance is given by:

$$L_e = L_1 + L_2 \left(\frac{V_1}{V_2}\right)^2.$$

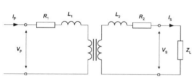

Figure 4.1.85 *Transformer equivalent circuit showing the effect of leakage flux and winding resistance*

The effective impedance, Z_e, appearing in series with the primary circuit (with both R_2 and L_2 referred into the primary) is thus given by:

$$Z_e = R_e + j\omega L_e$$

(Note that $|Z_e| = \sqrt{(R_e^2 + (\omega L_e)^2)}$ – see Example 4.1.12 on page 232).

Magnetising current

In a real transformer, a small current is required in order to magnetise the transformer core. This current is present regardless of whether or not the transformer is connected to a load. This magnetising current is equivalent to the current that would flow in an inductance, L_m, connected in parallel with the primary winding.

Core losses

A real transformer also suffers from losses in its core. These losses are attributable to *eddy currents* (small currents induced in the laminated core material) and *hysteresis* (energy loss in the core due to an imperfect *B–H* characteristic). These losses can be combined together and represented by a resistance, R_{cl}, connected in parallel with the primary winding. The complete equivalent circuit for a transformer is shown in Figure 4.1.86.

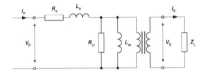

Figure 4.1.86 *Transformer equivalent circuit showing the effect of leakage flux, winding resistance, and core losses*

Example 4.1.32

A transformer designed for operation at 50 Hz has 220 primary turns and 110 secondary turns. The primary and secondary resistances are 1.5 Ω and 0.5 Ω, respectively, whilst the primary and secondary leakage inductances are 20 mH and 10 mH, respectively. Assuming core losses are negligible, determine the equivalent impedance referred to the primary circuit.

Now $Z_e = \sqrt{(R_e^2 + (\omega L_e)^2)}$

where $R_e = R_1 + R_2 \left(\dfrac{V_1}{V_2}\right)^2$ and $L_e = L_1 + L_2 \left(\dfrac{V_1}{V_2}\right)^2$.

Also $\left(\dfrac{V_1}{V_2}\right)^2 = \left(\dfrac{N_1}{N_2}\right)^2 = \left(\dfrac{220}{110}\right)^2 = 4$

thus $R_e = 1.5 + (0.5 \times 4) = 3.5\ \Omega$ and $L_e = 0.02 + (0.01 \times 4) = 0.06$ H.

Finally, $Z_e = R_e + j\omega L_e = 3.5 + j\,(2\pi \times 50 \times 0.06) = (3.5 + j18.84)\ \Omega$

$|Z_e| = \sqrt{(3.5^2 + 18.84^2)} = \sqrt{(3.5^2 + 18.84^2)} = 19.16\ \Omega$.

Questions 4.1.5

(1)   A flux of 60 μWb is produced in a coil. A second coil is wound over the first coil. Determine the flux that will be induced in this coil if the coefficient of coupling is 0.3.

(2) Two coils exhibit a mutual inductance of 40 mH. Determine the e.m.f. produced in one coil when the current in the other coil is increased at a uniform rate from 100 mA to 600 mA in 0.50 ms.

(3) An e.m.f. of 0.2 V is induced in a coil when the current in a second coil is varied at a rate of 1.5 A/s. Determine the mutual inductance that exists between the two coils.

(4) Two coils have self-inductances of 60 mH and 100 mH. If the two coils exhibit a mutual inductance of 20 mH, determine the coefficient of coupling.

(5) Two coupled coils have negligible resistance and mutual inductance of 20 mH. If a supply of 22 $\angle 90°$V is applied to one coil and a current of 0.5 A flows in the other coil, determine the current in the first coil if it is found to have an inductance of 60 mH.

(6) A transformer operates from a 110 V 60 Hz supply. If the transformer has 400 primary turns, determine the maximum flux present. By how much will the flux be increased if the transformer was to be used on a 50 Hz supply?

(7) A 'loss-free' transformer has 900 primary turns and 225 secondary turns. If the primary winding is connected to a 220 V supply and the secondary is connected to an 11 Ω resistive load, determine:
   (a) the secondary voltage;
   (b) the secondary current;
   (c) the supply power.

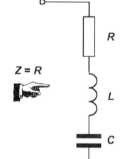

Figure 4.1.87   *A series R–L–C circuit*

Resonant circuits

In Chapter 3 we introduced series and parallel resonance and derived the expressions for resonant frequency, dynamic resistance, and Q-factor. This chapter considers the behaviour of resonant circuits at frequencies above and below resonance and the effects of loading on tuned circuit performance.

The impedance of the *R–L–C* series circuit shown in Figure 4.1.87 is given by the expression:

$$Z = R + j\omega L - j/\omega C = R + j(\omega L - 1/\omega C).$$

At resonance, $\omega L = 1/\omega C$, thus $Z = R + j(0) = R$.

In this condition, the circuit behaves like a pure resistor and thus the supply current and voltage are in-phase.

At any frequency, *f*, we can determine the current flowing in a series resonant circuit and also the voltage dropped across each of the circuit elements present, as shown in the following example:

Example 4.1.33

Determine the voltage dropped across each component in the circuit shown in Figure 4.1.88.

At 100 kHz, the inductive reactance is given by:

$$j\omega L = j \times 2\pi f \times L = j \times 6.28 \times 100 \times 10^3 \times 200 \times 10^{-6} = j125.6\,\Omega.$$

cont.

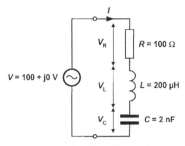

Figure 4.1.88 *See Example 4.1.33*

Similarly, the capacitive reactance is given by:

$$-\frac{j}{\omega C} = -j/(2\pi f \times C) = -j/(6.28 \times 100 \times 10^3 \times 2 \times 10^{-9})$$

$$= -j796.2\ \Omega.$$

The impedance of the circuit is given by:

$$Z = R + j(\omega L - 1/\omega C) = 100 + j(125.6 - 796.2) = (100 - j670.6)\ \Omega.$$

The current flowing in the circuit will be given by:

$$I = V/Z = (100 + j0)/(100 - j670.6) = 0.022 + j0.146\ \text{A}.$$

The voltage dropped across the capacitor will be given by:

$$V_C = I \times \frac{j}{\omega C} = (0.022 + j0.146) \times -j796.2$$

$$= (116.25 - j17.52)\ \text{V}.$$

The voltage dropped across the inductor will be given by:

$$V_L = I \times j\omega L = (0.022 + j0.146) \times j125.6 = (-18.33 + j2.76)\ \text{V}.$$

The voltage dropped across the resistor will be given by:

$$V_R = I \times R = (0.022 + j0.146) \times 100 = (2.2 + j14.6)\ \text{V}.$$

(You might like to check these answers by adding the three individual component voltages to see if they are equal to the supply voltage – they should be!)

The universal resonance curve

The resonance curves shown in Figure 4.1.89 are graphs that show the ratio of actual current (I) to current at resonance (I_0) plotted against the proportion of the resonant frequency (i.e. the ratio f/f_0) for a set of different Q-factors. Provided that we know the Q-factor of a tuned circuit *and it just happened to be the same* as one of those shown in Figure 4.1.89 we could use one of the graphs to determine the current flowing in the circuit at any frequency.

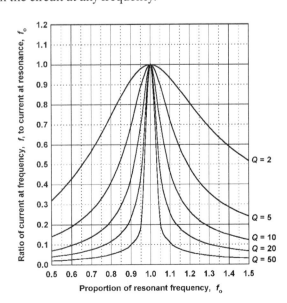

Figure 4.1.89 *Resonance curves for various Q-factors*

Figure 4.1.90   *Universal resonance curve*

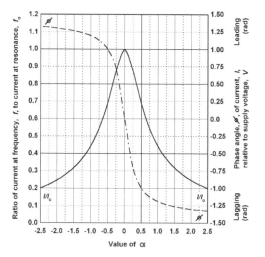

An alternative type of resonance curve (see Figure 4.1.90) can be used for *any* value of Q-factor. In this case, the horizontal axis shows values of α where:

$$\alpha = Q \times \frac{\Delta f}{f_0}$$

where Q is the Q-factor of the circuit ($Q = \omega L/R$), Δf is the deviation from the resonant frequency, f_0. This graph is known as a *universal resonance curve* since it can be made to apply to *any* resonant circuit.

Example 4.1.34

A tuned circuit has a Q-factor of 10 and a resonant frequency of 100 kHz. If the current at resonance is 100 mA, use the universal resonance curve to determine the current and phase angle at 102.5 kHz.

At 102.5 kHz, $\Delta f = f - f_0 = 102.5 - 100 = 2.5$ kHz.

Since $Q = 10$ and $f_0 = 100$ kHz, $\alpha = 10 \times (2.5/100) = 0.25$.

cont.

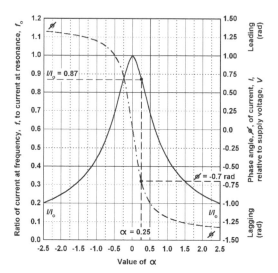

Figure 4.1.91   *See Example 4.1.34*

Referring to the universal resonance curve shown in Figure 4.1.91 shows that when $\alpha = 0.25$, $I/I_0 = 0.87$ and $\phi = -0.7$ rad.

Thus the current flowing, I, will be given by: $I = 100 \times 0.87 = 87$ mA.

and the phase angle, $\phi = -0.7$ rad. $= -40.1°$.

Example 4.1.35

At resonance, a current of 2.5 A flows in a circuit that is series resonant at a frequency of 400 Hz. At 440 Hz a current of 1.25 A flows in the same circuit. Use the universal resonance curve to determine the Q-factor of the circuit.

Now, $I/I_0 = 1.25/2.5 = 0.5$.

From the universal resonance curve (see Figure 4.1.90), when $I/I_0 = 0.5$, $\alpha = 0.7$.

Since $\alpha = Q \times \dfrac{\Delta f}{f_0}$, $Q = \alpha \times \dfrac{f_0}{\Delta f}$.

Now $f_0 = 400$ Hz and $\Delta f = (440 - 400) = 40$ Hz.

Thus $Q = 0.7 \times \dfrac{400}{40} = 7$.

Example 4.1.36

Use the universal resonance curve to determine the bandwidth of a series tuned circuit having a Q-factor of 20 and a resonant frequency of 1 MHz at the points on the response curve where the current has fallen to 55% of its maximum.

Now, $I/I_0 = 0.55$.

From the universal resonance curve (see Figure 4.1.92), the points at which $I/I_0 = 0.55$ correspond to values of α of $+0.75$ and -0.75.

cont.

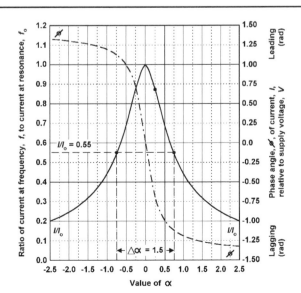

Figure 4.1.92   *See Example 4.1.36*

$$\text{Since } \alpha = Q \times \frac{\Delta f}{f_0}, \Delta f = \alpha \times \frac{f_0}{Q}.$$

$$\text{Thus } \Delta f = 0.75 \times \frac{1 \text{ MHz}}{20} = 37.5 \text{ kHz}.$$

The total bandwidth is thus equal to $(2 \times 37.5) = 75$ kHz.

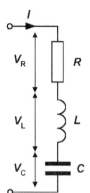

Figure 4.1.93 *Series resonant circuit*

Q-factor

In Chapter 3 we introduced the notion of the 'goodness' of a resonant circuit and we showed how this was related to the amount of 'magnification' that was produced by a circuit. We shall now develop this idea a little further.

For the series resonant circuit shown in Figure 4.1.93, Q-factor is the *voltage magnification factor* and it can be defined as:

$$Q = \frac{\text{voltage across } L}{\text{voltage across } R} = \frac{V_L}{V_R} = \frac{IX_L}{IR} = \frac{\omega L}{R}$$

where V_L and V_R are the voltages dropped across L and R at resonance and I is the current flowing in the circuit at resonance.

For the parallel resonant circuit shown in Figure 4.1.94, Q-factor is the *current magnification factor* and it can be defined as:

$$Q = \frac{\text{current in } L}{\text{current in } R} = \frac{I_L}{I_R} = \frac{\left(\frac{V}{X_L}\right)}{\left(\frac{V}{R}\right)} = \frac{VR}{VX_L} = \frac{R}{X_L} = \frac{R}{\omega L}$$

where I_L and I_R are the currents in L and R at resonance and V is the voltage developed across the circuit.

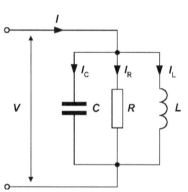

Figure 4.1.94 *Parallel resonant circuit*

Example 4.1.37

A parallel resonant circuit consists of $L = 10$ μH, $C = 20$ pF and $R = 10$ kΩ. Determine the Q-factor of the circuit at resonance.

First we need to find the resonant frequency from:

$$f_0 = \frac{1}{2\pi\sqrt{(LC)}} = \frac{1}{2\pi\sqrt{10 \times 10^{-6} \times 20 \times 10^{-12}}} = \frac{1}{2\pi\sqrt{200 \times 10^{-18}}}$$

$$f_0 = \frac{0.159}{1.414 \times 10^{-8}} = 11.24 \times 10^6 = 11.24 \text{ MHz}.$$

At resonance,

$$X_L = 2\pi f_0 L = 6.28 \times 11.24 \times 10^6 \times 10 \times 10^{-6} = 705.9 \text{ Ω}.$$

Now $Q = R/\omega L = 10\,000/705.9 = 14.2$.

Loading and damping

Loading of a series tuned circuit occurs whenever another circuit is coupled to it. This causes the tuned circuit to be *damped* and the result is a reduction in Q-factor together with a corresponding increase in bandwidth (recall that bandwidth $= f_0/Q$).

In some cases, we might wish to deliberately introduce damping into a circuit in order to make it less selective, reducing the Q-factor and increasing the bandwidth. For a series tuned circuit, we must introduce the damping resistance in series with the existing components (L, C and R) whereas, for a parallel tuned circuit, the damping resistance must be connected in parallel with the existing components.

Questions 4.1.6

(1) A series tuned circuit consists of $L = 450$ μH, $C = 1$ nF, and $R = 30$ Ω. Determine:
    (a) The frequency at which maximum current will flow in the circuit.
    (b) The Q-factor of the circuit at resonance.
    (c) The bandwidth of the tuned circuit.
(2) If the tuned circuit in Question 1 is connected to a voltage source of $(2 + j0)$V at a frequency of 500 kHz, determine the current that will flow and the voltage that will be developed across each component.
(3) A series tuned circuit has a Q-factor of 15 and is resonant at a frequency of 45 kHz. If a current of 50 mA flows in the circuit at resonance, use the universal resonance curve to determine:
    (a) The current that will flow at a frequency of 51 kHz.
    (b) The frequency at which the phase angle will be 72° (leading).
(4) A current of 10 mA flows in a series L–C–R circuit at its resonant frequency of 455 kHz. If the current falls to 3.7 mA at a frequency of 480 kHz, determine the Q-factor of the tuned circuit.
(5) A parallel resonant circuit comprises $L = 200$ μH, $C = 700$ pF, and $R = 75$ kΩ.
    (a) Determine the resonant frequency, Q-factor and bandwidth of the tuned circuit.
    (b) If the bandwidth is to be increased to 5 kHz, determine the value of additional parallel damping resistance required.

4.2 Networks

In the previous section we introduced some simple networks (see Figure 4.1.37 on page 224). We also showed how it was possible to replace one type of network with another without affecting its electrical characteristics. In this section we shall develop this work further so that you can gain an understanding of the effect that networks have on signals.

Basic network models

The simple networks that we met in the last section can exist in two basic forms, *unbalanced* and *balanced*. In the latter case, none of the network's input and output terminals are connected directly to

Figure 4.2.1 *Unbalanced and balanced forms of basic T and π-networks*

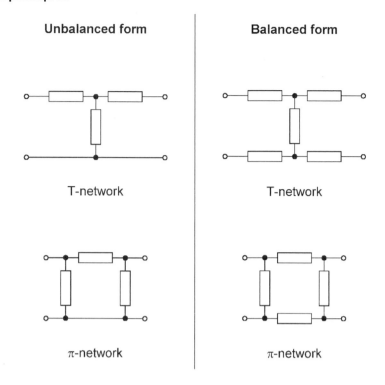

Unbalanced form	Balanced form
T-network	T-network
π-network	π-network

common or ground. The unbalanced and balanced forms of the basic T and π-networks are shown in Figure 4.2.1.

The networks shown in Figure 4.2.1 all have two *ports*. One port (i.e. pair of terminals) is connected to the input whilst the other is connected to the output. For convenience, many *two port networks* are made symmetrical and they perform exactly the same function and have the same characteristics regardless of which way round they are connected.

Characteristic impedance

It is often convenient to analyse the behaviour of a signal transmission path in terms of a number of identical series connected networks. One important feature of any network is that, when an infinite number of identical symmetrical networks are connected in series, the resistance (or impedance) seen looking into the network will have a definite value. This value is known as the *characteristic impedance* of the network.

Take a look at Figure 4.2.2. In Figure 4.2.2(a) an infinite number of identical networks are connected in series. By definition, the impedance seen looking into this arrangement will be equal to the characteristic impedance, Z_0. Now suppose that we remove the first network in the chain, as shown in Figure 4.2.2(b). To all intents and purposes, we will still be looking into an infinite number of series-connected networks. Thus, once again, we will see an impedance equal to Z_0 when we look into the network.

Finally, suppose that we place an impedance of Z_0 across the output terminals of the single network that we removed earlier. This *terminated* network (see Figure 4.2.2(c) will behave *exactly* the same way as the arrangement in Figure 4.2.2(a). In other words, by correctly terminating the network in its characteristic impedance, we have made one single network section appear the same as a series of identical networks stretching to infinity.

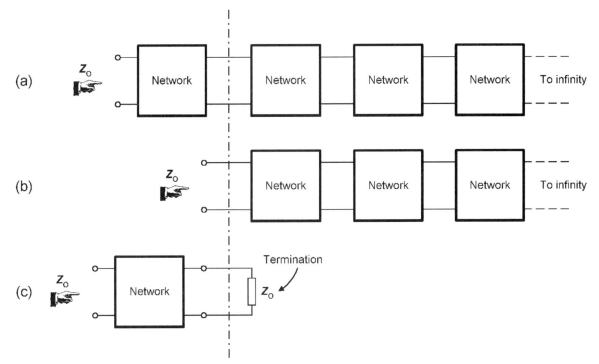

(a)

(b)

(c)

Termination

Figure 4.2.2 *Characteristic impedance seen looking into an infinite number of series-connected networks*

It should be fairly obvious that the characteristic impedance of a network is determined by the values of resistance (or impedance) within the network. We shall now prove that this is the case and determine values of Z_0 for the T and π-networks that we met in the previous section.

The T-network

Take a look at the network shown in Figure 4.2.3.

Now $Z_0 = Z_1 + (Z_2 \mid\mid (Z_1 + Z_0))$ (where $\mid\mid$ means 'in parallel with').

Thus $Z_0 = Z_1 + \dfrac{Z_2 \times (Z_1 + Z_0)}{Z_0 + Z_1 + Z_2}$

$$Z_0 = Z_1 + \frac{Z_1 Z_2 + Z_0 Z_2}{Z_0 + Z_1 + Z_2} = Z_1 + \frac{Z_1 Z_2}{Z_0 + Z_1 + Z_2} + \frac{Z_0 Z_2}{Z_0 + Z_1 + Z_2}$$

$$Z_0 = \frac{Z_1(Z_0 + Z_1 + Z_2) + Z_1 Z_2 + Z_0 Z_2}{Z_0 + Z_1 + Z_2} = \frac{Z_1 Z_0 + Z_1^2 + Z_1 Z_2 + Z_1 Z_2 + Z_0 Z_2}{Z_0 + Z_1 + Z_2}$$

$$Z_0 \times (Z_0 + Z_1 + Z_2) = Z_1^2 + 2Z_1 Z_2 + Z_1 Z_0 + Z_0 Z_2$$

or $Z_0^2 + Z_0 Z_1 + Z_0 Z_2 = Z_1^2 + 2Z_1 Z_2 + Z_1 Z_0 + Z_0 Z_2$.

Hence $Z_0^2 = Z_1^2 + 2Z_1 Z_2$ or $Z_0 = \sqrt{(Z_1^2 + 2Z_1 Z_2)}$.   (i)

The impedance seen looking into the T-network with the output terminals left open-circuit (see Figure 4.2.4) is simply given by:

$Z_{OC} = Z_1 + Z_2$.

The impedance seen looking into the T-network with the output

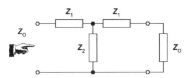

Figure 4.2.3 *T-network*

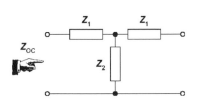

Figure 4.2.4 *Impedance seen looking into a T-network with the output terminals left open-circuit*

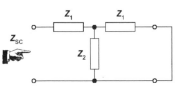

Figure 4.2.5 *Impedance seen looking into a T-network with the output terminals short-circuited*

terminals linked together by a short-circuit (see Figure 4.2.5) is given by:

$Z_{SC} = Z_1 + (Z_1 || Z_2)$ (where $||$ means 'in parallel with').

$$Z_{SC} = Z_1 + \frac{Z_1 Z_2}{Z_1 + Z_2} = Z_1 \frac{(Z_1 + Z_2)}{(Z_1 + Z_2)} + \frac{Z_1 Z_2}{(Z_1 + Z_2)} = \frac{Z_1^2 + Z_1 Z_2 + Z_1 Z_2}{(Z_1 + Z_2)}$$

$$= \frac{Z_1^2 + 2Z_1 Z_2}{(Z_1 + Z_2)}.$$

Now $Z_{OC} \times Z_{SC} = (Z_1 + Z_2) \times \dfrac{Z_1^2 + 2Z_1 Z_2}{(Z_1 + Z_2)} = Z_1^2 + 2Z_1 Z_2.$ (iii)

Combining equations (i) and (ii) gives: $Z_0 = \sqrt{Z_{OC} Z_{SC}}$.

The π-network

Take a look at the network shown in Figure 4.2.6.

Now $Z_0 = Z_2 || \left(Z_1 + \left(\dfrac{Z_0 Z_2}{Z_0 + Z_2} \right) \right)$ (where $||$ means 'in parallel with')

or $Z_0 = Z_2 || \left(\dfrac{Z_1 Z_0 + Z_1 Z_2 + Z_0 Z_2}{Z_0 + Z_2} \right)$

thus $Z_0 = \dfrac{Z_2 \times (Z_1 Z_0 + Z_1 Z_2 + Z_0 Z_2)/(Z_0 + Z_2)}{Z_2 + ((Z_1 Z_0 + Z_1 Z_2 + Z_0 Z_2)/(Z_0 + Z_2))}$

From which $Z_0^2 = \dfrac{Z_1 Z_2^2}{Z_1 + 2Z_2}$ or $Z_0 = \sqrt{\dfrac{Z_1 Z_2^2}{Z_1 + 2Z_2}}$. (i)

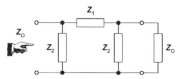

Figure 4.2.6 π-network

The impedance seen looking into the π-network with the output terminals left open-circuit (see Figure 4.2.7) is given by:

$Z_{OC} = Z_2 || (Z_1 + Z_2)$ (where $||$ means 'in parallel with')

$$Z_{OC} = \frac{Z_2 (Z_1 + Z_2)}{Z_2 + Z_1 + Z_2} = \frac{Z_2 (Z_1 + Z_2)}{Z_1 + 2Z_2}.$$

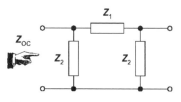

Figure 4.2.7 *Impedance seen looking into a π-network with the output terminal left open-circuit*

The impedance seen looking into the T-network with the output terminals linked together by a short-circuit (see Figure 4.2.8) is given by:

$Z_{SC} = Z_2 || Z_1$ (where $||$ means 'in parallel with')

$$Z_{SC} = \frac{Z_1 Z_2}{Z_1 + Z_2}.$$

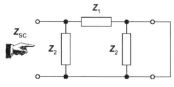

Figure 4.2.8 *Impedance seen looking into a π-network with the output terminal short-circuited*

Now $Z_{OC} \times Z_{SC} = \left(\dfrac{Z_2(Z_1 + Z_2)}{Z_1 + 2Z_2} \right) \left(\dfrac{Z_1 Z_2}{Z_1 + Z_2} \right) = \dfrac{Z_1 Z_2^2}{Z_1 + 2Z_2}.$ (ii)

Figure 4.2.9 *See Example 4.2.1*

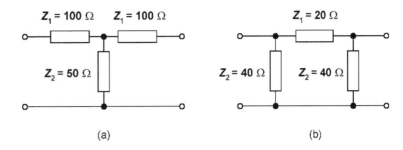

(a)   (b)

Combining equations (i) and (ii) once again gives: $Z_0 = \sqrt{(Z_{OC}Z_{SC})}$.

Example 4.2.1

Determine the characteristic impedance of each of the networks shown in Figure 4.2.9.

Figure 4.2.9(a) shows a T-network in which $Z_1 = 100\ \Omega$ and $Z_2 = 50\ \Omega$.

For a T-network, $Z_0 = \sqrt{Z_1^2 + 2Z_1Z_2}$.

Thus $Z_0 = \sqrt{(100^2 + (2 \times 100 \times 50))} = \sqrt{20\,000} = 141.4\ \Omega$.

Figure 4.2.9(b) shows a π-network in which $Z_1 = 20\ \Omega$ and $Z_2 = 40\ \Omega$.

For a π-network, $Z_0 = \sqrt{\dfrac{Z_1 Z_2^2}{Z_1 + 2Z_2}}$.

Thus $Z_0 = \sqrt{\dfrac{20 \times 40^2}{20 + (2 \times 40)}} = \sqrt{\dfrac{20 \times 1600}{100}} = \sqrt{320} = 17.9\ \Omega$.

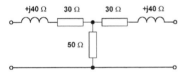

Figure 4.2.10 *See Example 4.2.2*

Example 4.2.2

Determine the characteristic impedance of the T-network shown in Figure 4.2.10.

Comparing this with the T-network shown in Figure 4.2.3 reveals that:

$Z_1 = 30 + j40\ \Omega$ and that $Z_2 = 50\ \Omega$.

For a T-network, $Z_0 = \sqrt{Z_1^2 + 2Z_1Z_2}$.

Thus $Z_0 = \sqrt{(30 + j40)^2 + (2 \times (30 + j40) \times 50)} = \sqrt{8000 + j6400}\ \Omega$.

In order to find the square root of this complex quantity we shall convert it to polar form:

Thus $Z_0 = \sqrt{(10\,245\,\angle 38.66°} = 101.2\,\angle 19.33\ \Omega$.

Converting this back to rectangular form gives:

$Z_0 = 95.5 + j33.5\ \Omega$.

Attenuators

One of the most common applications of the basic T and π-networks is to reduce the amplitude of signals present in a matched system by a

Figure 4.2.11 *Correctly matched attenuator*

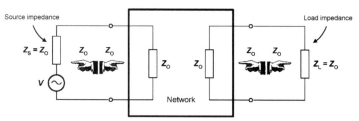

precise amount. The network is then said to *attenuate* the signal and the circuit is referred to as an *attenuator*. In order to work correctly (i.e. provide the required amount of attenuation) an attenuator needs to be *matched* to the system in which it is used. This simply means ensuring that the impedance of the source, as well as that of the load, matches the characteristic impedance of the attenuator. In this condition, we say that an attenuator is correctly *terminated*. Figure 4.2.11 illustrates this concept.

Before we take a look at the operation of two simple forms of attenuator, it is worth pointing out that the impedances used in attenuators are always pure resistances. The reason for this is that an attenuator must provide the same attenuation at all frequencies and the inclusion of reactive components (inductors and/or capacitors) would produce a non-linear attenuation/frequency characteristic.

The T-network attenuator

The circuit of a correctly terminated T-network attenuator is shown in Figure 4.2.12.

From Figure 4.2.12:

$$V_3 = V_1 - I_1 R_1$$

but $I_1 = V_1/R_0$ thus $V_3 = V_1 - \left(\dfrac{V_1}{R_0}\right) \times R_1$

or $V_3 = V_1\left(1 - \dfrac{R_1}{R_0}\right).$ (i)

Also $V_2 = V_3 \times \dfrac{R_0}{R_0 + R_1}$ or $V_3 = V_2\left(\dfrac{R_0 + R_1}{R_0}\right).$ (ii)

Equating (i) and (ii) gives: $V_1\left(1 - \dfrac{R_1}{R_0}\right) = V_2\left(\dfrac{R_0 + R_1}{R_0}\right).$

Thus $V_1\left(\dfrac{R_0 - R_1}{R_0}\right) = V_2\left(\dfrac{R_0 + R_1}{R_0}\right).$

The attenuation, α, provided by the attenuator is given by V_1/V_2, hence:

$$\alpha = \dfrac{V_1}{V_2} = \left(\dfrac{R_0}{R_0 - R_1}\right) \times \left(\dfrac{R_0 + R_1}{R_0}\right) = \dfrac{R_0 + R_1}{R_0 - R_1}.$$

Thus $\alpha = \dfrac{R_0 + R_1}{R_0 - R_1}.$

Now $R_1 + R_0 = \alpha\,(R_0 - R_1) = \alpha R_0 - \alpha R_1.$

Thus $R_1 + \alpha R_1 = \alpha R_0 - R_0$

Figure 4.2.12 *Correctly terminated T-network attenuator*

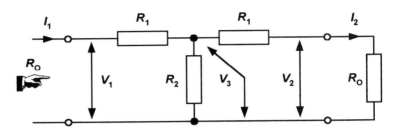

and $R_1 (1 + \alpha) = R_0 (\alpha - 1)$

thus $R_1 = R_0 \left(\dfrac{\alpha - 1}{\alpha + 1} \right).$

From equation (i) on page 261, $R_0 = \sqrt{(R_1^2 + 2R_1 R_2)}.$

Thus $R_0 = \sqrt{\left(R_0 \left(\dfrac{\alpha - 1}{\alpha + 1} \right) \right)^2 + 2R_0 \left(\dfrac{\alpha - 1}{\alpha + 1} \right) R_2}$

and $R_0^2 = \left(R_0 \left(\dfrac{\alpha - 1}{\alpha + 1} \right) \right)^2 + 2R_0 \left(\dfrac{\alpha - 1}{\alpha + 1} \right) R_2 = R_0^2 \left(\dfrac{\alpha - 1}{\alpha + 1} \right) + 2R_0 \left(\dfrac{\alpha - 1}{\alpha + 1} \right) R_2$

thus $R_0^2 = R_0^2 \left(\dfrac{\alpha - 1}{\alpha + 1} \right)^2 + 2R_0 R_2 \left(\dfrac{\alpha - 1}{\alpha + 1} \right).$

Dividing both sides by R_0 gives:

$R_0 = R_0 \left(\dfrac{\alpha - 1}{\alpha + 1} \right)^2 + 2R_2 \left(\dfrac{\alpha - 1}{\alpha + 1} \right)$

thus $2R_2 \left(\dfrac{\alpha - 1}{\alpha + 1} \right) = R_0 - R_0 \left(\dfrac{\alpha - 1}{\alpha + 1} \right)^2$

or $2R_2 = R_0 \left(\dfrac{\alpha + 1}{\alpha - 1} \right) - R_0 \left(\dfrac{\alpha - 1}{\alpha + 1} \right)^2 \left(\dfrac{\alpha + 1}{\alpha - 1} \right) = R_0 \left(\dfrac{\alpha + 1}{\alpha - 1} \right) - R_0 \left(\dfrac{\alpha - 1}{\alpha + 1} \right).$

Hence $2R_2 = R_0 \left(\left(\dfrac{\alpha + 1}{\alpha - 1} \right) - \left(\dfrac{\alpha - 1}{\alpha + 1} \right) \right)$

and $R_2 = \dfrac{R_0}{2} \left(\left(\dfrac{\alpha + 1}{\alpha - 1} \right) - \left(\dfrac{\alpha - 1}{\alpha + 1} \right) \right) = \dfrac{R_0}{2} \left(\dfrac{(\alpha + 1)(\alpha + 1) - (\alpha - 1)(\alpha - 1)}{(\alpha + 1)(\alpha - 1)} \right)$

thus $R_2 = \dfrac{R_0}{2} \left(\dfrac{(\alpha^2 + 2\alpha + 1) - (\alpha^2 - 2\alpha + 1)}{(\alpha^2 - 1)} \right) = \dfrac{R_0}{2} \left(\dfrac{4\alpha}{\alpha^2 - 1} \right)$

hence $R_2 = R_0 \left(\dfrac{2\alpha}{\alpha^2 - 1} \right).$

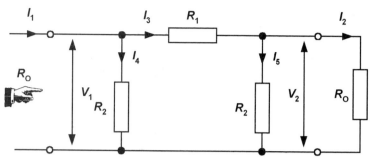

Figure 4.2.13 *Correctly terminated π-network attenuator*

The π-network attenuator

The circuit of a correctly terminated π-network attenuator is shown in Figure 4.2.13.

From Figure 4.2.13:

$I_1 = V_1/R_0$, $I_1 = I_3 + I_4$ and $I_3 = I_2 + I_5$.

Thus $I_1 = I_2 + I_4 + I_5$

and since $I_2 = V_2/R_0$, $I_4 = V_0/R_2$, and $I_5 = V_2/R_2$

$$\frac{V_1}{R_0} = \frac{V_2}{R_0} + \frac{V_1}{R_2} + \frac{V_2}{R_2}.$$

As before, $\alpha = V_1/V_2$.

Thus $V_2 = V_1/\alpha$.

Hence $\dfrac{V_1}{R_0} = \dfrac{V_1}{\alpha R_0} + \dfrac{V_1}{R_2} + \dfrac{V_1}{\alpha R_2}$

or $\dfrac{1}{R_0} - \dfrac{1}{\alpha R_0} = \dfrac{1}{R_2} + \dfrac{1}{\alpha R_2}$

and $\dfrac{1}{R_0}\left(1 - \dfrac{1}{\alpha}\right) = \dfrac{1}{R_2}\left(1 + \dfrac{1}{\alpha}\right)$

thus $\dfrac{1}{R_0}\left(\dfrac{\alpha - 1}{\alpha}\right) = \dfrac{1}{R_2}\left(\dfrac{\alpha + 1}{\alpha}\right)$

from which $R_2 = R_0\left(\dfrac{\alpha + 1}{\alpha - 1}\right)$. 　　　　　(i)

Now $I_1 = I_3 + I_4$ thus:

$$\frac{V_1}{R_0} = \frac{V_1 - V_2}{R_1} + \frac{V_1}{R_2}$$

or $\dfrac{V_1}{R_0} = \dfrac{V_1}{R_2} - \dfrac{V_2}{R_1} + \dfrac{V_1}{R_1}$

or $\dfrac{V_1}{R_0} = \dfrac{V_1}{R_2} - \dfrac{V_1}{\alpha R_1} + \dfrac{V_1}{R_1}$

or $\dfrac{1}{R_0} = \dfrac{1}{R_2} - \dfrac{1}{\alpha R_1} + \dfrac{1}{R_1}$

and $\dfrac{1}{R_0} - \dfrac{1}{R_2} = \dfrac{1}{R_1}\left(1 - \dfrac{1}{\alpha}\right)$.

Substituting R_2 from equation (i) gives:

$$\dfrac{1}{R_0} - \dfrac{1}{R_0}\left(\dfrac{\alpha - 1}{\alpha + 1}\right) = \dfrac{1}{R_1}\left(\dfrac{\alpha - 1}{\alpha}\right)$$

from which $\dfrac{1}{R_0}\left(1 - \left(\dfrac{\alpha - 1}{\alpha + 1}\right)\right) = \dfrac{1}{R_1}\left(\dfrac{\alpha - 1}{\alpha}\right)$

and $\dfrac{1}{R_0}\left(\dfrac{(\alpha + 1) - (\alpha - 1)}{(\alpha + 1)}\right) = \dfrac{1}{R_1}\left(\dfrac{\alpha - 1}{\alpha}\right)$

or $\dfrac{1}{R_0}\left(\dfrac{2}{\alpha + 1}\right) = \dfrac{1}{R_1}\left(\dfrac{\alpha - 1}{\alpha}\right)$

Hence $R_1 = R_0\left(\dfrac{\alpha - 1}{\alpha}\right)\left(\dfrac{\alpha + 1}{2}\right)$

or $R_1 = R_0\left(\dfrac{\alpha^2 - 1}{2\alpha}\right)$.

Propagation coefficient

The infinite series of symmetrical networks that we met on page 261 has another important characteristic. This is the relationship that exists between the input and output currents and voltages. Take a look at the arrangement shown in Figure 4.2.14.

Since each section is identical we can infer that the ratio of input to output current will be the same for each network section.

Thus $I_0/I_1 = I_1/I_2 = I_2/I_3$ = a constant.

The value of the constant will be determined by amount of attenuation and phase shift produced by each network section. It is convenient to express this as a complex quantity, e^{γ}, where γ is known as the *propagation coefficient.*

Thus, for a single section:

Figure 4.2.14 *Decreasing current in an infinite number of series connected networks*

$e^{\gamma} = I_0/I_1 = I_1/I_2 = I_2/I_3$

hence for the nth section, $e^{\gamma} = I_{n-1}/I_n$.

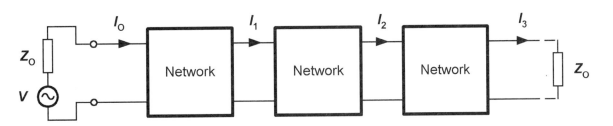

For two, correctly terminated, sections the relationship between the output and input currents will be given by:

$$I_0/I_2 = (I_1/I_2) \times (I_0/I_1) = e^\gamma \times e^\gamma = e^{2\gamma}.$$

Similarly, for three, correctly terminated, sections the relationship between the output and input currents will be given by:

$$I_0/I_3 = (I_2/I_3) \times (I_1/I_2) \times (I_0/I_1) = e^\gamma \times e^\gamma \times e^\gamma = e^{3\gamma}.$$

Thus, for n sections correctly terminated we can infer that the current

$$I_0/I_n = e^{n\gamma}.$$

Since γ is a complex number, we can infer that it has both real and imaginary parts. Thus:

$$\gamma = \alpha + j\beta$$

where α is the *attenuation coefficient* and β is the *phase change coefficient*.
α is measured in *nepers* and β is measured in *radians*.

For a single network section, $I_0/I_1 = e^\gamma$.

Ignoring phase shift, the real part of the current ratio (i.e. the *modulus*) will be given by:

$$|I_0/I_1| = e^\gamma.$$

Taking logs to the base e gives:

$$\ln |I_0/I_1| = \alpha.$$

The attenuation produced by n sections will thus be given by:

$$\ln |I_0/I_n| = n\alpha \text{ nepers}.$$

Finally, the phase shift produced by n sections will simply be given by $n \times \beta$ radians.

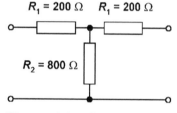

$R_1 = 200\ \Omega$ $R_1 = 200\ \Omega$

$R_2 = 800\ \Omega$

Figure 4.2.15 *See Example 4.2.3*

Example 4.2.3

Determine the characteristic impedance and attenuation provided by the T-network attenuator shown in Figure 4.2.15.
 Comparing this with the T-network shown in Figure 4.2.3 reveals that:

$Z_1 = R_1 = 200\ \Omega$ and that $Z_2 = R_2 = 800\ \Omega$.

For a T-network, $Z_0 = \sqrt{(Z_1^2 + 2Z_1Z_2)}$.

Thus $Z_0 = \sqrt{(200^2 + (2 \times 200 \times 800))} = \sqrt{3600} = 600\ \Omega$.

For a T-network attenuator, $\alpha = \dfrac{R_0 + R_1}{R_0 - R_1}$.

Thus $\alpha = \dfrac{600 + 200}{600 - 200} = \dfrac{800}{400} = 2$.

Relationship between nepers and decibels

In Chapter 3 we introduced the use of decibels (dB) for specifying voltage, current and power ratios. You should recall that decibels are calculated using logarithms to a base of 10. Nepers, on the other hand, are determined using the natural logarithm of a ratio of either voltage, current or power.

If the resistive components of the impedances at the input and output of the network are equal (this would be the case for a symmetrical network), attenuation can be easily converted from nepers to decibels and vice versa.

For a given current ratio, I_0/I_1, the attenuation, m, expressed using decibels would be given by:

$$m = 20 \log_{10}(I_0/I_1).$$

Using standard laws of logarithms gives:

$$m = 20 \log_e (I_0/I_1) \times \log_{10} e$$

$$m = 20 \times \log_{10} e \times \log_e (I_0/I_1)$$

$$m = 20 \times 0.4343 \times \log_e (I_0/I_1) = 8.686 \times \log_e (I_0/I_1).$$

But $\log_e (I_0/I_1)$ is the attenuation expressed in nepers. Thus, we can infer the following rules:

- to convert from nepers to decibels, multiply the value in nepers by 8.686;
- to convert from decibels to nepers, multiply the value in decibels by 0.1151.

Mathematics in action

Changing the base of logarithms

The change of base formula can be proved using the following simple steps:

If $y = \log_a(x)$ by definition, $x = a^y$

Now $\log_b(x) = \log_b(a^y) = y \log_b(a)$.

Thus $\log_b(x) = \log_a(x) \times \log_b(a)$.

Questions 4.2.1

(1) Determine the characteristic impedance of each of the networks shown in Figure 4.2.16.

(2) Design a T-network attenuator that will have a characteristic impedance of 50 Ω and an attenuation of 10.

(3) Design a π-network attenuator that will have a characteristic impedance of 600 Ω and an attenuation of 13 dB.

(4) Three identical symmetrical two-port networks provide a total

Figure 4.2.16 *See Questions 4.2.1*

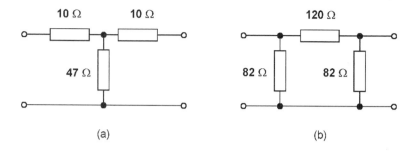

(a)

(b)

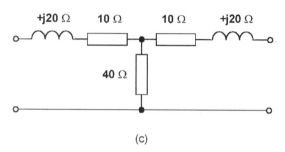

(c)

attenuation of 7.5 and a phase change of +15°. Determine the propagation coefficient for a single two-port network.

(5) The input and output voltage of a correctly terminated attenuator are 1.8 V and 225 mV, respectively. Determine the output current supplied to a matched 75 Ω load and the attenuation expressed in both decibels and nepers.

4.3 Complex waves

In Chapter 3 we introduced complex waveforms and showed how a complex wave could be expressed by an equation of the form:

$$v = V_1 \sin(\omega t) + V_2 \sin(2\omega t \pm \phi_2) + V_3 \sin(3\omega t \pm \phi_3) + V_4 \sin(4\omega t \pm \phi_4) + \ldots \text{(i)}$$

In this section we shall develop this idea further by introducing a powerful technique called Fourier analysis.

Fourier analysis

Fourier analysis is based on the concept that all waveforms, whether continuous or discontinuous, can be expressed in terms of a convergent series of the form:

$$v = A_0 + A_1 \cos t + A_2 \cos 2t + A_3 \cos 3t + \ldots + B_1 \sin t + B_2 \sin 2t + B_3 \sin 3t + \ldots \quad \text{(ii)}$$

where A_0, A_1, ... B_1, B_2, ... are the amplitudes of the individual components that make up the complex waveform. Essentially, this formula is the same as the simplified relationship in equation (i) but with the introduction of cosine as well as sine components. Note also that we have made the assumption that one complete cycle occurs in a time equal to 2π seconds (recall that $\omega = 2\pi f$ and one complete cycle requires 2π radians).

Another way of writing equation (ii) is:

$$v = A_0 + \sum_{n=1}^{n=\infty} (A_n \sin nt + B_n \cos nt). \tag{iii}$$

For the range $-\pi$ to $+\pi$, the value of the constant terms (the amplitudes of the individual components), $A_0, A_1, \ldots B_1, B_2 \ldots$ etc. can be determined as follows:

$$A_0 = \frac{1}{2\pi} \int_{-\pi}^{\pi} v \, dt = \frac{1}{2\pi} \int_{-\pi}^{\pi} \left(A_0 + \sum_{n=1}^{n=\infty} (A_n \cos nt + B_n \sin nt) \right) dt \tag{iv}$$

$$A_n = \frac{1}{\pi} \int_{-\pi}^{\pi} v \cos \, nt \, dt \text{ (where } n = 1, 2, 3, \ldots) \tag{v}$$

$$B_n = \frac{1}{\pi} \int_{-\pi}^{\pi} v \sin \, nt \, dt \text{ (where } n = 1, 2, 3, \ldots). \tag{vi}$$

The values, $A_0, A_1, \ldots B_1, B_2, \ldots$ etc. are called the *Fourier coefficients* of the *Fourier series* defined by equation (iii).

Just in case this is all beginning to sound a little too complex in terms of mathematics, let's take a look at some examples based on waveforms that you should immediately recognise:

Another view

If you are puzzled by equations (iv) to (vi), it may help to put into words what the values of the Fourier coefficients actually represent:

- A_0 is the mean value of v over one complete cycle of the waveform (i.e. from $-\pi$ to $+\pi$ or from 0 to 2π).
- A_1 is twice the mean value of $v \cos t$ over one complete cycle.
- B_1 is twice the mean value of $v \sin t$ over one complete cycle.
- A_2 is twice the mean value of $v \cos 2t$ over one complete cycle.
- B_2 is twice the mean value of $v \sin 2t$ over one complete cycle, and so on. We shall use this fact later when we describe an alternative method of determining the Fourier series for a particular waveform.

Example 4.3.1

Determine the Fourier coefficients of the sine wave superimposed on the constant DC level shown in Figure 4.3.1.

By inspection, the equation of the wave (over the range 0 to 2π) is:

$v = 10 + 5 \sin t.$

By comparison with equation (ii), we can deduce that:

$A_0 = 10$ (in other words, the mean voltage over the range 0–2 π is 10 V).

$B_1 = 5$ (this is simply the coefficient of sin t).

cont.

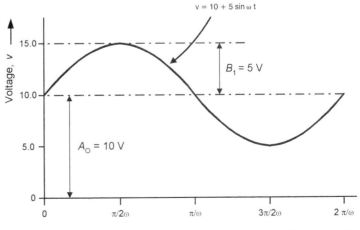

Figure 4.3.1 *See Example 4.3.1*

There are no other terms and thus all other Fourier coefficients (A_1, A_2, B_2, etc.) are all zero.

Now let's see if we can arrive at the same results using equations (iv), (v) and (vi):

The value of A_0 is found from:

$$A_0 = \frac{1}{2\pi} \int_0^{2\pi} v\, dt = \frac{1}{2\pi} \int_0^{2\pi} (10 + 5 \sin t)dt$$

$$A_0 = \frac{1}{2\pi} [10\, t]_{t=0}^{t=2\pi} + \frac{1}{2\pi} \int_0^{2\pi} 5 \sin t\, dt$$

$$A_0 = \left[\frac{20\pi}{2\pi} - 0 \right] + 0 = 10.$$

The value of A_1 is found from:

$$A_1 = \frac{1}{\pi} \int_0^{2\pi} v \cos t\, dt = \frac{1}{\pi} \int_0^{2\pi} (10 + 5 \sin t) \cos t\, dt$$

$$A_1 = \frac{1}{\pi} \int_0^{2\pi} (10 \cos t + 5 \sin t \cos t)\, dt$$

$$A_1 = \frac{1}{\pi} \int_0^{2\pi} 10 \cos t\, dt + \frac{1}{\pi} \int_0^{2\pi} 5 \sin t \cos t\, dt$$

$$A_1 = \frac{1}{\pi} \int_0^{2\pi} 10 \cos t\, dt + 0$$

$$A_1 = \frac{1}{\pi} [10 \sin t]_{t=0}^{t=\pi} + 0 = 0.$$

The value of B_1 is found from:

$$B_1 = \frac{1}{\pi} \int_0^{2\pi} v \sin t\, dt = \frac{1}{\pi} \int_0^{2\pi} (10 + 5 \sin t) \sin t\, dt$$

$$B_1 = \frac{1}{\pi} \int_0^{2\pi} (10 \sin t + 5 \sin^2 t)\, dt$$

$$B_1 = \frac{1}{\pi} \int_0^{2\pi} 10 \sin t\, dt + \frac{1}{\pi} \int_0^{2\pi} 5 \sin^2 t\, dt$$

$$B_1 = 0 + \frac{1}{\pi} \left[\frac{5}{2} \left(t - \frac{1}{2} \sin 2t \right) \right]_{t=0}^{t=2\pi}$$

cont.

$$B_1 = 0 + \frac{10\,\pi}{2\,\pi} = 5.$$

Example 4.3.2

Determine the Fourier series for the rectangular pulse shown in Figure 4.3.2.

Unlike the waveform in the previous example, this voltage is discontinuous and we must therefore deal with it in two parts; that from 0 to 1.5π and that from 1.5π to 2π.
The equation of the voltage is as follows:

$v = 2$ (over the range $0 < t < 1.5\pi$)

$v = 0$ (over the range $1.5\pi < t < 2\pi$).

To find A_0 we simply need to find the mean value of the waveform over the range $0–2\pi$. By considering the area under the waveform it should be obvious that this is 1.5.

Hence $A_0 = 1.5$.

Next we need to find the value of A_n for $n = 1, 2, 3, \ldots$ and so on. From equation (v):

$$A_n = \frac{1}{\pi}\int_0^{2\pi} v \cos n\,t\,\mathrm{d}t.$$

But since there are two distinct parts to the waveform, we need to integrate separately over the two time periods, $0–1.5\pi$ and $1.5\pi–2\pi$.

$$A_n = \frac{1}{\pi}\int_0^{1.5\pi} 2 \cos n\,t\,\mathrm{d}t + \frac{1}{\pi}\int_{1.5\pi}^{2\pi} 0 \cos n\,t\,\mathrm{d}t$$

$$A_n = \frac{1}{\pi}\int_0^{1.5} 2 \cos n\,t\,\mathrm{d}t + 0$$

cont.

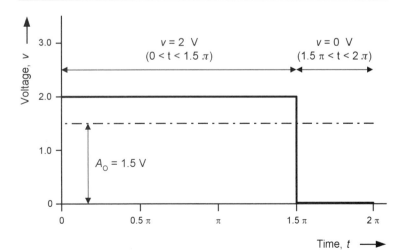

Figure 4.3.2 *See Example 4.3.2*

$$A_n = \frac{2}{\pi n} \left[\sin n\,t\right]_{t=0}^{t=1.5\pi} + 0 = \frac{2}{\pi n} \left[\sin 1.5\,n\,\pi - \sin 0\right].$$

Thus $A_n = \dfrac{2}{\pi n} \sin 1.5\,n\,\pi$.

Now, when $n = 1$, $A_1 = \dfrac{2}{\pi} \sin 1.5\pi = -\dfrac{2}{\pi}$

when $n = 2$, $A_2 = \dfrac{1}{\pi} \sin 3\pi = 0$

when $n = 3$, $A_3 = \dfrac{2}{3\pi} \sin 4.5\,\pi = +\dfrac{2}{3\pi}$

when $n = 4$, $A_4 = \dfrac{1}{2\pi} \sin 6\pi = 0$

when $n = 5$, $A_5 = \dfrac{2}{5\pi} \sin 7.5\,\pi = -\dfrac{2}{5\pi}$.

The cosine terms in the Fourier series will thus be:

$$-\frac{2}{\pi} \cos t + \frac{2}{3\pi} \cos 3t - \frac{2}{5\pi} \cos 5t \ldots$$

The sine terms can be similarly found:

$$B_n = \frac{1}{\pi} \int_0^{1.5\pi} 2\sin n\,t\, dt + \frac{1}{\pi} \int_{1.5\pi}^{2\pi} 0\sin n\,t\, dt$$

$$B_n = \frac{1}{\pi} \int_0^{1.5\pi} 2\sin n\,t\, dt + 0$$

$$B_n = \frac{-2}{\pi n} \left[\cos n\,t\right]_{t=0}^{t=1.5\pi} + 0 = \frac{-2}{\pi n} \left[\cos 1.5\,n\,\pi - \cos 0\right].$$

Thus $B_n = \dfrac{-2}{\pi n} (\cos 1.5\,n\,\pi - 1)$.

Now, when $n = 1$, $B_1 = \dfrac{-2}{\pi} (\cos 1.5\,\pi - 1) = \dfrac{2}{\pi}$

when $n = 2$, $B_2 = \dfrac{-2}{2\pi} (\cos 3\pi - 1) = \dfrac{2}{\pi}$

when $n = 3$, $B_3 = \dfrac{-2}{3\pi} (\cos 4.5\,\pi - 1) = \dfrac{2}{3\pi}$

when $n = 4$, $B_4 = \dfrac{-2}{4\pi} (\cos 6\pi - 1) = 0$

when $n = 5$, $B_5 = \dfrac{-2}{\pi} (\cos 7.5\,\pi - 1) = \dfrac{2}{5\pi}$.

cont.

Thus, the sine terms in the Fourier series will be:

$$\frac{2}{\pi}\sin t + \frac{2}{\pi}\sin 2t + \frac{2}{5\pi}\sin 5t \ldots$$

We are now in a position to develop the expression for the pulse. This will be similar to that shown in equation (ii) but with:

$$A_0 = 1.5, A_1 = -\frac{2}{\pi}, A_2 = 0, A_3 = \frac{2}{3\pi}, \text{ and so on.}$$

The expression for the pulse is thus:

$$v = 1.5 - \frac{2}{\pi}\cos t + \frac{2}{3\pi}\cos 3t - \frac{2}{5\pi}\cos 5t \ldots + \frac{2}{\pi}\sin t + \frac{2}{\pi}\sin 2t +$$

$$\frac{2}{3\pi}\sin 3t + \frac{2}{5\pi}\sin 5t \ldots$$

Using this function in a standard spreadsheet program (up to and including the ninth harmonic terms) produces the *synthesised* pulse shown in Figure 4.3.3.

Question 4.3.1

Derive the Fourier series for each of the waveforms shown in Figure 4.3.4.

An alternative method

An alternative tabular method can be used to determine Fourier coefficients. This method is based on an application of the trapezoidal rule and it is particularly useful when a number of voltage readings

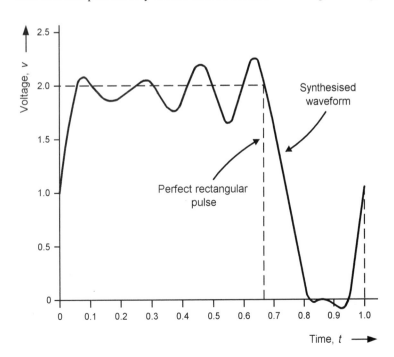

Figure 4.3.3 *A pulse waveform synthesised from harmonic components up to the ninth harmonic*

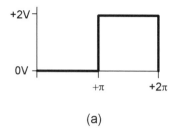

(a)

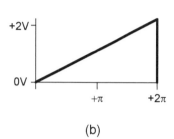

(b)

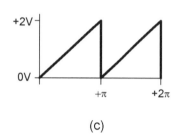

(c)

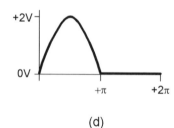

(d)

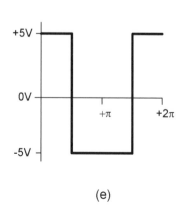

(e)

Figure 4.3.4 *See Question 4.3.1*

are available at regular time intervals over the period of one complete cycle of a waveform. Once again, we shall show how this method works by using an example:

Example 4.3.3

The values of voltage over a complete cycle of a waveform (see Figure 4.3.5) are as follows:

Angle, θ	30	60	90	120	150	180	210	240	270	300	330	360
Voltage, v	8.9	8.1	7.5	6.5	5.5	5.0	4.5	3.5	2.5	1.9	1.1	5

First we need to determine the value of A_0. We can do this by calculating the mean value of voltage over the complete cycle. Applying the trapezoidal rule (i.e. adding together the values of voltage and dividing by the number of intervals) gives:

$$A_0 = (8.9 + 8.1 + 7.5 + 6.5 + \dots 5)/12 = 5.$$

Similarly, to determine the value of A_1 we multiply each value of voltage by $\cos \theta$, before adding them together and dividing by *half* the number of intervals (recall that A_1 is *twice* the mean value of $v \cos t$ over one complete cycle). Thus:

$$A_1 = (7.707 + 4.049 - 0.002 - 3.252 + \dots 5)/6 = 0.$$

The same method can be used to determine the remaining Fourier coefficients. The use of a spreadsheet for this calculation is highly recommended (see Figure 4.3.6).

Having determined the Fourier coefficients, we can write down the expression for the complex waveform. In this case it is:

$$v = 5 + 2.894 \sin \theta + 1.444 \sin 2\theta + 0.635 \sin 3\theta + 0.522 \sin 4\theta + \dots$$

(note that there are no terms in $\cos \theta$ in the Fourier series for this waveform).

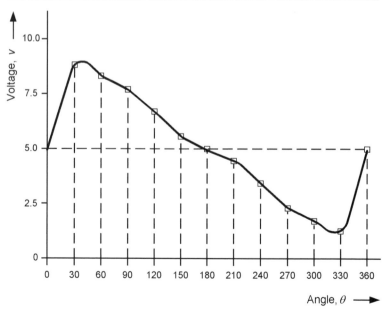

Figure 4.3.5 *See Example 4.3.3*

Fourier analysis - tabular method

y	θ	v	cos θ	v cos θ	sin θ	v sin θ	cos 2θ	v cos 2θ	sin 2θ	v sin 2θ	cos 3θ	v cos 3θ	sin 3θ	v sin 3θ	cos 4θ	v cos 4θ	sin 4θ	v sin 4θ
y1	30	8.9	0.866	7.707	0.500	4.451	0.500	4.449	0.866	7.708	0.000	-0.002	1.000	8.900	-0.500	-4.452	0.866	7.706
y2	60	8.1	0.500	4.049	0.866	7.015	-0.500	-4.052	0.866	7.014	-1.000	-8.100	0.000	-0.003	-0.500	-4.046	-0.866	-7.017
y3	90	7.5	0.000	-0.002	1.000	7.500	-1.000	-7.500	0.000	-0.003	0.001	0.005	-1.000	-7.500	1.000	7.500	0.001	0.006
y4	120	6.5	-0.500	-3.252	0.866	5.628	-0.500	-3.247	-0.866	-5.631	1.000	6.500	0.001	0.005	-0.501	-3.256	0.865	5.626
y5	150	5.5	-0.866	-4.764	0.500	2.748	0.501	2.753	-0.866	-4.761	-0.001	-0.006	1.000	5.500	-0.499	-2.744	-0.867	-4.767
y6	180	5	-1.000	-5.000	0.000	-0.002	1.000	5.000	0.001	0.004	-1.000	-5.000	-0.001	-0.006	1.000	5.000	0.002	0.008
y7	210	4.5	-0.866	-3.896	-0.500	-2.252	0.499	2.246	0.867	3.899	0.001	0.006	-1.000	-4.500	-0.502	-2.257	0.865	3.893
y8	240	3.5	-0.500	-1.748	-0.866	-3.032	-0.501	-1.753	0.865	3.029	1.000	3.500	0.002	0.006	-0.498	-1.743	-0.867	-3.035
y9	270	2.5	0.001	0.002	-1.000	-2.500	-1.000	-2.500	-0.001	-0.003	-0.002	-0.005	1.000	2.500	1.000	2.500	0.002	0.006
y10	300	1.9	0.501	0.951	-0.866	-1.645	-0.499	-0.948	-0.867	-1.647	-1.000	-1.900	-0.002	-0.004	-0.502	-0.954	0.865	1.643
y11	330	1.1	0.866	0.953	-0.499	-0.549	0.501	0.551	-0.865	-0.952	0.002	0.002	-1.000	-1.100	-0.497	-0.547	-0.868	-0.954
y12	360	5	1.000	5.000	0.001	0.004	1.000	5.000	0.002	0.008	1.000	5.000	0.002	0.012	1.000	5.000	0.003	0.016
	Sum	60	Sum	0.001	Sum	17.367	Sum	0.000	Sum	8.666	Sum	0.001	Sum	3.810	Sum	0.000	Sum	3.131
	A_0 =	5	A_1 =	0.000	B_1 =	2.894	A_2 =	0.000	B_2 =	1.444	A_3 =	0.000	B_3 =	0.635	A_4 =	0.000	B_4 =	0.522

Figure 4.3.6 *Spreadsheet table of Fourier coefficients*

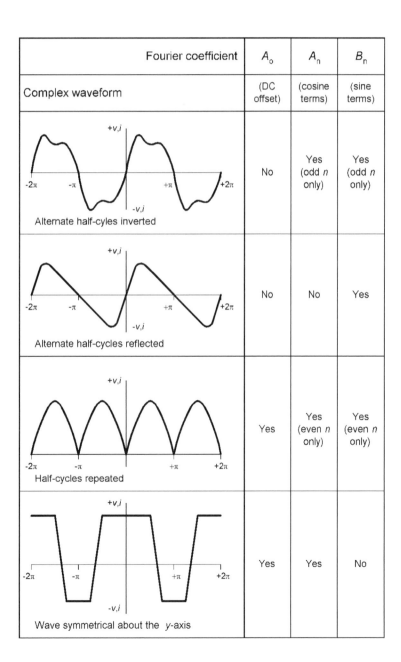

Figure 4.3.7 *Effect of symmetry and reflection on Fourier coefficients for different types of waveform*

Mathematics in action
The trapezoidal rule

In order to determine the mean value of $y = f(t)$ over a given period, we can divide the period into a number of smaller intervals and determine the y ordinate value for each.

If the y values are $y_1, y_2, y_3, \ldots y_m$, and there are m equal values in a time interval, t, the mean value of y, $\bar{y}$, will be given by:

$$\bar{y} = \frac{y_1 + y_2 + y_3 + \ldots y_m}{m} = \frac{\sum\limits_{n=0}^{n=m} y_n}{m}$$

Question 4.3.2

The following voltages are taken over one complete cycle of a complex waveform:

Angle, θ	0	30	60	90	120	150	180	210	240	270	300	330	360
Voltage, v	0	3.5	8.7	9.1	9.5	9.5	9.1	8.7	3.5	0	−3.5	−3.5	0

Use a tabular method to determine the Fourier series for the waveform.

In Chapter 3 we discussed the effect of adding harmonics to a fundamental on the shape of the complex wave produced. Revisiting this in the light of what we now know about Fourier series allows us to relate waveshape to the corresponding Fourier series. Figure 4.3.7 shows the effect symmetry (and reflected symmetry) has on the values of $A_0, A_1, A_2, \ldots B_1, B_2$, etc. This information can allow us to take a few short-cuts when analysing a waveform (there is no point trying to evaluate Fourier coefficients that are just not present!).

RMS value of a waveform

The root mean square (RMS) value of a waveform is the effective value of the current or voltage concerned. It is defined as the value of direct current or voltage that would produce the same power in a pure resistive load.

Assume that we are dealing with a voltage given by:

$$v = V \cos \omega t.$$

If this voltage is applied to a pure resistance, R, the current flowing will be given by:

$$i = \frac{V}{R} \cos \omega t.$$

The instantaneous power dissipated in the resistor, p, will be given by the product of i and v. Thus:

$$p = i\,v = \frac{V}{R}\cos \omega t \times V \cos \omega t = \frac{V^2}{R}\cos^2 \omega t = \frac{V^2}{R}(0.5 + 0.5 \cos 2\omega\,t).$$

Mathematics in action
Trigonometric identities

In the previous text, we used a trigonometric identity to replace the $\cos^2 \omega$ term with something that is easier to cope with. The reasoning is as follows:

$\cos (A + B) = \cos A \cos B - \sin A \sin B.$

Thus $\cos 2A = \cos^2 A - \sin^2 A.$

But $\sin^2 A + \cos^2 A = 1$ thus $\sin^2 A = 1 - \cos^2 A.$

Thus $\cos 2A = \cos^2 A - (1 - \cos^2 A) = 2 \cos^2 A - 1.$

Hence $\cos^2 A = 0.5(1 + \cos 2A) = 0.5 + 0.5 \cos 2A.$

Hence, $p = 0.5\dfrac{V^2}{R} + 0.5\dfrac{V^2}{R}\cos 2\omega\,t$

or, $p = 0.5\dfrac{V^2}{R} + 0.5\dfrac{V^2}{R}\cos 2\omega\,t = \dfrac{V^2}{2R} + \dfrac{V^2}{2R}\cos 2\omega\,t.$ \hfill (i)

This is an important result. From equation (i), we can infer that the power waveform (i.e. the waveform of p plotted against t) will comprise a cosine waveform at *twice* the frequency of the voltage (see Figure 4.3.8). Furthermore, if we apply our recently acquired knowledge of Fourier series, we can infer that the mean value of the power waveform (over a complete cycle of the voltage or current) will be the same as its amplitude. The value of the Fourier coefficients being:

$A_0 = \dfrac{V^2}{2R}$ \quad (i.e. the mean value of the power waveform over one cycle of voltage)

$A_2 = \dfrac{V^2}{2R}$ \quad (i.e. the amplitude of the term in $\cos 2\omega t$)

(note that no other components are present).

The mean power, over one cycle, is thus given by A_0.

Thus $P = \dfrac{V^2}{2R}.$

Thus $V = \sqrt{2PR}.$

Now, let V_{RMS} be the equivalent DC voltage that will produce the same power.

Thus $P = \dfrac{V_{RMS}^2}{R}.$

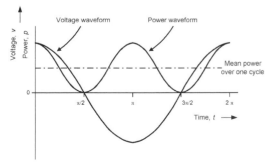

Figure 4.3.8 *Voltage and corresponding power waveform*

and $V_{RMS} = \sqrt{(PR)}$.

Hence $V = \sqrt{2}V_{RMS}$.

In other words, a waveform given by $v = 1.414 \cos \omega t$ V will produce the same power in a load as a direct voltage of 1V. A similar relationship can be obtained for current:

$$I = \sqrt{2}I_{RMS}$$

where I is the amplitude of the current wave.

When a wave is complex, it follows that its root mean square value can be found by adding together the values of each individual harmonic component present. Thus:

$$V_{RMS} = \sqrt{V_0^2 + \frac{V_1^2 + V_2^2 + V_3^2 + \ldots V_m^2}{2}} = \sqrt{V_0^2 + \frac{\sum_{n=1}^{n=m} V_n^2}{2}}$$

where V_0 is the value of any DC component that may be present, and $V_1, V_2, V_3 \ldots V_m$ are the amplitudes of the harmonic components present in the wave.

Power factor

When harmonic components are present, the power factor can be determined by adding together the power supplied by each harmonic before dividing by the product of root mean square voltage and current. Hence:

$$\text{Power factor} = \frac{\text{Total power supplied}}{\text{True power}} = \frac{P_{TOT}}{V_{RMS} I_{RMS}}$$

where P_{TOT} is found by adding the power due to each harmonic component present. Thus:

$$\text{Power factor} = \frac{\dfrac{V_1 I_1}{\sqrt{2}\sqrt{2}} \cos \phi_1 + \dfrac{V_2 I_2}{\sqrt{2}\sqrt{2}} \cos \phi_2 + \dfrac{V_3 I_3}{\sqrt{2}\sqrt{2}} \cos \phi_3 + \ldots + \dfrac{V_n I_n}{\sqrt{2}\sqrt{2}} \cos \phi_n}{V_{RMS} I_{RMS}}$$

$$\text{Power factor} = \frac{\dfrac{V_1 I_1 \cos \phi_1}{2} + \dfrac{V_2 I_2 \cos \phi_2}{2} + \dfrac{V_3 I_3 \cos \phi_3}{2} + \ldots + \dfrac{V_n I_n \cos \phi_n}{2}}{V_{RMS} I_{RMS}}$$

$$\text{Power factor} = \frac{\sum_{n=1}^{n=m} \dfrac{V_n I_n \cos \phi_n}{2}}{V_{RMS} I_{RMS}}.$$

Example 4.3.4

Determine the effective voltage and average power in a 10 Ω resistor when the following voltage is applied to it:

$v = 10 \sin \omega t + 5 \sin 2\omega t + 2 \sin 3\omega t$.

cont.

Now $V_{RMS} = \sqrt{V_0{}^2 + \dfrac{V_1{}^2 + V_2{}^2 + V_3{}^2 + \ldots V_n{}^2}{2}}.$

In this case, $V_0 = 0$ (there is no DC component), $V_1 = 10$, $V_2 = 5$ and $V_3 = 2$.

Thus $= V_{RMS} = \sqrt{\dfrac{10^2 + 5^2 + 2^2}{2}} = \sqrt{\dfrac{129}{2}} = 8.03$ V

$P = \dfrac{8.03^2}{10} = 6.45$ W.

Example 4.3.5

The voltage and current present in an AC circuit is as follows:

$v = 50 \sin \omega t + 10 \sin (3\omega t + \pi/2)$

$i = 3.54 \sin (\omega t + \pi/4) + 0.316 \sin (3\omega t + 0.321).$

Determine the total power supplied and the power factor.
The total power supplied, P_{TOT}, is given by:

$P_{TOT} = \dfrac{50 \times 3.54}{2} \cos (-\pi/4) + \dfrac{10 \times 0.316}{2} \cos (\pi/2 - 0.321)$

$P_{TOT} = (88.5 \times 0.707) + (1.58 \times 0.32) = 62.6 + 0.51 = 63.11$ W.
The root mean square voltage, V_{RMS}, is given by:

$V_{RMS} = \sqrt{\dfrac{50^2 + 10^2}{2}} = \sqrt{\dfrac{2600}{2}} = 36.1$ V.

The root mean square current, I_{RMS}, is given by:

$I_{RMS} = \sqrt{\dfrac{3.54^2 + 0.316^2}{2}} = \sqrt{\dfrac{12.63}{2}} = 2.51$ A.

The true power, P, is thus:

$P = V_{RMS} I_{RMS} = 36.1 \times 2.51 = 90.6$ W.

Finally, to find the power factor we simply divide the total power, P_{TOT}, by the true power, P:

Power factor $= \dfrac{P_{TOT}}{P} = \dfrac{63.11}{90.6} = 0.7.$

Questions 4.3.3

(1) A voltage is given by the expression:

$v = 100 \sin \omega t + 40 \sin (2\omega t - \pi/2) + 20 \sin (4\omega t - \pi/2).$

If this voltage is applied to a 50 Ω resistor:
(a) determine the power in the resistor due to the fundamental and each harmonic component;

(b)   derive an expression for the current flowing in the resistor;

(c)   calculate the root mean square voltage and current;

(d)   determine the total power dissipated.

(2)   The voltage, v, and current, i, in a circuit are defined by the following expressions:

$$v = 30 \sin \omega t + 7.5 \sin (3\omega t + \pi/4)$$
$$i = \sin \omega t + 0.15 \sin (3\omega t - \pi/12).$$

Determine:

(a)   the root mean square voltage;

(b)   the root mean square current;

(c)   the total power;

(d)   the true power;

(e)   the power factor of the circuit.

(3)   A voltage given by the expression:

$$v = 100 \sin 314\,t + 50 \sin 942\,t - 40 \sin 1570\,t$$

is applied to an impedance of $(40 + j\,30)\ \Omega$.

Determine:

(a)   an expression for the instantaneous current flowing;

(b)   the effective (RMS) voltage;

(c)   the effective (RMS) current;

(d)   the total power supplied;

(e)   the true power;

(f)   the power factor.

4.4 Transients in R–L–C circuits

Earlier in this Chapter, and in Chapter 3, we described how circuits containing resistance, inductance and capacitance, behave when steady alternating currents are applied to them. In this final section of Chapter 4, we investigate the behaviour of R–L–C circuits when subjected to a sudden change in voltage or current, known as a *transient*. To develop an understanding of how these circuits behave, we shall treat them as systems.

There are two types of system that we shall be concerned with; first order and second order systems:

- A *first order system* is one that can be modelled using a first order differential equation. First order electrical systems involve combinations of C and R or L and R.
- A *second order system* is one that can be modelled using a second order differential equation. Second order electrical systems involve combinations of all three types of component; L, C and R.

You are probably already familiar with the time response of simple C–R and L–R systems. In the first case, the voltage developed across the capacitor in a C–R circuit will grow exponentially (and the current will decay exponentially) as the capacitor is charged from a constant voltage supply (see Figure 4.4.1). In the second case, the inductor voltage will decay exponentially (and the current will grow exponentially) as the inductor is charged from a constant voltage supply (see Figure 4.4.2).

The mathematical relationships between voltage, current and time for the simple series C–R and L–R circuits are as follows:

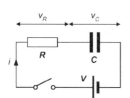

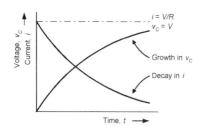

Figure 4.4.1 *Time response of a series C–R network*

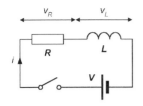

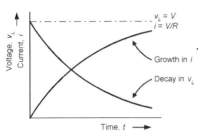

Figure 4.4.2 *Time response of a series L–R network*

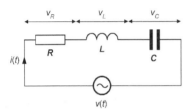

Figure 4.4.3 *Series L–C–R circuit showing forcing function (voltage) and forced function (current)*

For Figure 4.4.1:

$$i = \frac{V}{R}e^{-\frac{t}{CR}} \quad \text{and} \quad v_C = V\left(1 - e^{\frac{-t}{CR}}\right).$$

For Figure 4.4.2:

$$i = \frac{V}{R}\left(1 - e^{\frac{-Rt}{L}}\right) \text{and} \quad v_L = V\left(1 - e^{\frac{-Rt}{L}}\right).$$

The voltage impressed on a first or second order circuit is often referred to as the *forcing function* whilst the current is known as the *forced response*. In a simple case (as in Figures 4.4.1 and 4.4.2), once the switch has been closed, the forcing function is constant. This is not always the case, however, as the forcing function may *itself* be a function of time. This added degree of complexity requires a different approach to the solution of these, apparently simple, circuits.

Consider the following type of voltage that we met earlier in this chapter:

$$v = 100 + 50 \sin \omega t \text{ V}.$$

When used as a forcing function, this voltage can be thought of as constant step function (of amplitude 100 V) onto which is superimposed a sinusoidal variation (of amplitude 50 V).

In general, we can describe a forcing function as a voltage, $v(t)$, where the brackets and the t remind us that the voltage is a *function of time*. The forced function, $i(t)$, would similarly be a function of time (again denoted by the brackets and the t).

Now take a look at Figure 4.4.3. Here a forcing function, $v(t)$, is applied to a simple $L–C–R$ circuit. By applying Kirchhoff's Voltage Law we can deduce that:

$$v(t) = v_R + v_L + v_C$$

where $v_R = i(t)R$

and $v_L = L \times (\text{rate of change of current with time}) = L\dfrac{di(t)}{dt}$

and $v_C = \dfrac{1}{C} \times (\text{area under the current–time graph}) = \dfrac{1}{C}\displaystyle\int i(t)dt.$

Thus $v(t) = i(t)R + L\dfrac{di(t)}{dt} + \dfrac{1}{C}\displaystyle\int i(t)dt.$

This is an important relationship and you should have spotted that it is a second order differential equation. Solving the equation can be problematic but is much simplified by using a technique called the Laplace transform.

Laplace transforms

The Laplace transform provides us with a means of transforming differential equations into straightforward algebraic equations, thus making the solution of expressions involving several differential or integral terms relatively simple. The technique is as follows:

(1)   Write down the basic expression for the circuit in terms of voltage, current and component values.
(2)   Transform the basic equation using the table of standard Laplace transforms (each term in the expression is transformed separately).
(3)   Simplify the transformed expression as far as possible. Insert initial values. Also insert component values where these are provided.
(4)   When you simplify the expression, try to arrange it into the same form as one of the standard forms in the table of standard Laplace transforms. Failure to do this will prevent you from performing the inverse transformation (i.e. converting the expression back into the time domain). Note that you may have to use partial fractions to produce an equation containing terms that conform to the standard form.
(5)   Use the table of standard Laplace transforms in reverse to obtain the inverse Laplace transform.

Don't panic if this is beginning to sound very complicated – the three examples that follow will take you through the process on a step by step basis!

Mathematics in action
Laplace transforms

The Laplace transform of a function of time, $f(t)$, is found by multiplying the function by e^{-st} and integrating the product between the limits of zero and infinity. The result (if it exists) is known as the Laplace transform of $f(t)$.

Thus: $F(s) = \mathcal{L}\{f(t)\} = \int_0^\infty e^{-st} f(t)\, \mathrm{d}t.$

Frequently we need only to refer to a table of standard Laplace transforms. Some of the most useful of these are summarised in the table on page 285:

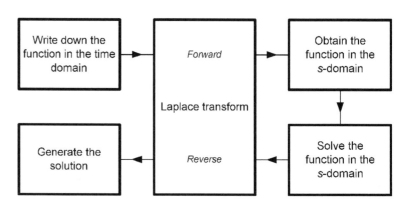

Figure 4.4.4 *Laplace transformation process*

$f(t)$	$F(s) = \mathcal{L}\{f(t)\}$	Comment
a	$\dfrac{1}{s}$	a constant
t	$\dfrac{1}{s^2}$	ramp
e^{at}	$\dfrac{1}{s-a}$	exponential growth
e^{-at}	$\dfrac{1}{s+a}$	exponential decay
$e^{-at}\sin(\omega t)$	$\dfrac{\omega}{(s+a)^2 + \omega^2}$	decaying sine
$e^{-at}\cos(\omega t)$	$\dfrac{s+a}{(s+a)^2 + \omega^2}$	decaying cosine
$\sin(\omega t + \phi)$	$\dfrac{s\sin\phi + \omega\cos\phi}{s^2 + \omega^2}$	sine plus phase angle
$\sin(\omega t)$	$\dfrac{\omega}{s^2 + \omega^2}$	sine
$\cos(\omega t)$	$\dfrac{s}{s^2 + \omega^2}$	cosine
$\dfrac{df(t)}{dt}$	$sF(s) - f(0)$	first differential
$\dfrac{d^2 f(t)}{dt^2}$	$s^2 F(s) - sf(0) - \dfrac{df(0)}{dt}$	second differential
$\int f(t)\,dt$	$\dfrac{1}{s}F(s) + \dfrac{1}{s}F(0)$	integral

Another view

You can think of the Laplace transform as a device that translates a given function in the time domain, $f(t)$, into an equivalent function in the s-domain, $F(s)$. We use the technique to translate a differential equation into an equation in the s-domain, solve the equation, and then use it in reverse to convert the s-domain solution to an equivalent solution in the time domain. Figure 4.4.4 illustrates this process.

Example 4.4.1

Use Laplace transforms to derive an expression for the current flowing in the circuit shown in Figure 4.4.5 given that $i = 0$ at $t = 0$.

The voltage, V, will be the sum of the two voltages, v_R and v_C, where $v_R = Ri(t)$ and

$$v_C = \frac{1}{C}\int i(t)\,dt.$$

Note that to remind us that the current is a function of time we have used $i(t)$ to denote current rather than just i on its own.

Thus $V = v_R + v_C, = Ri(t) + \dfrac{1}{C}\int i(t)\,dt.$

Applying the Laplace transform gives:

$$\mathcal{L}\{V\} = \mathcal{L}\left\{Ri(t) + \frac{1}{C}\int i(t)\,dt\right\} = \mathcal{L}\{Ri(t)\} + \mathcal{L}\left\{\frac{1}{C}\int i(t)\,dt\right\}$$

cont.

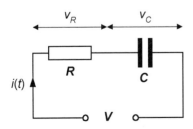

Figure 4.4.5 *See Example 4.4.1*

$$\mathscr{L}\{V\} = R\mathscr{L}\{i(t)\} + \frac{1}{C}\mathscr{L}\left\{\int i(t)\,dt\right\}.$$

Using the table of standard Laplace transforms on page 285 gives:

$$\frac{V}{s} = RI(s) + \frac{1}{sC}[I(s) + I(0)].$$

Note that we are now in the s-domain (s replaces t) and the large I has replaced the small i (just as F replaces f in the general expression).

Now $i = 0$ at $t = 0$ thus $I(0) = 0$, hence:

$$\frac{V}{s} = RI(s) + \frac{1}{sC}[I(s) - 0] = RI(s) + \frac{1}{sC}I(s) = I(s) \times (R + \frac{1}{sC})$$

or $I(s) = \dfrac{V}{s(R + \dfrac{1}{sC})} = \dfrac{V}{Rs + \dfrac{1}{C}} = \dfrac{V}{R}\left(\dfrac{1}{s + \dfrac{1}{CR}}\right)$

Now $i(t) = \mathscr{L}^{-1}\{I(s)\}$ thus:

$$i(t) = \mathscr{L}^{-1}\{I(s)\} = \mathscr{L}^{-1}\left\{\frac{V}{R}\left(\frac{1}{s + \dfrac{1}{CR}}\right)\right\} = \frac{V}{R}\mathscr{L}^{-1}\left\{\frac{1}{s + \dfrac{1}{CR}}\right\}.$$

We now need to find the inverse transform of the foregoing equation. We can do this by examining the table of standard Laplace transforms on page nn looking for $F(s)$ of the form

$$\frac{1}{s + a} \quad (\text{where } a = \frac{1}{CR}).$$

Notice, from the table, that the inverse transform of $\dfrac{1}{s + a}$ is e^{-at}, i.e.:

$$\mathscr{L}^{-1}\left\{\frac{1}{s + a}\right\} = e^{-at}.$$

Hence $i(t) = \mathscr{L}^{-1}\{I(s)\} = \dfrac{V}{R}\mathscr{L}^{-1}\left\{\dfrac{1}{s + \dfrac{1}{CR}}\right\} = \dfrac{V}{R}\left(e^{\frac{-t}{CR}}\right).$

Thus the current in the circuit is given by:

$$i(t) = \frac{V}{R}\left(e^{\frac{-t}{CR}}\right) \quad (\text{exponential growth}).$$

Example 4.4.2

Use Laplace transforms to derive an expression for the current flowing in the circuit shown in Figure 4.4.6 given that $i = 0$ at $t = 0$. The voltage, V, will be the sum of the two voltages, v_R and v_L,

cont.

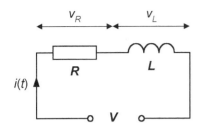

Figure 4.4.6 *See Example 4.4.2*

where $v_R = Ri(t)$ and

$v_L = L\dfrac{di(t)}{dt}$ and $i(t)$ reminds us that the current, i, is a function of time.

Thus $V = v_R + v_L, = Ri(t) + L\dfrac{di(t)}{dt}$.

Applying the Laplace transform gives:

$$\mathscr{L}\{V\} = \mathscr{L}\{Ri(t) + L\dfrac{di(t)}{dt}\} = \mathscr{L}\{Ri(t)\} + \mathscr{L}\{L\dfrac{di(t)}{dt}\}$$

$$\mathscr{L}\{V\} = R\mathscr{L}\{i(t)\} + L\mathscr{L}\{\dfrac{di(t)}{dt}\}.$$

Using the table of standard Laplace transforms on page 285 gives:

$$\frac{V}{s} = RI(s) + sL[I(s) - i(0)].$$

Note that we are now in the s-domain (s replaces t) and the large I has replaced the small i (just as F replaces f in the general expression).
Now $i = 0$ at $t = 0$ thus $i(0) = 0$, hence:

$$\frac{V}{s} = RI(s) + sL[I(s) - 0] = RI(s) + sLI(s) = I(s) \times (R + sL)$$

or $I(s) = \dfrac{V}{s(R + sL)}$.

We can simplify this expression by using partial fractions, thus:

$$\frac{V}{s(R + sL)} = \frac{A}{s} + \frac{B}{R + sL} = \frac{A(R + sL) + Bs}{s(R + sL)}$$

hence $V = A(R + sL) + Bs = AR + AsL + Bs$.

We now need to find the values of A, B and C
when $s = 0$, $V = AR + 0$ thus $A = \dfrac{V}{R}$

when $s = \dfrac{-R}{L}$, $V = AR + A\left(\dfrac{-R}{L}\right) + B\left(\dfrac{-R}{L}\right) = AR - AR - B\dfrac{R}{L}$

thus $V = \dfrac{-BR}{L}$ or $B = \dfrac{-VL}{R}$.

Replacing A and B with $\dfrac{V}{R}$ and $\dfrac{-VL}{R}$, respectively gives:

$$\frac{V}{s(R + sL)} = \frac{A}{s} + \frac{B}{R + sL} = \frac{V}{Rs} + \frac{\left(\dfrac{-VL}{R}\right)}{R + sL} = \frac{V}{Rs} - \frac{VL}{R(R + sL)}$$

cont.

Thus $I(s) = \dfrac{V}{Rs} - \dfrac{VL}{R(R + sL)}$.

We need to find the inverse of the above equation, since:

$$i(t) = \mathcal{L}^{-1}\{I(s)\} = \mathcal{L}^{-1}\left\{\dfrac{V}{Rs} - \dfrac{VL}{R(R + sL)}\right\}$$

thus $i(t) = \mathcal{L}^{-1}\left\{\dfrac{V}{Rs} - \dfrac{VL}{R(R + sL)}\right\} = \dfrac{V}{R}\mathcal{L}^{-1}\left\{\dfrac{1}{s} - \dfrac{L}{R + sL}\right\}$

hence $i(t) = \dfrac{V}{R}\mathcal{L}^{-1}\left\{\dfrac{1}{s} - \dfrac{1}{s + \dfrac{R}{L}}\right\}$.

Referring once again to the table on page 285 gives:

$$i(t) = \dfrac{V}{R}\left(1 - e^{\frac{-Rt}{L}}\right) \quad \text{(exponential decay)}.$$

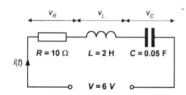

v_R v_L v_C

$R = 10\,\Omega$ $L = 2\,H$ $C = 0.05\,F$

$i(t)$

$V = 6\,V$

Figure 4.4.7   *See Example 4.4.3*

Example 4.4.3

Use Laplace transforms to derive an expression for the current flowing in the circuit shown in Figure 4.4.7 given that $i = 0$ at $t = 0$. Here, the voltage, V, will be the sum of three voltages, v_R, v_L, and v_C, where:

$$v_R = Ri(t), \quad v_L = L\dfrac{di(t)}{dt} \quad \text{and} \quad v_C = \dfrac{1}{C}\int i(t)dt.$$

Yet again $i(t)$ reminds us that the current, i, is a function of time.

Thus $V = v_R + v_L + v_C = Ri(t) + L\dfrac{di(t)}{dt} + \dfrac{1}{C}\int i(t)dt.$

Applying the Laplace transform gives:

$$\mathcal{L}\{V\} = \mathcal{L}\left\{Ri(t) + L\dfrac{di(t)}{dt} + \dfrac{1}{C}\int i(t)dt\right\}$$

$$\mathcal{L}\{V\} = R\mathcal{L}\{i(t)\} + L\mathcal{L}\left\{\dfrac{di(t)}{dt}\right\} + \mathcal{L}\left\{\dfrac{1}{C}\int i(t)dt\right\}$$

Using the table of standard Laplace transforms on page 285 gives:

$$\dfrac{V}{s} = RI(s) + sL[I(s) - i(0)] + \dfrac{1}{sC}[I(s) + I(0)].$$

Now $i = 0$ at $t = 0$ thus $i(0) = 0$ and $I(0) = 0$, hence:

$$\dfrac{V}{s} = RI(s) + sL[I(s) - 0] + \dfrac{1}{sC}[I(s) + I(0)]$$

$$\dfrac{V}{s} = RI(s) + sLI(s) + \dfrac{1}{sC}I(s) = I(s) \times (R + sL + \dfrac{1}{sC})$$

cont.

or $I(s) = \dfrac{V}{s(R + sL + \dfrac{1}{sC})} = \dfrac{V}{sR + s^2L + \dfrac{1}{C}}$.

Now $V = 6\,V$, $L = 2\,H$, $R = 12\,\Omega$ and $C = 0.05\,F$, thus:

$$I(s) = \frac{V}{s^2L + sR + \dfrac{1}{C}} = \frac{6}{2s^2 + 12s + \dfrac{1}{0.05}} = \frac{3}{s^2 + 6s + 10}. \quad \text{(i)}$$

In order to perform the inverse Laplace transformation we need to express equation (i) in a form that resembles one of the standard forms listed in the table on page 285. The nearest form is as follows:

$F(s) = \dfrac{\omega}{(s + a)^2 + \omega^2}$ which has $f(t) = e^{-at}\sin\,\omega t$ as its reverse transform.

Fortunately, it's not too difficult to rearrange equation (i) into this form:

$$I(s) = 3 \times \frac{1}{(s + 3)^2 + 1}$$

from which $a = 3$ and $\omega = 1$, thus the reverse transform is:

$i(t) = e^{-3t}\sin\,t$ (a decaying sine wave).

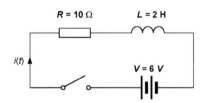

Figure 4.4.8 *See Questions 4.4.1*

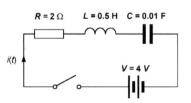

Figure 4.4.9 *See Questions 4.4.1*

Questions 4.4.1

(1) A voltage given by $V = 50\sin(100t + \pi/4)$ is applied to a resistor of $100\,\Omega$. Use the table of standard Laplace transforms to write down an expression for the current flowing in the resistor in the s-domain.

(2) The current flowing in a circuit in the s-domain is given by:

$$I(s) = \frac{0.025}{0.025s + 1}\ mA.$$

Use the inverse Laplace transform to write down an expression for the current in the time domain. Hence determine the current flowing in the circuit when $t = 12.5\,ms$.

(3) Use Laplace transforms to derive an expression for the current flowing in the circuit shown in Figure 4.4.8 given that $i = 0$ at $t = 0$.

(4) Use Laplace transforms to derive an expression for the current flowing in the circuit shown in Figure 4.4.9 given that $i = 0$ at $t = 0$.

5 Mechanical principles

Summary

In this chapter we will cover a range of mechanical principles which underpin the design and operation of mechanical engineering systems. The material is intended to extend the knowledge gained from Engineering Science (Chapter 3), which you should have studied first. It includes coverage of the strength of materials and mechanics of machines. The primary aim is to provide a firm foundation for work in engineering design and for the material to provide a basis for more advanced study.

5.1 Complex loading systems

You may remember from Chapter 3 that we defined Poisson's ratio and very briefly looked at bulk modulus. In this section we are going to consider these subjects in a little more detail, by determining the behaviour of materials under complex loading conditions. We start by looking once again at Poisson's ratio, which is valid providing the material is only subjected to loads within the elastic range.

$$\text{Poisson's ratio } (\nu) = -\frac{\text{lateral strain}}{\text{axial strain}}. \tag{5.1.1}$$

Look back at thermal stress and strain in Chapter 3, if you cannot remember the significance of the minus sign.

For a three-dimensional solid, such as a bar in tension, the lateral strain represents a decrease in width (negative lateral strain) and the axial strain represents elongation (positive longitudinal strain). In tables it is normal to show only the magnitudes of the strains considered, so tabulated values of Poisson's ratio are positive. Table 5.1.1 shows typical values of Poisson's ratio, for a variety of materials. Note that for most metals, when there is no other available information, Poisson's ratio may be taken to be 0.3.

Table 5.1.1 *Typical values of Poisson's ratio for a variety of materials*

Aluminium	0.33
Manganese Bronze	0.34
Cast iron	0.2–0.3
Concrete (non-reinforced)	0.2–0.3
Marble	0.2–0.3
Nickel	0.31
Nylon	0.4
Rubber	0.4–0.5
Steel	0.27–0.3
Titanium	0.33
Wrought iron	0.3

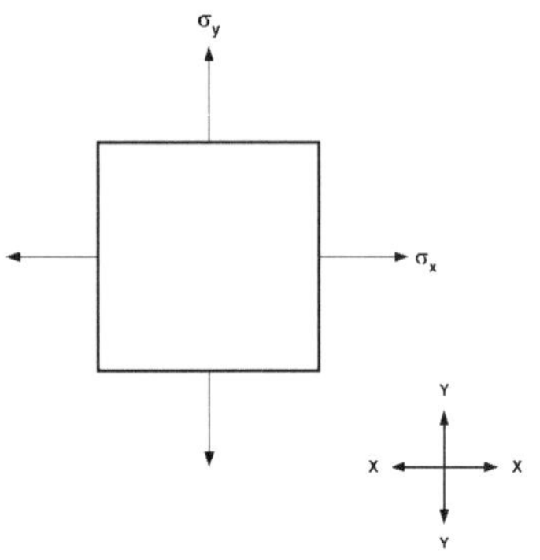

Figure 5.1.1 *Two-dimensional loading system*

Poisson's ratio and two-dimensional loading

A *two-dimensional stress system* is one in which all the stresses lie within one plane, such as the plane of this paper the *xy* plane. Consider the flat plate (Figure 5.1.1) which is subjected to two dimensional stress loading, as shown.

We know from our previous work that for a material within its elastic limit, we have from Hooke's law that:

Elastic modulus $E = \dfrac{\text{stress}}{\text{strain}} = \dfrac{\sigma}{\epsilon}$ and so strain $= \dfrac{\sigma}{E}$.

Also: lateral strain $= \nu$ (axial or longitudinal strain) from *Poisson's ratio*. Then in the *XX* direction we have:

strain due to the stress acting along the *XX* direction $= \dfrac{\sigma_x}{E}$ (tensile)

and strain due to stress acting along the *YY* direction $= -\dfrac{\nu\sigma_y}{E}$ (compressive)

(note minus sign for compressive strain in accordance with convention)
so combined strain due to the stresses acting perpendicular to each other in the *x* and *y* directions is:

combined strain in *XX* direction $\epsilon_x = \dfrac{\sigma_x}{E} - \dfrac{\nu\sigma_y}{E}$ $\qquad$ (5.1.2)

Similarly:

combined strain in *YY* direction $\epsilon_y = \dfrac{\sigma_y}{E} - \dfrac{\nu\sigma_x}{E}$ $\qquad$ (5.1.3)

Multiplying Equations 5.1.2 and 5.1.3 by *E* and solving simultaneously for the stresses σ_x and σ_y yields the following equations which relate stress, strain and Poisson's ratio:

$$\sigma_x = \frac{E}{(1 - \nu^2)}(\epsilon_x + \nu\epsilon_y) \qquad (5.1.4)$$

and $\sigma_y = \dfrac{E}{(1 - \nu^2)}(\epsilon_y + \nu\epsilon_x)$ $\qquad$ (5.1.5)

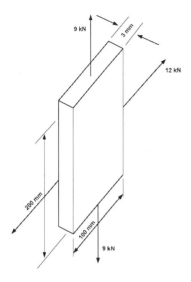

Figure 5.1.2 *Example 5.1.1 – plate loaded in two dimensions*

Example 5.1.1

If the plate loaded as shown in Figure 5.1.2, is made from a steel with $v = 0.3$ and $E = 205$ GPa.
Determine the changes in dimension in both the x and y directions.
In the x direction the plate is subject to a 12 kN load acting over an area of 600 mm².
Thus the stress in the x direction = L/A = 20 N/mm².
Similarly, the stress in the y direction = 9000/300 N/mm² = 30 N/mm².
Then from Equation 5.1.2 the *total strain in the x direction* is:

$$\varepsilon_x = \frac{20 - 0.3(30)}{205 \times 10^3} \frac{\text{N/mm}^2}{\text{N/mm}^2} \quad \text{(notice the careful manipulation of units)}$$

$$\varepsilon_x = 5.37 \times 10^{-5}$$

and since strain = change in length/original length then:

change in length = (strain) × (original length)

$$= (5.37 \times 10^{-5}) \times (100) \text{ mm} = 0.00537 \text{ mm}.$$

So change in dimension in x direction is an extension = 0.00537 mm.

Similarly from Equation 5.1.3 the *total strain in the y direction* is:

$$\varepsilon_y = \frac{30 - 0.3(20)}{205 \times 10^3} \quad \varepsilon_y = 1.17 \times 10^{-4}$$

and so change in length = $(1.17 \times 10^{-4}) \times (200)$ mm = 0.0234 mm.

So change in dimension in y direction is an extension = 0.0234 mm.

Questions 5.1.1

(1) A metal bar 250 mm long has a rectangular cross-section of 60 mm × 25 mm. It is subjected to an axial tensile force of 60 kN. Find the change in dimensions if the metal has an elastic modulus $E = 200$ GPa and Poisson's ratio is 0.3.

(2) A rectangular steel plate 200 mm long and 50 mm wide is subject to a tensile load along its length. If the width of the plate contracts by 0.005 mm find the change in length. Take Poisson's ratio for the steel as 0.28.

(3) A flat aluminium plate is acted on by mutually perpendicular stresses σ_1 and σ_2 as shown in Figure 5.1.3. The corresponding strains resulting from these stresses are $\varepsilon_1 = 4.2 \times 10^{-4}$ and $\varepsilon_2 = 9.0 \times 10^{-5}$. Find the values of the stress if E for aluminium = 70 GPa and the value of Poisson's ratio is that given for aluminium in Table 5.1.1.

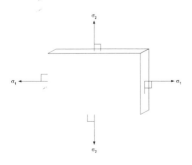

Figure 5.1.3 *Figure for Question 3 (Questions 5.1.1)*

Figure 5.1.4 *One-dimensional stress leading to three-dimensional strain*

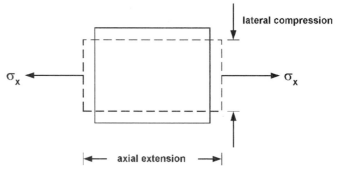

Figure 5.1.4a

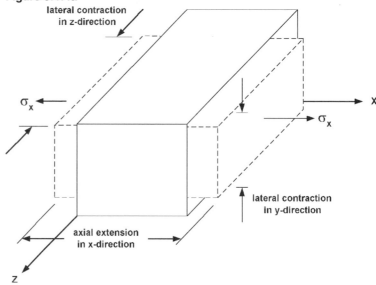

Figure 5.1.4b

Poisson's ratio and three-dimensional loading

You know from our work on two-dimensional strain that if, for example, a flat plate is subject to an axial tensile stress (Figure 5.1.4a) then there is an axial extension and lateral contraction of the plate. This relationship is given by Equation 5.1.2, assuming that the axial strain is in the *x* direction.

If we extend the argument to three dimensions, you can see (Figure 5.1.4b) that an axial extension of the bar (*x* direction) results in a lateral contraction in both the mutually perpendicular *y* and *z* directions.

Then by analogy, three-dimensional strain resulting from a tensile stress in the *x* direction is given by:

$$\varepsilon_x = \frac{\sigma_x}{E} - \frac{\nu\sigma_y}{E} - \frac{\nu\sigma_z}{E} \quad \text{or} \quad \varepsilon_x = \frac{1}{E}(\sigma_x - \nu\sigma_y - \nu\sigma_z). \tag{5.1.6}$$

Similarly the strains in the *y* and *z* directions in terms of the mutually perpendicular stresses are represented by:

$$\varepsilon_y = \frac{1}{E}(\sigma_y - \nu\sigma_x - \nu\sigma_z) \tag{5.1.7}$$

$$\varepsilon_z = \frac{1}{E}(\sigma_z - \nu\sigma_x - \nu\sigma_y). \tag{5.1.8}$$

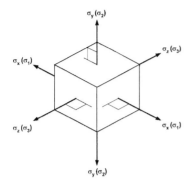

Figure 5.1.5 *Principal stresses*

Principal strains in terms of stresses

In the absence of shear stresses on the faces of the element shown in Figure 5.1.5, the stresses σ_x, σ_y, σ_z are in fact principal stresses. So the principal strain in a given direction may be obtained from the principal stresses.

For principal stresses and strains the suffix, x, y and z are replaced by 1, 2 and 3, respectively. So we may rewrite the above equations in terms of principal stresses and strains as

$$\varepsilon_1 = \frac{1}{E}(\sigma_1 - v\sigma_2 - v\sigma_3) \tag{5.1.6a}$$

$$\varepsilon_2 = \frac{1}{E}(\sigma_2 - v\sigma_1 - v\sigma_3) \tag{5.1.7a}$$

$$\varepsilon_3 = \frac{1}{E}(\sigma_3 - v\sigma_1 - v\sigma_2). \tag{5.1.8a}$$

Bulk modulus

You have already met the elastic constants, where Young's modulus E and the shear modulus G are defined as the ratio of stress/strain for direct and shear loading, respectively.

Also in Chapter 3 the bulk modulus was defined in Equation 3.1.15

as: $\text{K} = \dfrac{p}{\delta V/V}$

where p was the uniform pressure or *volumetric stress* and $\delta V/V$ was the *volumetric strain*. Equation 3.1.15 may therefore be written as:

$$\text{Bulk modulus (K)} = \frac{\text{volumetric stress } (\sigma)}{\text{volumetric strain } (\varepsilon_v)}. \tag{5.1.9}$$

Volumetric strain

We know from the definition, that volumetric strain

$$= \frac{\text{change in volume } (\delta V)}{\text{original volume } (V)}.$$

Now it can be shown that:

volumetric strain = the sum of the linear strains in the x, y *and* z *directions*

or $\quad \varepsilon_v = \varepsilon_x + \varepsilon_y + \varepsilon_z. \tag{5.1.10}$

Now substituting Equations 5.1.6, 5.1.7 and 5.1.8 into Equation 5.1.10 we produce an equation for the stresses in the x, y and z directions in terms of the volumetric strain, that is:

$$\varepsilon_v = \frac{1}{E}(\sigma_x - v\sigma_y - v\sigma_z) + \frac{1}{E}(\sigma_y - v\sigma_x - v\sigma_z) + \frac{1}{E}(\sigma_z - v\sigma_x - v\sigma_y)$$

and on rearrangement we get:

$$\varepsilon_v = \frac{1}{E}(\sigma_x + \sigma_y + \sigma_z)(1 - 2v). \tag{5.1.11}$$

Now let us consider the circular bar shown (Figure 5.1.6) which is subject to a uniaxial stress σ_x, then the stresses in the y and z planes are zero, therefore from Equation 5.1.11, we have:

$$\varepsilon_v = \frac{\sigma_x}{E}(1 - 2v) \text{ since } \sigma_y = \sigma_z = 0.$$

Figure 5.1.6 *Bar subject to uniaxial stress*

Circular bar subject to uniaxial tensile stress σ_x, causing
lateral contraction and axial elongation with $\sigma_y = \sigma_z = 0$

We also know that the volumetric strain $\varepsilon_v = dV/V$ and letting $\sigma_x = \sigma$, then the general equation for calculating the volumetric strain or the change in volume of a circular bar is given as:

$$\varepsilon_v = \frac{\delta V}{V} = \frac{\sigma}{E}(1 - 2\,v). \tag{5.1.12}$$

Example 5.1.2

A bar of circular cross-section (Figure 5.1.7) is subject to a tensile load of 100 kN, which is within the elastic range. The bar is made from a steel with $E = 210$ GPa and Poisson's ratio $= 0.3$, it has a diameter of 25 mm and is 1.5 m long. Determine the extension of the bar, the decrease in diameter and the increase in volume of the bar.

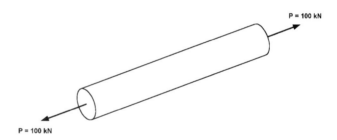

P = 100 kN

P = 100 kN

Figure 5.1.7 *Circular section bar subject to a tensile load of 100 kN*

Since we are told that the loaded bar is in the elastic range then we may find the axial strain from:

$\varepsilon = \dfrac{\sigma}{E}$ where $\sigma = 10^5/490.87)$ N/mm² $= 203.7$ MN/m²

then the *axial strain* $(\varepsilon) = \dfrac{203.7 \text{ MPa}}{210 \text{ GPa}} = 0.00097.$

Now the extension (change in length) is $=$ strain $\times$ original length

extension $= (0.00097 \times 1.5) = 1.455$ mm.

To find the decrease in diameter we first find the lateral strain, from Poisson's ratio:

$\varepsilon_{\text{lat}} = -v\varepsilon = (0.3 \times 0.00097) = -0.000291$

since decrease in diameter $=$ lateral strain $\times$ original diameter then:

decrease in diameter $= (-0.000291 \times 25) = 0.007275$ mm.

The change in volume is given by Equation 5.1.12, that is:

$\delta V = V\dfrac{\sigma}{E}(1 - 2v)$ or $\delta V = V\varepsilon(1 - 2v).$

So $\delta V = (490.87 \times 10^{-6} \times 1.5)(0.00097)(1 - 0.6) = 285.0$ mm³.

Relationship between the elastic constants and Poisson's ratio

You will remember from our review of shear stress and strain in Chapter 3 (page 87) that the modulus of rigidity, the *shear modulus* G is given by:

$$G = \frac{\tau}{\gamma}. \qquad (5.1.13)$$

Mathematics in action

We will use Equation 5.1.13 and Figure 5.1.8 to help us obtain an important relationship between the shear modulus G, Young's modulus E, and Poisson's ratio v.

The square section stress element ABCD (Figure 5.1.8a) is subject to pure shear by the application of stresses τ, the faces are then distorted into a rhombus as shown in Figure 5.1.8b. The diagonal *bd* has lengthened, while the diagonal *ac* has shortened, as a result of the shear stresses (note that the corresponding strains would in practice be very small, they are exaggerated in Figure 5.1.8b for the purpose of illustration).

Now since the strains are small, angles *abd* and *bde* are each approximately 45° and so the triangle *abd* has sides with the ratios; $1:1:\sqrt{2}$. Now by simple geometry:

the length of side $bd = s\sqrt{2}(1+\varepsilon_{max})$

where ε_{max} is the normal strain in the 45° direction along *bd*. In Figure 5.1.8b the shear strain is given by the angle of deformation in radians. You should now see that the angles *bda* and *dba* are both equal to $\pi/4 - \gamma/2$, and also due to the extension of diagonal *bd* the angle *dab* is equal to $\pi/2 + \gamma$, as shown.

Now using the *cosine rule* for triangle *dab* we have:

$$s\sqrt{2}(1+\varepsilon_{max})^2 = s^2 + s^2 - 2s^2 \cos\left(\frac{\pi}{2} + \gamma\right).$$

So $2s^2(1+\varepsilon_{max})^2 = 2s^2 - 2s^2 \cos\left(\frac{\pi}{2} + \gamma\right)$

and on division by $2s^2$ we have:

$$(1+\varepsilon_{max})^2 = 1 - \cos\left(\frac{\pi}{2} + \gamma\right).$$

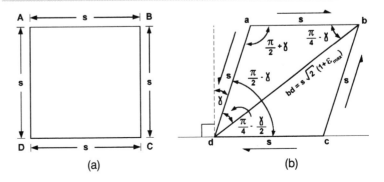

Figure 5.1.8 *Square section stress element subject to pure shear*

(a)   (b)

Now using the trigonometric identity $\cos(\pi/2+x) = -\sin x$ and expanding LHS gives:

$$1 + 2\varepsilon_{max} + (\varepsilon_{max})^2 = 1+\sin \gamma.$$

Also since ε_{max} and γ are very small strains we may use the approximation that $\sin \gamma = \gamma$ and also ignore $(\varepsilon_{max})^2$ when compared to $2\varepsilon_{max}$.

So now we have:

$$1 + 2\varepsilon_{max} = 1 + \gamma$$

or $\varepsilon_{max} = \dfrac{\gamma}{2}$. (5.1.14)

This condition shows the relationship is for pure shear between the normal strain acting in the 45° direction.

Equation 5.1.14 shows the relationship between the normal strain in the 45° direction ε_{max} and the shear strain γ.

Now consider the strains that occur in the stress element orientated at zero degrees, shown in Figure 5.1.9a. This can be replaced (from Hooke's law) by a system of *direct stresses* orientated at 45° as shown in Figure 5.1.9b.

Then using Equation 5.1.2 for the direct stress system along the diagonals and, noting the sign convention for the stresses we have:

The strain in the direction of positive normal diagonal,

$$\varepsilon_{max} = \frac{\sigma_1}{E} - \frac{\nu\sigma_2}{E}$$

$$= \frac{\tau}{E} - \frac{\nu}{E}(-\tau)$$

$$\varepsilon_{max} = \frac{\tau}{E}(1+\nu). \tag{5.1.15}$$

Also from Equation 5.1.13 we have: $\quad \gamma = \dfrac{\tau}{G}.$

Now substituting this expression for γ and the expression for ε_{max} (Equation 5.1.15) into Equation 5.1.14 gives:

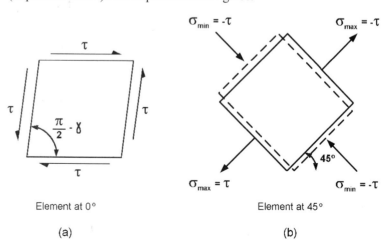

Element at 0°

(a)

Element at 45°

(b)

Figure 5.1.9 *Stresses resulting from pure shear*

$$\frac{\tau}{E}(1+v) = \frac{\tau/G}{2} \text{ and so } G = \frac{E}{2(1+v)} \qquad (5.1.16)$$

Do not worry too much if you were unable to follow the mathematical argument, much more important is that you understand, and are able to use, the relationships we have found. In determining the relationship given in Equation 5.1.16, we have only considered an element of material under *pure shear* conditions. The stresses imposed under these conditions are shown in Figure 5.1.9.

We will return to the relationships between shear stress and shear strain in the next section when we look at the loading of thin walled cylinders, such as pressure vessels. Before we leave our study of Poisson's ratio and the elastic constants, let us look at one or two examples.

Example 5.1.3

Assume that the cube is immersed in a fluid (Figure 5.1.10), which subjects the vessel to volumetric (hydrostatic) stress. Show, stating all assumptions, that the elastic constants E and K and Poisson's ratio are related, by the expression:

$$K = \frac{E}{3(1-2v)}.$$

In order to find this relationship we need to remember one very important fact which is:

volumetric strain = the sum of the strains in the x,y *and* z *directions.*

Now for hydrostatic stress the stresses in the *x*, *y* and *z* directions are equal so:

volumetric strain = $3 \times$ the linear strain.

Assumption: that all surfaces of the pressure vessel are subject to equal pressure and have the same material properties. Then, side wall stresses are equal and volumetric strain is equal to three times the linear strain.
So from Equation 5.1.6 where:

$\varepsilon_x = \frac{1}{E}(\sigma_x - v\sigma_y - v\sigma_z)$ the volumetric strain, on rearrangement,

is given by: $\quad \varepsilon_v = \frac{3\sigma}{E}(1-2v) \qquad (5.1.17)$

where $\sigma_x = \sigma_y = \sigma_z = \sigma$.

Now we need an expression involving the bulk modulus K, if we are to relate the elastic constants E and K with Poisson's ratio.
We know that the bulk modulus K = volumetric stress/ volumetric strain therefore:

volumetric strain $\varepsilon_v = \frac{\sigma}{K} \qquad (5.1.18)$

and from Equations (5.1.17) and (5.1.18) we have:

cont.

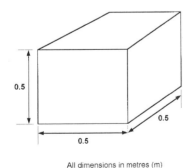

All dimensions in metres (m)

Figure 5.1.10 *Figure for Example 5.1.3*

$$\frac{\sigma}{K} = \frac{3\sigma}{E}(1 - 2v) \text{ and on rearrangement:}$$

$$K = \frac{E}{3(1 - 2v)}. \tag{5.1.19}$$

Equation 5.1.19 is the relationship we require.

Example 5.1.4

A cylindrical compressed air cylinder 1.5 m long and 0.75 m internal diameter has a wall thickness of 8 mm. Find the increase in volume when the cylinder is subject to an internal pressure of 3.5 MPa, also find the values for the constants G and K. Given that E = 207 GPa and Poisson's ratio v = 0.3.

For this problem, the increase in volume can be found directly using Equation (5.2.6), which you will meet in the next section. Then:

$$\text{change in internal volume} = \frac{pd}{4tE}(5 - 4v)V$$

where:
p = internal pressure
v = Poisson's ratio
d = internal diameter
V = internal volume
t = wall thickness.

So change in volume:

$$= \frac{(3.5 \times 10^6)(0.75)(5 - 4 \times 0.3)(0.6627)}{(4)(0.008)(207 \times 10^9)}$$

change in volume = 0.000998 m³
Also using Equations (5.1.16) and (5.1.19) then:

$$G = \frac{207 \times 10^9}{2(1 + 0.3)} = 79.6 \text{ GPa}$$

$$\text{and } K = \frac{207 \times 10^9}{3(1 - 0.6)} = 172.5 \text{ GPa}$$

Questions 5.1.2

(1) The principal strains (ε_1 and ε_2) at a point on a loaded steel sheet are found to be 350×10^{-6} and 210×10^{-6}, respectively. If the modulus of elasticity E = 210 GPa and Poisson's ratio for the steel is 0.3. Determine the corresponding principal stresses.

(2) An aluminium alloy has a modulus of elasticity of 80 GPa and a modulus of rigidity of 33 GPa. Determine the value of Poisson's ratio and the bulk modulus.

(3)   A rod 30 mm in diameter and 0.5 m in length is subjected to an axial tensile load of 90 kN. If the rod extends in length by 1.1 mm and there is a decrease in diameter of 0.02mm. Determine the values of Poisson's ratio and the values of E, G and K.

5.2 Loaded cylinders and beams

We continue with our theme of complex loading by considering thin-walled pressure vessels and thick-walled cylinders. Pressure vessels subject to both internal and external pressure include, compressed air receivers, boilers, submarine hulls, aircraft fuselages, hydraulic reservoirs and condenser casings, to name but a few. Thick-walled cylinders subject to pressure include, hydraulic linear actuators, extrusion dies, gun barrels and high pressure gas bottles.

Thin-walled pressure vessels

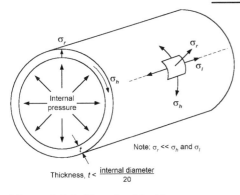

Figure 5.2.1 shows the note: $\sigma_r \ll \sigma_h$ and σ_l

Thickness, $t < \dfrac{\text{internal diameter}}{20}$

Figure 5.2.1 *Vessel subject to internal pressure*

When a thin-walled pressure vessel is subject to internal pressure, three mutually perpendicular principal stresses are set up in the vessel material. These are the *hoop stress* (often referred to as the circumferential stress), the *radial stress* and the *longitudinal stress*. We may define a *thin-walled pressure vessel* as one in which the *ratio* of the wall thickness to that of the inside diameter of the vessel is *greater* than 1:20, that is *less* than the fraction 1/20. Under these circumstances it is reasonable to assume that the hoop and longitudinal stresses are uniform across the wall thickness and that the radial stress is so small in comparison with the hoop and longitudinal stresses that for the purpose of our calculations it may be ignored.

Figure 5.2.1 shows a vessel subject to internal pressure and the nature of the resulting, hoop, longitudinal and radial stresses.

Hoop and longitudinal stress

Hoop or *circumferential stress* is the stress which resists the bursting effect of the applied internal pressure. If the internal diameter is (d) the applied pressure (p), the length (L) and the wall thickness (t). Then the force tending to burst or separate the vessel (Figure 5.2.2a) is given by pressure × internal diameter × length or, *force = pdL*. This is resisted by the hoop stress (σ_h) acting on an area $2tL$, so for equilibrium these values may be equated i.e.

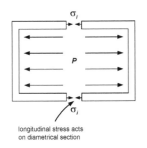

Internal pressure resisted by hoop stress

(a)

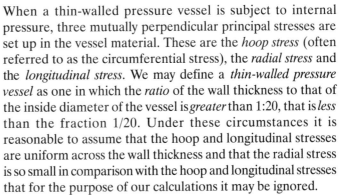

longitudinal stress acts on diametrical section

Internal pressure resisted by longitudinal stress

(b)

Figure 5.2.2 *Hoop and longitudinal stress resulting from internal pressure*

hoop stress $= F/A$ so $F = \sigma_h \times A$ or $F = (\sigma_h)(2tL)$

also from above $F = pdL$

then: $pdL = 2\sigma_h tL$ so $\sigma_h = \dfrac{pd}{2t}$.   (5.2.1)

In a similar manner the force trying to separate the pressure vessel across its diameter, that is longitudinally (Figure 5.2.2b) is given by:

pressure × area $= \dfrac{p\pi d^2}{4}$.

This force is resisted by the longitudinal stress (σ_L) acting on the area pdt so for equilibrium we have:

$\dfrac{p\pi d^2}{4} = \sigma_L \pi dt$ so $\sigma_L = \dfrac{pd}{4t}$.   (5.2.2)

In developing the formulae for hoop stress and longitudinal stress we have ignored the efficiency of the joints that go to make-up a pressure vessel. You will note that the hoop stress acts on a longitudinal section and that the longitudinal stress acts on a diametrical section (Figure 5.2.2). It therefore follows that, the efficiency of the longitudinal joints which go to make-up the pressure vessel directly affects the hoop stress at the joint. Similarly the longitudinal stress is affected by the efficiency of the diametrical joints.

The joint efficiency (η) is often quoted as; the strength of the joint/ strength of the undrilled plate. So taking this into account for hoop stress we may write:

$$\sigma_h = \frac{pd}{2t\eta_l} . \qquad (5.2.3)$$

Similarly for longitudinal stress we may write:

$$\sigma_L = \frac{pd}{4t\eta_h} . \qquad (5.2.4)$$

Example 5.2.1

A cylindrical compressed air vessel has an internal diameter of 50 cm and is manufactured from steel plate 4 mm thick. The longitudinal plate joints have an efficiency of 90% and the circumferential joints have an efficiency of 55%. Determine the maximum air pressure which may be accommodated by the vessel if the maximum permissible tensile stress of the steel is 100 MPa.

Using Equations 5.2.3 and 5.2.4 we have:

(i) $\quad \sigma_h = \dfrac{pd}{2\eta_l t} = \dfrac{0.5p}{2 \times 0.9 \times 0.004}$

and since maximum stress = 100 MPa

then the maximum air pressure accommodated by the longitudinal joints is 1440 kPa

(ii) similarly: $\sigma_L = \dfrac{pd}{4\eta_h t} = \dfrac{0.5p}{4 \times 0.55 \times 0.004}$

then maximum air pressure accommodated by circumferential joints is 1760 kPa.

So the maximum air pressure is determined by the longitudinal joints and is 1440 kPa.

Dimensional change in thin cylinders

The change in length of a thin cylinder subject to internal pressure may be determined from the longitudinal strain, providing we ignore the radial stress. The longitudinal strain may be found from Equation 5.1.2, i.e.

$$\varepsilon_l = \frac{\sigma_l}{E} - \frac{\nu\sigma_h}{E}$$

and since change in length = longitudinal strain × original length

then change in length $= \dfrac{1}{E}(\sigma_L - \nu\sigma_h)L$

so on substituting Equations 5.2.1 and 5.2.2 for hoop stress and longitudinal stress into the above equation for change in length, then:

$$\text{change in length} = \frac{pd}{4tE}(1-2v)L \qquad (5.2.5)$$

We can also find a relationship for the change in internal volume of a thin-walled cylinder by considering some of our earlier work on volumetric strain. You will remember that volumetric strain is equal to the sum of the mutually perpendicular linear strains (Equation 5.1.10). For a cylinder we have longitudinal strain (ε_L) and two perpendicular diametrical strains, so from:

$$\varepsilon_v = \varepsilon_x + \varepsilon_y + \varepsilon_z \text{ (Equation 5.1.10)}$$

on substituting $2\varepsilon_D$ (diametrical strain for $\varepsilon_y + \varepsilon_z$) then:

the volumetric strain for a cylinder is $= \varepsilon_L + 2\varepsilon_D$

and knowing that longitudinal strain $= 1/E\,(\sigma_L - v\sigma_h)$ and substituting into above equation for volumetric strain, then:

$$\varepsilon_v = \frac{1}{E}(\sigma_L - v\sigma_h) + \frac{2}{E}(\sigma_h - v\sigma_L)$$

and finally, substituting Equation 5.2.5 into above equation after simplification, gives:

$$\text{change in internal volume} = \frac{pd}{4tE}(5-4v)V \qquad (5.2.6)$$

The full derivation of Equation 5.2.6 together with the determination of a relationship for the change in diameter, is left as an exercise at the end of this section.

Example 5.2.2

A thin cylinder has an internal diameter of 60 mm, and is 300 mm in length. If the cylinder has a wall thickness of 4 mm and, is subject to an internal pressure of 80 kPa. Then assuming Poisson's ratio is 0.3 and $E = 210$ GPa, determine the:

(a)　change in length of the cylinder;
(b)　hoop stress;
(c)　longitudinal stress.

(a)　Using Equation 5.2.5 then, change in length $= \dfrac{pd}{4tE}(1-2v)L$

$$= \frac{80\times10^5 \times 60\times10^{-3}\times 300\times10^{-3}}{4\times4\times10^{-3}\times 210\times10^9}(1-0.6)$$

$$= 17.14\times10^{-6}\text{ m} \qquad \text{or} \qquad 17.14\ \mu\text{m}.$$

(b)　Hoop stress is given by $\sigma_h = \dfrac{pd}{2t} = \dfrac{80\times10^5\times 60\times10^{-3}}{2\times4\times10^{-3}}$

$$= 60 \text{ MPa}.$$

(c)　Longitudinal stress given by $\sigma_L = \dfrac{pd}{4t} = \dfrac{80\times10^5\times 60\times10^{-3}}{4\times4\times10^{-3}}.$

Pressure vessel applications

An application of the theory of thin cylinders and other thin- and thick-walled shells may be applied to the design and manufacture of pressure vessels. These vessels could be used for food or chemical processing, the storage of gases and liquids under pressure or even the fuselage of a pressurised aircraft. Whatever the industrial use, formulae can be simply derived from some of the relationships we have already found.

Let us first consider some useful relationships which aid the design of cylinders and spherical shells for the plant and process industry.

For a thin-walled cylindrical shell the minimum thickness required to resist internal pressure can be determined from Equation 5.2.1. We use this equation because the hoop stress is greater than the longitudinal stress and so, must always be considered first.

Let (d) be the internal diameter and (e) be the minimum thickness required, then the mean diameter will be given by:

$d + (2 \times e/2)$ or simply $(d+e)$

and substituting this into Equation 5.2.1 gives:

$$e = \frac{p(d+e)}{2\sigma_d}$$

where σ_d is the design stress and p the internal pressure. Rearranging the above formula gives:

$$e = \frac{pd}{2\sigma_d - p} \tag{5.2.7}$$

which is the form the equation takes in British Standards 5500, the standards manual used by pressure vessel designers.

The relationship for the principal stresses of a spherical shell is similar to those we have found for the hoop stress and longitudinal stress (principal stresses) of a thin-walled cylinder.

That is, the principal stresses for a spherical shell are given by:

$$\sigma_1 = \sigma_2 = \frac{pd}{4t}. \tag{5.2.8}$$

Then in a similar manner to above, an equation for minimum thickness of a sphere can be obtained from Equation 5.2.8, this is:

$$e = \frac{pd}{4\sigma_d - p}. \tag{5.2.9}$$

Equation 5.2.9 differs slightly from that given in BS 5500, because it is derived from thick-walled shell theory, which we will be looking at next. The formula for determining the minimum thickness of material required to resist internal pressure within a sphere is given in BS 5500 as:

$$e = \frac{pd}{4\sigma_d - 1.2p}. \tag{5.2.10}$$

If our pressure vessel has welded joints then we need to take into account the integrity of these joints, by using joint efficiency factors (η). The efficiency of welded joints are dependent on the heat treatment received after welding and, on the amount of non-destructive examination used to ascertain their integrity. If the joint integrity can be totally relied upon as a result of heat treatment and non-destructive examination, then the joint may be considered to be 100% efficient. In practice welded joint efficiency factors vary from about 0.65 up to 1.0. In other words the efficiency of welded joints normally ranges from around 65% up to 100%.

If we wish to consider cylindrical and spherical welded vessels then Equations 5.2.7 and 5.2.10 may be written as follows:

for a cylinder $\quad e = \dfrac{pd}{2\sigma_d\eta - p}$ (5.2.11)

for a sphere $\quad e = \dfrac{pd}{4\sigma_d\eta - p}$. (5.2.12)

When designing cylindrical pressure vessels due consideration must be given to the way in which the vessels are closed at their ends. For example, flat plates or domes of various shape could be used. There are in fact four principal methods used to close cylindrical pressure vessels, these are: flat plates and formed flat heads, hemispherical heads, torispherical and ellipsoidal heads.

Flat plates may be plain or flanged, and then bolted or welded in position (Figure 5.2.3). Torispherical domed heads, which are often used to close cylindrical pressure vessels are formed from part of a torus and part of a sphere (Figure 5.2.4). Torispherical heads are generally preferred to their elliptical counterparts (Figure 5.2.5), because they are easier and cheaper to fabricate. Hemispherical heads (Figure 5.2.6) are the strongest form of closure that can be used with cylindrical pressure vessels and are generally used for high pressure applications. They are however costly, so their use is often restricted to applications where pressure containment and safety take priority over all other considerations, including costs.

Minimum thickness formulae for the methods of closure can be determined in a similar manner to those we found earlier for cylinders and spheres, by analysing their geometry, stresses and method of constraint at their periphery. Time does not permit us to derive these relationships but three typical formulae for flat, ellipsoidal and torispherical heads are given below.

The equation for determining the thickness of flat heads is:

$$e = c_p d_e \sqrt{\dfrac{p}{\sigma_d}}$$ (5.2.13)

where c_p = a design constant, dependent on edge constraint,
σ_d = design stress,
d_e = nominal diameter.

The minimum thickness required for ellipsoidal heads is given by:

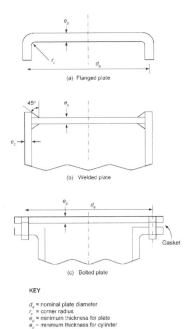

(a) Flanged plate

(b) Welded plate

(c) Bolted plate

KEY

d_e = nominal plate diameter
r_c = corner radius
e_p = minimum thickness for plate
e_c = minimum thickness for cylinder

Figure 5.2.3 *Flat plate closures for cylindrical pressure vessel*

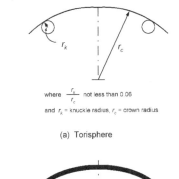

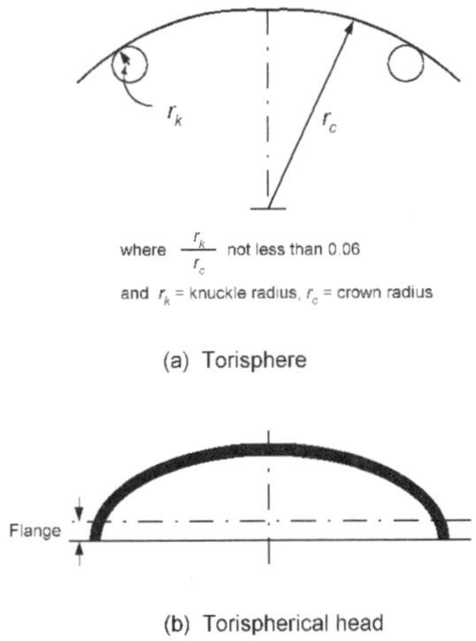

where $\dfrac{r_k}{r_c}$ not less than 0.06

and r_k = knuckle radius, r_c = crown radius

(a) Torisphere

(b) Torispherical head

Figure 5.2.4 *Torispherical domed head for cylindrical pressure vessel*

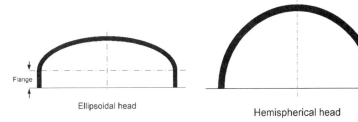

Ellipsoidal head

Figure 5.2.5 *Ellipsoidal head for cylindrical pressure vessel*

Hemispherical head

Figure 5.2.6 *Hemispherical head for cylindrical pressure vessel*

$$e = \frac{pd}{2\eta\sigma_d - 0.2p} \qquad (5.2.14)$$

where the symbols have their usual meaning.

Finally the minimum thickness required for torispherical heads is given by:

$$e = \frac{pR_cC_s}{2\eta\sigma_d + p(C_s - 0.2)} \qquad (5.2.15)$$

where C_s = stress concentration factor for torispherical heads and is found from: $C_s = \frac{1}{4}\left(3 + \sqrt{(R_c/R_k)}\right)$

R_c = crown radius,
R_k = knuckle radius.

We have spent some time looking at formulae which can be used in determining minimum thickness of materials for pressure vessels. For a full account of all aspects of pressure vessel design you should refer to BS 5500.

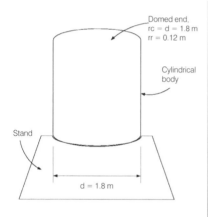

Domed end,
rc = d = 1.8 m
rr = 0.12 m

Cylindrical body

Stand

d = 1.8 m

Figure 5.2.7 *Figure for Example 5.2.3*

Example 5.2.3

Consider the pressure vessel (Figure 5.2.7). Estimate the thickness of material required for the cylindrical section and domed ends. Assume that the pressure vessel is to be used in the food processing industry and so the material for construction is to be a stainless steel. The vessel is to operate at a design temperature of 250°C and a design pressure of 1.6 MPa so, we will use a typical design stress of 115 MPa.

(a)   The cylindrical section
Using Equation 5.2.7 we have:

$$e = \frac{1.6 \times 10^6 \times 1.8}{(2 \times 115 \times 10^6) - (1.6 \times 10^6)} = 0.0126 \text{ m.}$$

Now we need to consider any relevant safety factors. The design pressure will have a safety factor built-in, so it will already be set at a higher value than normal operating pressure. Therefore we need only consider an allowance for corrosion and fabrication problems. Stainless steel is highly resistant to corrosion, but as a result of the fabrication process, may have 'locked-in' stresses which we are not aware of, therefore it would be prudent to make some allowance for the unknown without incurring too much unnecessary expense or material wastage. In practice, further detail design will determine the necessary allowances.

Let us assume a further 2 mm is the appropriate allowance, then:

$e = 12.6 + 2 = 14.6$ mm.

Now we are unlikely to obtain 14.6 mm SS plate as stock, so we use 15 mm plate.

(b)   The domed head
We shall first consider a standard dished torispherical head and assume that the head is formed by pressing, therefore it has no welded joints and so η may be taken as 1.0. Let us assume that the torisphere has a crown radius equal to the diameter of the

cont.

cylindrical section (its maximum value) and a knuckle radius of 0.12 m.

Then using Equation 5.2.15 we have for a torispherical head:

$$e = \frac{pR_cC_s}{2\sigma_d\eta + p(C_s - 0.2)}$$

where $C_s = \frac{1}{4}(3 + \sqrt{(R_c/R_k)})$

$C_s = \frac{1}{4}(3 + \sqrt{(1.8/0.12)})$

$C_s = 1.72$ and $\eta = 1.0$

So $e = \dfrac{1.6 \times 10^6 \times 1.8 \times 1.72}{(2 \times 115 \times 10^6) + 1.6 \times 10^6(1.72 - 0.2)}$

therefore, $e = 0.0213$ m or 21.3 mm.

Let us now try an ellipsoidal head.
From Equation 5.2.14 we get:

$$e = \frac{1.6 \times 10^6 \times 1.8}{(2 \times 115 \times 10^6) - (0.2 \times 1.6 \times 10^6)}$$

$e = 0.0125$ m or 12.5 mm.

Since the fabrication of both the torispherical and ellipsoidal head incur similar effort, then the most economical head would be ellipsoidal with a minimum thickness of 15.0 mm, after the normal allowances are added.

Thick-walled pressure vessels

When dealing with the theoretical analysis of thin cylinders we made the assumption that the hoop stress was constant across the thickness of the cylinder (Figure 5.2.2a) and that there was no difference in pressure across the cylinder wall. When dealing with thick pressure vessels where the wall thickness is substantial we can no longer accept either of the above two assumptions. So with thick-walled vessels, where the wall thickness is normally greater than one tenth of the diameter, both the hoop (σ_h) stress and the radial (σ_r) stress vary across the wall.

If thick-walled vessels are subject to pressure, then the value of these stresses may be determined by use of the *Lamé equations* given below.

$$\sigma_r = a - \frac{b}{r^2} \tag{5.2.16}$$

$$\text{and } \sigma_h = a + \frac{b}{r^2}. \tag{5.2.17}$$

These equations may be used to determine the radial stress and hoop stress at any radius r in terms of constants a and b. For any pressure condition there will always be two known stress conditions, that enable us to find values for the constants. These known stress conditions are often referred to as 'boundary conditions'.

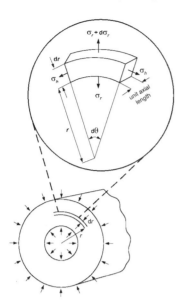

Figure 5.2.8 *Element of cylinder wall subject to internal and external pressure*

Mathematics in action

Consider a cylinder which is long in comparison with its diameter, where the longitudinal stress and strain are assumed to be constant across the thickness of the cylinder wall. If we take an element of the cylinder wall (Figure 5.2.8) when the cylinder is subject to internal and external pressure and, where the axial length is considered to be unity.

Then for radial equilibrium, applying elementary trigonometry and remembering that for small angles $\sin \theta$ approximately equals θ (in radians), we may write

$$(\sigma_r + d\sigma_r)(r + dr)d\theta = \sigma_r r d\theta + 2\sigma_h dr \frac{d\theta}{2}.$$

From which, after simplification, neglecting second-order small quantities we may write: $\sigma_r dr + r d\sigma_r = \sigma_h dr$

which on division by (dr) gives

$$\sigma_r + r \frac{d\sigma_r}{dr} = \sigma_h. \tag{5.2.18}$$

Now if we let the longitudinal stress be σ_L and the longitudinal strain be ε_L, then following our assumption that the longitudinal stress and strain are constant across the cylinder walls we may write:

$$\varepsilon_L = \frac{1}{E}(\sigma_L - v(\sigma_r + \sigma_h)). \tag{5.2.19}$$

Compare this equation with (5.1.16).

Also, with σ_L and ε_L constant then $\sigma_r + \sigma_h$ = constant (say $2a$)

or $\sigma_h = 2a - \sigma_r$. \hfill (5.2.20)

Then on substituting Equation 5.2.20 for σ_h into Equation 5.2.18 and simplifying, we get

$$2ar - r^2 \frac{d\sigma_r}{dr} - 2\sigma_r r = 0$$

which on integration gives

$$ar^2 - r^2 \sigma_r = \text{constant (say } b)$$

and on rearrangement and division by r we get

$$\sigma_r = a - \frac{b}{r^2} \tag{5.2.21}$$

and substituting Equation 5.2.21 for σ_r into Equation 5.5.20 gives

$$\sigma_h = a + \frac{b}{r^2}$$

which are Equations 5.2.16 and 5.2.17, the Lamé equations for internal components.

Thick cylinders subject to internal pressure

Consider a thick cylinder which is only subject to internal pressure (p), the external pressure being zero. For the situation shown in Figure 5.2.9,

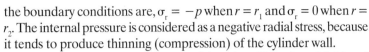

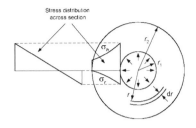

Figure 5.2.9 *Thick cylinder subject to internal pressure*

the boundary conditions are, $\sigma_r = -p$ when $r = r_1$ and $\sigma_r = 0$ when $r = r_2$. The internal pressure is considered as a negative radial stress, because it tends to produce thinning (compression) of the cylinder wall.

By substituting the boundary conditions into Equations 5.2.16 and 5.2.17 and determining expressions for the constants a and b, it can be shown that for thick cylinders subject to internal pressure only the:

$$\text{radial stress } \sigma_r = \frac{pr_1^2}{r_2^2 - r_1^2}\left(\frac{r^2 - r_2^2}{r^2}\right) \tag{5.2.22}$$

$$\text{and the hoop stress } \sigma_h = \frac{pr_1^2}{r_2^2 - r_1^2}\left(\frac{r^2 + r_2^2}{r^2}\right). \tag{5.2.23}$$

The maximum values for the radial stress and hoop stress across the section can be seen in Figure 5.2.9 and may be determined using:

$$\sigma_{r_{max}} = -p \tag{5.2.24}$$

$$\sigma_{h_{max}} = p\left(\frac{r_1^2 + r_2^2}{r_2^2 - r_1^2}\right). \tag{5.2.25}$$

Longitudinal and shear stress

So far we have established that the longitudinal stress in thick-walled pressure vessels, may be assumed to be uniform. This constant longitudinal stress (σ_l) is determined by considering the longitudinal equilibrium at the end of the cylinder. Figure 5.2.10 shows the arrangement for the simple case where the cylinder is subject to internal pressure.

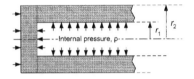

Figure 5.2.10 *Longitudinal equilibrium at end of cylinder*

So assuming equilibrium conditions and resolving forces horizontally we have:

$$p \times \pi r_1^2 = \sigma_l \times \pi(r_2^2 - r_1^2)$$

$$\text{and } \sigma_l = \frac{pr_1^2}{r_2^2 - r_1^2}. \tag{5.2.26}$$

In a similar manner it can be shown that for a thick-walled cylinder subject to *both internal and external pressure* the constant longitudinal stress is given by:

$$\sigma_l = \frac{p_1 r_1^2 - p_2 r_2^2}{r_2^2 - r_1^2}. \tag{5.2.27}$$

The radial stress (σ_r), hoop stress (σ_h) and longitudinal stress (σ_l) are all principal stresses. The maximum shear stress (τ) is equal to half the difference between the maximum and minimum principal stresses. So if we consider the case for *internal pressure only*, we have already seen that radial stress is compressive, the hoop stress and longitudinal stress are tensile. Therefore, remembering the sign convention, and noting the fact that the hoop stress is larger than the longitudinal stress. Then the maximum difference is given by $(\sigma_h - \sigma_r)$ and so the maximum shear stress is:

$$\tau_{max} = \frac{\sigma_h - \sigma_r}{2}$$

and substituting for σ_h and σ_r from the Lamé equations we have:

$$\tau_{max} = \frac{\left(\dfrac{a+b}{r^2}\right) - \left(\dfrac{a-b}{r^2}\right)}{2} = \frac{b}{r^2}.$$

Now applying the boundary conditions to the Lamé equation for radial stress where the cylinder is subject to internal pressure only gives:

$$-p = a - \frac{b}{r_1^2} \text{ and } o = a - \frac{b}{r_2^2}$$

from which $b = \dfrac{pr_1^2 r_2^2}{r_2^2 - r_1^2}$

and so $\tau_{max} = \dfrac{pr_1^2 r_2^2}{(r_2^2 - r_1^2)r^2} = \dfrac{b}{r^2}.$

So the maximum shear stress (τ_{max}) occurs at the inside wall of the cylinder where $r = r_1$

or $\tau_{max} = \dfrac{pr_2^2}{r_2^2 - r_1^2}.$ (5.2.28)

Example 5.2.4

The cylinder of a hydraulic actuator has a bore of 100 mm and is required to operate up to a pressure of 12 MPa. Determine the required wall thickness for a limiting tensile stress of 36 MPa.

In this example the boundary conditions are that at $r = 50$ mm, $\sigma_r = -12$ MPa and since the maximum tensile hoop stress occurs at the inner surface then, at $r = 50$ mm, $\sigma_h = 36$ MPa. In order to simplify the arithmetic when calculating the constants in the Lamé equations we will use the relationship that 1 MPa = 1 N/mm² and work in N/mm².

Therefore using $\sigma_h = a + b/r^2$, $\sigma_r = a - b/r^2$ (inner surface)

Where $\sigma_r = -12$ N/mm² $r = 50$ mm

$\sigma_h = 36$ N/mm²

we have $-12 = a - \dfrac{b}{2500}$

and $36 = a + \dfrac{b}{2500}.$

Adding the two equations:

$24 = 2a$, so $a = 12$

from which on substitution

$-12 = 12 - \dfrac{b}{2500}$ and $b = 60\,000$.

At the outer surface, $\sigma_r = 0$; therefore

$0 = a - \dfrac{b}{r^2} = 12 - \dfrac{60\,000}{r^2}$

and $r = 70.7$ mm.

Questions 5.2.1

(1) A thin cylinder has an internal diameter of 50 mm, a wall thickness of 3 mm and, is 200 mm long. If the cylinder is subject to an internal pressure of 10 MPa, determine the change in internal diameter and the change in length. Take $E = 200$ GPa and Poisson's ratio $v = 0.3$.

(2) A cylindrical reservoir used to store compressed gas is 1.8 m in diameter and assembled from plates 12.5 mm thick. The longitudinal joints have an efficiency of 90% and the circumferential joints have an efficiency of 50%. If the reservoir plating is to be limited to a tensile stress of 110 MPa, determine the maximum safe pressure of the gas.

(3) A thin cylinder has an internal diameter of 240 mm, a wall thickness of 1.5 mm and is 1.2 m long. The cylinder material has $E = 210$ GPa and Poisson's ratio $v = 0.3$. If the internal volume change is found to be 14×10^{-6} m^3 when pressurised. Find
  (i) the value of this pressure,
  (ii) the value of the hoop and longitudinal stresses.

(4) A steel pipe of internal diameter 50 mm and external diameter 90 mm is subject to an internal pressure of 15 MPa. Determine the radial and hoop stress at the inner surface.

(5) A thick cylinder has an internal diameter 200 mm and external diameter of 300 mm. If the cylinder is subject to an internal pressure of 70 MPa and an external pressure of 40 MPa. Determine:
  (i) the hoop and radial stress at the inner and outer surfaces of the cylinder,
  (ii) the longitudinal stress, if the cylinder is assumed to have closed ends.

Beams

Introduction

You will recall from your previous work that a beam is a structural member which is subject to loads that act normal (laterally) to their longitudinal axis. The loads create internal actions in the form of shear stresses and bending moments. The lateral loads that act on the beam cause it to bend, or flex, thereby deforming the axis of the beam into a curve (Figure 5.2.11) called the *deflection curve* of the beam.

We will only consider beams in this section that are symmetric about the *xy* plane, which means that the *y*-axis is an axis of symmetry of the cross-sections. In addition, all loads are assumed to act in the *xy* plane. As a consequence bending deflections occur in this same plane, which is known as the *plane of bending*. Thus the deflection curve AB of the beam (Figure 5.2.11) is a plane curve lying within the plane of bending.

Beams are normally classified by the way in which they are supported. The beams which will concern us are illustrated in Figure 5.2.12.

The *cantilever* (5.2.12a), is a beam which is rigidly supported at one end. The *simply supported* beam is either supported at its ends on rollers, or smooth surfaces (Figure 5.2.12b) or by one of the ends being pin-jointed and the other resting on a roller or smooth surface

Figure 5.2.11 *Cantilever subject to symmetric bending*

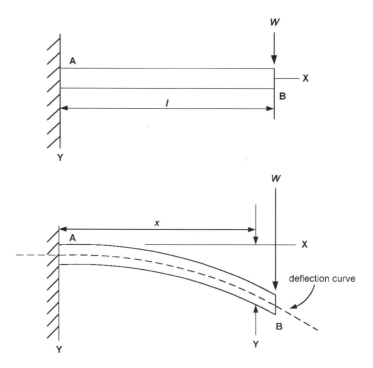

(Figure 5.2.12c). A built-in or encastre beam (Figure 5.2.12d) has both its ends rigidly fixed. Finally, Figure 5.2.12e, illustrates a simply supported beam with overhang, where the supports are positioned some distance from its ends.

The loads that are carried by beams are often classified as concentrated loads, where the load is deemed to act at a point and distributed loads, which are applied over a length of the beam. You should already be familiar with the concept of concentrated and uniformly distributed loads, illustrated in Figure 5.2.13.

When a beam is loaded reactions and resisting moments occur at the supports, and shear forces and bending moments of varying magnitude and direction occur along the length of the beam.

Shear force and bending moments

In order to determine the nature of the *shear forces* and *bending moments*, that act under plane bending conditions, it is necessary to take a cut normal to the axis of a uniform cross-section beam, to form a *free body* diagram. This technique is best illustrated using an example.

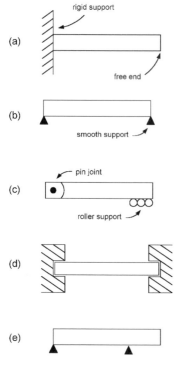

Figure 5.2.12 *Types of beam support*

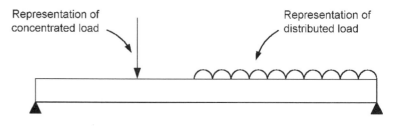

Figure 5.2.13 *Simply supported beam subject to concentrated and uniformly distributed loads*

Figure 5.2.14 *Figure for Example 5.2.5*

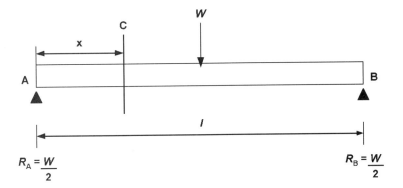

Figure 5.2.14 *Figure for Example 5.2.5*

Example 5.2.5

Consider the simply supported uniform beam AB of length l subject to concentrated load W, at its mid-point (Figure 5.2.14). We wish to determine the shear force F and the bending moment M at C, a distance x from A.

Then the method we adopt is as follows:

(1) Find the reactions at A and B.

These are determined by taking moments about either A or B to eliminate them from the calculation. In this particular case since the load W acts at the centre of the beam then:

$R_A = R_B = W/2$.

(2) Make a cut at C (Figure 5.2.15) to form the free body diagram, and add the internal shear force F and bending moment M. Note the direction of the shear force and bending moments at the cut.

(3) Then vertical equilibrium of AC gives: $\dfrac{W}{2} - F = 0$

or vertical equilibrium of BC gives: $F - W + \dfrac{W}{2} = 0$.

(4) For rotational equilibrium of AC about C gives:
$\dfrac{Wx}{2} - M = 0$ so $M = \dfrac{Wx}{2}$

or rotational equilibrium of BC about C gives:

$M + W\left(\dfrac{l}{2} - x\right) - \dfrac{W}{2}(l - x) = 0$

so $M = \dfrac{Wx}{2}$.

You could also have taken moments about A, for rotational equilibrium of AC, to give:

cont.

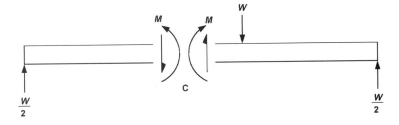

Figure 5.2.15 *Free body diagram shear force and bending moments*

$Fx - M = 0$.

From this it can be seen that M varies linearly with x from $x = 0$ to $x = \frac{1}{2}$.

You should also note that because M is already a moment its effect is the same irrespective of the point we take moments about. Also we get the same value of F and M regardless of whether we consider the left-hand part or right-hand part of the beam. In practice it is easier to consider the part which has fewest external forces acting on it.

When determining shear forces and bending moments it is convenient to use a *sign convention*. The convention we shall adopt is, *positive for clockwise shear acting on the element and sagging moments*, this convention is illustrated in Figure 5.2.16.

Shear force and bending moment diagrams for concentrated loads

Frequently we will need diagrams of the variation along the beam of shear force and bending moment. Using our sign convention, we again use an example to illustrate the method needed to formulate shear force and bending moment diagrams.

Figure 5.2.16 *Sign convention for shear force and bending moments*

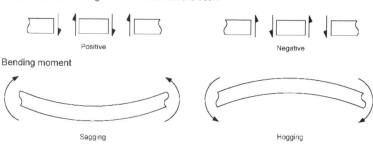

Shear force considering a small element of the beam

Positive   Negative

Bending moment

Sagging   Hogging

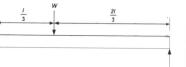

Figure 5.2.17 *Figure for Example 5.2.6 – simply supported beam with concentrated load*

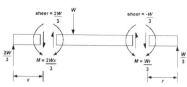

Figure 5.2.18 *Figure for Example 5.2.6 – free body diagram for cuts either side of the load*

Example 5.2.6

Draw the shear force and bending moment diagrams for the beam shown in Figure 5.2.17.

We have already been given the reactions at the left-hand and right-hand supports, they are $2W/3$ and $W/3$, respectively. So we may now determine the shear force and bending moment values by producing our free body diagram, making cuts either side of the load at the arbitrary points x and r (Figure 5.2.18).

For vertical equilibrium, using our sign convention, then at left-hand cut the shear force $F = 2W/3$ and at right-hand cut $F = -W/3$, note also the direction of the shear force pairs at the cuts, either side of the transverse load.

At the left-hand end of beam the reaction is $2W/3$ (positive), at the left hand cut the shear force is again $2W/3$, but this changes on the right-hand side of the load to $-W/3$. So at the load the shear force on the beam is the sum of these two shear forces, as shown in Figure 5.2.19a.

Note that the shear force only changes where a transverse load acts. Figure 5.2.19b shows the resulting bending moment diagram, where the bending moment under the load is obtained by putting

$x = \dfrac{l}{3}$ or $r = \dfrac{2l}{3}$ into the equation $M = \dfrac{2Wx}{3}$ or $M = \dfrac{Wr}{3}$, respectively.

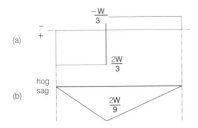

Figure 5.2.19 *Figure for Example 5.2.6 – shear force and bending moment diagram*

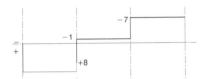

Figure 5.2.20 *Figure for Example 5.2.7 – simply supported beam with two concentrated loads*

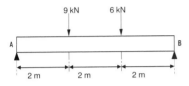

Figure 5.2.21 *Figure example Figure 5.2.7 – shear force distribution*

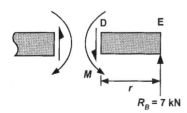

Figure 5.2.22a *Figure for Example 5.2.7 – free body diagram*

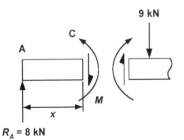

Figure 5.2.22b *Figure for Example 5.2.7 – bending moment diagram*

Example 5.2.7

Draw the shear force and bending moment diagram for the beam shown in Figure 5.2.20.

The first step is to calculate the reactions R_A and R_B.

The reactions may be found in the normal manner by considering rotational and vertical equilibrium. Then taking moments about R_A gives:

$(9 \times 2) + (6 \times 4) = 6R_B$ and $R_B = 7$ kN.

Also for vertical equilibrium $R_A + 7 = 15$ and so $R_A = 8$ kN.

Now considering the shear forces at the 9 kN load, and remembering that the shear force only changes where a transverse loads acts. Then on the left-hand side of the 9 kN load we have a positive shear force $R_A = +8$ kN. This is opposed at the load by the 9 kN force acting in a negative direction, so the shear force to the right of this load is $= -1$ kN. This shear force continues until the next transverse load is met, in this case the 6 kN load, again acting in a negative direction, producing a net load of -7 kN which continues to R_B. This shear force distribution is clearly illustrated in Figure 5.2.21.

The bending moment diagram can be easily determined from the shear force diagram by following the argument given in example 5.2.6. If we consider the free body diagram for AC (Figure 5.2.22a), then taking moments about C gives:

$R_A x - M = 0$ and at the 9 kN load where $x = 2$ m we have $(8 \times 2) = M = 16$ Nm (sagging).

Similarly, considering the free body diagram for DE (Figure 5.2.22b) and taking moments about D, when $r = 2$, then from $-R_B r + M = 0$, we have $-(7 \times 2) = -M = 14$ Nm.

These results can now be displayed on the bending moment diagram (Figure 5.2.23).

Note that the loads produce a sagging moment downwards, which intuitively you would expect.

Shear force and bending moment diagrams for uniformly distributed loads

We will now consider drawing shear force (SF) and bending moment (BM) diagrams for a beam subject to a uniformly distributed load. These loads are normally expressed as force per unit length.

The beam shown in Figure 5.2.24 is subject to a uniform distributed load (UDL) of 10 kN/m in addition to having a point load of 40 kN acting 2 m from A.

As usual we first find the reaction at A and B. To do this you will remember from your previous work, to treat the UDL as a point load acting at the centre of the uniform beam. This gives a load of 80 kN acting 4 m from A. Then the reactions may be calculated as $R_A = 70$ kN and $R_B = 50$ kN. We now cut the beam and, consider the free body on the left of the cut (Figure 5.2.25).

Our first cut is made at a distance x from the left-hand support and to the left of the point load. We consider the left-hand part because it is the simpler of the two (you would of course get the same values for SF and BM if you considered the right-hand part).

Then, from the convention used in Figure 5.2.25 we have:

Figure 5.2.23   *Figure for Example 5.2.7 – bending moment diagram*

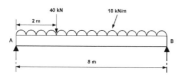

Figure 5.2.24 *Beam subject to concentrated load and UDL*

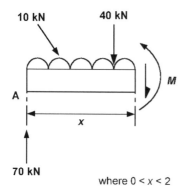

Figure 5.2.25 *Free body diagram of beam to left of cut*

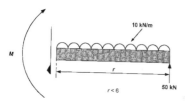

Figure 5.2.26 *Free body diagram of beam to right of cut*

(i)   SF = $(70 - 10x)$ kN

(ii)   and taking moments about the cut gives a BM = $(70x - \dfrac{10x^2}{2})$ kN m sagging.

Next consider a cut to the right of the point load (Figure 5.2.26). Again using the convention shown, we have:

(i)   SF = $(10r - 50)$ kN.

(ii) $(50r - \dfrac{10r^2}{2})$ kN m sagging.

You will note that the curve given by the expression for BM is a parabola. This is true for any section of a beam carrying only a *uniformly* distributed load.

If we plot the SF at A and then subsequently, at 1 m intervals for the whole length of the beam we will obtain the SF and BM diagrams illustrated in Figure 5.2.27.

You should ensure that you follow the above argument by verifying one or two of these SF and BM values for yourself.

Plotting diagrams point by point like this is a tedious business! Fortunately there are ways around it, based on relations between load, shear force and bending moment that we have not yet considered.

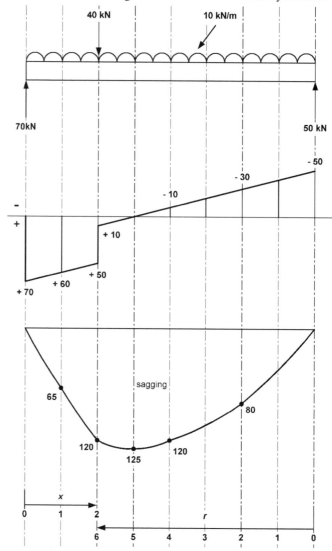

Figure 5.2.27 *Shear force and bending moment diagram for UDL*

Figure 5.2.28 *Small element from beam carrying a UDL*

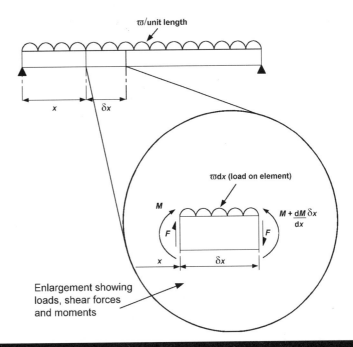

Mathematics in action

Take a small element, of length dx from a beam carrying a load w (downwards) per unit length (Figure 5.2.28).

Then we can see that if F is the shear force due to load at the point x, then at the point $x + \delta x$ the shear is $F + \dfrac{ds}{dx} \times \delta x$

and similarly if M is the bending moment at x, $M + \dfrac{dM}{dx} \times \delta x$ is the bending moment at the point $x + dx$.

Now for vertical equilibrium we have (adding downward forces)

$$-F + \omega \delta x + \left(\frac{F + ds\, \delta x}{dx} \right) = 0$$

(where ω is the distributed load) giving

$$\frac{dF}{dx} = -\omega. \tag{5.2.29}$$

This indicates that a downward load ω/unit length gives a negative rate of shear.

Next we consider the tendency of the element to rotate, i.e. equilibrium of moments (taken clockwise about the point x).

Then:

$$M + \omega \delta x.\frac{\delta x}{2} + \left(\frac{F + ds.\delta x}{dx} \right) \delta x - \left(\frac{M + dM.\delta x}{dx} \right) = 0$$

giving

$$M + \omega \frac{(\delta x)^2}{2} + F.\delta x + \frac{ds}{dx}(\delta x)^2 - M - \frac{dM}{dx}.\delta x = 0$$

neglecting second order terms in δx this gives:

$$F.\delta x - \frac{\mathrm{d}M}{\mathrm{d}x}.\delta x = 0$$

$$\text{or } \frac{\mathrm{d}M}{\mathrm{d}x} = F. \tag{5.2.30}$$

Check that you follow the above argument! This mathematical argument leads to two important results.

(i) For a maximum value of M, $\mathrm{d}M/\mathrm{d}x = 0$, thus $F = 0$ (from Equation 5.2.29), so if we have already drawn an SF diagram we know that the position of the maximum BM is at the point where the SF is zero.

(ii) If we integrate Equation 5.2.30 we get $M = \int F \, \mathrm{d}x$, so we can find the values for the BM diagram by adding numerically the areas of the SF diagram (i.e. integrating) and this may well be quicker than plotting numerical values found from the algebraic expression for the BM. Also, from Equation 5.2.29, $S = -\int \omega \mathrm{d}x$ and so substituting this into Equation 5.2.30 we get

$$\int \omega \mathrm{d}x = -\frac{\mathrm{d}M}{\mathrm{d}x}$$

$$\text{or } \omega = \frac{\mathrm{d}^2 M}{\mathrm{d}x^2}. \tag{5.2.31}$$

Example 5.2.8

Consider a beam with loads as shown in Figure 5.2.29 and draw the SF and BM diagrams for the beam.

Again the reactions may be found in the usual way, where R_A = 90 kN and R_B = 30 kN both acting upwards.

To obtain the SF diagram we start at the left-hand end of beam where R_A = +90 kN. Moving along to point B and adding up the loads we have, $+90 - \int w \mathrm{d}x$ which becomes $90 - 20 = +70$. At point B a downward load of 80 acts, giving $+90 - 20 - 80 = -10$. At D we have an SF of -30, resulting from the reaction R_B, this is constant until it meets the distributed load at C. Now moving from C to B the SF changes constantly from -30 to -10, as indicated on the diagram. You will note that on the SF diagram distributed load gives a straight sloping line between A and B, and between B and C.

To obtain the BM diagram we again start at A, where the BM is zero. *Adding up the area of the shear force diagram* between A and B gives a BM at B of +160. Continuing we may say that the sum of the areas between A and C is given by $\int_A^C F \mathrm{d}x = 160 - 40 = 120$. At D it can easily be seen that the BM $= 160 - 160 = 0$, by again summing the areas on the SF diagram between A and D.

Note that the SF is zero at point B, so the maximum BM occurs at B. The slope of the BM diagram is everywhere given by the ordinate of the SF diagram and because the SF varies linearly, the BM diagram is parabolic (between A and B, B and C), while between C and D where the SF is constant the BM is a straight line, remembering your integration!

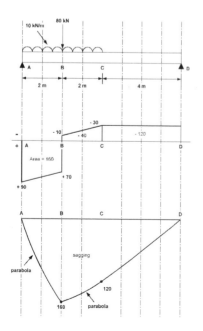

Figure 5.2.29 *Figure for Example 5.2.8*

Questions 5.2.2

(1) Draw the SF and BM diagrams for the following beams, following the techniques given in this section.

(a)

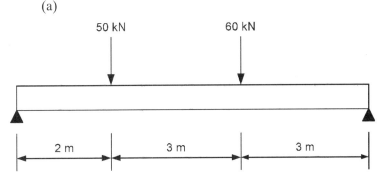

Figure 5.2.30 *Figure for Question 5.2.2.1a*

(b)

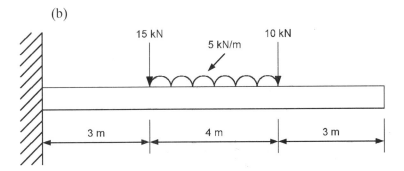

Figure 5.2.31 *Figure for Question 5.2.2.1b*

(2) A beam 10 m long is simply supported over a span of 8 m and overhangs the right-hand support by 2 m. The whole beam carries a uniformly distributed load of 10 kN/m together with a point load of 120 kN, 2.5 m from the left-hand support. Draw the SF and BM diagrams showing appropriate numerical values.

Engineers' theory of bending

When a transverse load is applied to a beam, we have seen that bending moments are set up which are resisted internally by the beam material. If the bending moment resulting from the load is sufficient, the beam will deform, creating tensile and compressive stresses to be set up within the beam (Figure 5.2.32). Between the areas of tension and compression there is a layer within the beam, which is unstressed termed the *neutral layer* and its intersection with the cross-section is termed the *neutral axis*.

Engineers' theory of bending is concerned with the relationships between the stresses, the beam geometry and the curvature of the applied bending moment. In order to formulate such relationships we make several assumptions, which simplifies the mathematical modelling. These assumptions include:

(i) the beam section is symmetrical across the plane of bending and remains within the plane after bending (refer back to the introduction);

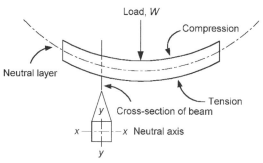

(ii) the beam is straight prior to bending and the radius of bend is large compared with the beam cross-section;

(iii) the beam material is uniform (homogenous) and has the same elastic modulus in tension and compression (this is not the case, for example, with ceramic materials).

Deformation in pure bending

Consider the beam subject to pure bending (Figure 5.2.33), which shows an exploded view of the beam element where the neutral axis, discussed above, is clearly seen. Figure 5.2.33a shows the general arrangement of the beam before and after bending, where we consider the deformations that occur between the sections *ac* and *bd*, which are δx apart, when the beam is straight. The longitudinal fibre of the beam material *ef* which is at a distance *y* from the neutral axis has the same initial length as the fibre *gh* at the neutral axis, prior to bending.

Now during bending we can see from Figure 5.2.33 that *ef* stretches to become *EF*. However, *gh*, which is at the neutral axis is not strained (since stress is deemed to be zero) and so it has the same dimensions when it becomes *GH*. So, by simple trigonometry, when R equals the radius of curvature of *GH* we may write:

$GH = gh = \delta x$ where $\delta x = R\mathrm{d}\theta$ (θ in radians) and $EF = (R+y)\mathrm{d}\theta$.

We also see from the figure that the longitudinal strain of *EF* is given by:

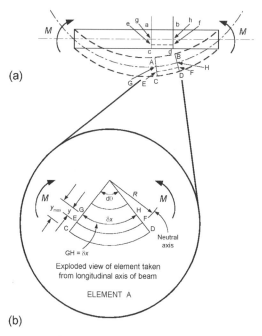

Figure 5.2.33 *Deformation in pure bending*

$$\varepsilon_x = \frac{EF - ef}{ef}.$$

Now since we know that prior to bending $ef = gh$ and that after bending because gh is on the neutral axis $gh = GH$, then GH must equal $ef = rd\theta$ (from above). Also, since $EF = (R+y)d\theta$ we have:

$$\varepsilon_x = \frac{(R+y)d\theta - Rd\theta}{Rd\theta}$$

$$\text{or } \varepsilon_x = \frac{Rd\theta + yd\theta - Rd\theta}{Rd\theta}$$

and on division by $d\theta$

$$\varepsilon_x = \frac{y}{R}. \tag{5.2.32}$$

We also showed that $dx = Rd\theta$ or $R = dx/d\theta$, and on substitution into Equation 5.2.32 we get:

$$\varepsilon_x = \frac{d\theta}{dx}y. \tag{5.2.33}$$

From our earlier work in this chapter on Poisson's ratio we established relationships for three-dimensional strains. So if we consider the strains ε_y and ε_z which are transverse to the longitudinal strain ε_x we can show that they may be determined from the relationship:

$$\varepsilon_y = \varepsilon_z = -\frac{\nu\sigma_x}{E}$$

Where ν, you will remember, is Poisson's ratio.

Stresses due to bending

We have already shown (Equation 5.2.32) that $\varepsilon_x = \frac{y}{R}$

and knowing that $\varepsilon = \frac{\sigma}{E}$ then substituting into the above equation yields the relationship:

$$\frac{\sigma}{E} = \frac{y}{R}. \tag{5.2.34}$$

We can also show, by considering the cross-section of element A in Figure 5.2.33 that the bending moment M imposed on the beam at the element is related to the stress on the element, the distance y from the neutral axis and the second moment of area I by:

$$\frac{M}{I} = \frac{\sigma}{y}. \tag{5.2.35}$$

The derivation of this relationship is left as an exercise at the end of this section. You have already met the Polar Second Moment of Area (J) in Chapter 3, and therefore, should have little difficulty in understanding the concept of Second Moment of Area (I). The derivation of I will be found in Chapter 6 under the applications of the calculus, which you should consult if you have difficulties.

If we combine Equations 5.2.34 and 5.2.35 we can write:

$$\frac{M}{I} = \frac{\sigma}{y} = \frac{E}{R}. \tag{5.2.36}$$

The above relationship is known as the general bending formula or *engineers' theory of bending*.

For a beam the maximum bending stress (s_{max}) will occur where the distance from the neutral axis is a maximum (y_{max}) and from Equation 5.2.36 we may write:

$$M = \frac{I\sigma_{max}}{y_{max}}$$

the quantity I/y_{max} depends only on the cross-sectional area of the beam under consideration, and is known as the section modulus Z, therefore:

$$M = Z\sigma_{max}.$$

Note that in some reference sources the section modulus is given the symbol S, we will not use this symbol since it may be confused with shear (see Equation 5.2.29). Many standard texts give values of the section modulus for a variety of commonly used beams (See for example Appendix A in Mechanics of Materials by Gere and Timoshenko).

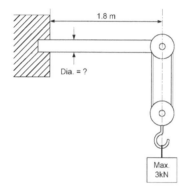

Figure 5.2.34 *Cantilever lifting gantry*

Example 5.2.9

Figure 5.2.34 shows a simplified diagram of a cantilever lifting gantry, 1.8 m long, designed to raise loads of up to 3 kN . If the maximum design stress of the cantilever beam material is 250 MPa. Determine a suitable diameter for the bar.

Now the design load is given as 3 kN, however, in design work we should always consider a factor of safety, in view of the fact that the suspended loads are involved we must ensure the complete safety of the operators so we will choose a factor of safety of 2. Thus the maximum design load = 6 kN.

The maximum bending moment for the cantilever beam is furthest from the support, in this case the maximum BM = 6×1.8 = 10.8 kN m.

Now the diagram shows a circular cross-section beam. We need to calculate the second moment of area *I* for this beam. This is given by the formula:

$$I = \frac{\pi d^4}{64} \text{ (refer to Chapter 6).}$$

The maximum allowable design stress (σ_{max} = 250 MN/m²) on the beam occurs at the maximum distance from the neutral axis (y_{max}), in this case at distance $d/2$.

So using engineers' theory of bending where $\dfrac{M}{I} = \dfrac{\sigma}{y}$ then:

$$\frac{10\,800}{\pi d^4/64} = \frac{250 \times 10^6}{d/2}$$

from which we get d = 76 mm.

Questions 5.2.3

(1) A simply supported beam (Figure 5.2.35) carries the UDL as shown. The maximum allowable bending stress for the beam is 120 MPa.

  (i) Construct the shear force diagram and so determine the maximum bending moment.

Figure 5.2.35 *Figure for Question 5.2.3*

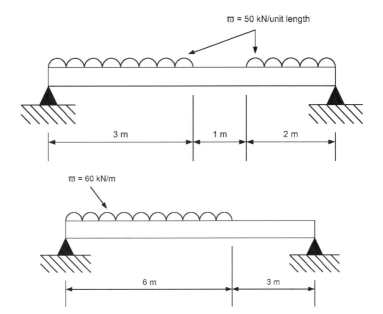

Figure 5.2.36 *Figure for Question 5.2.3*

(ii)   Calculate the required section modulus Z.
Ignore the weight of the beam.

(2)   A simply supported beam (Figure 5.2.36) carries the UDL as shown:
(i)   Construct the shear force diagram.
(ii)   Sketch the bending moment diagram.
(iii)   Determine the maximum bending moment.
(iv)   Given that the section modulus for the beam is 3500 cm³, determine the allowable bending stress.
Ignore the weight of the beam.

(3)   Below, in Figure 5.2.37, are shown two simplified views of element A (taken from Figure 5.2.32), cut to show a section BC, with neutral axis XX acting through the centroid of this section. Also shown is an infinitely small layer at distance y from the neutral axis which is assumed to have area dA.

If a longitudinal stress σ, that acts on the layer, produces a longitudinal force that is equal to σdA and, the element of the beam shown is in equilibrium.

Show that:
(i) $\int y dA = 0$.
(ii)   The total moment about the neutral axis
$$M = \frac{E}{R} \int y^2 dA.$$
(iii) $\dfrac{M}{I_{xx}} = \dfrac{E}{R}$ if $I_{xx} = \int y^2 dA$.

Figure 5.2.37 *Element taken from beam (Figure 5.2.32) cut to show section*

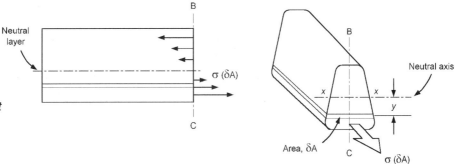

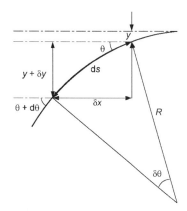

Figure 5.2.38 *Element of deflection curve of beam*

Deflection of beams

You have already been introduced to the deflection curve, at the beginning of this section. In order that engineers may estimate the loading characteristics of beams in situations where, for example, they are used as building support lintels or supports for engineering machinery. It is important to establish mathematical relationships which enable the appropriately dimensioned beam of the correct material to be chosen.

We have already established one very important relation, engineers theory of bending, from which other important equations may be found.

From Equation 5.2.36 we know that

$$\frac{M}{I} = \frac{E}{R}$$ and on rearrangement we may write

$$\frac{1}{R} = \frac{M}{EI}. \tag{5.2.37}$$

Now consider Figure 5.2.38 which shows a cantilever beam being deflected downwards (compare with Figure 5.2.11). At A the beam has been deflected by an amount y and at B the deflection is $y + dy$.

The deflection of beams in practice, is very small, therefore from simple geometry for very small angles we may assume that the arc distance ds is approximately equal to dx. So as dθ approaches zero then

$$dx = R d\theta \text{ from which } \frac{1}{R} = \frac{d\theta}{dx}.$$

Also, since the arc distance AB is very small, we may approximate it to a straight line which has a gradient (*slope*) equal to the tangent of the angle θ, therefore

$$\theta = \frac{dy}{dx} \text{ and so on differentiation } \frac{d\theta}{dx} = \frac{d^2y}{dx^2}.$$

Then from the above two equations $\frac{1}{R} = \frac{d^2y}{dx^2}.$ \hfill (5.2.38)

Finally combining Equations 5.2.37 and 5.2.38 yields

$$\frac{d^2y}{dx^2} = \frac{M}{EI}. \tag{5.2.39}$$

Deflections by integration

The above bending moment differential equation provides the fundamental relationship for the deflection curve of a loaded beam. Further manipulation of this equation using integration produces relationships in terms of the *slope* and *deflection* of the beam, as can be seen below.

Mathematics in action

Deflection by integration

When the beam is of uniform cross-section and M is able to be expressed mathematically as a function of x, we may integrate the equation $d^2y/dx^2 = M/EI$ with respect to x and get

$$\frac{dy}{dx} = \frac{1}{EI} \int M dx + A \qquad (5.2.40)$$

where A is the constant of integration, and dy/dx is the 'slope' of the beam.

Integrating a second time yields

$$Y = \frac{1}{EI} \iint M dx dx + Ax + B \qquad (5.2.41)$$

Which produces the expression for the 'deflection' of the beam y.

Also from equations 5.2.29 and 5.2.30 we have

$$\frac{dF}{dx} = -\omega \text{ and } \frac{dM}{dx} = F$$

where ω is distributed load and F the shear force due to point load.

Then integrating $\frac{d^2y}{dx^2} = \frac{M}{EI}$ once more gives

$$\frac{d^3y}{dx^3} = \frac{dM}{dx}. \frac{1}{EI}$$

and substituting $dM/dx = F$ gives

$$\frac{d^3y}{dx^3} = \frac{F}{EI}. \qquad (5.2.42)$$

Finally, using $\frac{dF}{dx} = -\omega$

and integrating $\frac{d^3y}{dx^3} = \frac{F}{EI}$ again, we get

$$\frac{d^4y}{dx^4} = \frac{dF}{dx}. \frac{1}{EI} \text{ and so}$$

$$\frac{d^4y}{dx^4} = -\frac{\omega}{EI}. \qquad (5.2.43)$$

Note that the slope and deflection equations, we have just found all contain $1/EI$, where the modulus (E) multiplied by the second moment of area (I) provides a measure of the *flexural rigidity* of the beam. Since the elastic modulus is a measure of the stiffness of the beam material and the second moment of area (sometimes known as the moment of inertia) is a measure of the beam's resistance to bending, about a particular axis. The *signs* associated with the equations we have just integrated, adopt the convention that the *deflection* (y) is *positive upwards* and *bending moments (M) causing sagging are positive*.

The integration method can be applied to standard cases, which include cantilevers and simply supported beams having a variety of combinations of concentrated loads, distributed loads, or both. Space does not permit us to consider all possible cases, so just one example of these applications is given in full below. This is followed by a table of some of the more important cases, that may be solved by use of the *integrating method*.

When applying the integrating method to the slop and deflection of beams we will use the following symbols, which matches with our previous theory: x as the distance measured from left-hand support of beam or from the fixed end of a cantilever; l for length; y for deflection w/per unit length for distributed loads and, F for shear force due to point load W.

Note: that in some textbooks v is used for y and P for W.

Example 5.2.10

To determine the deflection and slope of a cantilever subject to a point load consider the cantilever shown in Figure 5.2.11, subject to a single point load at its free end. Now the load causes a 'hogging' bending moment in the beam which, according to our convention, is negative so

$$-M = W(l-x) = Wl - Wx.$$

Using our fundamental Equation 5.2.39, where $\frac{d^2y}{dx^2} = -\frac{M}{EI}$ then on substitution $EI\frac{d^2y}{dx^2} = -W(l-x)$

so on integration we have

$$EI\frac{dy}{dx} = -W\left(lx - \frac{x^2}{2}\right) + A.$$

The constant of integration can be found from the 'boundary conditions'. The slope of the beam at the fixed end is zero, since the beam is deemed to be level at this point, in other words when $x = 0$, the slope $dy/dx = 0$. This implies that $A = 0$, so our equation becomes

$$EI\frac{dy}{dx} = -W\left(lx - \frac{x^2}{2}\right)$$

Or the 'slope' of the beam is given by

$$\frac{dy}{dx} = \frac{W}{EI}\left(lx - \frac{x^2}{2}\right). \tag{5.2.44}$$

If we integrate the above equation again then we get

$$y = -\frac{W}{EI}\left(\frac{lx^2}{2} - \frac{x^3}{6}\right) + B.$$

Again the constant of integration B can be found from the boundary conditions. These are that $y = 0$ at $x = 0$, this infers that $B = 0$.

So our equation for deflection y may be written as:

$$y = -\frac{W}{EI}\left(\frac{x^2}{2} - \frac{x^3}{6}\right). \tag{5.2.45}$$

We can see, quite clearly, from Figure 5.2.11 that the maximum deflection occurs at the free end, where $x = l$, then

$$y_{max} = -\frac{W}{EI}\left(\frac{l^3}{2} - \frac{l^3}{6}\right)$$

and so $y_{max} = -\frac{Wl^3}{3EI}.$ \hfill (5.2.46)

Also the maximum slope occurs at the free end and from Equation 5.2.44 when $x = l$ we have

$$\left(\frac{dy}{dx}\right)_{max} = -\frac{Wl^2}{2EI}. \tag{5.2.47}$$

Note the use of boundary conditions to establish values for the constants. You should ensure that you follow the logic of the

cont.

argument when establishing such conditions, which vary from case to case.

Now let us assume that the cantilever discussed above is made from a steel with an elastic modulus $E = 210$ Gpa, and the point load $W = 40$ kN. If the maximum deflection $y_{max} = 1.5$ mm, determine:

(i)   the second moment of area for the beam;
(ii) the maximum slope.

(i) The second moment of area is easily calculated using Equation 5.2.46, where

$$I = \frac{Wl^3}{3Ey_{max}} = \frac{(40 \times 10^3)(2.5)^3}{(3)(210 \times 10^9)(1.5 \times 10^{-3})} = 661.3 \times 10^{-6} \text{ m}^4$$

or $I = 661.3 \times 10^{-6} \times 10^{12} = 661.3 \times 10^6 \text{ mm}^4$.

Note the careful use of units!

(ii) The maximum slope may now be determined using equation 5.2.47

$$\left(\frac{dy}{dx}\right)_{max} = \frac{Wl^2}{2EI} = \frac{(40 \times 10^3)(2.5)^2}{(2)(210 \times 10^9)(661.3 \times 10^{-6})} = 0.0022 \text{ m/m}.$$

So maximum slope is 0.0022 metres per metre or 2.2 mm/m.

The process shown in the above example, where successive integration is applied to the fundamental equations to establish expressions for the slope and deflection, may be used for many standard cases involving plane bending. Remembering that the constants of integration may be found by applying the unique boundary conditions which exist, for each individual case.

Laid out in Table 5.2.1 are some standard cases for slopes and deflection associated with simply supported beams and cantilevers.

Principle of superposition

This principle states that if, one relationship connecting the bending moment, slope and deflection for a particular loading has been determined, for example a cantilever subject to a concentrated load at its free end (Figure 5.2.39). Then we may add algebraically, any other known relationship to it for a given loading situation.

The cantilever shown in Figure 5.2.39 is subject to both a concentrated load at its free end and a distributed load over its whole

Table 5.2.1   *Some standard cases for slope and deflection of beams*

Situation	Slope $\left(\frac{dy}{dx}\right)$	Deflection (y)	$\left(\frac{dy}{dx}\right)$	y_{max}
Cantilever with end couple	$\frac{-Mx}{EI}$	$\frac{-Mx^2}{2EI}$	$\frac{-Ml}{EI}$ at B	$\frac{-Mx^2}{EI}$ at B
Cantilever with concentrated end load	$-\frac{W}{EI}\left(lx - \frac{x^2}{2}\right)$	$-\frac{W}{EI}\left(\frac{x^2}{2} - \frac{x^3}{6}\right)$	$-\frac{Wl^2}{2EI}$ at B	$-\frac{Wl^3}{3EI}$ at B

Table 5.2.1 (cont'd)

Situation	Slope $\left(\dfrac{dy}{dx}\right)$	Deflection (y)	$\left(\dfrac{dy}{dx}\right)$	y_{max}
Cantilever with distributed load	$-\dfrac{w}{2EI}\left(l^2x - lx^2 + \dfrac{x^3}{3}\right)$	$-\dfrac{w}{2EI}\left(\dfrac{l^2x^2}{2} - \dfrac{lx^3}{3} + \dfrac{x^4}{12}\right)$	$-\dfrac{wl^3}{6EI}$ at B	$\dfrac{-wl^4}{8EI}$ at B
Simply supported beam with point load	$\dfrac{W}{EI}\left(\dfrac{lx}{2} - \dfrac{x^2}{2}\right)$	$\dfrac{W}{2EI}\left(\dfrac{lx^2}{4} - \dfrac{x^3}{6} - \dfrac{l^3}{24}\right)$	$\dfrac{Wl^2}{16\,EI}$ at A and B	$\dfrac{-Wl^3}{48EI}$ at C
Simply supported beam with distributed load	$\dfrac{w}{2EI}\left(\dfrac{l^2x}{4} - \dfrac{x^3}{3}\right)$	$\dfrac{w}{2EI}\left(\dfrac{l^2x^2}{8} - \dfrac{x^4}{12} - \dfrac{5l^4}{192}\right)$	$\dfrac{wl^3}{24EI}$ at A and B	$\dfrac{-5wl^4}{384EI}$ at C

Cantilever with distributed load: A, ϖ/unit length, B

Simply supported beam with point load: W, C, A, B, $\dfrac{l}{2}$, $\dfrac{l}{2}$

Simply supported beam with distributed load: ϖ/unit length, A, B, C

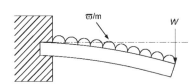

Figure 5.2.39 *Cantilever beam subject to loads causing deflection*

length so, for example, using the expressions for maximum deflections in Table 5.2.1 and, the principle of superposition. We may write

$$y_{max} = -\left[\frac{Wl^3}{3EI} + \frac{Wl^4}{8EI}\right].$$

Macaulay's method

No doubt if you have attempted Question 2 you will realise that the method of successive integration to determine the slope and deflection equations for each separate loading situation, is tedious and time consuming. We are required to find values for the constants of integration by considering boundary conditions on each separate occasion and, apart from the most simple of cases, this can involve quite complex algebraic processes. All is not lost however, since a much simpler method exists which enables us to formulate one equation which takes into account the bending moment expression for the whole beam and also enables us to find the necessary boundary conditions much more easily. This simplified process is known as *Macaulay's method*.

To assist us in formulating the Macaulay expression for a given loading situation of a beam, we need to use Macaulay functions. The graph of the Macaulay unit ramp function (Figure 5.2.40a) will help us to understand its nature.

Graph of Macaulay function F_1 (unit ramp function)

$$F_1(x) = \langle x - a \rangle^1 \text{ where } F = \text{function}$$

At point 1, $M = 0$
At point 2, $M = W(x - a)$ (a)
So Macaulay expression for all values of x is

$$M = W\langle x - a \rangle$$

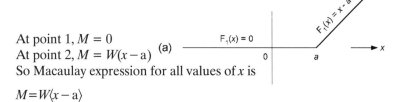

Figure 5.2.40 *Macaulay function application to beams*

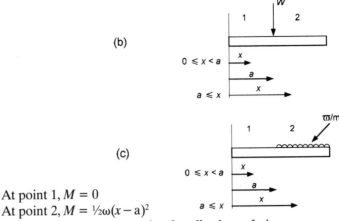

(b)

$0 \leqslant x < a$

$a \leqslant x$

(c)

$0 \leqslant x < a$

$a \leqslant x$

At point 1, $M = 0$

At point 2, $M = \frac{1}{2}\omega(x-a)^2$

So again Macaulay expression for all values of x is

$$M = \frac{1}{2}\omega\langle x-a\rangle^2$$

Macaulay functions are used to represent quantities that begin at some particular point on the x-axis. In our case for the unit ramp function, this is at $x = a$. To the left of the point a the function has the value zero, to the right of a, the function denoted by F_1 has the value $(x-a)$. This relationship may be written mathematically as:

$$F_1(x) = \langle x-a\rangle^1 = \begin{cases} 0 \text{ when } x \leqslant a \\ x-a \text{ when } x \geqslant a \end{cases}.$$

Now the pointed brackets indicate that the function is a *discontinuity function*. At zero our function can take *two* values, zero or one. This may be expressed mathematically as:

$$F_0(x) = \langle x-a\rangle^0 = \begin{cases} 0 \text{ when } x \leqslant a \\ 1 \text{ when } x \geqslant a \end{cases}.$$

So what does all this mean in practice? If we apply our function to a beam where a point load W is applied some way along the beam (Figure 5.2.40b) at a distance a. We can determine the bending moments in the normal way, by first considering forces to the left of x, which gives

$$M = 0 \text{ for } x \geqslant 0 < a.$$

Also for forces to the right of a, we may write the bending moment as

$$M = W(x-a) \text{ for } a \leqslant x.$$

Then the Macaulay expression for the bending moment for all values of x is given as:

$$M = W\langle x-a\rangle.$$

Now in using the Macaulay expression we must remember that we are applying the unit ramp function and step-function and different rules apply. In particular, all terms which make the value inside the bracket negative, are given the value zero.

Figure 5.2.40c shows our beam subject to a distributed load commencing at distance a. Again the bending moment at a distance x along the beam, may be determined in the normal manner. Considering forces to the left of x, we have

$M = 0$ when $0 \leqslant x \leqslant a$

and $M = \frac{1}{2}\omega(x-a)^2$.

Thus the Macaulay expression for the bending moment for all values of x is

$M = \frac{1}{2}\omega\langle x-a\rangle^2$.

Do remember the constraints that apply when considering the expression inside the pointed brackets.

The method you should adopt in determining the Macaulay expression for the whole beam is illustrated by the next example.

We will now consider an example, where the total procedure for obtaining the Macaulay expression for the entire beam is illustrated.

Example 5.2.11

Determine the Macaulay expressions for the slope and deflection of the simply supported beam shown in Figure 5.2.41.

Now we take the origin from the extreme left-hand end, as shown in Figure 5.2.4l and consider a section at the extreme right-hand end of the beam from which we write down the Macaulay expression. Once written the ramp function rules apply.

So taking moments in the normal way gives

$M = R_A x - W_1(x-a) - W_2(x-b)$

therefore the Macaulay expression for the entire beam is

$M = R_A x - W_1\langle x-a\rangle - W_2\langle x-b\rangle$

and using Equation 5.2.39 where $\dfrac{d^2y}{dx^2} = \dfrac{M}{EI}$ gives

$EI\dfrac{d^2y}{dx^2} = (R_A x - W_1\langle x-a\rangle - W_2\langle x-b\rangle)$

and integrating in the normal way gives

$\dfrac{dy}{dx} = \dfrac{1}{EI}\left(\dfrac{R_A x^2}{2} - W_1\dfrac{\langle x-a\rangle^2}{2} - W_2\dfrac{\langle x-b\rangle^2}{2} + A\right)$

cont.

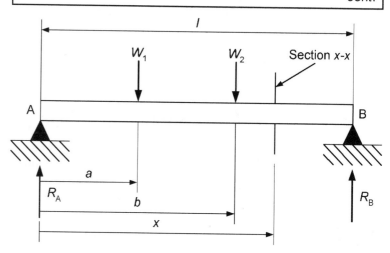

Figure 5.2.41 *Figure for Example 5.2.11*

and integrating a second time gives the expression for the deflection

$$y = \frac{1}{EI}\left(\frac{R_A x^3}{6} - W_1\frac{\langle x-a\rangle^3}{6} - W_2\frac{\langle x-a\rangle^3}{6} + Ax + B\right).$$

The constraints of integration can be found by applying the boundary conditions. At the supports the deflection is zero so when $x=0$, $y=0$ and when $x=l$, $y=0$.

From the above equation it can be seen that when $x=0$, $y=0$ and so $B=0$.

Also, when $x=l$, $y=0$ then from the slope equation, the value of A can be found by evaluating the expression

$$A = \frac{R_1 l^2}{6} - \frac{F_1(l-a)^3}{6l} - \frac{F_2(l-b)^3}{6l}.$$

Therefore, we have all that is necessary to determine the slope and deflection for any point along the beam.

Do remember that all negative Macaulay brackets should be treated as zero.

Using the Macaulay method for uniformly distributed loads

In using the Macaulay ramp function for a uniformly distributed load we started from a point a, an arbitrary distance from the zero datum, this was in order to comply with the requirements of the function. We then made the assumption that the UDL continued to the right-hand extremity of the beam. This has two consequences for use of the Macaulay method.

(i) For a UDL that covers the complete span of the beam, the Macaulay method is not appropriate. If we wish to find the slope and deflection for uniformly distributed loads, which span the entire beam, then we need to use the successive integration methods, discussed earlier.

(ii) For a UDL that starts at a, but does not continue for the entire length of the beam (Figure 5.2.42a) we cannot directly apply the Macaulay terms to our bending moment equation, for the reason given above. We need to modify our method in some way to accommodate this type of loading. In effect, what we do is allow our UDL to continue to the extremity of the beam (Figure 5.2.42b) and counter this additional loading by introducing an equal and opposite loading into the bending moment expression.

Figure 5.2.42a shows a simply supported beam with a UDL positioned between point a and point b. If we imagine that this load is extended to the end of the beam (Figure 5.2.42b) then our bending moment equation, as previously explained, would be

$$BM = R_A x - \omega\left\langle\frac{(x-a)^2}{2}\right\rangle.$$

This of course is incorrect, but if we now counter-balance the increase in the UDL by a corresponding decrease over the distance $(x-b)$,

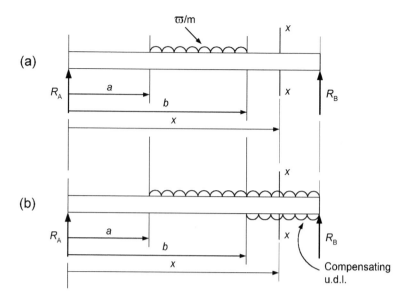

and introduce this as an additional Macaulay term, we obtain the correct Macaulay equation for a beam subject to a partial length UDL. The corrected equation is:

$$BM = R_{A}x - \omega\left\langle\frac{(x-a)^2}{2}\right\rangle + \omega\left\langle\frac{(x-b)^2}{2}\right\rangle.$$

Example 5.2.12

For the beam shown in Figure 5.2.43. Determine the deflection of the beam at its centre, if $EI = 100$ MNm².

We first find the reactions at the supports by taking moments about R_{B}, then

$$10R_{A} = (30 \times 7) + (80 \times 5) + (50 \times 3)$$

therefore $R_{A} = 76$ kN and so $R_{B} = 84$ kN.

Taking the origin as R_{A} we apply the Macaulay method shown previously. So by considering section $x-x$ to the extreme right of the beam and taking moments of the forces to the left of $x-x$ we get:

$$BM = 10^3[76x - 30\langle x-3\rangle - 50\langle x-7\rangle - 20/2\langle x-3\rangle^2 + 20/2\langle x-7\rangle^2]$$

and since $\dfrac{d^2y}{dx^2} = \dfrac{BM}{EI}$

then

$$\frac{d^2y}{dx^2} = \frac{10^3}{EI}\left[76x - 30\langle x-3\rangle - 50\langle x-7\rangle - \frac{20}{2}\langle x-3\rangle^2 + \frac{20}{2}\langle x-7\rangle^2\right]$$

cont.

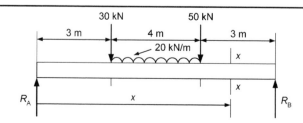

Figure 5.2.43 *Figure for Example 5.2.12*

therefore

$$\frac{dy}{dx} = \frac{10^3}{EI}\left[\frac{76x^2}{2} - \frac{30}{2}\langle x-3\rangle^2 - \frac{50}{2}\langle x-7\rangle^2 - \frac{20}{6}\langle x-3\rangle^3 + \frac{20}{6}\langle x-7\rangle^3 + A\right]$$

and

$$y =$$

$$\frac{10^3}{EI}\left[\frac{76x^3}{6} - \frac{30}{6}\langle x-3\rangle^3 - \frac{50}{6}\langle x-7\rangle^3 - \frac{20}{24}\langle x-3\rangle^4 + \frac{20}{24}\langle x-7\rangle^4 + Ax + B\right]$$

Now the boundary conditions yield the following values.
When $x=0$, $y=0$, so that $B=0$, since all the Macaulay terms are negative and so equate to zero.
  When $x=10$, $y=0$, now all the Macaulay terms are positive and so, we find $A=-879.3$.
When $x = 5$, then

$$y = \frac{10^3}{EI}\left[\frac{(76)(5)^3}{6} - \frac{30}{6}(2)^3 - \frac{20}{24}(2)^4 - 4396.7\right]$$

$$y = \frac{(10^3)(-2866.7)}{100 \times 10^6}$$

giving $y = -0.0287$ m or -28.7mm.

The exercises set out below should help you to consolidate, this rather difficult subject. You will need to refer to Chapter 6, if you are unclear on the mathematics associated with establishing (I) the second moment of area or the slope and deflection equations.

Questions 5.2.4

(1) Use the free body approach to draw the shear force and bending moment diagrams for the simply supported beams (Figure 5.2.44) shown below, giving values for the maximum bending moment in each case.

(2) A circular section cantilever beam 2 m long and 80 mm in diameter, supports a concentrated load of 1.2 kN acting 1.5 m from its fixed end. If the elastic modulus for the beam is $E = 210$ GPa, calculate the deflection at the load.

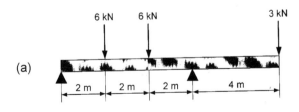

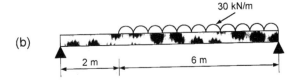

Figure 5.2.44 Figure for Question 5.2.4.1

(3) A beam simply supported at its ends, is 4 m long and carries point loads of 60 kN and 30 kN at distances 1 m and 2.5 m from the left-hand end of the beam. Determine the position and magnitude of the maximum deflection of the beam, given that $EI = 16 \times 10^6$ Nm^2.

(4) In a laboratory experiment, a steel beam 1.8 m long with rectangular cross-section 25 mm wide and 5 mm deep is simply supported on two knife edges 0.8 m apart spanning the centre section of the beam. A load of 30 N is hung at either end of the beam. Neglecting the mass of the beam and given that the steel has an elastic modulus $E = 200$ GN/m^2:
    (i) show that the bending moment is constant between the supports and find the value of this bending moment;
    (ii) find the maximum stress due to bending;
    (iii) calculate the radius of curvature between the supports.

(5) A beam of length 6 m is simply supported at its ends and carries a distributed load ($\omega = 10$ kN/m) which commences 1 m from the left-hand end of the beam and continues to the right-hand end of the beam. Two point loads of 30 kN and 40 kN act 1 m and 5 m, respectively, from the left-hand end. If the beam material is made of a steel with $EI = 20 \times 10^6$ Nm^2 determine the deflection at the centre.

5.3 Power transmission and rotational systems

In this final section we start with an investigation into power transmission between machines using belt drives, clutches and gear trains. We then turn our attention to the dynamics of rotating systems, looking particularly at static and dynamic balancing, flywheel dynamics and finally at the coupling of these components into rotational systems.

So, for example, the torque created by an electric motor may be converted into mechanical work by coupling the motor output to a gearbox, chain or belt drive. Motor vehicle transmission systems convert the torque from the engine into the torque at the wheels via a clutch, gearbox and drive shafts. The clutch is used to engage and disengage the drive from the engine, the gearbox assembly transmits the power to the wheel drive shafts, as dictated by the driving conditions.

Gears, shafts and bearings provide relatively rigid, but positive, power-transmission and may be used in many diverse applications ranging, for example, from power transmission for the aircraft landing flap system on a jumbo jet to the drive mechanism for a child's toy.

Belts, ropes, chains and other similar elastic or flexible machine elements are used in conveying systems and in the transmission of power over relatively long distances. In addition, since these elements are elastic and usually quite long, they play an important part in absorbing shock loads and in damping out and isolating the effects of unwanted vibration.

A classic example of the effects of a component being out of balance, may be felt through the steering wheel of a motor vehicle. Where a minor knock to a road wheel may be sufficient to put it out of balance and so cause rotational vibration which is felt back through the steering geometry. Any rotational machinery, which is incorrectly balanced or goes out of balance as a result of damage or a fault, will produce vibration, which is at best unacceptable and at worse may cause dangerous failure. In this section, you will be briefly introduced to the concept of balance and the techniques for determining out of balance moments.

We finish this section with a brief study of flywheels. These components provide a valuable source of rotational energy, for use in many engineering situations, such as machine presses, mills, cutting machines and prime movers for transport and agricultural machinery. We look, finally, at the inertia and energies involved when transmission and rotational components are coupled together, to form rotational systems.

We begin our study of power transmission by looking first at belt drives.

Belt drives

Power transmission from one drive shaft to another can be achieved by using drive pulleys attached to the shafts of machines being connected by belts of varying cross-section and materials.

A wide variety of belts are available, including flat, round, V, toothed and notched, the simplest of these and one of the most efficient is the flat belt. Belts may be made from leather or a variety of reinforced elastomer materials. Flat belts are used with crowned pulleys, round and V-belts are used with grooved pulleys, while timing belts require toothed wheels or sprockets.

To illustrate the set-up between a drive belt and pulleys a V-belt system is shown in Figure 5.3.1. Note that the pulley size is given by its pitch diameter and that in the case of a V-belt, it rides slightly higher than the external diameter of the pulley. The meaning of the symbols used in Figure 5.3.1 (if they are not obvious) are explained in the following paragraphs.

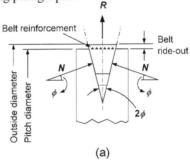

(a)

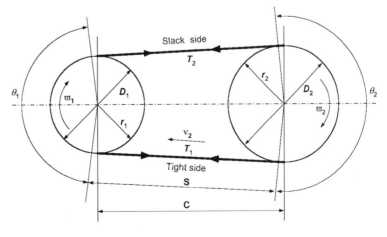

(b)

Figure 5.3.1 *Basic belt drive geometry*

The speed ratio between the driving and driven pulleys is inversely proportional to the ratio of the pulley pitch diameters. This follows from the observation that there is no slipping (under normal loads). Thus the linear velocity of the pitch line of both pulleys is the same and equal to the belt velocity (v), then,

$$v = r_1\omega_1 = r_2\omega_2 \tag{5.3.1}$$

or $v = D_1\omega_1/2 = D_2\omega_2/2$

so the angular velocity ratio is $\dfrac{\omega_1}{\omega_2} = \dfrac{D_2}{D_1}$. $\tag{5.3.2}$

The angles θ_1 and θ_2 are known as the *angle of lap* or *angle of contact* (Figure 5.3.1), this is measured between the contact length of the belt with the pulley and the angle subtended at the centre of the pulley.

The *torque* in belt drives and subsequent *transmitted power*, is influenced by the differences in tension that exist within the belt during operation. The differences in tension result in a tight side and slack side of the belt, as shown. These differences in tension influence the torque on the driving and driven pulley wheels, in the following way:

torque on driving pulley $= T_1r_1 - T_2r_2 = (T_1 - T_2)r_1$

torque on driven pulley $= T_1r_2 - T_2r_2 = (T_1 - T_2)r_2$.

Now we know that power is $=$ torque $\times$ angular velocity, and since angular velocity is given by the linear speed divided by the radius of action, then from Equation 5.3.1 we may write that

Transmitted power $(P) = (T_1 - T_2)V$. $\tag{5.3.3}$

The *stresses* set up in the drive belt are due to:

(i)   the tensile force in the belt, which is a maximum on the tight side of the belt;
(ii) the bending of the belt around the pulleys, which is a maximum as the tight side of the belt bends around the smaller diameter pulley; and,
(iii) the centrifugal forces created as the belt moves around the pulleys.

The maximum total stress occurs where the belt enters the smaller pulley and the bending stress is a major part. Thus there are recommended minimum pulley diameters for standard belts. Using smaller than recommended drastically reduces belt life.

Belt tension equations

Belts when assembled on pulleys are given an initial tension (T_0), you will be only too well aware of this fact if you have ever been involved in setting up the fan belt on a motor vehicle engine. If this initial tension is too slack the belt may slip below the maximum design power or, if too tight, undue stresses may be imposed on the belt causing premature failure. The initial pre-tension may be determined by using the formula:

$$T_0 = \frac{T_1 + T_2}{2}. \tag{5.3.4}$$

This follows from the argument that since the belt is made from an elastic material, when travelling around the pulleys, the belt length remains constant. Thus the increase in length on the tight side is equal to the reduction in length on the slack side, and so the corresponding tensions are given by, $T_1 - T_0 = T_0 - T_2$, from which Equation 5.3.4 results.

As already mentioned, when the belt passes over a pulley, there is an increase in the stresses imposed within the belt. These are due to an increase in centrifugal force, as the belt passes over the pulley, creating an increase in the tension of the belt.

Therefore, we have an increase in belt tension, additional to that of normal transmission power, which is due to the centrifugal forces set up in the belt as it passes over the pulleys. Our standard power Equation 5.3.3 needs to be modified to take the effects of centrifugal tension into account.

We can formulate this relationship, leading to the required power equation if we are prepared to undertake a little mathematical gymnastics! This exercise has been covered in Chapter 6, under Trigonometric Methods. The important results of this exercise *and*, the relationship between the coefficient of friction (μ) the angle of lap (θ) and the belt tensions (T_1, T_2); are given below.

$$T_c = m^1 v^2 \tag{5.3.5}$$

where (T_c) = the additional centrifugal tension, (m^1) is the mass per unit length of the belt and (v) is the belt velocity as indicated in Figure 5.3.1.

$$\frac{T_1 - T_c}{T_2 - T_c} = e^{\mu\theta}. \tag{5.3.6}$$

Under low speed condition the centrifugal tension term in Equation 5.3.6, is relatively small and may be neglected to give the following relationship

$$\frac{T_1}{T_2} = e^{\mu\theta}. \tag{5.3.7}$$

Also, as suggested we may modify our *power* equation to take into account the centripetal tension term, under high or low speed conditions, then substituting these terms into Equation 5.3.3 gives

$$P = (T_1 - T_c)(1 - e^{-\mu\theta})v \tag{5.3.8}$$

or, for low speed conditions $P = T_1(1 - e^{-\mu\theta})v.$ \qquad (5.3.9)

Modified equation for V-belt drives

The above equations have been formulated based on geometry related to flat belts travelling over flat drum pulleys (see Chapter 6). V-belt drives require us to replace μ by $\mu/\sin\phi$, where it appears in Equations 5.3.6 to 5.3.9.

Consider the V-pulley drive shown in Figure 5.3.1, where the groove angle is (2ϕ) and the reactions (N) act at angle ϕ as shown. The normal reaction (R), between the belt and pulley is given by simple trigonometry as

$R = 2N \sin\phi$

and so the *friction force* between the belt and pulley is

$$2\mu N = \frac{\mu R}{\sin\phi}. \tag{5.3.10}$$

So in effect, going from flat to V-grooved belt drives, transforms μ to $\mu/\sin\phi$.

Belt drive design requirements

In order to select a drive belt and its associated pulley assemblies for a specified use and, to ensure proper installation of the drive, several important factors need to be considered. The basic data required for drive selection is listed below. They are the:

- rated power of the driving motor or other prime mover;
- the service factor, which is dependent on the time in use, operating environment and the type of duty performed by the drive system;
- belt power rating;
- belt length;
- size of driving and driven pulleys and centre distance between pulleys;
- correction factors for belt length and lap angle on smaller pulley;
- number of belts;
- initial tension on belt.

Many design decisions depend on the required usage and space constraints. Provision must be made for belt adjustment and pre-tensioning, so the distance between pulley centres must be adjustable. The lap angle on the smaller pulley should be greater than 120°. Shaft centres should be parallel and carefully aligned so that the belts track smoothly, particularly if grooved belts are used.

The operating environment should be taken into consideration, where hostile chemicals, pollution or elevated temperatures may have an effect on the belt material. Consider other forms of drive, such as chains, if the operating environment presents difficulties.

From the above list, it can be seen that relationships between belt length, distance between centres, pulley diameters and lap angles, are required if the correct belt drive is to be selected. Equation 5.3.11 relates the belt pitch length, L; centre distance, C; and the pulley pitch diameters, D; by:

$$L = 2C + 1.57\,(D_2 - D_1) + \frac{(D_2 - D_1)^2}{4C}. \tag{5.3.11}$$

$$\text{Also } C = B + \frac{\sqrt{B^2 - 32\,(D_2 - D_1)^2}}{16} \tag{5.3.12}$$

where $B = 4L - 6.28(D_2 + D_1)$.

The contact angle (*lap angle*) of the belt on each pulley is

$$\left.\begin{array}{l} \theta_1 = 180° - 2\sin^{-1}\!\left[\dfrac{D_2 - D_1}{2C}\right] \\[2mm] \theta_2 = 180° + 2\sin^{-1}\!\left[\dfrac{D_2 - D_1}{2C}\right] \end{array}\right\} \tag{5.3.13}$$

One other useful formula is given below which enables us to find the span length S (see Figure 5.3.1), given the distance C between pulley centres and the pulley diameters.

$$S = \sqrt{C^2 - \left[\frac{D_2 - D_1}{2}\right]^2}. \tag{5.3.14}$$

This relationship is important since it enables us to check for correct belt tension, by measuring the force required to deflect the belt at mid-span. This length also has a direct bearing on the amount of vibration set up in the belt during operation therefore, we need to ensure that span length falls within required design limits.

Example 5.3.1

A flat belt drive system consists of two parallel pulleys of diameter 200 mm and 300 mm, which have a distance between centres of 500 mm. Given that the maximum belt tension is not to exceed 1.2 kN, the coefficient of friction between the belt and pulley is 0.4 and the larger pulley rotates at 30 rads/sec. Find:

(i)   the belt lap angle for the pulleys;
(ii) the power transmitted by the system;
(iii) the belt pitch length L;
(iv) the pulley system span length between centres.

(i) Lap angles are given by Equation 5.3.13.

So $\theta_1 = 180° - 2\sin^{-1}\left[\dfrac{300-200}{2\times500}\right]$

$= 180° - 2\sin^{-1}0.1$

$= 180° - 11.48° = 168.52°$

and $\theta_2 = 180° + 11.48° = 191.48°$.

(ii)   The maximum power transmitted will depend on the angle of lap of the smaller pulley. So, using Equation 5.3.9 and assuming low speed conditions, we get:

$P = T_1(1 - e^{-\mu\theta})v$ where $v = r\omega = (0.15)(30) = 4.5$.

$P = 1.2\times10^3\,(1 - e^{-(0.4)(2.94)})4.5$ note θ in radians.

$P = 1.2\times10^3(1 - 0.308)4.5$

$P = 3.74$ kW.

(ii) The belt pitch length is found using Equation 5.3.11.
$L = 2C + 1.57(D_2 - D_1) + \dfrac{(D_2 - D_1)^2}{4C}$.

The distance between pulley centres C is given as 500 mm. Then:
$L = (2)(0.5) + 1.57(0.3 - 0.2) + \dfrac{(0.3-0.2)^2}{(4)(0.5)}$

giving $L = 1.162$ m.

(iii) The span length S is given by Equation 5.3.14. Then:

$S = \sqrt{(0.5)^2 - \left(\dfrac{0.3-0.2}{2}\right)^2}$

Giving $S = 0.497$ m.

Friction clutches

With the sliding surfaces encountered in most machine components such as, bearings, gears, levers and cams it is desirable to minimise friction, so reducing energy loss, heat generation and wear. When considering friction in clutches and brakes, we try to maximise the friction between the driving and driven plates. This is achieved in dry clutches by choosing clutch plate materials with high coefficients of friction. Much research has been undertaken in materials designed to maximise the friction coefficient and minimise clutch disk wear, throughout a wide range of operating conditions during its service life.

The primary function of a clutch is to permit smooth transition between connection and disconnection of the driving and driven components and, when engaged, to ensure that the maximum possible amount of available torque is transmitted. Clutches are used where there is a constant rotational torque produced at the drive shaft, by a prime mover such as an engine. The drive shaft can then be engaged through the clutch to drive any load at the output, without the need to alter the power produced by the prime mover. The classic application of this transmission principle is readily seen in automobiles, where the engine provides the power at the drive shaft which is converted to torque at the wheels. When we wish to stop the vehicle, without switching off the engine, we disengage the drive using the clutch.

Disk clutches

A simple disk clutch, with one driving and one driven disk is illustrated in Figure 5.3.2. The principle of operation is dependent on the driving friction created between the two clutch disks when they are forced together. The means of applying pressure to the clutch disks may be mechanical or hydraulic.

As with other components that rely on friction, they are designed to operate either dry or wet using oil. Many types of clutch used in automobile transmissions are designed to operate wet. The oil provides an effective cooling medium for the heat generated during clutch engagement and instead of using a single disk (Figure 5.3.2.), multiple disks may be used to compensate for the inevitable drop in the coefficient of friction.

In a multi-disk clutch (Figure 5.3.3) the driving disks rotate with the input shaft, the clutch is engaged using hydraulic pressure which is generated from a master cylinder at the clutch pedal, or by some other means. The driving disks are constrained, normally by a key and keyway, to rotate with the input shaft. On engagement, the rotating drive disks are forced together with the driven disk to provide torque at the output. When the clutch is disengaged by releasing the hydraulic pressure, or through spring action, the disks are free to slide axially and so separate themselves.

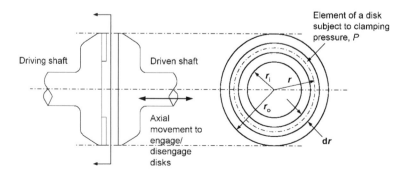

Figure 5.3.2 *Simple disk clutch*

Figure 5.3.3 *Hydraulically operated clutch assembly*

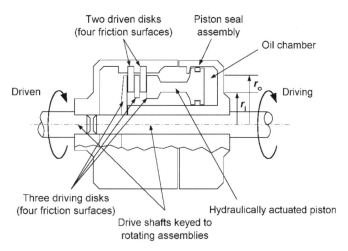

From our diagram you can see that the driving disks include the backing plates, which only provide one friction surface, whereas the middle driving disk and the two driven disks each provide two contact surfaces. So in our example the five disks provide a total of four driving and four driven contact surfaces.

Friction clutch equations

In order to estimate the torque and power from clutch driven shafts we need to use equations which relate factors such as: clutch size, friction coefficients, torque capacity, axial clamping force and pressure at the disks. When formulating such equations, we make two very important basic assumptions:

(i)   *we assume uniform distribution of pressure at the disk interface*;
(ii) *we assume a uniform rate of wear at the disk interface.*

Mathematics in action

The *first assumption* is valid for an unworn (new) accurately produced and installed clutch, with rigid outer disks. Consider a very small element of the disk surface (Figure 5.3.1), which is subject to an axial force normal to the friction surface. Then if *constant* pressure p acts on the element at radius r and the element has width dr, then the force on the element

$$= (2\pi r dr)p.$$

So the total normal force acting on the area of contact is

$$W = \int_{r_i}^{r_o} 2\pi p r \, dr = \pi p (r_o^2 - r_i^2) \tag{5.3.15}$$

Where W is the axial force damping the driving and driven plates together.

Now we know that the friction torque that can be developed on the disk 'element' is the product of the friction force and radius. The friction force is the product of the force on the element and the coefficient of friction. So we may write:

torque on disk element $= (2\pi p r dr)\mu r$

and the total torque that can be transmitted is

$$T = \int_{r_i}^{r_o} 2\pi p\mu r^2 dr = \frac{2}{3}\pi p\mu(r_o^3 - r_i^3). \tag{5.3.16}$$

The above equation represents the torque transmitted by one driving disk (plate) mating with one driven plate.

For clutches that have more than one set of disks, i.e. (N) disks, where N is an even number (see Figure 5.3.3), then:

$$T = \frac{2}{3}\pi p\mu(r_o^3 - r_i^3)N \tag{5.3.17}$$

If we transpose Equation 5.3.15 for p and substitute it into Equation 5.3.17 we get

$$T = \frac{2W\mu(r_o^3 - r_i^3)}{3(r_o^2 - r_i^2)}N. \tag{5.3.18}$$

This equation relates the torque capacity of the clutch with the axial clamping force.

If we now *assume a uniform rate of wear*. We know that the rate at which wear takes place between two rubbing surfaces is proportional to the rate at which it is rubbed (the rubbing velocity) multiplied by the pressure applied. So as a pair of clutch plates slide over each other, their relative velocities are linear, our clutch plates rotate with angular velocity, so the rubbing velocity is dependent on the radius of rotation. From the above argument we may write:

wear $\propto$ pressure $\times$ velocity (where velocity is dependent on radius)

so wear $\propto$ pressure $\times$ radius

and introducing the constant of proportionality gives

$pr = c$.

Now substituting the above expression into Equations 5.3.15 and 5.3.16 gives, respectively

$$W = \int_{r_i}^{r_o} 2\pi c dr = 2\pi c(r_o - r_i) \tag{5.3.19}$$

and

$$T = \int_{r_i}^{r_o} 2\pi c\mu r dr = \pi c\mu(r_o^2 - r_i^2)N \tag{5.3.20}$$

and on transposition of Equation 5.3.15 for p and then substitution into Equation 5.3.20 we get

$$T = W\mu\left(\frac{r_o + r_i}{2}\right)N. \tag{5.3.21}$$

We have said that $pr = c$, now it is know that the greatest pressure on the clutch plate friction surface occurs at the inside radius r_i. So for a maximum allowable pressure (p_{max}) for the friction surface, we may write

$$pr = c = p_{max}r_i. \tag{5.3.22}$$

Clutch plates are designed with a specified ratio between their inside and outside radius, to maximise the transmitted torque between surfaces. This is given below, without proof,

$r_i = 0.58r_o$.

In practice clutches are designed within the approximate range $0.5r_o \leqslant r_i \leqslant 0.8r_o$.

Figure 5.3.4 *Cone clutch*

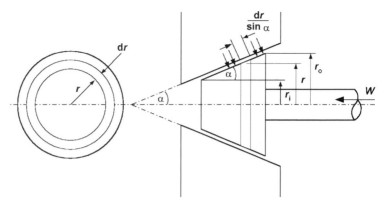

Cone clutches

A cone clutch has a *single pair* of conical friction surfaces (Figure 5.3.4) and is similar to a disk clutch, except that in the case of the disk clutch the coning angle take a specific value of 90°. So the cone clutch may be thought of as the generalised case, of which the disk clutch is a special case. One of the advantages of a cone clutch, apart from their simplicity, is that due to the wedging action the clamping force required is drastically reduced when compared to their disk clutch counterpart.

The formulae for cone clutches are derived in a very similar way to those you met earlier concerning the plate clutch and V-belts. We therefore adopt a very similar mathematical approach.

Mathematics in action

Conical clutches

Consider the element of width dr in Figure 5.3.4, from the geometry the area of this element is:

$$= \frac{2\pi r dr}{\sin \alpha}.$$

If p is again the normal pressure, then the normal force acting on the element is:

$$= \frac{(2\pi r dr)p}{\sin \alpha}$$

and the corresponding axial clamping force is

$$= (2\pi r dr)p$$

the above expression is exactly the same as for the disk clutch element, shown earlier. So the total axial force acting on the mating cone surfaces is

$$W = \int_{r_i}^{r_o} 2\pi p r dr \text{ (same as Equation 5.3.15)}$$

and assuming 'uniform pressure'

$$W = \pi p(r_o^2 - r_i^2), \text{ as before.}$$

We also know that the friction force is equal to the product of the coefficient of friction (μ) and the normal reaction force. If we again conside the normal reaction force in terms of pressure, we may write friction force $= \dfrac{(2\pi r \mu dr)p}{\sin \alpha}$

and the torque transmitted by the element is

$= \dfrac{(2\pi r \mu dr)}{\sin \alpha} pr.$

Then the total torque transmitted is

$T = \int_{r_i}^{r_o} 2\pi \mu p r^2 dr$

and assuming uniform pressure

$$T = \frac{2\pi \mu p}{3 \sin \alpha} (r_o^3 - r_i^3) \tag{5.3.23}$$

and as before eliminating p, using the force (W) equation, we have

$$T = \frac{2\mu W}{3 \sin \alpha} \left(\frac{r_o^3 - r_i^3}{r_o^2 - r_i^2} \right). \tag{5.3.24}$$

Finally assuming *uniform wear* we have

$$W = 2\pi p_{max} r_i (r_o - r_i) \tag{5.3.25}$$

$$T = \frac{\pi \mu p_{max} r_i}{\sin \alpha} (r_o^2 - r_i^2) \tag{5.3.26}$$

and in terms of W

$$T = \frac{\mu W}{2 \sin \alpha} (r_o + r_i). \tag{5.3.27}$$

Note that in Equations 5.3.25 and 5.3.26, the constant 'c' is replaced by $p_{max} r_i$, since $pr = c = p_{max} r_i$ (compare this with Equations 5.2.18 and 5.2.19).

Example 5.3.2

A multiple plate clutch, similar to that shown in Figure 5.3.3, needs to be able to transmit a torque of 160 Nm. The external and internal diameters of the friction plates are 100 mm and 60 mm, respectively. If the friction coefficient between rubbing surfaces is 0.25 and $p_{max} = 1.2$ MPa determine the total number of disks required, *and* the axial clamping force. State all assumptions made.

Since this is a design problem, we make the justifiable assumption that the wear rate will be uniform at the clutch plate rubbing surfaces. We must also assume that the torque is shared equally by all the clutch disks and, that the friction coefficient remains constant during operation.

This last assumption is unlikely, and contingencies in the design would need to be made to ensure the clutch met its service requirements. In practice, designers ensure that a new clutch is slightly over engineered to allow for *bedding in*. After the initial bedding-in period wear tends to be uniform and friction coefficients are able to meet the clutch design specification under varying conditions, so the clutch assembly performs satisfactorily.

cont.

Then using Equation 5.3.20, where $c = r_i p_{max}$, then

$$N = \frac{T}{\pi r_i p_{max} \mu (r_o^2 - r_i^2)}.$$

So $N = \dfrac{160}{(\pi)(30 \times 10^{-3})(1.2 \times 10^6)(0.25)(0.05^2 - 0.03^2)}$

$$N = \frac{160}{46.29} = 3.53.$$

Now since N must be an even integer, then the nearest N above design requirements is $N=4$. This is because the friction interfaces must transmit torque in parallel pairs. In Figure 5.3.3 there are four friction interfaces, which require 'five' disks, the two outer disks having only one friction surface.
We also need to find the axial clamping force (W).
This is given by Equation 5.3.21, where

$$T = W\mu \left(\frac{r_o + r_i}{2} \right) N.$$

So
$$W = \frac{T}{\mu N} \left(\frac{2}{r_o - r_i} \right) = \frac{160}{(0.25)(4)} \left(\frac{2}{0.05 - 0.03} \right)$$

$W = 16$ kN.
So we require a total of five disks and an axial clamping force of 16 kN.

Gear trains

We continue our study of power transmission by looking at gears. The gearbox is designed to transfer and modify rotational motion and torque. For example, electric motor output shafts may rotate at relatively high rates, producing relatively low torque. A reduction gearbox interposed between the motor output and load, apart from reducing the speed to that required, would also increase the torque at the gearbox output by an amount equivalent to the gearbox ratio.

The gear wheel itself is a toothed device designed also, to transmit rotary motion from one shaft to another. As mentioned in the introduction to transmission systems, gears are rugged, durable and efficient transmission devices, providing rigid drives. The simplest and most commonly used gear is the *spur gear*, they are designed to transmit motion between parallel shafts (Figure 5.3.5). Spur gears can be helical, providing smooth and efficient meshing, which provides a quieter action and tends to reduce wear. The gears used in automobile transmission systems are often helical, for the above reasons (synchronised meshing or synchromesh). If the gear teeth are bevelled, they enable the gears to transmit motion at right angles, the common hand drill is a good example of the use of bevelled gears.

Spur gears

Figure 5.3.6 shows the basic geometry of two meshing spur gears, there are several important points to note about this geometry, which

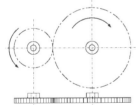

(a) Simple spur gears

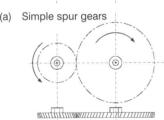

(b) Helical spur gears

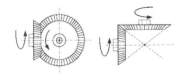

(c) Bevelled gears

Figure 5.3.5 *Gear wheels*

Figure 5.3.6 *Geometry for two gears in mesh*

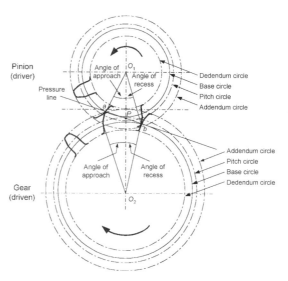

will help you when we consider gear trains. The gear teeth profile is *involute*, this geometric profile is the curve generated by any point on a taut thread as it unwinds from a *base circle*. Once the gear teeth have cut to this involute shape they will, if correctly spaced, mesh without jamming. It is interesting to note that this is the only shape where this can be achieved. The gear teeth involute profile is extended outwards beyond the *pitch circle* (the point of contact of inter-meshing teeth) by a distance called the addendum. Similarly the tooth profiles are extended inwards from the pitch circle by an identical distance called the dedendum. When we refer to the diameter of gear wheels, we are always referring to the *pitch diameter*.

Simple gear train

Meshing spur gears which transmit rotational motion from an input shaft to an output shaft are referred to as *gear trains*. Figure 5.3.7 shows a simple gear train consisting of just two spur gears of differing size, with their corresponding direction of rotation being given. If the larger gear *wheel* has 48 teeth (gear A) and the smaller *pinion* has 12 teeth (gear B). Then the smaller gear will have to complete four revolutions for each revolution of the larger gear. Now we know that the *velocity ratio* (VR) is defined as

$$\text{VR} = \frac{\text{distance moved by effort}}{\text{distance moved by load}}.$$

Figure 5.3.7 *Simple gear train*

The VR when related to our gear train, will depend on which of the gears is the *driver* (effort) and which the *driven* (load). The distance moved is dependent on the number of teeth on the gear wheel. In our example above, if the larger wheel is the driver then we have

$$\text{VR} = \frac{\text{distance moved by driver}}{\text{distance moved by driven}} = \frac{1}{4}.$$

What this shows is the relationship between the number of gear teeth and the velocity ratio or, when considering gears, the *gear ratio*.

Now when the gear train is in motion, the ratio of the angular velocities of the gears, is directly proportional to the angular distances travelled and as can be seen from above is the inverse ratio of the number of teeth on the gears. Thus the *gear ratio G* may be written formally as

Gear ratio $(G) \dfrac{t_B}{t_A} = \dfrac{\omega_A}{\omega_B}$ which for our gear pair gives $\dfrac{12}{48} = \dfrac{1}{4}$.

So the gear ratio in this case is 1:4.

The gear teeth on any meshing pair must be the same size, therefore the number of teeth on a gear is directly proportional to its *pitch circle diameter*. So, taking into account this fact the *gear ratio formula*, may be written as

$$\text{gear ratio } G = \dfrac{\omega_A}{\omega_B} = \dfrac{t_B}{t_A} = \dfrac{d_B}{d_A}. \tag{5.3.28}$$

So to summarise; the gear ratio is inversely proportional to the teeth ratio, and so inversely proportional to the pitch circle diameter ratio.

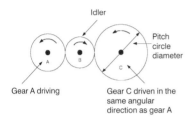

Idler

Pitch circle diameter

Gear A driving

Gear C driven in the same angular direction as gear A

Figure 5.3.8 *Gear train with idler*

Example 5.3.3

A gearwheel (A) rotates with an angular velocity of 32 rad/s, it has 24 teeth, it drives a gear B. If the gear ratio G is 8:1, determine the velocity of B and the number of teeth on B.
From Equation 5.3.28

$$G = \dfrac{\omega_A}{\omega_B} = \dfrac{t_B}{t_A} \text{ or } \dfrac{8}{1} = \dfrac{32}{4} = \dfrac{t_B}{24} \text{ and } t_B = 192.$$

Then the velocity of B = 4 rad/s and it has 192 teeth.

A simple gear train can, of course, carry more than two gears. When more than two gears are involved we still consider the gear train in terms of a combination of gear pairs. Figure 5.3.8 shows a simple gear train consisting of three gears.

The overall gear ratio is the ratio of the input velocity over the output velocity, in our example this is, ω_A/ω_C. By considering the velocity ratios of each gear pair sequentially from the input, we can determine the velocity ratio for the whole train. So for the first pair of meshing gears, AB, the velocity ratio is given in the normal manner as ω_A/ω_B. Now the second pair of meshing gears is ω_B/ω_C. So the overall gear ratio is

$$G = \dfrac{\omega_A}{\omega_B} \times \dfrac{\omega_B}{\omega_C} = \dfrac{\omega_A}{\omega_C}. \tag{5.3.28}$$

You will note that the middle gear, plays no part in the final gear ratio. This suggests that it is not really required. It is true that the size of the middle gear has no bearing on the final gear ratio, but you will note from Figure 5.3.8, that it does have the effect of reversing the direction of angular motion between the driving and driven gear, for this reason it is known as an *idler gear*.

Compound gear train

In a compound train at least one shaft carries a compound gear, that is two wheels which rotate at the same angular velocity. The advantage of this type of arrangement is that compound gear trains can produce higher gear ratios, without using the larger gear sizes that would be needed in a simple gear train. Examples of compound

Figure 5.3.9 *Compound gear train*

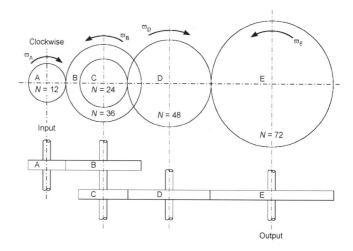

gear trains may be found in automobile gearboxes and in a variety of industrial machinery such as lathes and drilling machines.

The following example illustrates how the gear ratio may be determined by considering the angular velocities of the gears and their relationship with the number of teeth.

Example 5.3.4

The compound gear train illustrated in Figure 5.3.9, consists of five gears, of which two, gears B and C are on the same shaft. If gear A drives the system at 240 revs/min (clockwise) and the number of teeth on each gear (N), are as given. Determine the angular velocity of gear E at the output and the gear ratio for the system.

Again the system may be treated as gear pairs. Noting that gears B and C, being on the same shaft, rotate with the same angular velocity, then we proceed as follows.

From Equation 5.3.29, for a simple gear train, then the overall gear ratio (G) is given by

$$G = \frac{\omega_A}{\omega_B} \times \frac{\omega_B}{\omega_C} \times \frac{\omega_C}{\omega_D} \times \frac{\omega_D}{\omega_E} = \frac{\omega_A}{\omega_E} \text{ where } \omega_B = \omega_C.$$

The angular velocity of gear E may be found as follows:

$$\omega_E = \left(\frac{N_A}{N_B} \times \frac{N_B}{N_C} \times \frac{N_C}{N_D} \times \frac{N_D}{N_E} \right) \omega_A$$

$$\omega_E = \left(\frac{12}{36} \times \frac{36}{24} \times \frac{24}{48} \times \frac{48}{72} \right) \omega_A = \frac{1}{6}\omega_A$$

then $\omega_E = \dfrac{240}{6} = 40$ rev/min (counterclockwise).

The direction can be established as counterclockwise from the velocity arrows shown in Figure 5.3.9.

Now from above $G = \dfrac{\omega_A}{\omega_E} = \dfrac{240}{40} = \dfrac{6}{1} = 6{:}1.$

This is inversely proportional to the teeth ratio (1/6), which is correct. So we may write the general expression (G) for compound gear trains as

cont.

$$G = \frac{\omega_A}{\omega_B} \times \frac{\omega_B}{\omega_C} \times \frac{\omega_C}{\omega_D} \times \frac{\omega_D}{\omega_E} = \frac{\omega_A}{\omega_E} \dots \textit{etc.}$$

$$= \frac{t_B}{t_A} \times \frac{t_C}{t_B} \times \frac{t_D}{t_C} \times \frac{t_E}{t_D} = \frac{t_E}{t_A} \dots \textit{etc.}$$

$$= \frac{d_B}{d_A} \times \frac{d_C}{d_B} \times \frac{d_D}{d_C} \times \frac{d_E}{d_D} = \frac{d_E}{d_A} \dots \textit{etc.}$$

(5.3.30)

Note that

$$m = \frac{d_A}{t_A} = \frac{d_B}{t_B} \dots \textit{etc.}$$

where m is the module of the gear wheel, that is the number of millimetres of pitch circle diameter taken up per tooth.

Epicyclic gear trains

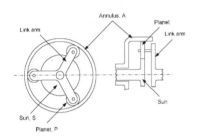

Figure 5.3.10 *Epicyclic gear train*

Unique gear ratios and movements can be obtained in a gear train by permitting some of the gear axes to rotate about others. Such gear combinations are called *planetary* or *epicyclic* gear trains. Epicyclic gear trains (Figure 5.3.10) always include a *sun* gear, a *planet* carrier or *link arm*, one or more *planet gears*. Often a ring is included which has internal gear teeth that mesh with the planet gear, this is known as the *annulus*. Epicyclic gear trains are unusual mechanisms in that they have two degrees of freedom. This means, for constrained motion, that any two of the elements in the train provide an input.

Epicylic gear trains are used for most automatic gearboxes and some complex steering mechanisms. The major advantages of epicylic mechanisms is the attainment of high gear ratios and the fact that they act as bearings as well as transmissions.

We will now attempt to analyse the motion of the gear train illustrated in Figure 5.3.10, the following procedure may be used.

(1) First imagine that the whole assembly is locked and rotated through one revolution, this has the effect of adding +1 revolutions to all elements.

(2) Next apply a rotation of −1 to the gear which is to be fixed, this has the effect of cancelling the rotation given initially. Now the rotations of the other elements can be compared with the rotation of the fixed element, by considering the number of teeth on each of the gears.

(3) Construct a table for each of the elements and add the results obtained from above actions. (When we sum the columns, using above procedure, we are able to ensure that the fixed link has zero revolutions.)

Example 5.3.5

An epicylic gear train (Figure 5.3.10) has a fixed annulus (A) with 180 teeth, and a sun wheel (S) with 80 teeth.

(i) Determine the gear ratio between the sun and the link arm.
(ii) Find the number of teeth on the planet gears and so determine the gear ratio between the planet and link arm.

cont.

So following our procedure, we first lock the whole assembly and give it +1 revolutions (so the annulus is zero when we sum). Next we apply a −1 revolution to the gear specified as being at rest, (so this element is zero when we sum), in this case the annulus. Then, to determine the arm in relation to the sun wheel, we fix the arm. Now, the number of revolutions of the sun wheel to that of the annulus equals the inverse ratio of their teeth. The results of these various actions are given below.

Action	Effect			
	Arm	A	S	P
(1) Lock assembly give all elements +1 revolutions	+1	+1	+1	+1
(2) Fix the link arm and give −1 to A	0	−1	$\dfrac{+180}{80}$	$\dfrac{-180}{50}$
(3) Sum 1 and 2	+1	0	3.25	−2.6

Then a rotation of +1 for the link arm results in +3.25 for the sun wheel. The gear ratio is 3.25:1.

Note that the number of teeth on the planet (P) was not required for part (1), because it is in effect an idler gear between the sun and the annulus.

(iii) Referring to Figure 5.3.10, we note that

$r_s = 2r_P = r_A$ and we know from Example 3.5.9 that

$$\frac{d_s}{d_p} = \frac{t_s}{t_p} \text{ and } \frac{d_P}{d_A} = \frac{t_P}{t_A} \qquad \text{(a)}$$

So $\dfrac{r_S}{r_P} = \dfrac{2r_P}{r_P} = \dfrac{r_A}{r_P}$ then since $2r = d$ we may write

$$\frac{d_s}{d_P} = 2 = \frac{d_A}{d_p} \text{ and so from (a)} \frac{t_S}{t_P} = 2 = \frac{t_A}{t_P}$$

which on multiplication by t_p gives $t_s = 2t_P = t_A$.

Then $t_P = \left(\dfrac{t_A - t_S}{2}\right) = \left(\dfrac{180 - 80}{2}\right) = 50.$

The required results for the gear ratio between the planet and the link arm is −2.6:1. The negative sign indicates that the angular velocity of the planet is in the opposite direction to that found for the sun. Since angular velocities determine the gear ratio, this would be expected. These results are shown in brackets in the table.

Torque in gear trains

The torques and angular velocities at the input and out of a gear train, are shown in Figure 5.3.11. The gear train could take the form of a gearbox. The input torque will be in the same direction as the rotation of the input shaft, the torque at the output will act in the opposite direction to the rotation of the output shaft. This is because, the torque created by the shaft driving the load at the output, acts in opposition to the rotation of the drive shaft.

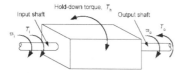

Figure 5.3.11 *Gearbox torques and angular velocities*

Under normal circumstances there will be a difference between the torque created at the input (T_i) and the torque created at the output (T_o). This difference produces a third torque which creates a positive or negative turning moment, which acts on the gear train casing. To counter this effect we need to ensure that the gear train assembly is held down securely, hence we refer to this third torque as the *hold down torque* (T_h). Under these circumstances the gear train assembly is in equilibrium and so the sum of the torques is zero, or

$$T_i + T_h + T_o = 0. \tag{5.3.31}$$

Also if there are no losses through the gear train then, the input power equals the output power. We already know that power (P) = $T\omega$, and for our system the output torque acts in opposition to the input torque (in a negative sense), so we may write:

$$T_i \omega_i = -T_o \omega_o.$$

The above assumes that no losses occur through the system, in practice there are always losses, due to bearing friction, viscous friction and gear drag. Now since the efficiency of a machine is defined as

$$\text{Efficiency } (\zeta) = \frac{\text{power at output}}{\text{power at input}}$$

then $\zeta = \dfrac{T_o \omega_o}{T_i \omega_i}.$ \hspace{1em} (5.3.32)

Example 5.3.6

A gearbox powered by a directly coupled electric motor, is required to provide a minimum of 12 kW of power at the output, to drive a load at 125 rpm. The gearbox chosen consists of a compound train (Figure 5.3.12), which has an efficiency of 95%. Determine the:

(i) gear ratio;
(ii) electric motor output shaft velocity;
(iii) power required from the electric motor;
(iv) hold down torque.

(i) $G = \dfrac{t_B}{t_A} \times \dfrac{t_C}{t_B} \times \dfrac{t_D}{t_C} = \dfrac{t_D}{t_A}$ (inverse of teeth ratios).

From the diagram
$$G = \frac{30}{12} \times \frac{40}{30} \times \frac{72}{40} = \frac{72}{12} = \frac{6}{1} = 6:1.$$

cont.

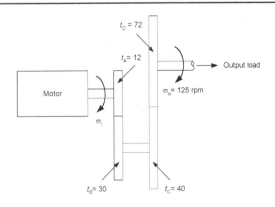

Figure 5.3.12 *Figure for Example 5.3.6*

(ii) $\omega_i = 6\omega_o = (6)(125) = 750$ r.p.m.
(iii) Power required at input is given by

$$\text{Efficiency} = \frac{\text{power at output}}{\text{power at input}}$$

so $P_i = \dfrac{12 \text{ kW}}{0.95} = 12.63$ kW.

(iv) The hold down torque may be found from Equation 5.3.31.
Then since $P = T\omega$ so $T_i = \dfrac{P_i}{\omega_i}$ and $\omega_i = 78.54$ rad/s

then $T_i = \dfrac{12.63 \times 10^3}{78.54} = 160.8$ Nm.

Similarly, at output $T_o = \dfrac{12 \times 10^3}{13.09} = -916.7$ Nm (T_o negative).

Then from Equation 5.3.31, where $T_o + T_h + T_i = 0$

$T_h = -T_o - T_i = 916.7 - 160.8 = 755.9$ Nm.

From the results you will note that the torque required from the motor is six times smaller than that required at the output. This is one of the reasons for placing a gearbox between the motor and the load, we are then able to substantially reduce the size of the motor!

Balancing

As already mentioned in the introduction to this section, the effects of components being out of balance can be severe. You have already met the concept of *inertia* applied to rotational systems, when you studied angular kinetic energy in Chapter 3. Inertia forces and accelerations occur in all rotational machinery and such forces must be taken into account by the engineering designer. Here, we are concerned with the alleviation of inertia effects on rotating masses.

If a mass m is rigidly attached to a shaft which rotates with an angular velocity ω about a fixed axis and the centre of the mass is at a radius r from the axis. Then we know from our previous work that the centrifugal force which acts on the shaft as a result of the rotating mass is given by

$$F_c = mr\omega^2. \tag{5.3.33}$$

When additional masses are added, which *act in the same plane* (Figure 5.3.13), they also produce centrifugal forces on the shaft. Then for equilibrium *to balance* the shaft, we require that the sum of all the centrifugal forces, $\Sigma m\omega^2 r = 0$. Since all the masses are rigidly fixed, they all rotate with the *same* angular velocity, so ω may be omitted from the above expression to give $\Sigma mr = 0$. Now to determine whether or not the shaft is in balance we can simply produce a vector polygon, where the vector product mr, is drawn to represent each mass at its radius, in the system.

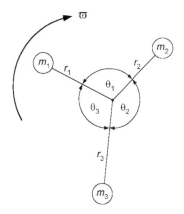

Figure 5.3.13 *Fixed masses about centre of rotating shaft*

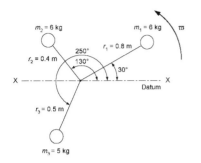

Figure 5.3.14 *Rotating mass system*

Example 5.3.7

A rotating system consists of three masses rigidly fixed to a shaft, which all act in the same plane, as shown in Figure 5.3.14. Determine the magnitude and direction of the mass required for balance, if it is to be set at a radius of 0.5 m from the centre of rotation of the shaft.

Then from the diagram, we draw our vector polygon, where each component is mr. Then $m_1 r_1 = 4.8$ kg m, $m_2 r_2 = 2.4$ kg m and $m_3 r_3 = 2.5$ kg m. Now constructing the vector polygon, we get

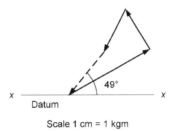

Scale 1 cm = 1 kgm

which gives a value for the balancing couple of 2.5 kg m acting at the angle shown above. So required mass at radius 0.5 m is

$$\frac{2.5 \text{ kg m}}{0.5 \text{ m}} = 5 \text{ kg}.$$

Note that the common angular velocity, ω, was not needed to determine the out of balance force.

If the masses rotate in *different planes*, then not only must the centrifugal forces be balanced but the moments of these forces about any plane of revolution must also be balanced. We are now considering rotating masses, in three dimensions.

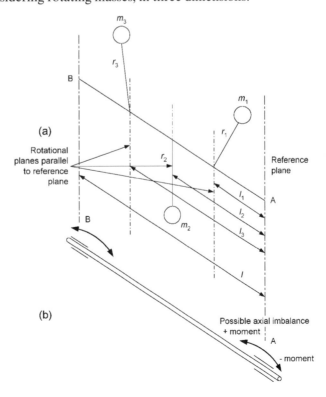

Figure 5.3.15 *Rotating masses in different planes*

This system is illustrated in Figure 5.3.15, where masses now rotate about our shaft, some distance (l) from a reference plane. The result of these masses rotating in *parallel* planes to the reference plane, is that they create moment effects. Quite clearly the magnitude of these moments will depend on the position chosen for the reference plane.

The technique we adopted for determining the balance state of the system is first, to transfer the out of balance *couple* to the reference plane, by adding the individual moment couples, by vector addition, using a *moment polygon*. Each couple is given by the product $mr\omega^2 l$ (where l is the distance of each rotating mass from the reference plane). We then solve the *force polygon*, at the plane of reference, in the same way as Example 5.3.7. Note, that as before, since all masses are rigidly attached to the shaft, they will all rotate with the same angular velocity, so this velocity may be ignored, when constructing the moment polygon.

So we have a *two stage* process which first involves the production of a *moment polygon* for the couples, from which we determine whether or not there is an out of balance moment. If there is, we will know the nature of this moment that is; its mass, radius of action, angular position and, distance from the reference plane (mrl). The *force polygon* may then be constructed, which will include the *mr* term from the moment polygon. We are then able to balance the system by placing the *balance* mass, derived from our polygons, at a position (l), angle (t), radius (r) to achieve *dynamic balance* of the rotating system.

If the shaft is supported by bearings, radial balance across the bearing will be achieved by adopting the above method. However, axial forces may still exist at the bearing if this is not chosen as the reference plane for the unknown mass. This would result in longitudinal in-balance of the shaft (Figure 5.3.15 b). There would then be a requirement for the addition of a counter-balance mass, to be placed in an appropriate position to one side of the shaft bearing. Thus there may be the need for two balance masses, to ensure total dynamic balance throughout the system.

The above procedure, should be readily understood, by considering the following example.

Example 5.3.8

A rotating shaft is supported by bearings at A and B. It carries three disk cams, which are represented by the equivalent concentrated masses shown in (Figure 5.3.16). Their relative angular positions are also shown.

For the given situation, calculate the magnitude of the reaction at bearing A resulting from dynamic imbalance when the shaft rotates at 3000 rev/min.

Using $F = mr\omega^2$ we first need to establish the $m_A r_A$ value, then find $m_A r_A l_A$ and so determine the reaction at bearing A, as required.

To assist us we will set up a table, and by taking moments about 'B', we eliminate it from the calculation, in the normal manner. Then from the diagram we find values as:

cont.

Figure 5.3.16 *Figure for Example 5.3.8*

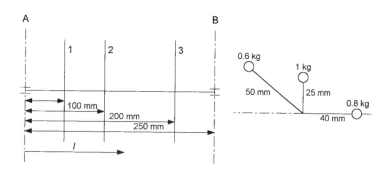

Force vector diagram

Scale 1 cm = 10 kg mm

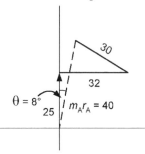

Moment vector diagram

Scale 1 cm = 1 kg mm
using $m \, r \, l_A = 10$

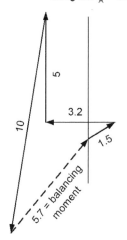

Plane	m (kg)	r (mm)	mr	l(m)	mrl
A	m_A	r_A	$m_A r_A$	0.25	$0.25 \, m_A r_A$
1	1	25	25	0.2	5
2	0.8	40	32	0.1	3.2
3	0.6	50	30	0.05	1.5
B	m_B	r_B	$m_B r_B$	0	0

Check that you arrive at the values in the table. Next we construct our force polygon as in the previous example.

Now the required force may be found since $m_A r_A = \dfrac{5.7}{0.25}$ kg mm = 22.8 kg mm.

Then $F_A = m_A r_A \omega^2 = (22.8 \times 10^{-3})\left(\dfrac{3000 \times 2\pi}{60}\right)^2 = 2250$ N.

So force on bearing A = 2250 N.
Note that reaction B may be found in a similar manner by taking moments about A.

Flywheels

A flywheel is an energy storage device. It absorbs mechanical energy by increasing its angular velocity and gives out energy by decreasing its angular velocity. The mechanical energy is given to the flywheel by a torque, which is often provided by a motor. It is accelerated to operating speed, thus increasing its inertial energy, which can be used to drive a load after the accelerating torque has been removed.

Mathematics in action

If a flywheel is given an input torque T_i, at angular velocity ω_i, this will cause the flywheel to increase speed. Then if a load or some other output acts on the flywheel, it will create an opposing torque T_o at angular velocity ω_o, which will cause the flywheel to reduce speed. The difference between these torques will cause the flywheel to accelerate or retard, according to the relationship.

$$T_i(\omega_i) - T_o(\omega_o) = I\omega'$$

where $\omega' = \dfrac{d\omega}{dt}$ which is the rate of change of velocity with respect to time, which of course is the angular acceleration α and I, you should recognise as the second moment of inertia (see Equation 3.1.44). We could have applied the torque from rest through some angle t, then both angular displacement and angular velocity become variables. Then the above relationship connecting input torque, output torque and acceleration would be written

$$T_i(\theta_i, \omega_i) - T_o(\theta_o, \omega_o) = I\theta''$$

where $\theta'' = \dfrac{d^2\theta}{dt}$ which is the second derivative of angular displacement with respect to time or again, acceleration α.

Now assuming that the flywheel is mounted on a rigid shaft, then $\theta_i = \theta_o = \theta$ and so $\omega_i = \omega_o = \omega$.

Then the above equation may now be written as

$$T_i(\theta, \omega) - T_o(\theta, \omega) = I\theta'' = I\alpha. \tag{5.3.34}$$

The above equation can be solved by direct integration, when the start values for angular displacement and angular velocity are known.

Now the *work input* to a flywheel $= T_i(\theta_1 - \theta_2)$, where the torque acts on the flywheel shaft through the angular displacement $(\theta_1 - \theta_2)$. Similarly the *work output* to a flywheel $= T_o(\theta_3 - \theta_4)$ where again the torque acts on the shaft through some arbitrary angular displacement.

We can write these relationships in terms of kinetic energy (see Equation 3.1.46). At $\theta = \theta_1$ the flywheel has a velocity ω_1 rads/sec, and so its kinetic energy is

$$E_1 = \tfrac{1}{2}I\omega_1^2$$

and similarly at, θ_2 with velocity ω_2

$$E_2 = \tfrac{1}{2}I\omega_2^2$$

So the change in the kinetic energy of the flywheel is given by

$$E_2 - E_1 = \tfrac{1}{2}I(\omega_2^2 - \omega_1^2). \tag{5.3.35}$$

Example 5.3.9

A steel cylindrical flywheel is 500 mm in diameter, has a width of 100 mm and it is free to rotate about its polar axis. It is uniformly accelerated from rest and takes 15 seconds to reach an angular velocity of 800 rev/min. Acting on the flywheel is a constant friction torque of 1.5 Nm. Taking the density of the steel to be 7800 kg/m³, determine the torque which must be applied to the flywheel to produce the motion.

To solve this problem we will need to use not only Equation 5.3.34, but also to refer back to Chapter 3 for the work we did on angular motion.

To find the angular acceleration, we may use Equation 3.1.38.

Where $\omega_f = \omega_i + \alpha t$ this gives

cont.

83.8 = 0 + 15α (where 800 r.p.m. = 83.8 rad/s)

then α = 5.58 rad/s².

Now the mass moment of inertia I, for a cylindrical disk, is given by

I = ½Mr^2 (see Figure 3.1.17).

In order to find I we first need to find the mass of the flywheel, i.e.

M = (0.25π)(0.5)²(0.1)(7800)

M = 153.15 kg.

Then I = (0.5)(153.15)(0.25)²

I = 4.79 kg m².

Now the net accelerating torque is given from Equation 5.3.34:

$T_N = T_i - T_o = I\alpha$

$T_N = (T_i - 1.5) = (4.79)(5.58)$.

Thus the accelerating torque T_i = 28.23 Nm.

A further example of an integrated system involving the use of a flywheel is given in the next section, when we deal with coupled systems.

Coupled systems

In the final part of this section we are going to look at *coupled systems*. For example, an electric motor may be used as the initial power source for a lathe. The rotary and linear motion required for machining operations being provided by a gearbox and selector mechanism, positioned between the motor and the workpiece. The motor vehicle provides another example where the engine is the prime mover, coupled through a clutch to a gearbox and drive shafts.

We are already armed with the necessary under pinning principles, from our previous work on belt drives, clutches, gear trains, rotating masses and flywheels. We may, however, also need to draw on our mechanical knowledge from Chapter 3, particularly with respect to angular motion.

We will study this mainly through examples, which are intended to draw together and consolidate our knowledge of power transmission systems.

Let us first consider the inertia characteristics of a rotational system, in this case a gearbox. Remembering our work on free-body diagrams from our study of shear force and bending moments!

Figure 5.3.17 shows a gearbox assembly with rotors attached to its input and output shaft. Let us assume that the gearbox ratio is a step-down, that is the input velocity is higher than the output velocity and these are related to the overall gear ratio G, by Equation 5.3.29.

Now each of the rotors will have a mass moment of inertia whic' depends on their radius of gyration, for the purpose of our anal' we will assume the rotors are cylindrical. Then if we draw the body diagram for the system, *with respect to the input* (Figure 5

Figure 5.3.17 *Gearbox assembly with rotors*

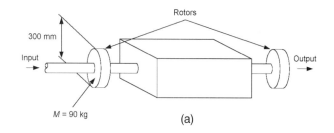

(a)

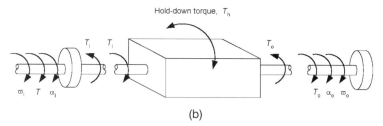

(b)

we are able to consider with respect to any component in the system, the resulting torques, velocities and accelerations. We will use these ideas in the following example.

Example 5.3.10

A gearbox has rotors attached to its input and output as shown in Figure 5.3.17a, where for the output rotor I_o = 25 kg m², the input rotor is cylindrical having a mass of 90 kg and diameter of 300 mm. The angular velocity of the rotor on the input shaft is higher than that of the rotor at the output. If a torque of 1.0 kN is applied to the high speed input shaft. Calculate the value of the angular acceleration of the input shaft, when G = 5 and the gearbox has an efficiency of 95%.

In order to solve this problem we first need to formulate the relationships, which enable us to determine the appropriate torques and accelerations for the system, from the point of view of the input.

The free body diagram in Figure 5.3.17b shows the isolated system elements (when the torque is applied to the input rotor shaft) as, the input rotor and drive shaft, the gearbox assembly and the output rotor and drive shaft. For the input rotor the torque, T, is opposed by the torque produced at the gearbox side of the shaft, so the equation of motion for I_i is given as:

$$T - T_i = I_i \alpha_i \qquad \text{(a)}$$

compare this with Equation 5.3.34!

For the gearbox the input and output torques are directly related to the gearbox ratio and the efficiency of the gearbox by

$$T_o = T_i G \zeta. \qquad \text{(b)}$$

The above relationship is true for a reduction gearbox, if you remember that one of the primary functions of a gearbox is to provide an increase in power at the output and if the velocity at the output is reduced, as in this case, the torque is raised (see Equation 5.3.32).

cont.

Now at the output rotor, from the free body diagram we have the relationship

$$T_o = I_o \alpha_o \text{ (as before)} \qquad \text{(c)}$$

$$\text{and } G\alpha_o = \alpha_i. \qquad \text{(d)}$$

Now combining equations (b) and (c) gives

$$T_i G \zeta = I_o \alpha_o$$

and substituting the above expression into equation (a) and at the same time from (d) substituting for α_o we get:

$$T - I_o \alpha_o = I_i \alpha_i$$

and so finally $\qquad T = \alpha_i [I_i + I_o/G^2 \zeta]. \quad$ (e)

So this equation tells us that the torque for the system, (when for a reduction gearbox the torque is assumed to act at the high velocity input side), is equal to the acceleration caused by this torque multiplied by the equivalent inertia (I_e) of the system, which is the term in the square brackets.

As you can see, most of this example has been concerned with formulating the required relationship, the above technique can be used when developing system equations for many similar situations.

The required acceleration at the input can be found by direct substitution of the variables into equation *e*, once we have determined I_i. Then mass moment of inertia for a cylinder gives

$$I_i = \tfrac{1}{2}Mr^2 = (0.5)(90)(0.15)^2 = 1.01 \text{ kg m}^2.$$

Now from equation (e), acceleration of input shaft is

$$\alpha_i = \frac{1.0}{(1.01 + 25/23.75)} = \frac{1.0}{(1.01 + 1.053)} = 0.48 \text{ rad/s}^2.$$

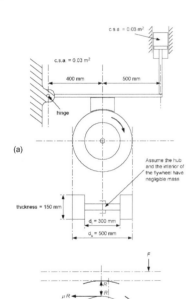

(a)

c.s.a = 0.03 m²

c.s.a. = 0.03 m²

400 mm

500 mm

hinge

Assume the hub and the interior of the flywheel have negligible mass

thickness = 150 mm

d_i = 300 mm

d_o = 500 mm

(b)

μR

R

F

Figure 5.3.18 *Flywheel and shoe brake system*

We will now consider another fairly simple system which consists of a shoe brake, being used to retard a flywheel. The flywheel may be thought of as an annulus, where the internal thickness is considered, as a first approximation, to be negligible (Figure 5.3.18).

Example 5.3.11

The steel flywheel shown in Figure 5.3.18 has an external diameter of 500 mm, an internal diameter of 300 mm, and the annulus formed by these diameters is 150 mm thick. The flywheel is retarded by means of a lever brake. Just prior to the application of the brake the flywheel is rotating with an angular velocity of 3500 r.p.m. and then the brake is applied for 50 seconds. A piston assembly is used to activate the brake assembly acting on the end of the lever. If the pressure applied by the piston is 3.5 kPa and the piston area is 0.03 m³. Find:

(i) the angular velocity of the flywheel at the end of the braking period;

cont.

(ii)   the power released from the flywheel during the braking period.

Take the flywheel density as 7800 kg/m³ and the coefficient of friction between the brake and the flywheel as 0.3.

Figure 5.3.18a shows the situation diagram with all relevant dimensions, Figure 5.3.18b shows the free body diagram, illustrating the braking reaction.

(i) The solution to this part of the problem relates to the equation $T = I\alpha$, which is a simplified form of Equation 5.3.34, where, in our case, $T = T_B =$ torque due to braking. We need to find this and the mass moment of inertia of the flywheel.

To find the braking torque we first determine the braking force, this is

$$F = P \times A = (3500)(0.03) = 105 \text{ N}.$$

Now considering the equilibrium of the brake lever, the reaction force R is found by taking moments about the hinge (see free body diagram).

So $R \times 400 = (105)(500) = 131.25 \text{ N}.$

So, assuming the friction is limiting friction and using the fact that braking force $F = \mu R$, then braking torque $T_B = (\mu R)r$ ($r =$ radius of flywheel).

So $T_B = (0.3)(131.25)(0.25) = 9.84 \text{ Nm}.$

Now the mass moment of inertia of a disk is given by

$$I = \tfrac{1}{2}Mr^2$$

Logic suggests that I for an annulus will have the form $\tfrac{1}{2}M(R^2 - r^2)$, which it does. However, it is normally expressed as

$$I = \pi \frac{(d_o^4 - d_i^4)t\rho}{32}$$

where $t =$ thickness of the flywheel and $\rho =$ density of flywheel material. The above form is often used as a first approximation where the hub and inner disk is considered insignificant, when compared with the bulk of the rim.

So $I_F = \dfrac{\pi(0.5^4 - 0.3^4)(0.15)(7800)}{32}$

giving $I_f = 6.25 \text{ kg m}^2.$

To find the angular retardation α, then

$T_B = I\alpha$ and $\alpha = \dfrac{9.84}{6.25} = 1.574 \text{ rad/s}^2.$

So final angular velocity is given by Equation 3.1.39

$\omega_f = \omega_i + \alpha_t$ and $\omega_f = (366.5 - 78.5)$

and $\omega_f = 288 \text{ rad/s}$ or 2750 rev/min.

(ii) To find the power given out by the flywheel during braking,

cont.

we first find the kinetic energy change. Then using Equation 5.3.35 where $E_2 - E_1 = \frac{1}{2}I(\omega_2^2 - \omega_1^2)$

then energy change $= (0.5)(6.25)(366.5^2 - 288^2)$

$= 160.588$ kJ

and so power $P = \dfrac{J}{S} = \dfrac{160\,558.59}{50} = 3.21$ kW.

We now discuss a system for the *operation of a crank pressing machine*, which consists of a prime mover, gearbox and flywheel, to drive the press crank.

Let us first consider the selection of a power unit for a punch pressing machine. What information do we require to help with our selection? Well, at the very least, we would need information about the service load, the frequency of the pressing operation, the torque requirements and rotational velocities of the system.

Figure 5.3.19 shows the torque requirements for a typical pressing operation. Note that the torque requirement fluctuates, very high torque being required for the actual pressing operation, which is of course what we would expect.

We first determine the torque versus angular displacement graph (Figure 5.3.19).

Let us assume that the press operates at 150 revs/min and the pressing operation is completed in 0.3 seconds and, takes place 30 times a minute. This is not an unreasonable assumption, many punch presses operate at much higher rates than this, obviously dependent on the intricacy and force requirements of the product. Assume that

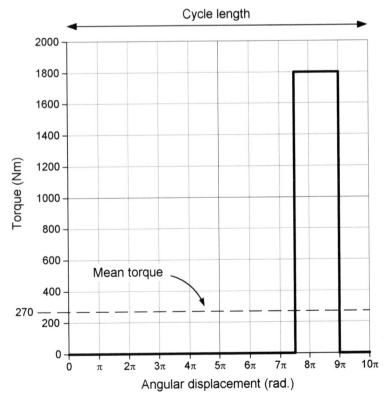

Figure 5.3.19 *Turning moment diagram for crank pressing machine*

Scale: 1 cm = π rad.
         1 cm = 200 Nm

the resisting torque during the pressing operation is 1800 Nm.

Based on the above information we can deduce the angular displacement of the crank for one complete cycle of events (Figure 5.3.19).

The punching operation takes place at 30 times a minute, therefore the punch press crank will rotate five revolutions, for each operation.

So the angular displacement for one complete operation is $5 \times 2\pi = 10\pi$ radians.

The angular displacement of the crankshaft during the actual pressing operation, is easily determined by comparing the time for one complete cycle of events (2 seconds), with the time for the actual pressing operating (0.3 seconds).

So angular displacement for pressing operation $= \dfrac{0.3 \times 10\pi}{2.0} = 1.5\pi$ radians (Figure 5.1.19).

We know that the high resisting torque of 1800 Nm acts for 15% of the time required for one complete cycle of events, this torque would need to be provided by the prime mover, if no flywheel was included in the system. So, what power would we require from our prime mover under these circumstances?

Using an electric motor (the most likely source of power) we would need to provide gearing between the motor output and the input to the press crank, for the following reasons.

A suitable electric motor for such a task is likely to be 3-phase ac. This type of motor is normally produced to run at synchronous speeds that depend on the number of pole pairs of the machine, 600, 900, 1200 rpm, etc. We will choose a motor that runs at 1200 rpm, which requires a reduction gearbox of 8 to 1. What is the power required from our motor to provide the maximum torque, assuming that this peak torque needs to be *continuously* available from the motor?

Without losses we know that the power provided by the motor is equal to the power at the crankshaft, to drive the press. Then from the diagram we have already seen that the peak torque we require at the crank is 1800 Nm. Then we know that

$$T_i \omega_i = T_o \omega_o$$

$$\frac{T_i}{T_o} = \frac{\omega_o}{\omega_i} \quad \text{so } T_i = \frac{(1800)(60)}{1200} = 90 \text{ Nm.}$$

Note that we may leave the angular velocities in rev/min because here they are a ratio.

Now if the motor has the capacity to deliver this torque *continuously*, then the *work* required from the motor in one complete cycle of events is $10\pi(90) = 900\pi$ Nm.

Now power is the rate of doing work, so in 1 second the motor shaft rotates 20 revolutions, the power required from our motor is

$$20 \times 900\pi = 56.5 \text{ kW!}$$

It is obviously wasteful to provide such a large motor when the maximum torque requirements are only needed for a small fraction of the time. This is why our system *needs* the addition of a flywheel.

Let us assume that we require a flywheel to be fitted to our system, which keeps the press crank velocity between 140 and 160 rev/min during pressing operations. In order to select such a flywheel we need to know the energy requirements of the system and, the total inertia that needs to be contributed by the flywheel during these pressing operations.

From our discussion so far I hope you recognise that the area under our crank torque–crank angle graph in Figure 5.3.19, represents the energy requirements for the press.

We assume as before that, the energy in, is equal to the energy out, and that there are no losses due to friction.

The total energy required during the pressing operation is given from our previous calculations as:

$(1800 \, \text{Nm})(1.5p) = 8482 \, \text{J}$ (area under graph for pressing operation).

This energy is required to be spread over the complete operation cycle, i.e. spread over 10π radians, so the uniform torque required from the motor is given by

$(1800 \times 1.5\pi) = (T_m \times 10\pi)$ where T_m = uniform motor torque

$T_m = 270 \, \text{Nm}$ (this is indicated on the diagram).

Then the *change* in energy is

$\Delta E = (T_{max} - T_{mean}) \times$ angular distance

$\Delta E = (1800 - 270)1.5\pi = 7210 \, \text{J}.$

So the flywheel must supply 7210 J and therefore the motor must supply approximately

$8482 - 7210 = 1272 \, \text{J}.$

Note the drastic reduction in the energy required by the motor!

We are now in a position to determine the inertia required by the flywheel, using Equation 5.3.34 we have

$7210 = \frac{1}{2}I(\omega^2_{max} - \omega^2_{min})$

where $\omega_{max} = 160 \, \text{rev/min} = 16.76 \, \text{rad/s}$

and $\omega_{min} = 140 \, \text{rev/min} = 14.66 \, \text{rad/s}.$

Then the inertia of the flywheel is given by

$I = \frac{2 \times 7210}{(280.9 - 214.9)} = 218.5 \, \text{kg m}^2.$

This is where we end our discussion of this system. We could go on to make a specific motor selection, determine the requirements of a pulley and belt drive system, and also determine the dimensional requirements of the flywheel, if we were so inclined!

We leave our discussion on the coupling of transmission and other systems, with one further example.

Example 5.3.12

An electric motor drives a cylindrical rotational load through a clutch (Figure 5.3.20). The armature of the motor has an external diameter of 400 mm and a mass of 60 kg. It may be treated as being cylindrical in shape. The rotor has a diameter of 500 mm and a mass of 480 kg. If the motor armature is revolving at 1200 rev/min and the rotor at 450 rev/min, both

Figure 5.3.20 *Figure for Example 5.3.12*

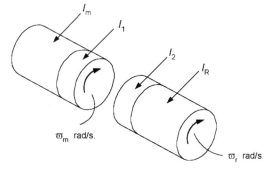

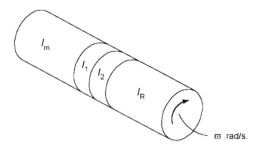

(a) before impact

(b) after impact

in the same direction. Assuming the moment of inertia of the driving clutch plate is 0.25 kg m² and that after engagement the moment of inertia of the whole clutch is 0.5 kg m². Ignoring all other inertia effects, find the loss of kinetic energy when the clutch is engaged.

This problem requires us to find the angular momentum change, as a result of the impact. Angular momentum is given by the product of moment of inertia and the angular velocity. Knowing this, we are able to find the angular velocity after impact, and then the loss of kinetic energy.

The mass moment of inertia for a cylinder is determined simply from the $I = \tfrac{1}{2}Mr^2$ in the normal way.

Then moment of inertia of motor armature and driving clutch assembly, where I_i = driving clutch assembly: $= I_m + I_i$

$= (0.5)(60)(0.2)^2 + 0.25$

$= 1.45$ kg m².

Moment of inertia of rotor end driven clutch assembly, where I_2 = driven clutch assembly $= I_R + I_2$

$= (0.5)(480)(0.25)^2 + 0.5$

$= 15.5$ kg m².

Also the angular velocity of the motor armature and driving clutch plate (same shaft) = 125.66 rad/s and the angular velocity of the rotor end driven clutch assembly (same shaft) = 47.12 rad/s.

Now from the conservation of angular moment we know that the total angular moment before impact equals the total angular moment after impact.

cont.

Therefore, $(I_m+I_i)\omega_m + (I_R+I_2)\omega_R = (I_M + I_1 + I_R + I_2)\omega_f$

Where ω_f = final velocity after impact.

So $(1.45)(125.66) + (15.5)(47.12) = (1.45 + 15.5)\,\omega_f$

$912.567 = 16.95\,\omega_f$

$\omega_f = 53.84$ rad/s.

Now to find the loss of kinetic energy at impact, then
Loss of KE = total KE before impact $-$ total KE after impact

$= $ (KE of $I_m + I_i$) + (KE of $I_R + I_2$) $-$ KE of $(I_M + I_1 + I_R + I_2)$

$= \tfrac{1}{2}(1.45)(125.66)^2 + \tfrac{1}{2}(15.5)(47.2)^2 - \tfrac{1}{2}(16.5)(53.84)^2$

$= 4088$ J.

So loss of KE = 4.088 kJ.

Questions 5.3.1

(1)   A flat belt connects two pulleys with diameters of 200 mm and 120 mm. The larger pulley has an angular velocity of 400 rpm. Given that the distance C between centres is 400 mm, the belt maximum tension is 1200 N and the coefficient of friction between belt and pulleys is 0.3. Determine:
(i)   the angle of lap of each pulley;
(ii)   the maximum power transmitted.

(2)   A single V-belt with $\phi = 18°$ and a mass of 0.25 kg/m is to be used to transmit 14 kW of power from a 180 mm diameter drive pulley rotating at 2000 rpm to a driven pulley rotating at 1500 rpm. The distance between the centres of the pulleys is 300 mm.
(i)   If the coefficient of friction is 0.2 and the initial belt tension is sufficient to prevent slippage, determine the values of T_1 and T_2.
(ii)   Determine the maximum stress in the belt, if its density is 1600 kg/m³. Note that the centripetal tension created by the centripetal force acting on the belt is given by $T_c = m'v^2$, where m' = the mass per unit length of the belt (see Equation 5.3.5).

(3)   A multiplate clutch consists of five disks, each with an internal and external diameter of 120 mm and 200 mm, respectively. The axial clamping force exerted on the disks is 1.4 kN. If the coefficient of friction between the rubbing surfaces is 0.25, and uniform wear conditions may be assumed, determine the torque that can be transmitted by the clutch.

(4)   A motorised winch is shown below (Figure 5.3.21). The motor supplies a torque of 10 Nm at 1400 rpm. The winch drum rotates at 70 rpm and has a diameter of 220 mm. The overall efficiency of the winch is 80%.
Determine the:
(i)   number of teeth on gear A;
(ii)   power input to the winch;
(iii)   power output at the drum;

Figure 5.3.21 *Figure for Question 5.3.4 – motorised winch*

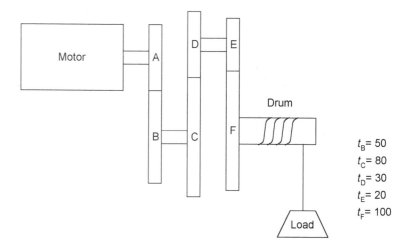

$t_B = 50$
$t_C = 80$
$t_D = 30$
$t_E = 20$
$t_F = 100$

(iv)   load being raised.

(5)   A flat belt 150 mm wide and 8 mm thick transmits 11.2 kW. The belt pulleys are parallel, and their axes are horizontally spaced 2 m apart. The driving pulley has a diameter of 150 mm and rotates at 1800 rev/min, turning so that the slack side of the belt is on top. The driven pulley has a diameter of 450 mm. If the density of the belt material is 1800 kg/m³. Determine:

(i)   the tension in the tight and slack sides of the belt if the coefficient of friction between the belt and the pulleys is 0.3;

(ii)   the length of the belt.

6 Analytical methods

Summary

The primary aim of this chapter is to introduce the mathematical principles, methods and applications that underpin the Higher National Engineering units that you will meet in this book. The fundamental techniques presented here, should also prepare you for further study of Mathematics and other Engineering units which you may encounter during your future studies. You will not only cover aspects from the core analytical methods unit, but also selected topics from some of the more advanced units where these are considered necessary to aid your understanding.

This chapter is not intended to be a mathematical treatise! There are many excellent text books (see Appendix) which cover in great detail the foundations of mathematics and, provide an in-depth treatment of all the material introduced in the mathematical units of Higher National Engineering programmes.

We will, however, be covering in detail three major areas of mathematics that are applicable to engineers; fundamental algebra, trigonometry, and the calculus. In addition we will briefly cover; complex numbers, differential equations, matrices, Fourier series and Laplace transforms, where their application is considered necessary to enhance the subject matter presented in this book.

Each topic will be presented in three parts. Starting with a formula sheet, the *formulae* being presented without proof. We will then look at the mathematical *methods* used for manipulating these formulae and establishing mathematical relationships. Finally, where appropriate, examples of their engineering *application* will be given drawn directly from the subject matter contained within this book. Since mathematics is a linear subject, some topics simply act as prerequisites for others, without having a direct engineering application. Under these circumstances only the appropriate formulae and methods will be presented.

In order to successfully study the subject matter presented in this way, you should be armed, already, with a number of fundamental mathematical tools. You *must* be completely familiar with elementary algebra, trigonometry and number. In particular, you should be able to manipulate algebraic formulae and fractions, equations, indices, and logarithms, constants of proportionality and inequalities. You should be able to calculate the areas and volumes of familiar solids and, solve problems involving the basic trigonometric ratios and their inverses. You should be familiar with estimation techniques and the manipulation and presentation of experimental data. Finally, you should be able to manipulate and use vulgar fractions!

Many people experience difficulties with the basics, identified above. You will notice as you study the analytical subjects presented in this book, that much of your time will be spent manipulating expressions *using these basics*, this is why they are so important. The intention now, is to build on these basics by studying a number of additional analytical techniques. You will then be in a position to provide the solution to engineering problems more quickly, and hopefully with less effort!

6.1 Algebra

In this section we will review those fundamental algebraic methods, which are considered essential to manipulate engineering expressions. These include, logarithms and logarithmic functions, exponential laws, and the solution of equations involving exponential and logarithmic functions.

New algebraic methods will be introduced primarily concerned with partial fractions, which we will need later to simplify expressions in order to *integrate* them more easily. Partial fractions will also be used when we find *Laplace* transforms and their inverses.

We will finally take a look at the binomial, exponential, logarithmic and hyperbolic series. These will be required for the power series and Fourier analysis which you will meet later.

Algebraic fundamentals

Formulae

(1) *Factors:*

$$(a+b)^2 = a^2 + 2ab + b^2$$
$$(a-b)^2 = a^2 - 2ab + b^2$$
$$a^3 + b^3 = (a+b)(a^2 - ab + b^2)$$
$$a^3 - b^3 = (a-b)(a^2 + ab + b^2)$$
$$(a+b)^3 = a^3 + 3a^2b + 3ab^2 + b^3$$
$$(a-b)^3 = a^3 - 3a^2b + 3ab^2 - b^2.$$

(2) *Indices:*

$$a^m \times a^n = a^{m+n} \tag{i}$$

$$\frac{a^m}{a^n} = a^{m-n} \tag{ii}$$

$$(a^m)^n = a^{mn} \tag{iii}$$

$$\frac{1}{a^n} = a^{-n} \tag{iv}$$

$$\sqrt[n]{a^m} = a^{m/n} \tag{v}$$

$$a^0 = 1. \tag{vi}$$

(3) *Logarithms:*

$$\text{If } a = b^c, \text{ then } c = \log_b a \tag{i}$$

$$\log_a MN = \log_a M + \log_a N \tag{ii}$$

$$\log_a \frac{M}{N} = \log_a M - \log_a N \tag{iii}$$

$$\log_a(M^n) = n \log_a M \tag{iv}$$

$$\log_b M = \frac{\log_a M}{\log_a b}. \tag{v}$$

(4) *Exponential and logarithmic functions:*

$$y = a^x \quad \text{where } x \text{ is the exponent or index}$$

$$\exp x = e^x$$

$$\log_e x = \ln x \quad (\text{where } \ln \text{ is the log to base e known as the}$$

$$\text{natural or Naperian logarithm})$$

$$y = \log_a x \quad (\text{logarithmic function to any base}).$$

(5) (i) *Remainder theorem:* When $f(x)$ is divided by $(x - a)$ the remainder is $f(a)$.
   (ii) *Factor theorem:* If $f(a) = 0$, $(x - a)$ is a factor of $f(x)$.

Methods

Simultaneous equations, and general algebraic manipulation have not been included in the fundamental algebraic methods that follow. However some methods, and engineering applications given after these methods, do require their use.

Logarithms

Manipulation of expressions and formulae where the variable you require is part of an index, requires us to use the laws of logarithms to find the variable. For example, if

$a = b^c$ make c the subject of the formula.

Then from laws (i) and (iv), and taking logs to base b gives

$$\log_b a = c \log_b b$$

so $\log_b a = c$ as required.

Example 6.1.1

If $U_2 = U_1 e^{\left(\frac{w}{pv}\right)}$ make w the subject of the formula then

$$\log_e\left(\frac{U_2}{U_1}\right) = \left(\frac{w}{pv}\right)\log_e e$$

(on division by U_1 and applying logs)

$$\log\left(\frac{U_2}{U_1}\right) = \frac{w}{pv} \quad (\text{since } \log_e e = 1)$$

and so $w = pv \log_e\left(\frac{U_2}{U_1}\right)$.

Note: It is normal to present the subject of a formula to the left of the equals sign.

Transposition of equations or expressions, when the new subject is part of a logarithmic expression, involves an extra step. This step is that of finding the antilogarithm of both sides of the equation.

Example 6.1.2

Transpose $b = \log_e t - a \log_e D$ to make t the subject:

$$b = \log_e t - \log_e D^a \quad \text{(law iv)}$$

$$b = \log_e \left(\frac{t}{D^a} \right) \quad \text{(from law iii)}$$

$$e^b = \frac{t}{D^a} \quad \text{(since } e^x \text{ is the inverse or antilogarithm}$$

$$\text{of } \log_e x = \ln x)$$

$$t = e^b D^a.$$

Let us consider one more example where we are required to determine a numerical answer.

Example 6.1.3

If $\theta_f - \theta_i = \dfrac{R}{J} \log_e \left(\dfrac{U_2}{U_1} \right)$ find the value of U_2 given that

$$\theta_f = 3.5, \quad \theta_i = 2.5, \quad R = 0.315, \quad J = 0.4, \quad U_1 = 50.$$

Placing the values directly into the formula does not get us very far, because we will end up trying to take the logarithm, of an unknown.

Transposing gives:

$$\frac{J(\theta_f - \theta_i)}{R} = \log_e \left(\frac{U_2}{U_1} \right)$$

and so $e^{\left(\frac{J(\theta_f - \theta_i)}{R} \right)} = \dfrac{U_2}{U_1}$ (antiloging)

then $U_2 = U_1 e^{\left(\frac{J(\theta_f - \theta_i)}{R} \right)}$ which on substitution of the values gives $U_2 = 178$ (check you get this value).

Exponential equations

These equations are simply an extension of our previous transposition methods where we used logarithms. They can sometimes be a little awkward, requiring a little algebraic manipulation.

Example 6.1.4

Find the value of x, when

$$\log 8^x = \log 10$$

$$x \log 8 = \log 10 \quad \text{(using rule (iv) of logarithms)}$$

$$x = \frac{\log 10}{\log 8} \quad \text{(and using calculator or tables)}$$

$$x = 1.107 \quad \text{correct to three decimal places.}$$

Example 6.1.5

Solve the equation $2(2^{2x}) - 5(2^x) + 2 = 0$ for x. Now at first glance taking logs appears to be no help! The trick is to simplify the equation using a substitution. In performing the substitution we need to be aware of the rules for indices as well as logs.

If we substitute $y = 2^x$, then we get

$2y^2 - 5y + 2 = 0$ (where $2^{2x} = 2^x \cdot 2^x = y \cdot y = y^2$).

Now we solve a simple quadratic equation, which you should be familiar with. The simplest method is to factorise, if all else fails revert to the quadratic formula!

$$\text{Then}\quad (2y - 1)(y - 2) = 0$$

from which $2y - 1 = 0$ giving $y = \frac{1}{2}$

or $y - 2 = 0$ giving $y = 2.$

So $2^x = \frac{1}{2}$ or $2^x = 2$

then by inspection $x = \pm 1$.

If you could not see this straightaway look carefully at the rules given for indices, particularly where

$a^0 = 1$ and $\dfrac{1}{a^n} = a^{-n}.$

Example 6.1.6

Solve the equations simultaneously for x and y:

$\log x - 2 \log y = 1$ (i)

$xy = 270.$ (ii)

Then from the laws of logs (i) becomes

$\log \dfrac{x}{y^2} = 1$ and antiloging we get

$\dfrac{x}{y^2} = 10$ (logs to base 10)

so $x = 10y^2$ (substituting in ii) we get

$10y^3 = 270$

$y^3 = 27$

$y = 3$

Then $x = 90$, $y = 3$.

Reduction of non-linear laws

We will now look at a technique for reducing non-linear relationships to linear form. This process enables us to verify experimental laws and analyse experimental data in a convenient way. Two examples serve to illustrate the process.

Example 6.1.7

The experimental values of p and s shown below, are believed to be related by the law $s = ap^2 + b$. By plotting a suitable graph verify this law and determine approximate values of a and b.

p	1	2	3	4	5
s	6.73	11.93	20.59	32.7	48.3

If s is plotted against p a curve results and we are unable to determine the value of the constants a and b. By comparing the above relationship, with the law for a straight line where

$$s = ap^2 + b$$

and $y = mx + c$

it can be seen that a plot of s against p^2, should produce a straight line enabling us to estimate the constants a, b and so verify or otherwise the suggested relationship.

To produce the graph (Figure 6.1.1) we draw up a table of values in the normal way.

p	1	2	3	4	5
p^2	1	4	9	16	25
s	6.73	11.93	20.59	32.7	48.3

The graph results in a straight line where slope (constant a) is shown to be 1.73 and the intercept with the s axis, (constant b) has a value close to 5.0.

So the law relating the data is $s = 1.73p^2 + 5.0$ (you should verify that this is correct).

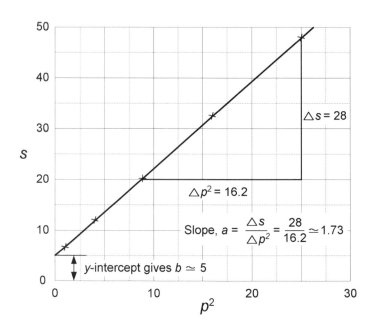

Figure 6.1.1 *Graph of p^2 against s*

You should remember that the slope of the graph enables us to find the constant m, in the straight line law and, the constant c is found where the graph crosses the y-intercept at zero. Occasionally it is not convenient to start the graph at zero, under these circumstances, simultaneous equations may be used to determine the constants. One final example should ensure you are able to follow the method.

Example 6.1.8

It is thought that the load (W) and the effort (E) for a machine are related by $E = \dfrac{aW}{6} + b$. The results of an experiment carried out on the machine are given below. Determine the constants a, b and verify the law of the machine.

Load W (N)	1000	2000	3000	4000	5000
Effort E (N)	580	660	750	833	920
$W/6$ (N)	166.7	333.3	500	666.7	833.3

Again, if we compare the supposed law of the machine with that of a straight line, then clearly all we need to do is plot effort E against $W/6$ from which a straight line graph should result which verifies the law. The above table includes the row for $E/6$, so we can plot the graph (Figure 6.1.2) straightaway.

From the graph it can be seen that constant $a = 0.5$ and constant $b = 500$. The law is verified because the graph yields a straight line and the law of the machine is given by

$$E = \frac{0.5W}{6} + 500.$$

Any set of values may be used to check that you have determined the constants correctly.

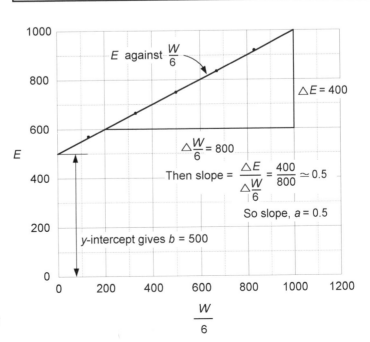

Figure 6.1.2 *Graph of w/6 against E*

The manipulation of experimental data and the verification of laws and other ideas, is an essential tool for experimental work. Sources of experimental error and error arithmetic are not covered here, since they are best learnt as part of your experimental work. However, it will certainly be in your best interests to become familiar with these topics if you have not already done so.

Engineering applications

We now turn our attention to the application of fundamental algebra to engineering situations. You will see that a lot of your time is spent manipulating logarithmic expressions, solving simple equations and making substitutions. All the engineering applications are in the form of examples, with all necessary explanation forming part of the example.

Example 6.1.9

In information theory (Chapter 3) it can be shown that the information content of a message is given by $I = \log_2(1/p)$. Show using the laws of logs that the information content may be expressed as $I = -\log_2(p)$ and find the information content of the message if the probability (p) of receiving the code is $1/16$.

We need to show that

$$I = \log_2\left(\frac{1}{p}\right) = -\log_2(p).$$

The left-hand side of the above expression may be written as $\log_2(p^{-1})$ and comparing with law (iv) where

$$\log_a(M^n) = n \log_a M$$

then in this case $M = p$ and $n = -1$ so

$$\log_2(p^{-1}) = -1 \log_2 p \quad \text{as required.}$$

Now substituting $\frac{1}{16}$ into: $\log_2\left(\dfrac{1}{p}\right)$ gives $\log_2(16)$.

So using law (v) we get

$$\log_2 16 = \frac{\log_{10} 16}{\log_{10} 2} = 4.$$

Then the information content of the message $= 4$.

Example 6.1.10

The Lamé equations relating radial and hoop stress in thick cylinders are given below (see Chapter 5.1). If the radial stress is $\sigma_r = -24\,\text{N/mm}^2$ and the hoop stress $\sigma_h = 54\,\text{N/mm}^2$, when the internal radius $r = 60\,\text{mm}$. Find the value of the constants a and b.

cont.

$$\sigma_r = a - \frac{b}{r^2} \tag{1}$$

$$\sigma_h = a + \frac{b}{r^2}. \tag{2}$$

Now in order to find the constants a and b, we must first substitute the given values into equations 1 and 2.

Then

$$-24 = a - \frac{b}{3600} \tag{1A}$$

$$54 = a + \frac{b}{3600}. \tag{2A}$$

So we need to solve these equations simultaneously. Multiplying both equations by 3600 to eliminate fractions gives

$$86\,400 = 3600a - b \tag{1B}$$

and $194\,400 = 3600a + b.$ $\tag{2B}$

'Adding' both equations eliminates b to give

$280\,800 = 7200a$ so $a = 39.$

Now substituting our value of 'a' into any convenient equation (1B directly above) gives

$86\,400 = (3600)(39) - b$

then $b = 54\,000.$

So required constants are $a = 39$, $b = 54\,000$.

Example 6.1.11

In a belt and pulley drive system (Chapter 5.3) the relationships between belt pitch length (L), distance between centres (C), the pulley pitch diameters (D) and the span length (S), are given by the formulae below. Determine the belt pitch length and the span length for the system when $D_1 = 0.2$, $D_2 = 0.3$ and $B = 4.02$.

$$L = 2C + 1.57(D_2 - D_1) + \frac{(D_2 - D_1)^2}{4C} \tag{1}$$

$$C = \frac{B + \sqrt{B^2 - 32(D_2 - D_1)^2}}{16} \tag{2}$$

$$S = \sqrt{C^2 - \left[\frac{D_2 - D_1}{2}\right]^2}. \tag{3}$$

We need to find the belt pitch length (L) from equation 1 above. Before we can do this we need to find the value of C. We can obtain C from equation 2 since we are given $B = 4.02$.

cont.

Then

$$C = \frac{4.02 + \sqrt{(4.02)^2 - 32(0.3 - 0.2)^2}}{16}$$

$$C = 0.5$$

and substituting into equation 1, to find L we have

$$L = (2)(0.5) + 1.57(0.1) + \frac{(0.1)^2}{(4)(0.5)}$$

so $L = 1.162$.

The span S is given by equation 3 as

$$S = \sqrt{(0.5)^2 - \left(\frac{0.1}{2}\right)^2}$$

$$S = 0.495.$$

This is relatively simple, but beware of making arithmetic errors!

Example 6.1.12

The resistors from two networks are connected by the equations given below. Use these equations to derive an equation for R_1 and find its value when $R_A = 20\,k\Omega$, $R_B = 45\,k\Omega$ and $R_C = 15\,k\Omega$.

$$R_1 + R_3 = \frac{R_A \times (R_B + R_C)}{R_A + R_B + R_C} \tag{1}$$

$$R_2 + R_3 = \frac{R_B \times (R_A + R_C)}{R_A + R_B + R_C} \tag{2}$$

$$R_1 + R_2 = \frac{R_C \times (R_A + R_B)}{R_A + R_B + R_C}. \tag{3}$$

By adding or subtracting the above equations from one another we should be able to leave R_1 on its own on the right-hand side of an equation. So our first task is to eliminate R_2 and R_3 from the right-hand side of the above equations. If we subtract equation 2 from equation 1, we first eliminate R_3. So:

$$(R_1 + R_3) - (R_2 + R_3) = \left(\frac{R_A \times (R_B + R_C)}{R_A + R_B + R_C}\right)$$
$$- \left(\frac{R_B \times (R_A + R_C)}{R_A + R_B + R_C}\right).$$

Now providing you know the rules for manipulating fractions, the above process is simple! We get

$$R_1 - R_2 = \frac{[R_A \times (R_B + R_C)] - [R_B \times (R_A + R_C)]}{R_A + R_B + R_C}.$$

cont.

If you did not understand this step you need to revise your knowledge of fractions.

Writing the above a little more succinctly gives

$$R_1 - R_2 = \frac{R_A(R_B + R_C) - R_B(R_A + R_C)}{R_A + R_B + R_C}. \tag{4}$$

Now I hope you can see that by adding equations 3 and 4, we eliminate R_2 and leave an expression for R_1, as follows

$$(R_1 + R_2) + (R_1 - R_2) = \left(\frac{R_C \times (R_A + R_B)}{R_A + R_B + R_C} \right)$$
$$+ \left(\frac{R_A(R_B + R_C) - R_B(R_A + R_C)}{R_A + R_B + R_C} \right).$$

Then

$$2R_1 = \frac{R_C(R_A + R_B) + R_A(R_B + R_C) - R_B(R_A + R_C)}{R_A + R_B + R_C}$$

and simplifying gives

$$2R_1 = \frac{R_C R_A + R_C R_B + R_A R_B + R_A R_C - R_B R_A - R_B R_C}{R_A + R_B + R_C}$$

or $\quad 2R_1 = \dfrac{R_A R_C + R_B R_C + R_A R_B + R_A R_C - R_A R_B - R_B R_C}{R_A + R_B + R_C}$

Note that putting multiplied terms into alphabetical order helps to identify like expressions.

Then $\quad 2R_1 = \dfrac{2R_A R_C}{R_A + R_B + R_C} \quad$ or $\quad R_1 = \dfrac{R_A R_C}{R_A + R_B + R_C}$

as required. Finally substituting values gives:

$$R_1 = \frac{(20)(15)}{20 + 45 + 15} = 3.75\,\text{k}\Omega.$$

Compare the above example with the equations given for T, π, star and delta networks in Chapter 4.

Example 6.1.13

The formula set out below relates the power (P), belt tension (T), angle of lap (θ), linear velocity (v) and the coefficient of friction (m), for a flat belt drive system (Chapter 5.3). Determine an expression for the coefficient of friction and find its value when, $P = 2500$, $T = 1200$, $v = 3$ and $\theta = 2.94$.

$$P = T(1 - e^{-\mu\theta})v \tag{1}$$

We are required to find an expression for 'μ'; in other words make 'μ' the subject of the formula. We proceed as follows, from equation 1.

$$\frac{P}{T} = (1 - e^{-\mu\theta})v \quad \text{or} \quad \frac{P}{Tv} = 1 - e^{-\mu\theta}$$

and so $\quad \dfrac{P}{Tv} - 1 = -e^{-\mu\theta} \quad$ or $\quad 1 - \dfrac{P}{Tv} = e^{-\mu\theta}$

cont.

Now taking logarithms to the base 'e' of both sides then

$$\ln\left(1 - \frac{P}{Tv}\right) = -\mu\theta.$$

so

$$\frac{-\ln\left(1 - \frac{P}{Tv}\right)}{\theta} = \mu.$$

Now substituting the given values, we get

$$\frac{-\ln\left(1 - \frac{2500}{(1200)(3)}\right)}{2.94} = \mu$$

$$\frac{-\ln(0.305)}{2.94} = \mu$$

Coefficient of friction $\mu = 0.4.$

There are many more engineering applications which require us to apply the fundamental rules of algebra, but space does not permit us to review them here. The problems set at the end of this section will give you the opportunity to put more of these basic algebraic skills into practice.

Partial Fractions

Formulae

Provided that the numerator $f(x)$ is second degree or lower

$$\frac{f(x)}{(x+a)(x+b)(x+c)} \equiv \frac{A}{(x+a)} + \frac{B}{(x+b)} + \frac{C}{(x+c)} \tag{1}$$

$$\frac{f(x)}{(x+d)(ax^2+bx+c)} \equiv \frac{A}{(x+d)} + \frac{Bx+C}{(ax^2+bx+c)} \tag{2}$$

$$\frac{f(x)}{(x+a)(x+b)^2} \equiv \frac{A}{(x+a)} + \frac{B}{(x+b)} + \frac{C}{(x+b)^2}. \tag{3}$$

Methods

Relatively complex algebraic fractions can often be broken down into a series of smaller fractions which are much more easily manipulated, these are known as *partial fractions*, the formulae given above show examples. Partial fractions (PFs), are very useful when trying to integrate complex expressions that are able to be broken down in this way. As mentioned previously, they are also useful when determining the inverse of non-standard Laplace transforms, which cannot be identified from tables.

You will be using PFs later, when we consider the calculus and the engineering applications of Laplace functions. In the mean time we will concentrate on the methods used to obtain them, covering their engineering applications at a later stage when we consider a number of more advanced topics.

The methods used for finding PFs, will depend on the types of factor that go to make up the original fraction.

Linear factors

If the denominator consists only of linear factors, where the degree of the unknown is one, as in formula 1 above. Then for every linear factor, e.g. $(x - a)$ in the denominator, there corresponds a partial fraction of the form $A/(x - a)$. Then in order to find the PFs we need to find the values of the *coefficients* in the numerator. Remembering that in this instance a coefficient is any number which is used as a multiple of the unknown, for example in the expression $2x$, 2 is the coefficient of x. The following example illustrates the process for finding PFs, where only linear factors are involved. You will see later that the process developed here is similar to that required for determining all PFs, irrespective of their form.

Example 6.1.14

Find the PFs for the expression

$$\frac{3x}{(x - 1)(x - 2)(x - 3)}.$$

From what has already been said and, following the format given in formula 1. Then we may write

$$\frac{3x}{(x - 1)(x - 2)(x - 3)} \equiv \frac{A}{(x - 1)} + \frac{B}{(x - 2)} + \frac{C}{(x - 3)}.$$

This is the required form for *linear* factors. Now multiplying *every* term by $(x - 1)(x - 2)(x - 3)$, that is the denominator and, cancelling where appropriate gives

$$3x \equiv A(x - 2)(x - 3) + B(x - 1)(x - 3) + C(x - 1)(x - 2).$$

The sign $\equiv$ means 'always equal to'.

Now we need to find values of the unknown 'x' which satisfies the above equation, in order to determine the coefficients A, B and C. This is, in fact, a lot easier than it looks! There are two methods often used, each has its merits. For linear factors the following simple short-cut method will be used. You will meet the other method next, when we consider quadratic factors.

All we need to do is substitute a value of 'x' into the bracketed expressions which, when possible, reduces them to zero. For example the brackets multiplied by the coefficient A are $(x - 2)(x - 3)$ so choosing a value of $x = 2$ or $x = 3$ would reduce the whole expression, including the coefficient A, to zero. Whatever value of 'x' is chosen, it must be applied to 'all x' throughout the whole expression.

So substituting $x = 2$, into the whole expression, gives

$$6 \equiv 0 + B(2 - 1)(2 - 3) + 0$$

then $6 = -B$ or $B = -6$.

cont.

We repeat this process by substituting appropriate values, then substituting $x = 3$, gives

$$9 \equiv 0 + 0 + C(3-1)(3-2)$$

then $\quad 9 \equiv 2C \quad$ or $\quad C = 4.5$.

One last substitution $x = 1$, gives

$$3 \equiv A(1-2)(1-3) + 0 + 0$$

then $\quad 3 \equiv 2A \quad$ or $\quad A = 1.5$.

So when $A = 1.5$, $B = -6$ and $C = 4.5$ we get

$$\frac{3x}{(x-1)(x-2)(x-3)} \equiv \frac{1.5}{(x-1)} - \frac{6}{(x-2)} + \frac{4.5}{(x-3)}.$$

This solution can be checked by adding together all the PFs, which should result in the original expression.
Check:

$$\frac{1.5(x-2)(x-3) - 6(x-1)(x-3) + 4.5(x-1)(x-2)}{(x-1)(x-2)(x-3)}$$

and on multiplication of brackets in the numerator

$$= \frac{1.5x^2 - 7.5x + 9 - 6x^2 + 24x - 18 + 4.5x^2 - 13.5x + 9}{(x-1)(x-2)(x-3)}$$

$$= \frac{3x}{(x-1)(x-2)(x-3)}$$

Correct, the left-hand side equals the right-hand side (LHS = RHS). Note the use of algebraic manipulation and fractions!

If the linear factors in the denominator are *repeated*, for example $(x+b)^2$, as in formula 3 above, then to every repeated factor there corresponds PFs to the value of the power of the repeated factor. So for example $(x+b)^2$ in the denominator will have two partial factors of the form $A/(x+b)$ and $B/(x+b)^2$. Similarly for factors like $(x-a)^3$ we would have three PFs, $A/(x-a)$, $B/(x-a)^2$ and $C/(x-a)^3$.

So the expression $\dfrac{(x-2)}{(x-1)(x-3)(x+1)^2}$ would have the following, corresponding PFs:

$$\frac{x-2}{(x-1)(x-3)(x+1)^2} \equiv \frac{A}{(x-1)} + \frac{B}{(x-3)} + \frac{C}{(x+1)} + \frac{D}{(x+1)^2}.$$

The coefficients in the above expression may be found in a similar way to the method described in Example 6.1.14, or by the method illustrated in Example 6.1.15 below.

Example 6.1.15

Determine the PFs for the following expression

$$\frac{1}{(x-2)^2(x-3)}.$$

In this case we have a 'repeated' linear factor, which requires two PFs. So following formula 3, above then

$$\frac{1}{(x-2)^2(x-3)} \equiv \frac{A}{(x-2)} + \frac{B}{(x-2)^2} + \frac{C}{(x-3)}.$$

Now substituting appropriate values of x, as in our previous method would provide values of the coefficients very easily. However, there is another method known as '*equating coefficients*', which can be used independently, or in conjunction with method 1. Although rather more long-winded, it has the advantage that it can *always* be applied. Method 1 becomes rather more hit and miss, when we deal with quadratic factors or higher. Here, we will combine method 1 with the new method of equating coefficients.

So clearing fractions we get:

$$1 \equiv A(x-2)(x-3) + B(x-3) + C(x-2)^2$$

or $\quad 1 \equiv A(x^2 - 5x + 6) + B(x-3) + C(x^2 - 4x + 4).$ (i)

Substituting an appropriate value of 'x', as before say $x = 2$ gives

$$1 = -B \quad \text{or} \quad B = -1.$$

Now, we will equate coefficients. This requires us to set-up a very simple table containing the unknown in ascending powers, as follows.

Considering equation (i) then the 'powers' of the unknown involved are: x^0, x^1, x^2. Note that $x^0 = 1$ and this allows us to consider the *constants* in the equation, remembering, for example, that $6x^0 = 6 \times 1 = 6$. So equating the coefficients of these powers of x, gives

$$x^0 \quad 1 = \quad 6A - 3B + 4C \tag{1}$$

$$x^1 \quad 0 = -5A + \quad B - 4C \tag{2}$$

$$x^2 \quad 0 = \quad A \quad\quad + C \tag{3}$$

So for row (1) where we equate x^0, we note that on the right-hand side of the equation we have '1' $= 1 \times x^0$, this is why it appears as a coefficient of x^0 in (1). Also multiplying the first expression $A(x^2 - 5x + 6)$ to get $Ax^2 - 5Ax + 6A$, shows why $6A$, appears as the coefficient of x^0. This process is repeated for coefficients of x^1 and x^2, up to the highest power in the expression. What we are left with is a set of equations in which we know the value of B. So treating the equations simultaneously then, adding (1) to (2) gives:

$$1 = A - 2B \quad \text{(eliminating } C\text{)}$$

cont.

and since we know that $B = -1$, this immediately gives A as

$1 = A + 2$ and $A = -1$.

Then from equation (3) $A = -C$ or $C = 1$.
  Then:

$$\frac{1}{(x-2)^2(x-3)} \equiv -\frac{1}{(x-2)} - \frac{1}{(x-2)^2} + \frac{1}{x-3}.$$

You should check that the PFs are correct, by multiplication, as in the previous example.

Quadratic factors

Formula 2 above, shows the way in which an expression involving linear and quadratic factors, should be split into partial fractions. To every *quadratic factor* like $x^2 + ax + b$ there corresponds a PF $Cx + D/(x^2 + ax + b)$. Repeated quadratic factors require additional PFs in the same way as before. Thus a factor $(ax^2 + ax + b)^2$ would require PFs $(Cx + D/(x^2 + +ax + b)$ and $(Ex + F)/(x^2 + ax + b)^2$.

Our final example concerning partial fractions, illustrates the method you will need to adopt when dealing with both linear and quadratic factors, in the denominator of an expression. The system of linear equations which results from equating coefficients, is in this example, fairly easily solved. More complex systems require the use of matrix methods for their solution.

Example 6.1.16

Express $\dfrac{x}{(x-1)^2(x^2+5x+1)}$ as a sum of partial fractions.
Then

$$\frac{x}{(x-1)^2(x^2+5x+1)} \equiv \frac{A}{(x-1)} + \frac{B}{(x-1)^2} + \frac{Cx+D}{x^2+5x+1}$$

and multiplying every term by the left-hand side denominator and cancelling gives

$$x = A(x-1)(x^2+5x+1) + B(x^2+5x+1)$$
$$+ (Cx+D)(x-1)^2. \tag{i}$$

Using the method of equating coefficients of different powers of 'x' on *both* sides of the equation, gives:

x^0: $0 = \quad\quad -A + \; B \quad\quad + D$

x^1: $1 = (A - 5A) + 5B + \; C + 2D$

x^2: $0 = \quad\quad 4A + \; B - 2C + \; D$

x^3: $0 = A \quad\quad\quad\quad + \; C$

Make sure you see how these values were obtained. If you find this difficult multiply out the brackets in (i) above, and try again.

cont.

Now we need to solve this system of linear equations to find coefficients. In *this case*, this is relatively easy. Numbering the equations then

$$0 = -A + B \qquad\qquad D \qquad\qquad\qquad (1)$$

$$1 = -4A + 5B + \quad C \quad 2D \qquad\qquad (2)$$

$$0 = \quad 4A + \quad B - 2C \quad D \qquad\qquad (3)$$

$$0 = \quad A \qquad\qquad C \qquad\qquad\qquad\qquad (4)$$

Starting with the simplest equation we see from (4) that $A = -C$ and substituting $A = -C$ into equations (2) and (3) we get

$$1 = -5A + 5B + 2D \qquad\qquad\qquad (2')$$

$$0 = \quad 6A + \quad B + \quad D \qquad\qquad\qquad (3')$$

Now subtracting equation (1) from (3')

$$0 = 6A + B + D$$
$$-(0 = -A + B + D)$$
$$\overline{0 = 7A}$$

This implies that $A = 0$.

Now substituting this value of A into (2') and (3') gives

$$1 = 5B + 2D$$

$$0 = \quad B + \quad D$$

and solving simultaneously

$$1 = 7B \quad \text{so} \quad B = \tfrac{1}{7} \quad \text{and so} \quad D = -\tfrac{1}{7}.$$

Then required PFs are:

$$\frac{x}{(x-1)^2(x^2+5x+1)} = \frac{1}{7(x-1)^2} - \frac{1}{7(x^2+5x+1)}.$$

Note that we could have found 'B' initially using method (1). This would have simplified the manipulation of the equations.

If the numerator of an expression is of higher degree than the denominator then we have an *improper algebraic fraction*. To convert an improper fraction to a proper fraction, we need to divide the denominator into the numerator, for algebraic expressions this requires the use of long division of algebra, with which you should be familiar.

Series

As already mentioned, the algebraic manipulation of sequences and series has few direct applications to engineering, but does enable us to gain mathematical skills which will be of use later, when we study complex numbers, numerical integration and Fourier analysis. For this reason only methods, not applications, will be covered in this section.

Formulae

(1) *Sigma notation:* $\sum$ = 'sum of all terms such as' Range of summation depends on the highest and lowest terms above and below the sigma sign, respectively.

$$\sum_{n=1}^{n=\infty} a^n = \text{Sum of all terms } a^n \text{ between } n = 1 \text{ and } n = \infty$$

(an infinite series).

(2) *Permutations and combinations*

$$^n P_r = \frac{n!}{(n-r)!} \tag{i}$$

$$^n C_r = \frac{n!}{(n-r)!\, r!} = \binom{n}{r} \tag{ii}$$

(3) *Binomial theorem*

$$(a \pm b)^n = {}^n C_0 a^n + {}^n C_1 a^{n-1} b + {}^n C_2 a^{n-2} b^2 + \cdots \tag{i}$$

$$(a + b)^n = a^2 + n a^{n-1} b + \frac{n(n-1)a^{n-2}b^2}{2!} + \cdots b^n \tag{ii}$$

$$(1 + x)^n = 1 + nx + \frac{n(n-1)}{2!} x^2 + \frac{n(n-1)(n-2)}{3!} x^3 + \cdots \tag{iii}$$

If 'n' is a positive integer, the series terminates after $(n+1)$ terms and is valid for all x.

If 'n' is not a positive integer the series terminates and is valid only for values of x such that $-1 < x < 1$.

(4) *Hyperbolic functions*

$$e^x = \cosh x + \sinh x \tag{i}$$

$$e^{-x} = \cosh x - \sinh x \tag{ii}$$

$$\sinh x = \frac{e^x - e^{-x}}{2} \tag{iii}$$

$$\cosh x = \frac{e^x + e^{-x}}{2} \tag{iv}$$

$$\tanh x = \frac{e^{2x} - 1}{e^{2x} + 1} \tag{v}$$

$$\sinh^{-1} x = \ln(x + \sqrt{x^2 + 1}) \quad \text{(for all } x) \tag{vi}$$

$$\cosh^{-1} x = \pm\ln(x + \sqrt{x^2 - 1}) \quad (|x| > 1) \tag{vii}$$

$$\tanh^{-1} x = \tfrac{1}{2} \ln \frac{1 + x}{1 - x} \tag{viii}$$

(5) *Series*

$$\exp x = 1 + x + \frac{x^2}{2!} + \frac{x^3}{3!} + \qquad \text{(i)}$$

valid for all x

$$\ln(1 + x) = x - \frac{x^2}{2} + \frac{x^3}{3} - \frac{x^4}{4} + \qquad \text{(ii)}$$

valid for $-1 < x \leqslant 1$

$$\ln(1 - x) = -x - \frac{x^2}{2} - \frac{x^3}{3} - \frac{x^4}{4} - \qquad \text{(iii)}$$

valid for $-1 \leqslant x < 1$

$$\sin x = x - \frac{x^3}{3!} + \frac{x^5}{5!} - \qquad \text{(iv)}$$

x in radians, valid for all x

$$\cos x = 1 - \frac{x^2}{2!} + \frac{x^4}{4!} - \qquad \text{(v)}$$

x in radians, valid for all x

$$\sinh x = x + \frac{x^3}{3!} + \frac{x^5}{5!} + \qquad \text{(vi)}$$

$$\cosh x = 1 + \frac{x^2}{2!} + \frac{x^4}{4!} + \qquad \text{(vii)}$$

Methods

In this final section on algebra we will concentrate on exponential, logarithmic and hyperbolic functions and their related series, in addition to investigating the concept of the limit and, the use of the Binomial theorem as a mathematical tool.

You have already met the exponential function and its inverse (the natural logarithm), in our work on fundamentals. In this section we will be looking at their series and those for trigonometric and hyperbolic functions, to see how they interrelate. We start by considering a little notation.

Sigma notation

The notation given in formula 1 above, requires a little explanation, particularly for those who may not be familiar with the way in which we can represent a series using this notation. Also, the following explanation should be useful for those who have not met the concept of a *limit*.

Using *Sigma* notation (symbol $\sum$) for summing the terms of a series enables us to define the series more concisely.

The terms of a *sequence* when added form a *series*. For example the set of numbers: $1, 2, 4, 8, 16$ are in a set order and form a sequence because each number may be found from the previous number by applying an obvious rule. When this sequence of numbers is added a series is formed. This series *can be bounded* by stipulating the number of terms in the series using Sigma notation, i.e.

$$\sum_{n=1}^{n=5} 2^n \quad \text{writing this out in full we get:}$$

$$\sum_{n=0}^{n=5} 2^n = 2^0 + 2^1 + 2^2 + 2^3 + 2^4 + 2^5$$

$$= 1 + 2 + 4 + 8 + 16 + 32.$$

So we have a succinct way of writing this series, which tells us the precise number of terms and in this case gives us a precise numerical value. It provides a lower boundary (1) and upper boundary (63) for the sum of the series.

Consider the following series, written in Sigma notation

$$\sum_{n=0}^{n=\infty} \frac{1}{2^n}$$ Does this series have a limit?

In other words does it have an upper and lower boundary. Expanding the series we get:

$$\sum_{n=0}^{n=\infty} \frac{1}{2^n} = \frac{1}{2^0} + \frac{1}{2^1} + \frac{1}{2^2} + \frac{1}{2^3} + \frac{1}{2^4} \cdots \frac{1}{2^n}$$

So $$\sum_{n=0}^{n=\infty} \frac{1}{2^n} = \frac{1}{1} + \frac{1}{2} + \frac{1}{4} + \frac{1}{8} + \frac{1}{16} \cdots \frac{1}{2^n}.$$

I am sure you can see that if we sum just the first term of the series the lower bound would be 1. Also that as $n \to \infty$ then $\frac{1}{2^n} \to 0$, so there must also be an upper bound, because successive terms approach zero.

The above series is known as *Geometric* because each successive term is obtained from the preceding term by multiplication of a common ratio. The common ratio (r) for the above series is $r = \frac{1}{2}$. This series is known as a *convergent* series because $r < 1$.

The previous series $\sum_{n=1}^{n=\infty} 2^n$ is known as a *divergent* series as written above it has no *upper bound*, the common ratio for successive terms is 2.

If a geometric series is *convergent* then an upper bound can be found from

$$S_\infty = \frac{a}{1 - r} \quad \text{where } |r| < 1.$$

$|r|$ means the modulus of r or the positive value of r. Then for our series the sum to infinity S_∞ is equal to $S_\infty = \frac{1}{1 - \frac{1}{2}} = \frac{1}{\frac{1}{2}} = 2$ (where $a = $ the first term). So this series has a *limit*.

Example 6.1.17

Find the upper and lower limits of x so that the series

$$\sum_{n=0}^{n=\infty} \frac{(x - 1)^n}{2^n} \quad converges$$

and determine the sum to infinity of the series when $x = 2$.

cont.

To see how the series might behave, we need to expand the series for the first few terms. Then

$$\sum_{n=0}^{n=\infty} \frac{(x-1)^n}{2^n} = 1 + \frac{x-1}{2} + \frac{(x-1)^2}{2^2} + \frac{(x-1)^3}{2^3}. \qquad \text{(i)}$$

Then the common ratio (r) for the series can be seen as $\left(\dfrac{x-1}{2}\right)$, so the condition for convergence is $|r| < 1$ or in this case

$$\left|\frac{x-1}{2}\right| < 1 \qquad \text{(ii)}$$

then removing modulus sign, allows us to consider negative limit, i.e.

$$-1 < \frac{x-1}{2} < 1$$

which on multiplication by 2 gives:

$$-2 < x - 1 < 2 \quad \text{and} \quad x - 1 < 2 \quad \text{means} \quad x < 3.$$

So $-1 < x < 3$ in other words x must lie between a lower value greater than -1 and less than 3.

Then for $x = 2$, the series will converge. Using our formula for sum to infinity $S_\infty = \dfrac{a}{1-r}$ when $x = 2$, $r = \frac{1}{2}$, from (ii). So $S_\infty = \dfrac{1}{1 - \frac{1}{2}} = 2$. (Note that if you put $x = 2$ into the series (i) we obtain the series $1 + \frac{1}{2} + \frac{1}{4} \cdots$, etc. as before!)

The binomial theorem

Before we look at the methods for using the binomial theorem, we must be clear on how to use the combination formula 2(ii). This formula tells us how many combinations there are of selecting (r) objects from (n) objects. For *combinations* the selection is made *irrespective of order*, for *permutations* formula 2(i) is the selection of (r) objects from (n) *in a specified order*.

We are primarily interested in combinations, because it is these that are used in the binomial theorem (3i).

So the selection of four objects from six, *irrespective of order* is given by formula (2ii) as

$$^6C_4 = \frac{n!}{(n-r)!\,r!} = \frac{6!}{(6-4)!\,4!} = \frac{6 \times 5 \times 4 \times 3 \times 2 \times 1}{(2 \times 1)(4 \times 3 \times 2 \times 1)} = \frac{720}{48} = 15.$$

Note that a number followed by an exclamation mark (!) is a *factorial* number, that is a number which multiplies itself by all the numerals down to one, as shown in the example above.

We will now use the form the binomial theorem given in formula 3i and 3ii.

Example 6.1.18

Expand $(x - y)^5$ using the binomial theorem, from formula (3ii), we get

$$(x - y)^5 = x^5 + (5)(x)^4(-y)^1 + \frac{(5)(4)(x)^3(-y)^2}{2!}$$

$$+ \frac{(5)(4)(3)(x)^2(-y)^3}{3!} + \frac{(5)(4)(3)(2)(x)^1(-y)^4}{4!}$$

$$+ \frac{(5)(4)(3)(2)(1)(-y)^5}{5!}$$

and simplifying gives:

$$(x - y)^5 = x^5 - 5x^4y + 10x^3y^2 - 10x^2y^3 + 5xy^4 - y^5.$$

The general interpretation of the binomial theorem considered in the previous example is restricted to an expansion which has integer powers of the index n. This restriction is removed when we consider the binomial expansion in *standard form*. If the expansion is modified to exclude the a^n term in formulae 3i and 3ii, replacing it with the number 1. Then the theorem may be written to include values of the index which are negative, fractional, or both, provided certain limitations are observed.

So the standard binomial expansion may be written as:

$$(1 + x)^n = 1 + (n)(x) + \frac{(n)(n-1)(x)^2}{2!} + \frac{(n)(n-1)(n-2)(x)^3}{3!} + \cdots$$

You will remember from our work on *limits*, that if we restrict values of x to $-1 < x < +1$, then the series will converge to a finite value. Under these circumstances the number of terms in the series is unrestricted for convergence. If the value of x is outside these limits, we still have no restriction on the number of terms in the series but the series may diverge.

If the binomial expansion contains multiples of x, such as $2x, 3x, \ldots$ then the limits imposed on 'x' must be adjusted accordingly. For example:

$$\text{if} \quad -1 < 2x < +1 \quad \text{then} \quad -\tfrac{1}{2} < x < \tfrac{1}{2}$$

$$\text{or if} \quad -1 < 3x < +1 \quad \text{then} \quad -\tfrac{1}{3} < x < \tfrac{1}{3} \quad \text{and so on.}$$

Example 6.1.19

Expand $(1 + 2x)^{-1/2}$ to 4 terms.

$$(1 + 2x)^{-1/2} = 1 + (-\tfrac{1}{2})(2x) + \frac{(-\tfrac{1}{2})(-\tfrac{1}{2} - 1)(2x)^2}{2!}$$

$$+ \frac{(-\tfrac{1}{2})(-\tfrac{1}{2} - 1)(-\tfrac{1}{2} - 2)(2x)^3}{3!} + \cdots$$

$$= 1 - x + \tfrac{3}{2}x^2 - \tfrac{5}{2}x^3 + \cdots$$

restriction $-\tfrac{1}{2} < x < +\tfrac{1}{2}$ for convergence.

It can be seen from the example 6.1.19 that a series can be produced for functions of the type $(1 + x)^n$, where n is a whole, fractional or negative number, by using the binomial theorem to the required degree of accuracy. It is this use of the binomial, where we are able to find *approximations* for algebraic expressions in the form of a *power series*, that will help us with our later work. Another, more subtle extension of the above approximation method may be used to calculate percentage changes.

Example 6.1.20

Given that $C = \sqrt{p}$, find the percentage change in C, caused by a 3% change in p.

Then new value of

$$C = \sqrt{1.03p}$$

$$= \sqrt{p}(1 + 0.03)^{1/2} \quad \text{(laws of indices!)}$$

and first approximation, using the binomial theorem (formula 3iii) gives

$$C = \sqrt{p}[1 + (\tfrac{1}{2})(0.03) + \cdots]$$

$$C = \sqrt{p}[1 + 0.015]$$

$$C = \sqrt{p}(1.015)$$

therefore there will be a 1.5% increase in C.

Example 6.1.21

The volume and temperature for air, when treated as a perfect gas subject to adiabatic expansion, may be related by the formula

$$\frac{V_1}{V_2} = \left(\frac{T_2}{T_1}\right)^{1.4}.$$

Find the percentage change in the volume ratio, given that there is a 2% change in T_1 and T_2.

Then

$$\left(\frac{V_1}{V_2}\right) + \text{change in} \left(\frac{V_1}{V_2}\right) = 2\% \text{ change in } (T_2^{1.4})(T_1^{-1.4})$$

therefore from binomial as a first approximation

$$\text{LHS} = [(1 - 0.02)^{1.4}(1 + 0.02)^{-1.4}] T_2^{1.4} \, T_1^{-1.4}$$

$$= [(0.972)(0.972)] T_2^{1.4} \, T_1^{-1.4}$$

$$= (1 - 0.028)(1 - 0.028) T_2^{1.4} \, T_1^{-1.4}$$

$$= (1 - 0.056) \frac{V_1}{V_2}.$$

Then 2% change in T_1 and T_2 gives 5.6% decrease in volume ratio $\left(\dfrac{V_1}{V_2}\right)$.

Series

We have already discussed the exponential and natural logarithm functions. If we use the binomial expansion we can derive a particularly useful series, the *Exponential* series (see Example 6.1.22). Estimates of exponential, logarithmic and hyperbolic functions using *power series* (a series where each term is a simple power of the independent variable), are particularly useful when considering the Calculus and other mathematical topics, which you will meet later.

Example 6.1.22

Expand $\left(1 + \dfrac{x}{n}\right)^n$ using the binomial theorem and deduce what happens to the series as $n \to \infty$. Then

$$\left(1 + \frac{1}{n}x\right)^n = 1 + x + \frac{n(n-1)}{2!}\left(\frac{1}{n}x\right)^2$$

$$+ \frac{n(n-1)(n-2)}{3!}\left(\frac{1}{n}x\right)^3 + \cdots$$

$$= 1 + x + \frac{\left(1 - \dfrac{1}{n}\right)}{2!}x^2 + \frac{\left(1 - \dfrac{1}{n}\right)\left(1 - \dfrac{2}{n}\right)}{3!}x^3 + \cdots$$

Now you already know that for the terms $\dfrac{1}{n}, \dfrac{2}{n}$, etc. as $n \to \infty$, these terms $\to 0$. Therefore the bracketed expressions, all approach 1, or *in the limit* equal 1.

So $\underset{n \to \infty}{\mathrm{Lt}}\left(1 + \dfrac{1}{n}x\right)^n = 1 + x + \dfrac{x^2}{2!} + \dfrac{x^3}{3!} + \cdots.$

I hope you recognise this series as the *Exponential* series, that is formula 5i, where

$$\exp(x) = e^x = 1 + x + \frac{x^2}{2!} + \frac{x^3}{3!} + \cdots.$$

This series provides us with *the base* for natural or Napierian logarithms (ln or $\log_e$), when $x = 1$.

i.e. $e^1 = 1 + 1 + \dfrac{1}{2!} + \dfrac{1}{3!} + \dfrac{1}{4!} + \dfrac{1}{5!} \cdots$

$= 1 + 1 + 0.5 + 0.166\,66' + 0.041\,666' + 0.008\,333'$

$e^1 \simeq 2.72$ correct to two decimal places.

Continuing the series gives us a better and better approximation for 'e', correct to five decimal places

$e \simeq 2.718\,28.$

Example 6.1.23

Formula 5ii and 5iii give the power series for the natural logarithm functions $\ln(1 + x)$ and $\ln(1 - x)$, respectively. Write down the expansion for the function

$$\ln\left(\frac{1 + x}{1 - x}\right).$$

Remembering our laws of logarithms, in particular law (iii), where

$$\log_a \frac{M}{N} = \log_a M - \log_a N,$$

then the above may be written as

$$\ln\left(\frac{1 + x}{1 - x}\right) = \log_e(1 + x) - \log_e(1 - x)$$

and since

$$\log_e(1 + x) = x - \frac{x^2}{2} + \frac{x^3}{3} - \frac{x^4}{4} + \frac{x^5}{5} + \cdots$$

$$\log_e(1 - x) = -x - \frac{x^2}{2} - \frac{x^3}{3} - \frac{x^4}{4} - \frac{x^5}{5} - \cdots$$

then $\log_e(1 + x) - \log_e(1 - x) = 2x + \frac{2x^3}{3} + \frac{2x^5}{5} + \cdots$

or $\ln\left(\frac{1 + x}{1 - x}\right) = 2\left(x + \frac{x^3}{3} + \frac{x^5}{5} + \cdots\right).$

Before we consider techniques for manipulating hyperbolic functions, it is important to understand what these functions represent. You are no doubt already familiar with the basic trigonometric functions; sine, cosine and tangent, and will be aware that they can be represented by the radius of a circle of unit length, rotating anti-clockwise from the horizontal. As this radius rotates values for the basic trigonometric ratios can be determined by construction, using suitable axis graduated in radians. For this reason the trigonometric functions are sometimes known as *circular functions*. Now, if instead of a circle we use a hyperbola then, in a similar way, we can use this geometrical method to produce the hyperbolic sine, cosine and tangent.

It is no coincidence that the power series for the sine and cosine functions, are very similar to their hyperbolic counterparts! Notice also, (formula 4iii and 4iv) that the hyperbolic functions can be represented in exponential form.

Example 6.1.24

Determine the value of cosh 1.932 using the definition given in formula 4iv and compare this value with that obtained by the first four terms of its power series (formula 5vii).

cont.

$$\cosh x = \frac{e^x + e^{-x}}{2} \quad \text{so} \quad \cosh(1.932) = \frac{e^{1.932} + e^{-1.932}}{2}$$

let $y = e^{1.932}$ then $\log_e y = 1.932$ and so $y = 6.9033$
therefore

$$\frac{1}{y} = e^{-1.932} = 0.1449$$

then $\cosh x = \dfrac{6.9033 + 0.1449}{2} = 3.524$ (3 decimal places).

Now using

$$\cosh x = 1 + \frac{x^2}{2!} + \frac{x^4}{4!} + \frac{x^6}{6!} + \cdots$$

$$\cosh(1.932) = 1 + \frac{(1.932)^2}{2} + \frac{(1.932)^4}{24} + \frac{(1.932)^6}{720} + \cdots$$

$$= 1 + \frac{3.7326}{2} + \frac{13.9325}{24} + \frac{52.0047}{720}$$

$$= 1 + 1.8663 + 0.580\,55 + 0.0722$$

$$= 3.52$$

An accuracy of three decimal places would require a minimum of five terms.

Example 6.1.25

Find the value of $\tanh^{-1} 0.825$ for *real* values of x. We could use formula (4viii) and solve directly, but we will take a more convoluted route to incorporate a little algebraic manipulation! The trigonometric identity $\dfrac{\sin x}{\cos x} = \tan x$ is well known, in a similar manner $\dfrac{\sinh x}{\cosh x} = \tanh x$

or $\quad \dfrac{\dfrac{e^x - e^{-x}}{2}}{\dfrac{e^x + e^{-x}}{2}} = \left(\dfrac{e^x - e^{-x}}{2}\right)\left(\dfrac{2}{e^x + e^{-x}}\right) = \dfrac{e^x - e^{-x}}{e^x + e^{-x}}$

We can say therefore that

$$\tanh x = \frac{e^x - e^{-x}}{e^x + e^{-x}} \tag{4ix}$$

Now if we let $x = \tan^{-1} 0.825$, then we find the value of x such that $\tanh x = 0.825$. Using 4ix we get:

$$\frac{e^x - e^{-x}}{e^x + e^{-x}} = 0.825$$

$$e^x - e^{-x} = 0.825(e^x + e^{-x})$$

$$e^x - 0.825e^x = e^{-x} + 0.825e^{-x}$$

$$0.175e^x = 1.825e^{-x} \quad \text{(and on multiplication by } e^x\text{)}$$

$$0.175(e^x)^2 = 1.825$$

cont.

$$\left(e^{x}\right)^{2} = \frac{1.825}{0.175}$$

$$e^{x} = \sqrt{\frac{1.825}{0.175}} = \pm 3.229$$

(e^{x} cannot be *negative* for real values of x) so $e^{x} = 3.229$
$\therefore$ $x = \ln 3.229 = 1.172$ then $\tanh^{-1} 0.825 = 1.172$. Check
this solution using formula (4viii)!

There are many other useful results which can be acquired by manipulating the above functions and their associated series, we will return to these series later.

We leave our study of fundamental algebra with a number of problems which will enable you to practice the methods presented here, and in some cases, to apply them to engineering situations.

Questions 6.1

(1) The thermal resistance of a composite wall is related by the equation

$$R_T = \frac{x_1}{k_1 A} + \frac{x_2}{k_2 A} + \frac{x_3}{k_3 A}.$$

Transpose the formula for 'A' and find the value of A when $k_1 = k_2 = 0.2$, $k_3 = 0.3$, $x_1 = 40 \times 10^{-3}$, $x_2 = 125 \times 10^{-3}$, $x_3 = 60 \times 10^{-3}$ and $R_T = 625$.

(2) The heat flow rate through a cylindrical wall is given by

$$Q = \frac{2k(t_1 - t_2)}{\ln(r_2/r_1)},$$

also the thermal resistance of a cylinder wall is given by

$$R = \frac{\ln(r_2/r_1)}{2\pi k}.$$

Obtain a formula for 'R' in terms of Q, and the temperature difference $(t_1 - t_2)$ and find its value when $t_1 = 250$, $t_2 = 20$, and $Q = 16.3$.

(3) The Bernoulli equation may be written as

$$\frac{P_1}{\gamma} + \frac{v_1^2}{2g} + h_1 = \frac{P_2}{\gamma} + \frac{v_2^2}{2g} + h_2$$

given that $(h_1 - h_2) = 2$, $(v_1^2 - v_2^2) = 8.4$, $P_1 = 350$, $\gamma = 10$, transpose the formula in a suitable way to find the value of the pressure P_2 ($g = 9.81$).

(4) The signal to noise ratio is given by the formula

$$\frac{S}{N} = 10 \log_{10}\left(\frac{P_{\text{signal}}}{P_{\text{noise}}}\right)$$

where P is signal and noise power is watts. Determine P noise given $\frac{S}{N} = 20$ and $P_{\text{signal}} = 2$.

cont.

(5) Values for a reciprocal motion problem result in the following simultaneous equation being formulated

$$12 = \omega\sqrt{r^2 - 0.08^2} \qquad (1)$$

$$3 = \omega\sqrt{r^2 - 0.24^2} \qquad (2)$$

Find the value for the radius 'r'.

(6) Solve the simultaneous equations

$$\log_x y = 2 \qquad (1)$$

$$xy = 8 \qquad (2)$$

(7) It is believed that the law connecting two variables H and V is of the form $H = aV^n$, where a and n are constants. The results of an experiment for V and H are given below. By plotting a graph using suitable axes, find the law relating H and V.

V	9	12	15	18	20	24
H	36.45	86.4	168.8	291.6	400	691.2

(8) Find the PFs for the expressions given below

(i) $\dfrac{1}{(x+2)(x-3)}$

(ii) $\dfrac{x}{(x+1)(x^2+2x+6)}$

(iii) $\dfrac{x^2}{(x-1)(x+2)^2}$

(9) The exponential function $f(x) = Ae^{-bx}$ satisfies the conditions $f(0) = 2$ and $f(1) = 0.5$. Find the constants A and b.

(10) If the general condition for 'convergence' of a series is

$$\underset{n \to \infty}{\text{Lt}} \left| \frac{u_{n+1}}{u_n} \right| < 1 \quad \text{where } u_n \text{ is } n\text{th term of a series.}$$

Determine whether or not the following series converge:

$$\sum_{1}^{\infty} \frac{1}{n^2 + 1} \qquad (i)$$

$$\sum_{1}^{\infty} \frac{n}{n^2 + 1} \qquad (ii)$$

(11) Expand the following binomial expressions and extract their coefficients:

$$(a+b)^0, (a+b)^1, (a+b)^2, (a+b)^3, (a+b)^4, (a+b)^5$$

By placing the coefficients of each expansion sequentially one underneath the other (starting with the lowest), study the triangular pattern formed and determine the coefficients of $(a+b)^6$, *without* carrying out the expansion.

cont.

(12) Given that $Q = \frac{1}{R}\left(\frac{L}{c}\right)^{1/2}$. Use the binomial theorem to find the percentage change in Q caused by a 2% increase in L, a 3% decrease in R and a 4% decrease in C.

(13) Find the first five terms in the expansion of $(x^2 - 1)e^{-x}$.

(14) Sketch the graphs of the hyperbolic functions given by formulae 4iii, 4iv and 4v, choosing a suitable scale. Comment on the characteristics of these functions when $x = 0$, $x = 1$, $x \rightarrow -\infty$ and $x \rightarrow +\infty$.

(15) Using appropriate formulae, find the value of

 (i) $\sinh x = 1.5$

 (ii) $\cosh x = 1.875$

 (iii) $\sinh^{-1} 1.375 = x$

 (iv) $\tanh x = 0.32$.

6.2 Trigonometry

In this short section on trigonometry, we will treat the subject matter in an identical way to the algebra you meet previously. We start with a review of some fundamental trigonometric methods concerned with radian measure and basic trigonometric functions and their associated graphs. In the second part of this section we will concentrate on the identification and use of trigonometric identities to simplify expressions and solve trigonometric equations, which will be of use when we study some of the more advanced mathematical topics.

Trigonometric fundamentals

Formulae

(1) Radian measure

 (i) π radians $= 180°$,
 (ii) if θ is small and measured in radians, then $\sin\theta = \theta$, $\cos\theta = 1$, $\tan\theta = \theta$,
 (iii) arc length $s = r\theta$ (θ in radians),
 (iv) area of sector $= \frac{1}{2}r^2\theta$ (θ in radians).

(2) Trigonometric ratios for angles of any magnitude

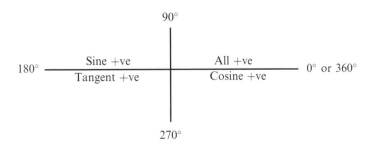

For all values of θ: $\sin(-\theta) = -\sin\theta$

$$\cos(-\theta) = \cos\theta$$

$$\tan(-\theta) = -\tan\theta$$

(3) Equivalent ratios

	$90° \le \theta \le 180°$	$180° \le \theta \le 270°$	$270° \le \theta \le 360°$
$\sin\theta =$	$\sin(180° - \theta)$	$-\sin(\theta - 180°)$	$-\sin(360° - \theta)$
$\cos\theta =$	$-\cos(180° - \theta)$	$-\cos(\theta - 180°)$	$\cos(360° - \theta)$
$\tan\theta =$	$-\tan(180° - \theta)$	$\tan(\theta - 180°)$	$-\tan(360° - \theta)$

(4) Triangle formulae

(i) Sine Rule $\quad \dfrac{a}{\sin A} = \dfrac{b}{\sin B} = \dfrac{c}{\sin C} = 2r$

(ii) Cosine Rule $\quad a^2 = b^2 + c^2 - 2bc\cos A$

(iii) Area $\quad = \frac{1}{2}bc\sin A$

$$= \sqrt{s(s-a)(s-b)(s-c)} \quad \text{where} \quad s = \frac{(a+b+c)}{2}.$$

(5) Polar and Cartesian co-ordinate system

(i) $x = r\cos\theta \qquad y = r\sin\theta$

(ii) $r = +\sqrt{x^2 + y^2}, \quad \theta = \tan^{-1}\left[\dfrac{y}{x}\right], \quad \tan\theta = \dfrac{y}{x}.$

(6) Equation of a circle, centre at the origin, radius (a)

$$x^2 + y^2 = a^2$$

$$x^2 + y^2 + 2gx + 2fy + x = 0$$

where the centre is at $(-g, -f)$ and radius $= \sqrt{g^2 + f^2 - c}$.

(7) Superposition

(i) $a\sin\omega x + b\cos\omega x = c\sin(\omega x + \phi)$ where point (a, b) has polar co-ordinates r, θ and $c = r = \sqrt{a^2 + b^2}$ and $\tan\phi = \dfrac{b}{a}.$

Methods

We will look at a variety of methods using examples, as we did previously. These will cover radian measure, polar and Cartesian co-ordinates, graphs of sinusoidal functions, and the solution of triangles using trigonometric ratios.

Radian measure

Circular measure using degrees has been with us since the days of the Babylonians, when they divided a circle into 360 equal parts, corresponding to what they believed were the days in a year. The *degree*, being an arbitrary form of circular measurement, has not always proved an appropriate unit for mathematical manipulation.

Another less arbitrary unit of measure has been introduced, which you will know as the *radian*, the advantage of this unit is its relationship with the arc length of a circle.

Example 6.2.1

An arc *AB* of length 4.5 cm is marked on a circle of radius 3 cm. Find the area of the sector bounded by this arc and the centre of the circle, using formulae 1iii and 1iv.

Then arc length $S = r\theta$

$$\theta = \frac{S}{r} = \frac{4.5}{3} = 1.5 \text{ radians.}$$

So area of sector $= \frac{1}{2}r^2\theta = (0.5)(3)^2(1.5)$

$$\text{Area} = 6.75 \text{ cm}^2.$$

Example 6.2.2

The area of a sector of a circle is 20 cm². If the circle has a diameter of 9.5 cm. What is the length of the arc of the sector?

Then again from $A = \frac{1}{2}r^2\theta$

$$\theta = \frac{2A}{r^2} = \frac{40}{4.75^2} = 1.77 \text{ radians.}$$

So required arc length $S = (4.75)(1.77)$

$$S = 8.42 \text{ cm.}$$

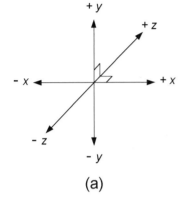

(a)

Figure 6.2.1   *Cartesian co-ordinates*

(b)

Polar and Cartesian co-ordinates

The ability to use both Polar and Cartesian co-ordinate systems and, to be able to change from one co-ordinate system to another is important for your later work, especially when we deal with complex numbers.

You will be familiar with the *Cartesian co-ordinate* system (Figure 6.2.1), where two or more mutually perpendicular axes are used to identify a point or position in space.

In Figure 6.2.1b the point *P* in the plane is located using the *xy* co-ordinates. The length of the line *OP* is found using Pythagoras i.e. $(OP)^2 = x^2 + y^2$ so $OP = \sqrt{x^2 + y^2}$.

The position of the point *P* in Figure 6.2.1b is thus defined by the *x* and *y* values. Equally if the angle *POQ* is known in conjunction with the length of *OP*, then point *P* can be defined in terms of the *x*-axis only. We then have a *polar co-ordinate* system (Figure 6.2.2), where in this case $OP = r$.

Example 6.2.3

Convert the point $(3, 4)$ to polar co-ordinates and convert the point $(3, 30°)$ to Cartesian co-ordinates.

(i) Using the formulae given in 5ii, we have for point $(3, 4)$,

cont.

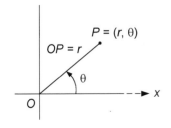

Figure 6.2.2   *Polar co-ordinates*

$$r = \sqrt{3^2 + 4^2} = \sqrt{25} = 5$$

and $\theta = \tan^{-1}\left(\dfrac{y}{x}\right) = \tan^{-1}\left(\dfrac{4}{3}\right) = \tan^{-1} 1.333 = 53.13°.$

So point $(3, 4)$ in polar form is $(5, 53.13°)$.

(ii) Using the formulae given in 5i, we have for point $(3, 30°)$,

$$x = r \cos \theta = 3 \cos 30 = 2.6$$

and $y = r \sin \theta = 3 \sin 30 = 1.5.$

Periodic functions

Circular sinusoidal functions such as $y = A \sin nx$ and $y = B \cos nx$ may be represented graphically as shown in Figure 6.2.3. The important points to notice are:

(i) These curves have a repeating pattern that occurs at period or wavelength $2\pi/n$.
(ii) They have amplitude, A or B.
(iii) The curves first reach their positive maximum value from a position or time zero at different points, this difference when measured as angular distance is known as *phase shift*. They are out of phase by some angular distance ϕ.
(iv) The general sinusoidal wave form may be written $y = A \sin(nx + \phi)$, where n is sometimes referred to as the *harmonic* and ϕ is the phase angle which may lead or lag.

When the above graphs take the form $y = A \sin \omega x$ or $y = B \cos \omega x$ (formula 7iv), then if the variable (x) is time, the time period is given by $2\pi\omega$, where ω is the *angular frequency measured in rad/s* and the reciprocal of the time period $\omega/2\pi$ is the *frequency measured in cycles per second or Hertz*. In this form we are able to use formula 7iv, which is known as the superposition formula, to sum two or more sinusoidal functions. This relationship is important in the study of waves and vibration analysis.

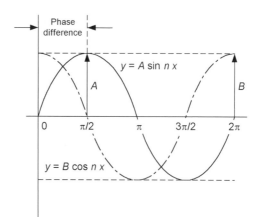

Figure 6.2.3 *Circular sinusoidal functions*

Example 6.2.4

Find the sum of the two trigonometric functions $y_1 = 4 \sin \omega x$, $y_2 = 3 \cos \omega x$, which both have the same period.

Then using the superposition formula (7i)

$$4 \sin \omega x + 3 \cos \omega x = c \sin(\omega x + \phi)$$

where $c = \sqrt{4^2 + 3^2} = \sqrt{25} = 5$

and $\tan \phi = \dfrac{B}{A} = \dfrac{3}{4}$ giving $\theta = 36.9°$

then $y_3 = 5 \sin(\omega x + 36.9)$

Example 6.2.5

(i) Write down the amplitude, order of harmonic and phase angle of the sine function

$$y = 30 \sin(3\theta + 20°)$$

then: Amplitude = 30

Harmonic = 3

Phase angle = 20° leading.

(ii) Find the instantaneous value of the sinusoidal function

$$y = 4 \sin\left(\omega t + \frac{\pi}{4}\right) \text{ given } \omega = 214 \text{ rad/s and } t = 7.5 \text{ ms}$$

$$y = 4 \sin\left(1.605 + \frac{\pi}{4}\right) \quad \text{remember} \quad 1.605 \text{ radians}$$

so $y = 4 \sin(2.39 \text{ rad})$

$y = 4(0.6825)$

$y = 2.73$.

Solution of triangles

This is very much revision to remind you of the use of the fundamental ratios for the solution of triangles.

Example 6.2.6

If the ratios of the sides of a triangle are such that $\sin \theta = \frac{3}{5}$ find the ratios for $\cos \theta$ and $\tan \theta$, without referring to tables or using a calculator. We can apply the fundamental trigonometric identity:

$$\sin^2 \theta + \cos^2 \theta = 1 \quad \text{as follows}$$

$$\cos^2 \theta = 1 - \sin^2 \theta$$

then $\cos \theta = \sqrt{1 - \sin^2 \theta}$

$$\cos \theta = \sqrt{1 - \left(\tfrac{3}{5}\right)^2} = \sqrt{\tfrac{16}{25}} = \tfrac{4}{5}.$$

cont.

We also know that

$$\tan \theta = \frac{\sin \theta}{\cos \theta}$$

therefore $\quad \tan \theta = \frac{\frac{3}{5}}{\frac{4}{5}} = \frac{3}{4}.$

Then required ratios are: $\sin \theta = \frac{3}{5}$, $\cos \theta = \frac{4}{5}$ and $\tan \theta = \frac{3}{4}.$

Example 6.2.7

In the triangle ABC, $BC = 8.2$ cm, $AC = 4.5$ cm and $\angle ACB = 60°$. Calculate:

(i) the other angles of the triangle;
(ii) the area of the triangle.

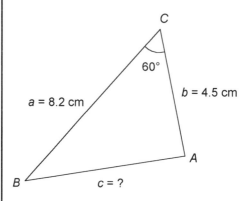

First we sketch the triangle to establish the sides a, b and c. Then using the cosine rule (formula 4ii), then

$$c^2 = a^2 + b^2 - 2ab \cos C.$$

So $\quad c^2 (8.2)^2 + (4.5)^2 - (2)(8.2)(4.5) \cos 60$

$\qquad c^2 = 67.24 + 20.25 - 36.9$

$\qquad c = 7.11$ cm.

Now using the sine rule

$$\frac{\sin A}{8.2} = \frac{\sin B}{4.5} = \frac{\sin 60}{7.11}$$

so $\quad \sin B = \frac{4.5 \sin 60}{7.11} = 0.3169$

and $\quad \angle B = 18.48°$ therefore $\angle A = 101.52°$

and then the area of the triangle $= \frac{1}{2} ab \sin C.$

So $\quad$ area $= (0.5)(8.2)(4.5) \sin 60$

required area $= 15.98$ cm^2.

Example 6.2.8

The diagram below shows a plot to be used as the footprint for a factory site. Determine the length of the diagonal QR.

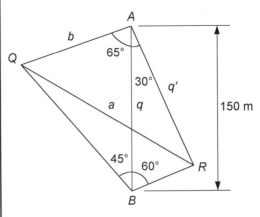

In triangle ARB

$\angle ARB = 90°$.

So length AR is given by:

$AR = 150 \cos 30°$

$\quad = 129.9\,\text{m}.$

Also in triangle AQB

$\angle AQB = 70°$

So $\dfrac{q}{\sin Q} = \dfrac{b}{\sin B}$ or $\dfrac{150}{\sin 70} = \dfrac{b}{\sin 45}$

and $b = \dfrac{(150)(\sin 45)}{\sin 70}$

side $b = 112.9\,\text{m}.$

Now considering triangle QAR and using cosine rule where sides are q, q', r, i.e. $AR = q'$, $AQ = r$, $a = QR$

$$a^2 = q'^2 + r^2 - 2q'r \cos 95$$

$$a^2 = (129.9)^2 + (112.9)^2 - (2)(129.9)(112.9)(-0.087)$$

$$a^2 = 168\,74.01 + 127\,46.41 + 2556.4$$

and $QR = a = 179.4\,\text{m}.$

Engineering applications

The applications of trigonometric fundamentals are numerous. We will look specifically at the use of trigonometry for manipulating formulae concerned with shear strain and simple harmonic motion, in addition to alternating current and signal wave forms. As before, each application is given in the form of an example.

Example 6.2.9

An element of material is subject to shear strain (refer to Figure 5.1.8). Using the figure and the cosine rule for triangle *dab*, derive the relationship:

$$1 + 2\varepsilon_{max} + (\varepsilon_{max})^2 = 1 + \sin \gamma.$$

Also, if ε_{max} and γ are very small show that $\varepsilon_{max} = \dfrac{\gamma}{2}$. Applying the cosine rule as suggested to triangle *dab*, where the longest side is written in terms of $s\sqrt{2}(1 + \varepsilon_{max})$ then

$$[s\sqrt{2}(1 + \varepsilon_{max})]^2 = s^2 + s^2 - 2s^2 \cos\left(\frac{\pi}{2} + \gamma\right)$$

So $2s^2(1 + \varepsilon_{max})^2 = 2s^2 - 2s^2 \cos\left(\dfrac{\pi}{2} + \gamma\right)$

which on division by $2s^2$ gives

$$(1 + \varepsilon_{max})^2 = 1 - \cos\left(\frac{\pi}{2} + \gamma\right).$$

Now using the trigonometric identity (formula 3) that $\cos 90 + x = -\sin x$ then $\cos\dfrac{\pi}{2} + \gamma = -\sin \gamma$. So we get

$$(1 + \varepsilon_{max})^2 = 1 + \sin \gamma$$

and on expansion of the left-hand side:

$$1 + 2\varepsilon_{max} + (\varepsilon_{max})^2 = 1 + \sin \gamma \quad \text{as required.}$$

Also if ε_{max} and γ are very small then $\sin \gamma \simeq \gamma$ (formula 1ii). We may also ignore the squared terms, i.e. $(\varepsilon_{max})^2$,

then $\quad 1 + 2\varepsilon_{max} = 1 + \gamma$

and $\quad \epsilon_{max} = \dfrac{\gamma}{2} \quad$ as required.

In the next example we use the trigonometric identity you have already met, i.e. $\sin^2 \theta + \cos^2 \theta = 1$.

Example 6.2.10

In simple harmonic motion the variation of velocity of a point Q is given by $Q = -\omega r \sin \omega t$. With reference to Figure 3.1.19, show that $Q = \pm\omega\sqrt{r^2 - x^2}$. You need to know that $\theta = \omega t$ and that the linear velocity v is equal to the product of the angular velocity ω and its radius from the centre, r. So $v = \omega r$.

Then $\quad Q = -\omega r \sin \omega t$

so $\quad Q = -\omega r \sin \theta \quad$ and from $\quad \sin^2 \theta = 1 - \cos^2 \theta$

then $\quad Q^2 = \omega^2 r^2 (1 - \cos^2 \theta)$

cont.

$$\text{or} \quad Q = \pm \omega r \sqrt{1 - \cos^2 \theta}$$

Now $\cos \theta = \dfrac{x}{r}$ (from Figure 3.1.19a)

$$\text{so} \quad Q = \pm \omega r \sqrt{1 - \left(\frac{x}{r}\right)^2} \quad \text{and placing } r \text{ inside } \sqrt{}$$

gives $Q = \pm \omega \sqrt{r^2 - x^2}$.

Note that in the previous example you should also be able to obtain the original expression for the velocity of Q, from the information given in Figure 3.1.19. The component of P's velocity which is parallel and equal to OQ is given in Figure 3.1.19c as, $\theta = -\omega r \sin \theta$, make sure you can derive this from the geometry of the situation. The minus sign is present, because velocity is a vector quantity (speed in a given direction) and acts in opposition to the velocity deemed to be positive.

In the following example we look at a method for graphically adding two sinusoidal waveforms and analysing the resultant waveform. This is similar to the technique used to solve question 3.3.5 (refer to waveforms in Chapter 3).

Example 6.2.11

Sketch the graphs of

$$y_1 = 10 \sin\left(100\pi t + \frac{\pi}{4}\right)$$

and $y_2 = 5 \sin\left(200\pi t - \frac{2\pi}{3}\right)$

on the same axis and hence sketch the graph of:

$$y_3 = 10 \sin\left(100\pi t + \frac{\pi}{4}\right) + 5 \sin\left(200\pi t - \frac{2\pi}{3}\right).$$

In order to sketch the resultant graph (y_3), the following procedure may help. Choose points:

(i) Where the original two graphs cross.
(ii) Where there is a maximum and minimum value for each waveform.
(iii) Where each waveform crosses the x-axis.

Now the format for each of the waves is slightly different because the velocity is given in cycle/sec not radians/sec. So general form is:

$y = a \sin(2\pi f t + \phi)$ where $2\pi f = $ frequency

in cycles per second. Taking $y_1 = 10 \sin\left(100\pi t + \frac{\pi}{4}\right)$, then the frequency $f = 50$ Hz (cycles per second). Therefore the time period $t = \frac{1}{50}$ s $= 20$ ms (that is the time to complete one cycle). We already know that the constant $a = $ the amplitude, which in the case of y, $a = 10$. So we have amplitude $= 10$,

cont.

Figure 6.2.4 *Graphical solution*

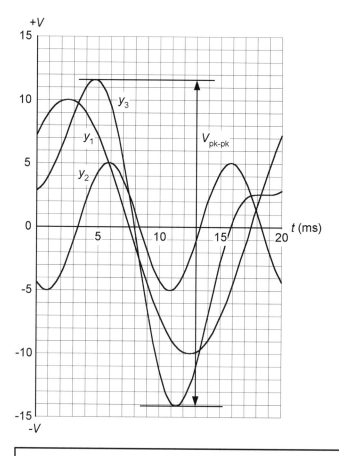

frequency $(f) = 50\,Hz$ and the periodic time $t = 20\,ms$ (milli-seconds).

In a similar manner for y_2 we have amplitude $= 5$, frequency $= 100\,Hz$, periodic time $= \frac{1}{100} = 10\,ms$. Now using these values we are able to set-up the axis of our graph, where y is amplitude (which may be voltage, current or power) and the x-axis is time (in this case).

The appropriate waveforms are shown in Figure 6.2.4. The waveform for y_3 is sketched by graphically adding y_1 and y_2 as suggested earlier. If the y-axis is graduated in volts, you can see that the peak-to-peak voltage (v_{pp}) for y_3 approximately $= 26\,V$.

You will find this technique very useful when looking at a.c. circuits and determining complex waveforms, which may be other than sinusoidal.

The final example in this section is rather contrived, and is concerned with determining the length of the bracing members used for the jib of a crane. It does however, give us the opportunity to use the sine rule.

Example 6.2.12

Figure 6.2.5 shows the jib of a crane, consisting of three members plus the pulley wire supporting the load. Find the

cont.

Figure 6.2.5 *Crane jib*

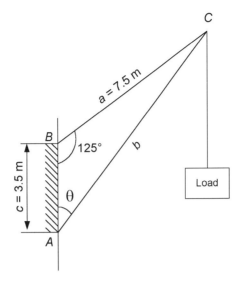

length of the strut *AC* and determine its angle of inclination *from* the vertical.

Now we have two sides and an included angle so we need to use the cosine rule, then

$b^2 = a^2 + c^2 - 2ac \cos B$ which gives

$b^2 = (7.5)^2 + (3.5)^2 - (2)(7.5)(3.5) \cos 125°$

$b^2 = 56.25 + 12.25 + 30.113$

$b = 9.93 \, \text{m}.$

The angle of inclination θ may now be found using the sine rule.

$$\frac{7.5}{\sin A} = \frac{9.93}{\sin 125°} \quad \text{or} \quad \frac{(7.5)(\sin 125°)}{9.93} = \sin A.$$

$$\sin A = 0.6187 \qquad \text{and so} \quad A = \theta = 38.2°.$$

There are numerous applications of fundamental trigonometry to engineering situations, space has permitted us to look at just a few. You will be able to gain further practice by attempting the problems which will be found at the end of this section on trigonometry.

Trigonometric identities

Formula

(1) General identities

(i) $\operatorname{cosec} \theta = \dfrac{1}{\sin \theta}; \quad \sec \theta = \dfrac{1}{\cos \theta}; \quad \cot \theta = \dfrac{1}{\tan \theta}.$

(ii) $\tan \theta = \dfrac{\sin \theta}{\cos \theta}; \quad \sin^2 \theta + \cos^2 \theta = 1.$

(iii) $\tan^2 \theta + 1 = \sec^2 \theta; \quad \cot^2 \theta + 1 = \operatorname{cosec}^2 \theta.$

(iv) $\sin(A \pm B) = \sin A \cos B \pm \cos A \sin B.$

(v) $\cos(A \pm B) = \cos A \cos B \mp \sin A \sin B$.

(vi) $\tan(A \pm B) = \dfrac{\tan A \pm \tan B}{1 \mp \tan A \tan B}$.

(2) Products to sums

(i) $\sin A \cos B = \frac{1}{2}[\sin(A + B) + \sin(A - B)]$.

(ii) $\cos A \sin B = \frac{1}{2}[\sin(A + B) - \sin(A - B)]$.

(iii) $\cos A \cos B = \frac{1}{2}[\cos(A + B) + \cos(A - B)]$.

(iv) $\sin A \sin B = -\frac{1}{2}[\cos(A + B) - \cos(A - B)]$.

(3) Sums to products

(i) $\sin A + \sin B = 2 \sin \dfrac{A + B}{2} \cos \dfrac{A - B}{2}$.

(ii) $\sin A - \sin B = 2 \cos \dfrac{A + B}{2} \sin \dfrac{A - B}{2}$.

(iii) $\cos A + \cos B = 2 \cos \dfrac{A + B}{2} \cos \dfrac{A - B}{2}$.

(iv) $\cos A - \cos B = -2 \sin \dfrac{A + B}{2} \sin \dfrac{A - B}{2}$ where $A > B$.

(4) Doubles and squares

(i) $\sin 2A = 2 \sin A \cos A$.

(ii) $\cos 2A = \cos^2 A - \sin^2 A = 2 \cos^2 A - 1 = 1 - 2 \sin^2 A$.

(iii) $\tan 2A = \dfrac{2 \tan A}{1 - \tan^2 A}$

(iv) $\cos^2 A + \sin^2 A = 1$.

(v) $\sec^2 A = 1$.

(vi) $\operatorname{cosec}^2 A = 1 + \cot^2 A$.

(5) Hyperbolic doubles and squares

(i) $\sinh 2A = 2 \sinh A \cosh A$.

(ii) $\cosh 2A = \cosh^2 A + \sinh^2 A = 2 \cosh^2 A - 1 = 1 + 2 \sinh^2 A$.

(iii) $\tanh 2A = \dfrac{2 \tanh A}{1 + \tanh^2 A}$.

(iv) $\cosh^2 A - \sinh^2 A = 1$.

(v) $\operatorname{sech}^2 A = 1 - \tanh^2 A$.

(vi) $\operatorname{cosech}^2 A = \coth^2 A - 1$.

(6) Negatives

(i) $\sin(-A) = -\sin A$.

(ii) $\cos(-A) = \cos A$.

(iii) $\tan(-A) = -\tan A$.

(iv) $\sinh(-A) = -\sinh A$.

(v) $\cosh(-A) = \cosh A$.

(vi) $\tanh(-A) = -\tanh A$.

(7) Halves
When $t = \tan \dfrac{\theta}{2}$ then:

(i) $\sin \theta = \dfrac{2t}{1 + t^2}$.

(ii) $\cos \theta = \dfrac{1 - t^2}{1 + t^2}$.

(iii) $\tan \theta = \dfrac{2t}{1 - t^2}$.

When $t = \tanh \dfrac{\theta}{2}$ then:

(iv) $\sinh Q = \dfrac{2t}{1 - t^2}$.

(v) $\cosh Q = \dfrac{1 + t^2}{1 - t^2}$.

(vi) $\tanh Q = \dfrac{2t}{1 + t^2}$.

Methods

We now look at ways in which expressions can be manipulated and simplified using trigonometric identities. One or two simple identities were needed to solve the application examples given previously. Our study here, is concerned with the simplification and rearrangement of expressions in order to obtain solutions using the calculus. Their engineering applications will form part of our later study.

Example 6.2.13

Solve the following trigonometric equations

(i) $4 \sin^2 \theta + 5 \cos \theta = 5$.

(ii) $3 \tan^2 \theta + 5 = 7 \sec \theta$.

(i) The most difficult problem when manipulating identities, is to know where to start! In this equation, we have two unknowns (sine and cosine) so the most logical approach is to try and get the equation in terms of one unknown, this leads us to the use of an appropriate identity. We can in this case use our old friend,

$\sin^2 \theta + \cos^2 \theta = 1$ from which $\sin^2 \theta = 1 - \cos^2 \theta$

which when substituted into equation (i) gives

$$4(1 - \cos^2 \theta) + 5 \cos \theta = 5$$

or $4 - 4 \cos^2 \theta + 5 \cos \theta = 5$.

This is a quadratic expression, which may be solved in a number of ways but it helps if you can factorise!

cont.

Then $(-4\cos\theta + 1)(\cos\theta - 1) = 0$

$\Rightarrow$ $-4\cos\theta = -1$ or $\cos\theta = 1$

$\Rightarrow$ $\cos\theta = \frac{1}{4}$ or $\cos\theta = 1$

so $\theta = 75.5°$ or $0°$.

(ii) Proceeding in a similar manner to (i), we need a trigonometric identity which relates $\tan\theta$ and $\sec\theta$ (look at formula 1vii).

then $3\tan^2\theta + 5 = 7\sec\theta$

and using $\sec^2\theta = 1 + \tan^2\theta$ or $\tan^2\theta = \sec^2\theta - 1$

we get $3(\sec^2\theta - 1) + 5 = 7\sec\theta$

$\Rightarrow$ $3\sec^2\theta - 3 + 5 = 7\sec\theta$

$\Rightarrow$ $3\sec^2\theta - 7\sec\theta + 2 = 0$

(again a quadratic expression) factorising gives

$(3\sec\theta - 1)(\sec\theta - 2) = 0$

so $3\sec\theta = 1$ or $\sec\theta = 2$

$\left(\text{remembering that } \sec\theta = \dfrac{1}{\cos\theta}\right)$

$\sec\theta = \frac{1}{3}$ or $\sec\theta = 2$

so $\cos\theta = 3$ or $\cos\theta = \frac{1}{2}$

$\cos\theta = 3$ (not permissible) so there is only one solution $\cos\theta = 0.5$ so $\theta = 60°$.

Example (6.2.14), is intended to show one or two techniques that may be used to verify trigonometric identities involving compound angle and double angle formulae. Example (6.2.15) shows how trigonometric identities may be used for evaluating trigonometric ratios.

Example 6.2.14

Verify the following identities by showing that each side of the equation is equal in all respects:

(i) $(\sin\theta + \cos\theta)^2 \equiv 1 + \sin 2\theta$

(ii) $\dfrac{\sin 3\theta - \sin\theta}{\cos\theta - \cos 3\theta} \equiv \cot 2\theta$.

(i) Simply requires the RHS to be manipulated algebraically to equal the LHS. So multiplying out gives

$(\sin\theta + \cos\theta)^2 \equiv 1 + \sin 2\theta$

$\sin^2\theta + 2\sin\theta\cos\theta + \cos^2\theta \equiv$

$\sin^2\theta + \cos^2\theta + 2\sin\theta\cos\theta \equiv$ (and from $\sin^2\theta + \cos^2\theta = 1$)

cont.

then $1 + 2 \sin \theta \cos \theta \equiv$ (and from formula 4i)

$$1 + \sin 2\theta \equiv 1 + \sin 2\theta \quad \text{as required.}$$

(ii) Again considering LHS and using sums to products (formulae 3)

where $(3\text{ii}) = \sin A - \sin B = 2 \cos \dfrac{A+B}{2} \sin \dfrac{A-B}{2}$

$$(3\text{iv}) = \cos A - \cos B = -2 \sin \dfrac{A+B}{2} \sin \dfrac{A-B}{2}$$

where $A > B$ then

$$\sin 3\theta - \sin \theta = 2 \cos \left(\frac{3+1}{2} \right) \theta \sin \left(\frac{3-1}{2} \right) \theta = 2 \cos 2\theta \sin \theta$$

and

$$\cos \theta - \cos 3\theta = -2 \sin \left(\frac{1-3}{2} \right) \theta \sin \left(\frac{1+3}{2} \right) \theta$$

$$\cos \theta - \cos 3\theta = -2 \sin(-\theta)(\sin 2\theta)$$

and from formula (6i) $\sin(-\theta) = -\sin \theta$

so $\cos \theta - \cos 3\theta = 2 \sin 2\theta \sin \theta$

therefore $\dfrac{\sin 3\theta - \sin \theta}{\cos \theta - \cos 3\theta} \equiv \dfrac{2 \cos 2\theta \sin \theta}{2 \sin 2\theta \sin \theta} \equiv \cot 2\theta.$

Example 6.2.15

If A is an acute angle and B is obtuse, where $\sin A = \frac{3}{5}$ and $\cos B = -\frac{5}{13}$, find the values of:

   (i) $\sin(A + B)$;

   (ii) $\sin 2A$;

   (iii) $\tan(A + B)$.

(i) $\sin(A + B) = \sin A \cos B + \cos A \sin B.$ \hfill (1)

In order to use this identity we need to find the ratios for $\sin B$ and $\cos A$. So again we need to choose an identity that allows us to find $\sin \theta$ or $\cos \theta$, in terms of each other.
   We know that $\sin^2 B + \cos^2 B = 1$

so $\sin^2 B = 1 - \cos^2 B$

$$= 1 - \left(-\tfrac{5}{13} \right)^2$$

$$\sin B = 1 - \tfrac{25}{169} = \tfrac{144}{169}$$

$$\sin B = \tfrac{12}{13}$$

(since B is obtuse $90° < B < 180°$ and sign ratio is positive in second quadrant. Then only positive values of ratio need be considered). Similarly:

cont.

$$\sin^2 A + \cos^2 A = 1$$

$$\text{so}\quad \cos^2 A = 1 - \sin^2 A$$

$$= 1 - \tfrac{9}{25}$$

$$\cos A = \tfrac{4}{5}$$

(since angle A is $< 90°$, i.e. acute only the positive value is considered).

Now using equation (1) above

$$\sin(A + B) = \sin A \cos B + \cos A \sin B$$

$$= (\tfrac{3}{5})(-\tfrac{5}{13}) + (\tfrac{4}{5})(\tfrac{12}{13})$$

$$= -\tfrac{15}{65} + \tfrac{48}{65}$$

$$\text{then}\quad \sin(A + B) = \tfrac{33}{65}.$$

Note the use of fractions to keep exact ratios!

(ii) Now to find the value for $\sin(2A)$, we again need an identity which relates 'double' and single angles formula (4i) provides just such a relationship.

$$\sin 2A = 2 \sin A \cos A$$

(for which we already have values)

$$\text{then}\quad \sin 2A = 2(\tfrac{3}{5})(\tfrac{4}{5}) = \tfrac{24}{25}.$$

(iii) For this part of the question we simply need to remember that $\dfrac{\sin A}{\cos A} = \tan A$ and use formula (1vi) in a similar way as before.

$$\text{then}\quad \tan A = \frac{\sin A}{\cos A} = \frac{\tfrac{3}{5}}{\tfrac{4}{5}} = (\tfrac{3}{5})(\tfrac{5}{4}) = \tfrac{3}{4}$$

$$\text{and}\quad \tan B = \frac{\sin B}{\cos B} = \frac{\tfrac{12}{13}}{-\tfrac{5}{13}} = (\tfrac{12}{13})(-\tfrac{13}{5}) = -\tfrac{12}{5}$$

and using formula (1vi),

$$\tan(A + B) = \frac{\tan A + \tan B}{1 - \tan A \tan B} = \frac{\tfrac{3}{4} - \tfrac{12}{5}}{1 - (\tfrac{3}{4})(-\tfrac{12}{5})}$$

multiplying *every* term in the expression by 20 (the lowest common multiple), we get

$$\tan(A + B) = \frac{15 - 48}{20 + 36} = -\frac{33}{56}.$$

The final example in this section requires the use of the half-angle identities, which we have not used up till now.

Example 6.2.16

Solve the equation $2\sin\theta + \cos\theta = 2$ for angles between 0 and 2π (radian). Then using formulae 7i and 7ii where $t = \tan\dfrac{\theta}{2}$

$$2\sin\theta + \cos\theta = 2$$

$$\frac{4t}{1+t^2} + \frac{1-t^2}{1+t^2} = 2 \quad \text{and on multiplication by } 1+t^2$$

$$4t + 1 - t^2 = 2 + 2t^2$$

or $\quad -3t^2 + 4t - 1 = 0 \quad$ (yet another quadratic!)

again factorising gives:

$$(-3t+1)(t-1) = 0$$

$\qquad \Rightarrow \quad -3t+1 = 0 \quad$ giving $\quad t = \frac{1}{3}$

or $\quad \Rightarrow \quad t-1 = 0 \quad$ giving $\quad t = 1 \quad$ where $\quad t = \dfrac{\tan\theta}{2}$

$\qquad$ so $\quad \tan\dfrac{\theta}{2} = \dfrac{1}{3} \quad$ or $\quad 1$

so we require values of $\dfrac{\theta}{2}$ between 0 and π

so $\quad \tan\dfrac{\theta}{2} = \dfrac{1}{3} \quad$ gives $\quad \dfrac{\theta}{2} = 18.43°$

or $\quad \tan\dfrac{\theta}{2} = 1 \quad$ gives $\quad \dfrac{\theta}{2} = 45°.$

So in our range 0–360°,

$$\theta = 36.87° \quad \text{or} \quad 90°.$$

Note the next values of $\dfrac{\theta}{2}$ corresponding to $\frac{1}{3}$ and 1 are 198.43° and 225° which when doubled are outside the required range.

This concludes our brief study of trigonometry, now try the problems!

Questions 6.2

(1) Find the length of arc and area of sector of a circle when:

(i) $\theta = 30°, r = 18$ cm; (ii) $\theta = 135°, r = 24$ cm.

(2) Convert the points $(-3, 4)$ and $(-2, -3)$ to polar co-ordinates.

(3) Convert the points $\left(4, \dfrac{\pi}{4}\right)$ and $(2, -30°)$ to Cartesian co-ordinates.

(4) Sketch the graphs of the following sinusoidal functions:

(i) $y_1 = 2\sin\left(2\theta + \dfrac{\pi}{2}\right)$

(ii) $y_2 = 10\cos(\theta - 60).$

cont.

What is the 'phase angle' between the functions y_1 and y_2?

(5) State the amplitude and phase angle for the following functions

  (i) $y_1 = 6 \sin\left(3\theta - \dfrac{\pi}{6}\right)$

  (ii) $y_2 = 4 \sin\left(20\theta - \dfrac{\pi}{3}\right)$.

(6) For each of the functions

  (i) $i = 3 \sin\left(200\pi t - \dfrac{\pi}{4}\right)$

  (ii) $v = 0.7 \sin\left(400\pi t + \dfrac{\pi}{3}\right)$.

  Calculate the amplitude, frequency, periodic time, phase angle and the time taken for each function to reach its first positive maximum value.

(7) On the same axes sketch the graphs

  $$y_1 = 3 \sin\left(x + \frac{\pi}{3}\right) \quad \text{and} \quad y_2 = 2 \sin\left(x - \frac{\pi}{3}\right),$$

  between 0 and 2π. On the same axes graphically sum these functions to produce the graph of

  $$y_3 = 3 \sin\left(x + \frac{\pi}{3}\right) + 2 \sin(x - 60).$$

(8) For an alternating current $i = 40 \sin(100\pi t - 0.32)$, find:

  (i) the amplitude, periodic time, frequency and phase angle with respect to $40 \sin 100\pi t$;
  (ii) i when $t = 0$ and i when $t = 4$;
  (iii) find t when the current is 25 amps.

(9) A complex waveform is described by the equation

  $$v = 100 \sin(100\pi t) + 60 \sin\left(200\pi t - \frac{\pi}{4}\right).$$

  Determine graphically the shape of the waveform and estimate the peak to peak voltage.

(10) The instantaneous current (i) and the instaneous voltage (v) in a pure resistance a.c. circuit is given by

  $$i = I_{max} \sin(\omega t) \quad \text{and} \quad v = V_{max} \sin(\omega t).$$

  Since power $P = IV$ show that an equation for instantaneous power is

  $$P = \frac{I_{max}^2 R}{2[1 - \cos(2\omega t)]}.$$

(11) For the situation shown in the force diagram (figure 6.2.6), find, using the resolution of forces and trigo-

cont.

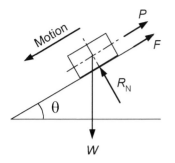

Figure 6.2.6   *Force diagram*

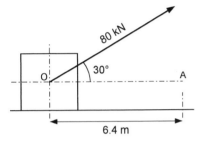

Figure 6.2.7   *Force diagram*

nometric ratios, equations that relate P, F, W and R_N and so show that

$$\frac{F}{R_N} = \tan \theta.$$

(12) Given that $\sin(\theta + \phi) = 0.6$ and $\cos(\theta + \phi) = 0.9$, find a value for '$\mu$' when $\mu = \tan \phi$.

(13) Calculate the component of force parallel to the horizontal axis (Figure 6.2.7) and the work done given that work = force × distance moved in direction of force.

(14) Verify the following identities:

(i) $\tan 3\theta = \dfrac{\sin \theta + \sin 3\theta + \sin 5\theta}{\cos \theta + \cos 3\theta + \cos 5\theta}$;

(ii) $\tan 2\theta = \dfrac{1}{1 - \tan \theta} - \dfrac{1}{1 + \tan \theta}$.

(15) Express the following as ratios of single angles:

(i) $\sin 5\theta \cos \theta + \cos 5\theta \sin \theta$;

(ii) $\cos 9t \cos 2t - \sin 9t \sin 2t$.

6.3 Calculus

Calculus is a branch of mathematics involved with the quantitative analysis of continually varying functions. It falls conveniently into two parts, *differential calculus*, which is used by engineers mainly to model instantaneous rates of change, while *integral calculus* is used primarily as a summation tool.

You will already have met some of the arithmetic of the calculus, when you *differentiated* (found the differential coefficient) of simple algebraic and trigonometric expressions. You may also be familiar with the *anti-derivative*, when you *integrated* similar expressions.

This section is not intended to offer a rigorous treatment of the derivation of the calculus but rather, to provide a vocabulary of methods needed for its engineering application. The methods are sub-divided into differentiation and integration, as discussed above. The engineering applications of the calculus will be dealt with separately, after our study of the methods. Algebraic and trigonometric techniques needed to manipulate the calculus will also be highlighted, as required.

Differential calculus

Formulae

(1) Standard derivatives

y	$\dfrac{\mathrm{d}y}{\mathrm{d}x}$
ax^n	nax^{n-1}
$\sin f(x)$	$f'(x) \cos f(x)$
$\cos f(x)$	$-f'(x) \sin f(x)$

(1) Standard derivatives (cont.)

$\tan f(x)$	$f'(x)\sec^2 f(x)$
$\operatorname{cosec} f(x)$	$-f'(x)\operatorname{cosec} f(x)\cot f(x)$
$\sec f(x)$	$f'(x)\sec f(x)\tan f(x)$
$\cot f(x)$	$-f'(x)\operatorname{cosec}^2 f(x)$
$\log_e f(x)$	$\dfrac{f'(x)}{f(x)}$
$e^{f(x)}$	$f'(x)\,e^{f(x)}$
a^x	$a^x \log_e a$
$\sin^{-1}\dfrac{x}{a}$	$\dfrac{1}{\sqrt{a^2 - x^2}}$
$\tan^{-1}\dfrac{x}{a}$	$\dfrac{a}{a^2 + x^2}$

(2) Rules of differentiation

(i) Function of a function rule $\dfrac{dy}{dx} = \dfrac{dy}{du}\cdot\dfrac{du}{dx}$.

Where u and v are functions of x:

(ii) Product rule $\dfrac{d}{dx}(uv) = u\dfrac{dv}{dx} + v\dfrac{du}{dx}$.

(iii) Quotient rule $\dfrac{d}{dx}\left[\dfrac{u}{v}\right] = \dfrac{v\dfrac{du}{dx} - u\dfrac{dv}{dx}}{v^2}$.

(3) Conditions for maxima and minima

$f(x)$ has a **maximum value** at $x = a$

if $f'(a) = 0$ **and** $f''(x)$ changes sign from +ve to −ve as x goes through the value a,

or if $f'(a) = 0$ **and** $f''(a)$ is negative.

$f(x)$ has a **minimum value** at $x = a$

if $f'(a) = 0$ **and** $f''(x)$ changes sign from −ve to +ve as x goes through the value a.

or if $f'(a) = 0$ **and** $f''(a)$ is positive.

Methods

We start with a review of the differing terminology used to describe the differential coefficient and then we review the standard derivatives, with which you should already be familiar. We will then look at the rules needed to differentiate more complex functions. These will include the use of the product, quotient and, function of a function rules. When applied to algebraic, trigonometric, exponential and hyperbolic functions.

Standard derivatives

The terminology used for determining the derivative of an expression, differs from textbook to textbook.

For example, we may say:

(a) find the derivative of ...
(b) find the differential coefficient for ...
(c) differentiate ...
(d) find the rate of change of ...
(e) find the tangent to the function ...
(f) find the gradient of the function at a point ...

This differing terminology is often confusing to beginners. It is further complicated by the fact that different symbols are used for the differentiation process, based on the convention chosen.

We can, for example, use Leibniz notation, where the differential coefficient for the function $y(x)$ is given as, $\frac{dy}{dx}$.

This is asking for the function 'y' (with independent variable x) to be differentiated once.

In this notation the *second derivative* is expressed as $\frac{d^2y}{dx^2}$, in this case we differentiate the function $y(x)$ twice, while $\frac{d^3y}{dx^3}$ is the differential coefficient of *degree three*.

Similarly, we may use *functional notation* where for a function $f(x)$ the first derivative is given as $f'(x)$, the second derivative $f''(x)$ and so on.

One other notation which is often used in mechanics is *dot* notation. For example $\dot{v}, \ddot{s}$, etc. which means the variable is differentiated once ($\dot{v}$) or twice ($\ddot{s}$) and so on.

The terminology in this chapter may vary according to the application. The standard derivative formulae are given in functional notation, $f(x)$, etc. but the rules for differentiation are given in Leibniz notation.

Remember also that the notation differs according to the variable! So $\frac{dy}{dx}, \frac{ds}{dt}, \frac{du}{dx}$, each require the first derivative of the functions y, s and u, respectively.

It is useful to aid understanding, to state in words the meaning of expressions like $\frac{ds}{dt}$. This is saying 'differentiate the function s with respect to the variable t.'

Let's look at a few examples of direct application of the standard derivatives, to remind you of the arithmetic process required.

Differentiate the following functions:

(i) $y = 6x^3$ then $\frac{dy}{dx} = (6)(3)x^{3-1} = 18x^2$

(ii) $y = e^{3x}$ $\frac{dy}{dx} = 3e^{3x}$

(iii) $y = \sin 4x$ $\frac{dy}{dx} = 4\cos 4x$

(iv) $y = 3\cos 3x$ $\frac{dy}{dx} = (3)(-3\sin 3x) = -9\sin 3x.$

Our first example will act as revision for differentiating using standard derivatives.

Example 6.3.1

Differentiate the following functions with respect to the variables given.

(i) $\dfrac{x^3}{3} - 6x^{-2} + \dfrac{12x^4}{32}$

(ii) $(\sqrt{x})^3 - (x^{-3})^2$

(iii) $(2x + 3)^2$

(iv) $\sin 4t - 6\cos 2t + e^{-3t}$.

The *golden rule* is to *simplify* whenever possible *before* differentiating.

So for (i) if $\quad y = \dfrac{x^3}{3} - 6x^{-2} + \dfrac{3x^4}{8}$

then $\quad \dfrac{dy}{dx} = \dfrac{3x^2}{3} + 12x^{-2} + \dfrac{12x^3}{8}$

$\dfrac{dy}{dx} = x^2 + 12x^{-2} + \dfrac{3x^3}{2}$.

(ii) Let the function

$$f(x) = (\sqrt{x})^3 - (x^{-3})^2$$

then simplifying $\quad f(x) = x^{3/2} - x^{-2/3}$ (laws of indices!)

so $\quad f'(x) = \frac{3}{2}x^{1/2} + \frac{2}{3}x^{-5/3}$.

(iii) $(2x + 3)^2$. We need to expand the bracket in order to apply a standard derivative

So $\quad (2x + 3)(2x + 3) = 4x^2 + 12x + 9$

and if $\quad y = 4x^2 + 12x + 9$

then $\quad \dfrac{dy}{dx} = 8x + 12$.

(iv) Here, we are required to use the standard derivatives for trigonometric and exponential functions. They are all differentiated with respect to the same variable t, then

if $\quad f(t) = \sin 4t - 6\cos 2t + e^{-3t}$

$\quad f'(t) = A\cos 4t + 12\sin 2t - 3e^{-3t}$.

Note also that the standard derivative for the Napierian logarithm ($\ln x$) or ($\log_e x$) is $\log_e f(x) = \dfrac{f'(x)}{f(x)}$ so for example if $y = \log_e f(t)$ and $f(t) = x^2 + 1$, then

$y = \log_e(x^2 + 1) \quad$ and $\quad \dfrac{dy}{dx} = \dfrac{2x}{x^2 + 1}$.

Function of a function rule

We will now look at examples of the use of the function of a function rule, which is really differentiation by substitution.

Example 6.3.2

Find $\dfrac{dy}{dx}$ if $y = (x^2 - x)^9$.

Now the function of a function rule

$$\frac{dy}{dx} = \frac{dy}{du} \times \frac{du}{dx}$$

is a rule that requires us to make a substitution, in the above function where $y = (x^2 - x)^9$ if we let the bracketed expression $= u$, then we would get

$$y = u^9 \quad (\text{where } u = x^2 - x).$$

We cannot, however, directly differentiate this expression because it involves a different variable to that required, i.e. we want to differentiate with respect to x, but we have y in terms of u.

So to complete the function of a function formula we need

$$\frac{dy}{du} \text{ and } \frac{du}{dx}$$

we have $\quad y = u^9 \quad$ and $\quad u = x^2 - x$

$$\text{So} \quad \frac{dy}{du} = 9u^8 \quad \text{and} \quad \frac{du}{dx} = 2x - 1$$

$$\text{Then} \quad \frac{dy}{dx} = \frac{dy}{du} \times \frac{du}{dx} = (9u^8)(2x - 1).$$

The differentiation is complete, all that is required is to put u in terms of x by using the original substitution.

$$\text{So} \quad \frac{dy}{dx} = 9(x^2 - x)^8(2x - 1).$$

To ensure that you are able to master the substitution method for differentiating functions. You should study the example given next, very carefully. Make sure that you are able to follow the standard procedure used for all the worked solutions.

Example 6.3.3

Differentiate the following functions with respect to the variables given using the function of a function (substitution) rule.

(i) $y = \sqrt{1 - 5x^3}$

(ii) $\dfrac{d}{d\theta}(\sin 7\theta)$

(iii) $\dfrac{d}{dt}\left[\cos\left(2t - \dfrac{3\pi}{2}\right)\right]$

(iv) $\dfrac{d}{dx}[\log_e(x^2 + 5)]$.

(i) Let $y = \sqrt{1 - 5x^3}$ i.e. $y = (1 - 5x^3)^{1/2}$. Then $y = u^{1/2}$ where $u = 1 - 5x^3$.

cont.

Therefore $\dfrac{dy}{du} = \tfrac{1}{2}u^{-1/2}$ and $\dfrac{du}{dx} = -15x^2$.

But $\dfrac{dy}{dx} = \dfrac{dy}{du} \times \dfrac{du}{dx}$,

therefore $\dfrac{dy}{dx} = \tfrac{1}{2}u^{-1/2} \times (-15x^2)$.

So $\dfrac{dy}{dx} = \tfrac{1}{2}(1 - 5x^3)^{-1/2}(-15x^2)$

$= -\tfrac{15}{2}x^2(1 - 5x^3)^{-1/2}$

$\dfrac{dy}{dx} = -\dfrac{15x^2}{2\sqrt{1 - 5x^3}}$.

Hence $\dfrac{d}{dx}(\sqrt{1 - 5x^3}) = -\dfrac{15x^2}{2\sqrt{1 - 5x^3}}$.

(ii) $\dfrac{d}{d\theta}(\sin 7\theta)$. Let $y = \sin 7\theta$.

Then $y = \sin u$ where $u = 7\theta$

$\dfrac{dy}{du} = \cos u$ and $\dfrac{du}{d\theta} = 7$.

But $\dfrac{dy}{d\theta} = \dfrac{dy}{du} \times \dfrac{du}{d\theta}$.

Therefore $\dfrac{dy}{d\theta} = \cos u \times 7 = 7 \cos u = 7 \cos 7\theta$

Therefore $\dfrac{d}{d\theta}(\sin 7\theta) = 7 \cos 7\theta$.

(iii) $\dfrac{d}{dt}\left(\cos\left(2t - \dfrac{3\pi}{2}\right)\right)$. Let $y = \cos\left(2t - \dfrac{3\pi}{2}\right)$.

Then $y = \cos u$ where $u = 2t - \dfrac{3\pi}{2}$.

Therefore $\dfrac{dy}{du} = -\sin u$ and $\dfrac{du}{dt} = 2$.

But $\dfrac{dy}{dt} = \dfrac{dy}{du} \times \dfrac{du}{dt}$,

therefore $\dfrac{dy}{dt} = (-\sin u) \times 2 = -2 \sin u$

$= -2 \sin\left(2t - \dfrac{3\pi}{2}\right)$

hence $\dfrac{d}{dt}\left(\cos\left(2t - \dfrac{3\pi}{2}\right)\right) = -2 \sin\left(2t - \dfrac{3\pi}{2}\right)$.

(iv) $\dfrac{d}{dx}(\log_e(x^2 + 5))$. Let $y = \log_e(x^2 + 5)$.

Then $y = \log_e u$ where $u = x^2 + 5$.

Therefore $\dfrac{dy}{du} = \dfrac{1}{u}$ and $\dfrac{du}{dx} = 2x$

But $\dfrac{dy}{dx} = \dfrac{dy}{du} \times \dfrac{du}{dx}$.

cont.

Therefore $\dfrac{dy}{dx} = \dfrac{1}{u} \times 2x = \dfrac{1}{x^2 + 5} \times 2x.$

Hence $\dfrac{d}{dx}(\log_e(x^2 + 5)) = \dfrac{2x}{x^2 + 5} = \dfrac{f'(x)}{f(x)}!$

With practice the analysis of a function becomes a mental process and with it the differentiation of a function of a function. For example:

If $y = (3x^3 - 3x)^4$

$$\frac{dy}{dx} = 4(3x^3 - 3x)^3(9x^2 - 3).$$

If $y = \cos^3 5x = (\cos 5x)^3$

$$\frac{dy}{dx} = (3\cos^2 5x)(-\sin 5x)5$$

$$= -15\cos^2 5x \sin 5x.$$

If $y = \log_e(3x^2 - 6x)^3$

$$\frac{dy}{dx} = \frac{1}{(3x^2 - 6x)^3} \cdot 3(3x^2 - 6x)^2(6x - 6)$$

$$= \frac{3(6x - 6)}{(3x^2 - 6x)}.$$

If this mental process is unclear, you should stick to the full process already covered for differentiation using the function of a function (substitution) rule.

Product and quotient rules

These rules (formula 2ii and 2iii) enable us to differentiate more complex functions which are made up, as their name suggests from products and quotients, for example the function,

$$y = x^3 \sin 2x \quad \text{is the product of } x^3 \text{ and } \sin 2x$$

while $y = \dfrac{e^{2x}}{x + 3}$ is e^{2x} divided by $x + 3$.

The rules for products and quotients require the use of similar algebraic manipulation to that required when using the function of a function rule. So if $y = u \times v$ where u and v are functions of any other variable (say x), then we must use the product rule (2ii) to find the differential of the function. Similarly, if $y = u/v$ where u and v are functions of any variable (say x) then we must use the quotient rule (2iii) to find the differential of the function.

The following examples illustrate the similarities between the use of the product and quotient rules. The quotient rule is considered slightly more difficult to use than the product rule, because it is dependent on order, as you will see. Quotients can often be represented as products by manipulating the function using the laws of indices. The quotient may then be treated as a product and differentiated using the product rule.

Example 6.3.4

(a) Find $\dfrac{d}{dx}(x^3 \sin 2x)$.

(b) Differentiate $(x^2 + 1) \log_e x$.

Both of the above functions are products so rule 2ii is appropriate.

(a) Let $y = x^3 \sin 2x$

then $y = u \times v$ where $u = x^3$ and $v = \sin 2x$.

Therefore $\dfrac{du}{dx} = 3x^2$ and $\dfrac{dv}{dx} = 2 \cos 2x$.

But from the rule:

$$\frac{dy}{dx} = v\frac{du}{dx} + u\frac{dv}{dx}.$$

So substituting our values from above:

$$\frac{dy}{dx} = (\sin 2x)(3x^2) + (x^3)(2 \cos 2x)$$

then $\dfrac{d}{dx}(x^3 \sin 2x) = x^2(3 \sin 2x + 2x \cos 2x)$.

(b) Again letting $y = (x^2 + 1) \log_e x$, then we are required to find $\dfrac{dy}{dx}$.

$$y = u \times v \quad \text{where} \quad u = x^2 + 1 \text{ and } v = \log_e x.$$

Therefore $\dfrac{du}{dx} = 2x$ and $\dfrac{dv}{dx} = \dfrac{1}{x}$

but $\dfrac{dy}{dx} = v\dfrac{du}{dx} + u\dfrac{dv}{dx}$ and on substitution:

$$\frac{dy}{dx} = (\log_e x)(2x) + (x^2 + 1)\left(\frac{1}{x}\right)$$

$$= 2x(\log_e x) + x + \frac{1}{x}.$$

Example 6.3.5

(a) Find $\dfrac{dy}{dx}$ if $y = \dfrac{e^{2x}}{x + 3}$.

(b) Find $\dfrac{d}{d\theta}(\tan \theta)$.

These are both quotients, although for *b* this is not obvious!

cont.

(a) $y = \dfrac{e^{2x}}{x+3}$ let $y = \dfrac{u}{v}$ where $u = e^{2x}$ and $v = x+3$.

Therefore $\dfrac{du}{dx} = 2e^{2x}$ and $\dfrac{dv}{dx} = 1$.

But $\dfrac{dy}{dx} = \dfrac{v\dfrac{du}{dx} - u\dfrac{dv}{dx}}{v^2}$

When selecting u and v this rule requires u to be the numerator and v to be the denominator of the function.

Then substitution gives

$$\frac{dy}{dx} = \frac{(x+3)(2e^{2x}) - (e^{2x} \times 1)}{(x+3)^2}$$

$$= \frac{e^{2x}(2x+6-1)}{(x+3)^2}$$

$$\frac{dy}{dx} = \frac{(2x+5)(e^{2x})}{(x+3)^2}.$$

(b) Let $y = \tan\theta$, we require $\dfrac{dy}{d\theta}$. Now to turn $\tan\theta$ into a quotient we use the trigonometric identity

$$\tan\theta = \frac{\sin\theta}{\cos\theta} \quad \text{so} \quad y = \frac{\sin\theta}{\cos\theta}$$

where $u = \sin\theta$ $\dfrac{du}{d\theta} = \cos\theta$

$v = \cos\theta$ $\dfrac{dv}{d\theta} = -\sin\theta$.

But $\dfrac{dy}{d\theta} = \dfrac{v\dfrac{du}{d\theta} - u\dfrac{dv}{d\theta}}{v^2}$

$$= \frac{(\cos\theta)(\cos\theta) - (\sin\theta)(-\sin\theta)}{\cos^2\theta}$$

$$= \frac{\cos^2\theta + \sin^2\theta}{\cos^2\theta} \quad \text{(our old friend!)}$$

$$\frac{dy}{d\theta} = \frac{1}{\cos^2\theta} = \sec^2\theta.$$

So $\dfrac{d(\tan\theta)}{d\theta} = \sec^2\theta.$

We will now consider two general methods for the differential calculus which enable us to find *rates of change* and *turning points*. These methods may then be adopted for specific engineering use.

The derivative and rate of change

You saw in the beginning of this section that we can define the differential coefficient dy/dx as the rate at which y changes as x changes. Suppose we have a function where both y and x depend on time (t), then we may use the *chain* (function of a function) *rule* to write this as:

$$\frac{dy}{dx} = \frac{dy}{dt} \cdot \frac{dt}{dx} \quad \text{(dot notation here means multiplication)}$$

then
$$\frac{dy}{dt} = \frac{dy}{dx} \cdot \frac{dx}{dt}$$

or
$$\frac{dx}{dt} = \frac{dy}{dt} \cdot \frac{dx}{dy}.$$

So from the above the rate of change of y with respect to t, or the rate of change of y with respect to x may be found, as required.

This method of calculating the rate of change of some variable or function with respect to time has wide practical use, since many situations that occur in real life depend upon time as their independent variable. The exponential growth of bacteria, the decay of charge in a capacitor, or the change in velocity of a car, are all dependent on time.

Example 6.3.6

Calculate the rate at which the area of a circle increases with respect to time (t), when the radius of the circle increases at 0.2 cm/s. Now from your previous work the area of a circle is

$$A = \frac{\pi d^2}{4} = \pi r^2.$$

Therefore, the rate of change of area, A, with respect to the radius, r is given by:

$$\frac{dA}{dr} \quad \text{so when} \quad A = \pi r^2, \frac{dA}{dr} = 2\pi r$$

and remembering that $\frac{dr}{dt} = 0.2$ cm/s, then using function of a function rule

$$\frac{dA}{dr} = \frac{dA}{dt} \cdot \frac{dt}{dr}$$

so $\frac{dA}{dt}$ (the change in area with respect to time) is given by:

$$\frac{dA}{dt} = \frac{dA}{dr} \cdot \frac{dr}{dt}$$

so $\frac{dA}{dt} = (2\pi r)(0.2) = 0.4\pi r \quad \text{cm}^2/\text{s}.$

Then for example when $r = 2$ cm, the rate of change of area $\frac{dA}{dt} = 0.8\pi \, \text{cm}^2/\text{s}.$

Example 6.3.7

Suppose an empty spherical vessel is filled with water. As the water level rises the radius of the water in the vessel and the volume of water will change. Now if the radius of water in the sphere increases at 0.5 cm/s, find the rate of change of volume, when the radius is 5 cm.

We know that

$$V = \tfrac{4}{3}\pi r^3 \quad \text{and} \quad \frac{dr}{dt} = 0.5 \,\text{cm/s},$$

so $\dfrac{dV}{dr} = 4\pi r^2$

and using function of a function rule:

$$\frac{dV}{dt} = \frac{dV}{dr} \cdot \frac{dr}{dt}$$

then $\dfrac{dV}{dt} = (4\pi r^2)(0.5) = 2\pi r^2$

and when $r = 5$, $\dfrac{dV}{dt} = 50\pi = 157 \,\text{cm}^3/\text{s}.$

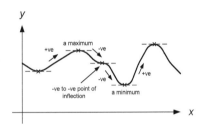

Figure 6.3.1 *Curve showing turning points*

Turning points

Consider the curve of the function $y = f(x)$, shown in Figure 6.3.1. Each cross represents a turning point for the function where, the gradient of the tangent at these points is zero. So a *turning point* is defined as *the point on any continuous function where the gradient of the tangent of the point is zero*. So to find turning points (TPs) values (independent variable x in this case), we solve the gradient equation

$$\frac{dy}{dx} = 0.$$

To find the corresponding y-ordinate of the TP, we substitute the x-values back into the original function, i.e. $y = f(x)$, these values are called *stationary values* (SVs). The following example illustrates the process.

Example 6.3.8

Find the TPs for the function $y = 2x^3 - 3x^2 - 12x + 4$. The requirement for a TP is that the rate of change of the function equals zero, that is:

$$\frac{dy}{dx} = 0 \quad \text{or} \quad \frac{dy}{dx} = 6x^2 - 6x - 12 = 0. \tag{a}$$

Solving this equation will produce the x value for which there is a turning point.

Dividing equation (a) by 6 gives $\dfrac{dy}{dx} = x^2 - x - 2 = 0$ which has factors $(x + 1)(x - 2) = 0$.

So TPs exist at $x = 2$ or -1.

cont.

Now substituting these values back into the original function y we get:

$$y = 2(2)^3 - 3(2)^2 - 12(2) + 4 = -16$$

or $y = 2(-1)^3 - 3(-1)^2 - 12(-1) + 4 = 11$.

So TPs and SVs are at the points $(2, -16)$ and $(-1, 11)$.

To decide on the nature of the TPs, i.e. whether they are a *maximum*, or *minimum* or *point of inflection*, two methods are often adopted (see formula 3).

Method 1

Determine the rate of change of the gradient function, in other word find the value for the second derivative of the function $\dfrac{d^2 y}{dx^2}$ at the turning point. If this value is *positive* the turning point is a *maximum*, if this value is *negative* then the turning point is a *minimum*.

Method 2

Consider the *gradient of the curve* close to the TPs, i.e. near to either side. Then for a *minimum* the gradient goes from *negative to positive* (Figure 6.3.1) and for a *maximum* the gradient goes from *positive to negative*. Both methods are illustrated in the following two examples.

Example 6.3.9

Using Method 1 find the maximum and minimum values of y given that

$$y = x^3 + 3x^2 - 9x + 6$$

then $\dfrac{dy}{dx} = 3x^2 + 6x - 9$ and $\dfrac{d^2 y}{dx^2} = 6x + 6$.

At a TP $\dfrac{dy}{dx} = 0$, therefore $3x^2 + 6x - 9 = 0$ or $x^2 + 2x - 3 = 0$ and $(x - 1)(x + 3) = 0$. Therefore, TPs at $x = 1$ and $x = -3$. Now we must test for maximum or minimum. So at TP where $x = 1$

$$\dfrac{d^2 y}{dx^2} = 6x + 6 = 6(1) + 6 = +12.$$

This is *positive* and so the TP at $x = 1$ is a *minimum and* the stationary value (SV) at this point is found by substituting $x = 1$ into the *original* equation for y,

i.e. $y_{min} = (1)^3 + 3(1)^2 - 9(1) + 6 = 1$.

So a minimum at point $(1, 1)$.

cont.

Similarly at the point where $x = -3$:

$$\frac{d^2y}{dx^2} = 6(-3) + 6 = -12$$

this is *negative* and so at $x = -3$ there is a *maximum* TP.

$$y_{max} = (-3)^3 + 3(-3)^2 - 9(-3) + 6 = +13.$$

So a maximum at point $(-3, 13)$.

Example 6.3.10

For the function $y = x^3 - 3x$ find the TPs and determine their nature.

For TPs we require

$$\frac{dy}{dx} = 3x^2 - 3 = 0$$

and so $3x^2 - 3 = 0$ or $x^2 - 1 = 0$

so $x = \pm 1$.

SV values corresponding to TP at $+1$ and -1 are found by substitution into original equation:

$$y = x^3 - 3x \quad \text{then} \quad y = (+1)^3 - 3(+1) = -2$$

$$\text{and also} \quad y = (-1)^3 - 3(-1) = +2.$$

So TPs at $(1, -2)$ and $(-1, +2)$.

Now for point $(1, -2)$ we consider values above and below the value of $x = 1$ so choose $x = 0$ and $x = 2$, then using the *gradient* equation

$$\frac{dx}{dy} = 3x^2 - 3,$$

when $x = 0$, $\frac{dy}{dx}$ is negative;

when $x = 2$, $\frac{dy}{dx}$ is positive.

Gradient function goes from $-$ve to $+$ve. Then point $(1, -2)$ is a minimum.

Similarly at point $(-1, 2)$ at values of $x = -2$ and $x = 0$,

then when $x = -2$, $\frac{dy}{dx} = 3x^2 - 3$ is $+$ve,

and when $x = 0$, $\frac{dy}{dx} = 3x^2 - 3$ is $-$ve.

Gradient function goes from $+$ve to $-$ve. Then point $(-1, -2)$ is a maximum.

This concludes our short study on methods for using the differential calculus, its application to engineering will be found later after our study of the integral calculus.

Integral calculus

Formulae

(1) Standard integrals (constant of integration omitted).

y	$\int y \, dx$
$x^a (a \neq -1)$	$\dfrac{x^{a+1}}{a+1}$
$\dfrac{1}{x}$	$\ln\lvert x \rvert$
$\ln x$	$x \ln x - x$
$a^x (a > 0)$	$\dfrac{a^x}{\ln a}$
$(ax+b)^n ; n \neq -1$	$\dfrac{1}{a} \dfrac{(ax+b)^{n+1}}{n+1}$
$\dfrac{1}{ax+b}$	$\dfrac{1}{a} \ln(ax+b)$
$\tan x$	$\sec^2 x$
$\sin(ax+b)$	$-\dfrac{1}{a} \cos(ax+b)$
$\cos(ax+b)$	$\dfrac{1}{a} \sin(ax+b)$
$\tan(ax+b)$	$\dfrac{1}{a} \ln[\sec(ax+b)]$
e^{ax+b}	$\dfrac{1}{a} e^{ax+b}$
$\sec x$	$\begin{cases} \ln\lvert \tan(\frac{1}{2}x + \frac{1}{4}\pi) \rvert \\ \ln\lvert \sec x + \tan x \rvert \end{cases}$
$\operatorname{cosec} x$	$\begin{cases} \ln\lvert \tan \frac{1}{2}x \rvert \\ \ln\lvert \operatorname{cosec} x - \cot x \rvert \end{cases}$
$\cot x$	$\ln\lvert \sin x \rvert$
$\sinh x$	$\cosh x$
$\cosh x$	$\sinh x$
$\tanh x$	$\ln \cosh x$
$\sin^2 x$	$\frac{1}{2}x - \frac{1}{4}\sin 2x$
$\cos^2 x$	$\frac{1}{2}x + \frac{1}{4}\sin 2x$
$\tan^2 x$	$\tan x - x$

(2) Integration by substitution (or change of variable)

$$\int f(x)dx = \int f(g(t))g'(t)dt \quad \text{when} \quad x = g(t).$$

(3) Integration by parts

$$\int u \frac{dv}{dx} dx = uv - \int v \frac{du}{dx} dx \quad \textbf{or} \quad \int uv' dx = uv - \int vu' dx.$$

(4) Numerical integration

Simpson's rule: If a plane area is divided into an even number of strips of equal width, then:

$$\text{Area} = \frac{\text{Common width}}{3} \left[\begin{array}{l} \text{sum of first and last ordinates} \\ + 4 \times (\text{sum of even ordinates}) \\ + 2 \times (\text{sum of remaining ordinates}) \end{array} \right]$$

(5) Trapezoidal rule

If a plane area is divided into strips of equal width, then the area = common width × (half the sum of the first and the last ordinates + the sum of the other ordinates).

Methods

In this section we will briefly review the standard integrals and their use in the determination of areas by direct integration. Next we will look at the techniques for integrating functions by substitution and, by parts. Finally we will use the Trapezoidal rule and Simpson's rule to illustrate the very powerful technique of numerical integration.

Standard integrals

Finding the prime function using standard integrals, having been given its derivative, is often referred to as *anti-differentiation*. A glance at the rule for finding the prime function (F) for a simple polynomial expression shows that we are finding the antiderivative. The following examples show the straightforward use of some standard integrals.

Example 6.3.11

Find the *integrals* of the following prime functions, using standard integrals.

(i) $3x + 2$;   (ii) $(x + 2)^6$;   (iii) $3e^{2x}$;

(iv) $3\cos 2x$; (v) $3\cosh 2x$; (vi) $(x + 2)^{-1} x > -2$;

(vii) $3e^{2x+1}$; (viii) $3\cos(2x + 1)$; (ix) $3\cosh(2x + 1)$;

(x) $(3x + 2)^{-1} x > -\frac{2}{3}$.

Using the table the *integrals* for the above functions are:

(i) $\frac{3}{2}x^2 + 2x + c$

The constant 'c', results from the process of anti-differentiation where it may or may not be present it is known as the *constant of integration*, with which you should already be familiar.

cont.

(ii) $\dfrac{(x+2)^7}{7};$ (iii) $\frac{3}{2}e^{2x};$ (iv) $\frac{3}{2}\sin 2x;$

(v) $\frac{3}{2}\sinh 2x;$ (vi) $\log(x+2);$ (vii) $\frac{3}{2}e^{2x+1};$

(viii) $\frac{3}{2}\sin(2x+1);$ (ix) $\frac{3}{2}\sinh(2x+1);$

(x) $\log(x+2).$

Make sure you can integrate elementary functions using standard integrals!

Example 6.3.12

Find: (i) $\displaystyle\int(x^2+2x-3)dx,$

(ii) $\displaystyle\int(\sin 7\theta + 2\cos 5\theta)d\theta;$

(iii) $\displaystyle\int\left(e^{6t}-\dfrac{1}{e^{3t}}\right)dt.$

(i) $\displaystyle\int(x^2+2x-3)dx = \dfrac{x^3}{3}+\dfrac{2x^2}{2}-3x+c$

$$=\dfrac{x^3}{3}+x^2-3x+c$$

(applying the rule to each term sequentially).

(ii) $\displaystyle\int(\sin 7\theta + 2\cos 5\theta)d\theta = -\dfrac{1}{7}\cos 7\theta + \dfrac{2}{5}\sin 5\theta + c.$

(iii) $\displaystyle\int\left(e^{6t}-\dfrac{1}{e^{3t}}\right)dt = \int(e^{6t}-e^{-3t})dt$ (indices!)

$$=\dfrac{1}{6}e^{6t}-\dfrac{1}{(-3)}e^{-3t}+c$$

$$=\dfrac{1}{6}e^{6t}+\dfrac{1}{3}e^{-3t}+c.$$

Example 6.3.13

Evaluate $\displaystyle\int_2^3(1+\cos 2\phi)d\phi.$

The above integral is known as a *definite integral* because it has definite *limits*. It is telling us to find the integral of the function and then find a numerical value, when the limiting values are substituted for the variable (ϕ in this case). So we first integrate in the normal manner.

$$\int_2^3(1+\cos 2\phi)d\phi = \left[\phi + \tfrac{1}{2}\sin 2\phi\right]_2^3$$

cont.

We put the integral in square brackets, with the upper and lower limit of ϕ, as indicated. Now the final numerical value for the function will be the difference between the upper and lower limits, when the values of ϕ are substituted. In other words:

$$= [3 + \tfrac{1}{2}\sin(2)(3)] - [2 + \tfrac{1}{2}\sin(2)(2)]$$

$$= 3 + \tfrac{1}{2}\sin 6 - 2 - \tfrac{1}{2}\sin 4.$$

Now remembering that when limits are substituted into trigonometrical functions, we deal in *radian* values. So evaluating, using a calculator, gives:

$$\int_2^3 (1 + \cos 2\phi)\,d\phi = 1.24.$$

The above examples were all straightforward, all we had to do was use standard integrals either referring to the formulae given in the table or rely on our memory. We will shortly be looking at methods to help us integrate some of the more complex expressions. Before we do, let us look at an example where we need to simplify the integral in some manner, prior to using standard integrals.

Example 6.3.14

Find: (i) $\displaystyle\int \frac{e^x + e^{6x} + e^{3x}}{e^{4x}}\,dx$;

   (ii) $\displaystyle\int \tan^2 x\,dx$.

(i) This is simple when you realise that all we need to do is divide each term by e^{4x} and then integrate each term sequentially. It does, once again, require us to use indices.

$$\int \frac{e^x + e^{6x} + e^{3x}}{e^{4x}}\,dx = \int e^{-3x} + e^{2x} + e^{-x}\,dx$$

$$= -\tfrac{1}{3}e^{-3x} + \tfrac{1}{2}e^{2x} - e^{-x} + c.$$

(ii) $\displaystyle\int \tan^2 x\,dx$.

This is fairly straightforward if we use a trigonometric identity, the problem is which one? A general rule is that whenever you see $\tan^2 x$, think of this as $\sec^2 x - 1$ (look at your standard integrals to see why).

   The above may now be written:

$$\int \tan^2 x\,dx = \int (\sec^2 x - 1)\,dx = \tan x - x + c.$$

This is why it is so useful to know your trigonometric identities!

The following example illustrates the technique for finding the area of a given function, when we know the rule of the function. In practice, this seldom occurs, and we would need to use one of the numerical methods, you will meet later.

Example 6.3.15

Determine the shaded area shown below (see figure), between the function and the x-axis.

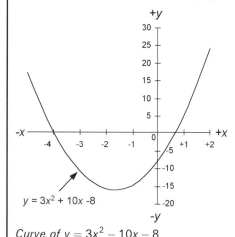

Curve of $y = 3x^2 - 10x - 8$

Now in order to find this area by integration, we first need to find the 'limits of integration'. The function crosses the x-axis when $y = 0$, so the required limits are at $y = 0$.

Then $3x^2 + 10x - 8 = 0$

and factorising we get:

$(3x - 2)(x + 4) = 0$,   so $x = \frac{2}{3}$ and $x = -4$.

So we have our limits. To obtain the required area, we need to integrate between the limits -4 to 0 and 0 to $\frac{2}{3}$. In other words the required area is given by:

$$\int_0^{-4} 3x^2 + 10x - 8 \, dx + \int_0^{2/3} 3x^2 + 10x - 8 \, dx$$

$$= \left[x^3 + 5x^2 - 8x + c \right]_0^{-4} + \left[x^3 + 5x^2 - 8x + c \right]_0^{\frac{2}{3}}$$

$$= \left[(-4)^3 + 5(-4)^2 - 8(-4) + c \right]_0^{-4}$$

$$+ \left[\left(\tfrac{2}{3}\right)^3 + (5)\left(\tfrac{2}{3}\right)^2 - (8)\left(\tfrac{2}{3}\right) + c \right]_0^{\frac{2}{3}}$$

$$= [(48 + c) - (0 + c)] + [(0.259 + c) - (0 + c)].$$

Notice the constants of integration are eliminated.

Then $= 48 + 0.259$.

So required area $= 48.259$ sq. units.

Example 6.3.16 shows how a relatively complex expression may be simplified using partial fractions and then the integral found with relative ease. You are now able to reap the benefits of your previous work on partial fractions!

Example 6.3.16

Find $\int \dfrac{dx}{(x-2)^2(x-3)}$.

At first sight, this looks a rather daunting problem, but reference to the table of standard integrals, gives us an idea of what is required. Functions of the type $\int \dfrac{1}{x+a}\,dx$ have solutions that involve the ln function.

If we can find terms similar to this for our integral, it can be solved. To do this we use *partial fractions* (PFs).

The above integral will have PFs of the form

$$\dfrac{A}{x-2} + \dfrac{B}{(x-2)^2} + \dfrac{C}{(x-3)}$$

In fact reference to Example 6.1.15 gives us the PFs.

Then $\displaystyle\int \dfrac{dx}{(x-2)^2(x-3)} = \int -\dfrac{1}{(x-2)} - \dfrac{1}{(x-2)^2} + \dfrac{1}{x-3}\,dx$

So $\displaystyle\int \dfrac{dx}{(x-2)^2(x-3)} = -\ln(x-2) + \dfrac{1}{(x-2)} + \ln(x-3) + C.$

Integration by substitution

You have already met a technique involving substitution, when you used the function of a function rule for differentiation. Unfortunately, unlike derivatives, when considering substitutions to simplify integrals we do not always know what to substitute. With practice and a little trial and error, the appropriate substitution can normally be found. The following examples illustrate some of the more important substitutions, that provide useful results.

Example 6.3.17

Find $\displaystyle\int \dfrac{\arctan x}{1+x^2}\,dx$.

The clue is in the arctan (the angle whose tangent is). We could let $u = 1 + x^2$, but then the integral still involves arctan x. A good course of action is to try and eliminate that which we know least about. So let's try $u = \arctan x$, then $x = \tan u$. If we substitute now, we have

$$\int \dfrac{u}{1+\tan^2 u}\,dx$$

(this is nonsense because we are told to integrate with respect to x, but the integral has another variable) so we look again at $x = \tan u$ and note that $\dfrac{dx}{du} = \sec^2 u$, so $dx = \sec^2 u\,du$ (now we are getting somewhere). We have a substitute for

cont.

dx in terms of u:

Then $\displaystyle\int \frac{\arctan x}{1 + x^2}\, dx = \int \frac{u}{1 + \tan^2 u}\, \sec^2 u\, du$

$$= \int u\, du \left(\begin{array}{l} \text{because} \\ \sec^2 u = 1 + \tan^2 u \end{array} \right)$$

$$= \frac{u^2}{2} \text{ and using our original substituition;}$$

then $\displaystyle\int \frac{\arctan x}{1 + x^2}\, dx = \frac{(\arctan x)^2}{2}.$

Look back carefully over this example and make sure you follow the logic of the process.

Example 6.3.18

Find $\displaystyle\int \frac{dx}{\sqrt{1 + x^2}}.$

Integrals of this type are best dealt with using a trigonometric or hyperbolic substitution, since the expression under the square root is of the form $a^2 + x^2$, or $a^2 - x^2$, or $x^2 - a^2$. In our case $a^2 + x^2$, we use $x = \sinh u$.

Substitutions for the other variations are given at the end of this example.

For our integral we have

$$x = \sinh u, \qquad \frac{dx}{du} = \cosh u \quad \text{or} \quad dx = \cosh u\, du.$$

So our integral becomes:

$$\int \frac{\cosh u}{\sqrt{1 + \sinh^2 u}}\, du$$

and remembering yet again our old friend in hyperbolic form

$$\cosh^2 u - \sinh^2 u = 1 \quad \text{then} \quad \cosh^2 u = 1 + \sinh^2 u$$

and so $\cosh u = \sqrt{1 + \sinh^2 u}$

so $\displaystyle\int \frac{\cosh u}{\cosh u}\, du = \int 1\, du = u = \sinh^{-1} x.$

For other integrals of this type, try the following substitutions.

For integrals involving:

(i) $\sqrt{a^2 - x^2}$ try $x = a \sin u$ or $x = a \cos u.$

(ii) $\sqrt{x^2 - a^2}$ try $x = a \cosh u$ or $x = a \sec u.$

For integrals involving odd power of trigonmetric ratios like:

$$\int \cos^n x\, dx\, (n \text{ odd}) \quad \text{try} \quad u = \sin x$$

$$\int \sin^n x\, dx\, (n \text{ odd}) \quad \text{try} \quad u = \cos x.$$

It would be useful to memorise these substitutions!

Integration by parts

This method (formula 3) is used mainly where the function under consideration is a *product*, in a similar way to the product rule for differentiation (the reverse process). In comparison with the previous substitutions methods, it is relatively straightforward to use. The following examples illustrate the method very well.

Example 6.3.19

Find $\int \ln x \, dx$.

On first sight it seems that this function is not a product and a substitution might be appropriate. Although the integral can be solved in this way, it is much easier to integrate by parts when we consider the integral as

$$\int \ln x(1) \, dx.$$

Now applying the rule (in functional notation)

$$\int uv' \, dx = uv - \int vu' \, dx$$

with $\quad u = \ln x \quad$ so $\quad u' = \dfrac{1}{x} \quad$ or $\quad u = \ln x \quad$ so $\quad \dfrac{du}{dx} = \dfrac{1}{x}$

and $\quad v' = 1 \quad$ so $\quad v = x \quad$ or $\quad \dfrac{dv}{dx} = 1 \quad$ so $\quad v = x$

then $\quad \displaystyle\int uv' \, dx = x \ln x - \int x \dfrac{1}{x} \, dx$

$$\int \ln x \, dx = x \ln x - x + c$$

(constant of integration is often assumed).

Check carefully that you can follow this process noting that we started with letting $\ln x = u$ and $1 = \dfrac{dv}{dx}$ or v'.

Example 6.3.20

Find $\int x \cos^2 x \, dx$.

This is quite clearly a product, however the integral of $\cos^2 x$ is not straightforward, but by using a trigonometric substitution we can simplify the integral before we start. Referring back to formula (4ii) in the trigonometry where

$$\cos 2A = 2\cos^2 A - 1 \quad \text{or} \quad \cos 2A + 1 = 2\cos^2 A.$$

Then for above integral

cont.

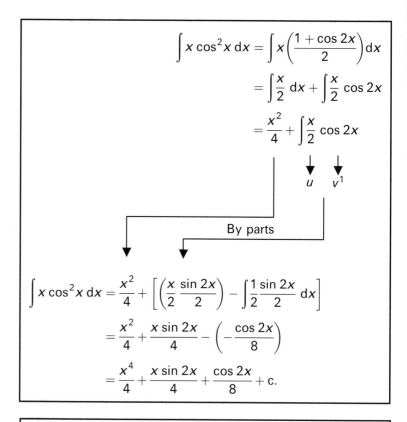

$$\int x\cos^2 x\,dx = \int x\left(\frac{1+\cos 2x}{2}\right)dx$$

$$= \int \frac{x}{2}\,dx + \int \frac{x}{2}\cos 2x$$

$$= \frac{x^2}{4} + \int \frac{x}{2}\cos 2x$$

By parts

$$\int x\cos^2 x\,dx = \frac{x^2}{4} + \left[\left(\frac{x}{2}\frac{\sin 2x}{2}\right) - \int \frac{1}{2}\frac{\sin 2x}{2}\,dx\right]$$

$$= \frac{x^2}{4} + \frac{x\sin 2x}{4} - \left(-\frac{\cos 2x}{8}\right)$$

$$= \frac{x^4}{4} + \frac{x\sin 2x}{4} + \frac{\cos 2x}{8} + c.$$

Example 6.3.21

Find $\int xe^x\,dx$.

Let $u = x$ and $v' = e^x$

then $u' = 1$ and $v = e^x$.

Using $\int uv'\,dx = uv - \int vu'\,dx$.

Then $\int xe^x\,dx = xe^x - \int (e^x)(1)\,dx$

$\int xe^x\,dx = xe^x - e^x + c.$

What about finding $\int x^2 e^x\,dx$?

Let's try the same process.

$u = x^2$ and $v' = e^x$ then $u' = 2x$ and $v = e^x$.

So by parts:

$$\int x^2 e^x\,dx = x^2 e^x - \int 2xe^x\,dx$$

the right-hand integral can be integrated by parts again! With $u = 2x, u' = 2, v' = e^x, v = e^x$

$$\int x^2 e^x\,dx = x^2 e^x - \left[2xe^x - 2\int e^x\,dx\right]$$

taking constant behind integral sign.

cont.

$$\text{so} \quad \int x^2 e^x dx = x^2 e^x - [2xe^x - 2e^x]$$

$$\int x^2 e^x dx = e^x (x^2 - 2x - 2)$$

constant assumed.

If you were able to follow the process illustrated in examples 6.3.19–6.3.21, you should be able to integrate products involving polynomial, trigonometric and exponential functions. There is not space here to show some of the more intricate techniques, which you may require. Integration, unlike differentiation, is often considered to be something of an art rather than a science. With *practice* you will be able to master the art!

Numerical integration

Up till now, the integrals we have been dealing with have all been clearly defined by a *rule*. In many engineering situations we are required to make sense of data which may result from experimentation or, be produced during system/component operation. If we have no clearly defined rule governing the summing process or the integral cannot be evaluated using analytical methods. Then, we need another method of solution, we use a *numerical summing process*. Two simple, but very powerful techniques of numerical integration (summing) are given next. These techniques involve the use of Simpson's rule and the Trapezoidal rule (formulae 4 and 5).

The Trapezoidal rule:

This rule is illustrated in Figure 6.3.2.

If the area represented by $\int_a^b f(x)\, dx$ is divided into strips (Figure 6.3.2), each of equal width w, then each strip is approximately a trapezium. The area of a trapezium is equal to half the sum of the parallel sides, multiplied by the distance between them. If we use the sum of these areas as an approximation for the actual value of the area under the integral sign, then

$$\int_a^b f(x)dx \simeq \tfrac{1}{2}\, d(h_0 + h_1) + \tfrac{1}{2}\, d(h_1 + h_2)$$

$$+ \tfrac{1}{2}\, d_2(h_2 + h_3) + \tfrac{1}{2}\, d\,(h_3 + h_4) + \tfrac{1}{2}\, d(h_4 + h_5)$$

$$\simeq \tfrac{1}{2}\, d[h_0 + 2h_1 + 2h_2 + 2h_3 + 2h_4 + h_5].$$

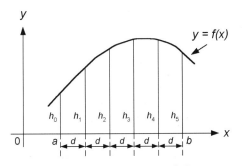

Figure 6.3.2 *Trapezoidal rule*

Now in general the rule depends on the number of ordinates chosen so the rule (formula 5) may be written as:

$$\int_a^b f(x)\,\mathrm{d}x \simeq \tfrac{1}{2}\,\mathrm{d}[h_0 + 2h_1 + \cdots 2h_{n-2} \cdots h_{n-1}].$$

Example 6.3.22

Use the trapezoidal rule, with five ordinates to evaluate

$$\int_0^1 e^{x^2}\,\mathrm{d}x.$$

This integral could be solved by integration, using a substitution, but it serves as a simple example of the process.

We have five ordinates h_0, h_1, h_2, h_3, h_4, evenly spaced between $h_0 = 0$ and $h_4 = 1$, so $d = 0.25$.

Then when $h = e^{x^2}$

$$h_0 = e^0 \quad\ = 1 \qquad\qquad \text{where}\quad d = 0$$

$$h_1 = e^{0.0625} = 1.0645 \quad \text{where}\quad d = 0.25$$

$$h_2 = e^{0.25} \quad = 1.2840 \quad \text{where}\quad d = 0.5$$

$$h_3 = e^{0.5625} = 1.7551 \quad \text{where}\quad d = 0.75$$

$$h_4 = e^1 \quad\ \ = 2.7183 \quad \text{where}\quad d = 1.$$

So the trapezoidal rule gives:

$$\int_0^1 e^{x^2} \simeq \tfrac{1}{2}\,\mathrm{d}[h_0 + 2h_1 + 2h_2 + 2h_3 + h_4]$$

$$\simeq (\tfrac{1}{2})(0.25)[1 + (2)(1.0645) + 2(1.284)$$
$$\qquad + 2(1.7551) + 2.7183]$$

$$\simeq 0.125(1 + 2.129 + 2.568 + 3.5102 + 2.7183)$$

$$\simeq 1.366.$$

Simpson's rule:

Simpson's rule will not be proved. It can however, be easily used by applying the version of the rule given in formula 4. 1.

Then for Simpson's rule use:

$$1/3\mathrm{d}\{(I + F) + 4(\text{ODD}) + 2(EVEN)\}$$

where $I =$ the initial ordinate (h_0),

$\qquad\quad F =$ the final ordinate (h_{2n}).

ODD $=$ the sum of the intervening odd ordinates, $h_1, h_3 \ldots$
EVEN $=$ the sum of the intervening even ordinates, $h_2, h_4 \ldots$

Example 6.3.23

Use Simpson's rule to evaluate the integral of example 6.3.22, with five ordinates.

cont.

Then we evaluate $\int_0^1 e^{x^2} dx$.

Simpson's rule written out in full is:

$$\int_u^b f(x)\,dx \simeq \tfrac{1}{3}\,d[h_0 + 4h_1 + 2h_2 + 4h_3 + 2h_4$$

$$+ \cdots 4h_{2n-1} + h_{2n}]$$

$$\int_0^1 e^{x^2} dx \simeq \tfrac{1}{3}\,d[(I + F) + 4(\text{ODD}) + 2(\text{EVEN})].$$

Now we have already calculated the values in example 6.3.22, they are

$h_0 = 1 \qquad d = 0$

$h_1 = 1.0645 \quad d = 0.25$

$h_2 = 1.2840 \quad d = 0.5$

$h_3 = 1.7551 \quad d = 0.75$

$h_4 = 2.7183 \quad d = 1.$

Then using Simpson's rule:

$$\int_0^1 e^{x^2} dx \simeq \tfrac{1}{3}(0.25)[(1 + 2.7183) + 4(2.8196) + 2(1.2840)]$$

$$\simeq 0.0833(3.7183 + 11.2784 + 2.568)$$

$$\simeq 1.463.$$

Compare this value with that obtained using the trapezoidal rule. Simpson's gives a more accurate solution.

Engineering Applications

We now turn our attention to the numerous engineering applications of the calculus. These will include examples on rates of change, centroids, second moment of area, second moment of mass, root mean square (rms) values of waves and numerical summation of engineering functions. Some important formulae concerned with engineering applications are also given in this section, for convenience.

Formulae

(1) Areas and volumes of revolution

Area $A = \int_a^b f(x)\,dx$

Volume of revolution obtained by rotating area A through four right-angles.

(i) about $0x = \pi \int_a^b [f(x)]^2 dx$

(ii) about $0y = 2\pi \int_a^b x f(x)\,dx.$

(2) Centroids of plane areas

For the figure, if $\bar{x}, \bar{y}$ denote the co-ordinates of the centroid of area A, then:

$$\bar{x} = \frac{\int_a^b xf(x)\,\mathrm{d}x}{\int_a^b f(x)\,\mathrm{d}x} ; \bar{y} = \frac{\frac{1}{2}\int_a^b [f(x)]^2\mathrm{d}x}{\int_a^b f(x)\,\mathrm{d}x}.$$

(3) Second moment of area

If I_x and I_y denote second moments of area A about $0x$ and 0_y respectively, then:

$$I_x = \frac{1}{3}\int_a^b y^3\,\mathrm{d}x \quad \text{and} \quad I_y = \int_a^b x^2 y\,\mathrm{d}x$$

(4) Second moment of mass (moment of inertia)

If the mass per unit volume of the volume of revolution generated by the rotation of area A about $0x$ is m, then the moment of inertia of the solid about $0x$ is:

$$I_{0x} = \frac{1}{2}\int_a^b m\pi y^4\,\mathrm{d}x.$$

(5) Theorem of Pappus

Volume of revolution obtained by rotating area A through four right-angles:

   (i)   about $0x = 2\pi y A$;

   (ii) about $0y = 2\pi x A$.

(6) Mean and rms values

If $y = f(x)$, then the **mean value** of y over the range $x = a$ to $x = b$ is given by:

$$\text{Mean value} \quad = \frac{1}{b-a}\int_a^b f(x)\,\mathrm{d}x$$

and the **rms value** over the same range is given by

$$\text{rms value} \quad = \sqrt{\frac{1}{b-a}\int_a^b [f(x)]^2\mathrm{d}x}.$$

Centroids of area

The *centroid* of an area is the point at which the total area is considered to be situated for calculation purposes. It is needed in the calculation of second moments of area, which follows. For simple shapes such as rectangles and circles, the centroid is easily found. There are however, many more complex shapes which cannot be divided into these standard shapes. For these areas, we need a mathematical summing technique to establish their centroid. The following two examples illustrate the process.

Note; that the centroid is found from conveniently positioned axes.

Example 6.3.24

Find by integration the value of $\bar{x}$ (i.e. the distance of the centroid from left-hand edge) for the rectangle shown below

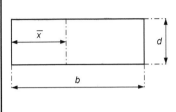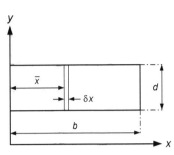

Determination of second moment of area of rectangle

The rectangle must be set up on suitable axes as shown above. We include a very small elemental strip.

Now the area of the rectangle $= \sum A$

where $\sum A =$ the sum of the elementary strip areas

$$= \sum_{x=0}^{x=b} d \cdot dx$$

$$= \int_0^b d\ dx = d \int_0^b 1 \cdot dx$$

$$= d[x]_0^b = d(b - 0) = b \cdot d.$$

Also the first moment of the rectangular area about the y-axis:

$$= \sum A \cdot x$$

$=$ sum of the first moment of area

of each of the elementary strips

$=$ sum of: (area of strip x distance of its

centroid from the y-axis)

$$= \sum_{x=0}^{x=b} d\ dx \cdot x$$

$$= \int_0^b dx\ dx \qquad = d \int_0^b x \cdot dx \qquad d \left[\frac{x^2}{2} \right]_0^b$$

$$= d \left(\frac{b^2}{2} - \frac{O^2}{2} \right) \qquad = \frac{b^2 \cdot d}{2}.$$

and so $\quad \bar{x} = \dfrac{\sum A.x}{\sum A} = \dfrac{b^2 d/2}{bd} = \dfrac{b}{2}.$

This of course is what we would expect from symmetry!

Example 6.3.25

Consider the area bounded by the curve $y = x(2 - x)$ and the x-axis.

Always draw a sketch and set up axes.

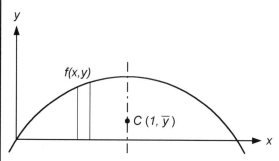

Area bounded by curve $y = x(2 - x)$ and the x-axis

From symmetry the centroid C lies on $x = 1$, i.e. $\bar{x} = 1$. To find $\bar{y}$ we consider the first moment about Ox. Taking a vertical strip through P as shown we have

$$\delta A \simeq y\,\delta x.$$

The centroid of the element is approximately a distance of $\frac{y}{2}$ from Ox. Therefore, the first moment of δA about $Ox \simeq (y\,\delta x)\frac{1}{2}y$

so $\bar{y} \simeq \dfrac{\displaystyle\sum_{x=0}^{2} \frac{1}{2}y^2\,\mathrm{d}x}{\displaystyle\sum_{x=0}^{2} y\,\mathrm{d}x}$

or $\bar{y} = \dfrac{\displaystyle\int_0^2 \frac{1}{2}y^2\,\mathrm{d}x}{\displaystyle\int_0^2 y\,\mathrm{d}x}$ where $y = x(2 - x)$

$$\bar{y} = \frac{\frac{1}{2}\displaystyle\int_0^2 x^2(2-x)^2\,\mathrm{d}x}{\displaystyle\int_0^2 x(2-x)\,\mathrm{d}x} = \frac{\left[\dfrac{4x^3}{3} - x^4 + \dfrac{x^5}{5}\right]_0^2}{\left[x^2 - \dfrac{x^3}{3}\right]_0^2}$$

$$= \frac{\dfrac{8}{15}}{\dfrac{4}{3}} = \frac{2}{5}$$

So centroid at $\bar{x} = 1$, $\bar{y} = \frac{2}{5}$.

Second moment of area

In engineers bending theory you will use the relationship:

$$\frac{M}{I} = \frac{\sigma}{y} = \frac{E}{R}$$

where I is known as the second moment of area about the cross-

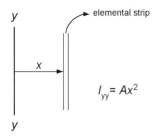

y

x

$I_{yy} = Ax^2$

y

Figure 6.3.3 *Method for calculating second moment of area*

section, it is in fact a measure of the resistance to the bending of a structural member, based on the geometry of its cross-section. It is an important engineering property and is calculated in the following way.

Figure 6.3.3 shows an elemental strip with very small width of area A, which is distance x from the *YY* axis.

Now from our work on centroids we know that the first moment of the area AA about YY is given by Ax. Then the *second moment* of the area A about YY is given by Ax^2. The second moment of area is always stated with reference to an axis or datum line. So in the above case $I_{yy} = Ax^2$.

Example 6.3.26

Find the second moment of area of the rectangle shown, about its base edge.

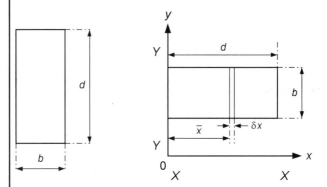

Second moment of area for rectangle

The rectangle is shown set up on suitable reference axes. It is in this case convenient to turn the rectangle through 90° and let the base edge lie on the *y*-axis (*YY*). The diagram shows a typical elemental strip area parallel to the reference axix (*YY*). Whose area is $b \cdot \delta x$.

Now second moment of the rectangular area:

$$\text{about the } y\text{-axis} = \sum A \cdot x^2 = \sum_{x=0}^{x=d} b \cdot \delta x \cdot x^2 = \int_0^d b \cdot dx \cdot x^2.$$

$$\text{So} \quad I_{yy} = b \int_0^d x^2 \, dx = b \left[\frac{x^3}{3} \right]_0^d = \frac{bd^3}{3}.$$

In engineers theory of torsion where:

$$\frac{T}{J} = \frac{\tau}{R} = \frac{G\theta}{L}$$

J is known as the *polar second moment of area* of the cross-section.

It is a measure of the resistance of a shaft to torsional loads, an expression for its value may be found in a similar manner to the method illustrated in example 6.3.26.

Example 6.3.27

Find the polar second moment of area (J) of the circular area shown in the figure.

As before we consider the second moment of area of the elemental strip, set-up on the polar axis as shown.

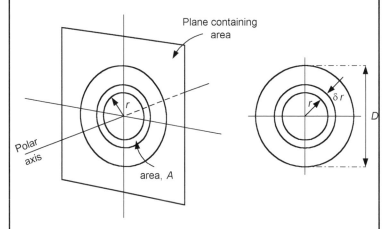

Polar second moment of area

Then J for the circular area $= \sum Ar^2$. Now approximate area A of the elemental strip $=$ (circumference of strip $\times$ width of strip).

$$\text{So} \quad A \simeq 2\pi r \times \delta r.$$

$$\text{Therefore} \quad J = \sum_{r=0}^{r=D/2} 2\pi r \cdot \delta r \cdot r^2$$

$$= \int_0^{D/2} 2\pi r \cdot dr \cdot r^2$$

$$= 2\pi \int_0^{D/2} r^3 \, dr$$

$$= 2\pi \left[\frac{r^4}{4} \right]_0^{D/2}$$

$$= \frac{\pi D^4}{32}.$$

Moment of inertia

The second moment of mass (*moment of inertia*) of a rotating body is found in a similar way to the second moment of area. You met the moment of inertia of rotating masses in Chapter 3 when you studied angular motion.

The moment of inertia, I, of a body about a given axis is the sum of the products of each element of mass and the square of its distance from a given axis (same as for moment of area except mass is taken into account). The procedure we adopt is therefore the same as that for second moment of area.

Example 6.3.28

Find the *moment of inertia* of a rectangular lamina (solid area without depth) of length 10 cm and breadth 5 cm about an axis parallel to the 10 cm side and 10 cm from it. Take the area density of the laminar to be ρ kg/cm^2.

The figure illustrates the situation.

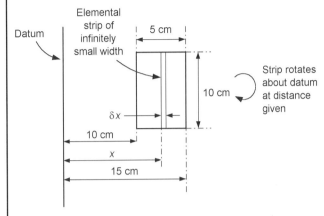

Determination of moment of inertia

$$\text{Then}\quad \text{mass of element} = 10\rho\,\mathrm{d}x \quad \text{kg}$$

$$I \text{ (second moment) for element} = 10\rho x^2\,\mathrm{d}x\,\text{kg m}^2$$

$$I \text{ for rectangle} = \int_{10}^{15} 10\rho x^2\,\mathrm{d}x$$

$$= 10\rho \left[\frac{x^3}{3}\right]_{10}^{15}$$

$$= 10\rho \left[\frac{15^3}{3} - \frac{10^3}{3}\right]$$

$$= 10\rho [1125 - 333\tfrac{1}{3}]$$

$$= 7916\tfrac{2}{3}\rho\,\text{kg cm}^2.$$

Mean and rms values

We have already used the integral calculus to find areas under curves, when you integrated between limits. The average or *mean value* of the height of such curves above the limits of integration may easily be found by dividing the area obtained (between the limits), by the distance between these limits (Figure 6.3.4). From this process we obtain the expression:

$$\text{Mean value of height} = \frac{\displaystyle\int_a^b f(x)\,\mathrm{d}x}{b - a} = \frac{\text{Area}}{\text{horizontal distance}}$$

$$\text{so}\quad \text{mean value} = \frac{1}{b-a}\int_a^b f(x)\,\mathrm{d}x.$$

Figure 6.3.4 *Mean value of curve*

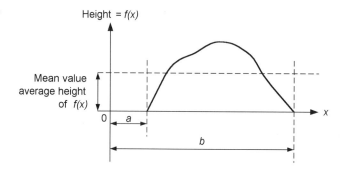

So, for example if the function represented in Figure 6.3.4 was given by:

$$\int_2^4 e^x \, dx$$

then the mean value would be given by:

$$\frac{1}{4-2} \int_2^4 e^x \, dx = \tfrac{1}{2} [e^x]_2^4$$

$$= \tfrac{1}{2} [e^4 - e^2]$$

$$= \tfrac{1}{2} [54.6 - 7.39]$$

$$= 23.6 \quad \text{units of length.}$$

When finding average values of sinusoidal wave forms, we need to consider the area above and below the *x*-axis. If we square the mean value of the sinusoidal wave its height (*y*) will always be positive and, if we then square root this value we obtain the average or the *root mean squared (rms) value*.

So considering our formula for the mean value of a function we get:

$$\text{mean value} \quad f(x) = \frac{1}{b-a} \int_a^b f(x) \, dx.$$

So squaring gives $\quad (f(x))^2 = \dfrac{1}{b-a} \displaystyle\int_a^b (f(x))^2 \, dx$

and taking the square root gives:

$$\text{rms value} = \sqrt{\frac{1}{b-a} \int_a^b (f(x))^2 \, dx}.$$

The rms value of sinusoidal waveforms is particularly useful when determining power and other parameters from a.c voltages and currents. The following example shows how we calculate the rms value for sinusoidal wave forms.

Example 6.3.29

Find the rms value of the function $f(x) = \sin x$ between 0 and π (radians).

cont.

Then using $\text{rms} = \sqrt{\dfrac{1}{b-a}\displaystyle\int_a^b (f(x))^2\,dx}$

then $\text{rms} = \sqrt{\dfrac{1}{\pi - 0}\displaystyle\int_0^\pi \sin^2\theta\,d\theta}$

and using the identity $\cos 2A = 1 - 2\sin^2 A$

then $\sin^2\theta = \frac{1}{2}(1 - \cos 2\theta)$.

So $\text{rms} = \sqrt{\dfrac{1}{\pi - 0}\displaystyle\int_0^\pi \frac{1}{2}(1 - \cos 2\theta)\,d\theta}$

$= \sqrt{\dfrac{1}{2\pi}\left[\theta - \dfrac{\sin 2\theta}{2}\right]_0^\pi}$ $\left(\begin{array}{l}\text{constant } \frac{1}{2}\text{, put outside} \\ \text{integral sign}\end{array}\right)$

$= \sqrt{\dfrac{1}{2\pi}\left[\left(\pi - \dfrac{\sin 2\pi}{2}\right) - \left(0 + \dfrac{\sin 0}{2}\right)\right]}$

$= \sqrt{\dfrac{1}{2\pi}(\pi)} = \sqrt{\dfrac{1}{2}} = 0.7071$.

So rms $= 0.7071$ (I hope this is what you expected!)

Since the sinusoidal function is squared and then square rooted, the wave repeats itself every π rad, so the rms value for the whole sinusoidal wave form is always 0.7071.

Numerical method for buoyancy

The distance and velocity of falling bodies in water, or other resistive mediums can be related by the integral calculus. The following example illustrates the use of the Trapezoidal rule and Simpson's rule to solve a problem, where the relationship has been established by experiment.

Example 6.3.30

The results of an experiment show that for a particular body falling in a resistive medium the distance that it has fallen when its velocity is 0.2 m/s, is given by:

$$\int_0^2 \frac{v}{9.81 - 0.2v^2}\,dv$$

Evaluate this distance using the Trapezoidal rule and Simpson's rule, take $d = 0.25$. For this integral, we will set up a table.

v_i	v initial and final	$\dfrac{v}{9.81 - 0.2v^2}$ ODD	$\dfrac{v}{9.81 - 0.2v^2}$ EVEN
$v_0 = 0$	$v_0 = 0$		
$v_1 = 0.25$		0.025 52	
$v_2 = 0.25$			0.051 23
$v_3 = 0.75$		0.077 34	
$v_4 = 1.0$			0.104 06

cont.

v_i	v initial and final	$\dfrac{v}{9.81 - 0.2v^2}$ ODD	$\dfrac{v}{9.81 - 0.2v^2}$ EVEN
$v_5 = 1.25$		0.131 61	
$v_6 = 1.5$			0.160 26
$v_7 = 1.75$		0.190 27	
$v_8 = 2.0$	$v_8 = 0.221\,98$		
Totals	0.221 98	0.424 74	0.315 55

Then using the abbreviated versions of both rules, we get:
(i) From Trapezoidal rule:

$$\int_0^2 \frac{v}{9.81 - 0.2v^2}\,dv \simeq d[\tfrac{1}{2}(I + F) + ODD + EVEN]$$

$$\simeq 0.25[0.110\,99 + 0.424\,74 + 0.315\,55]$$

$$\simeq 0.25(0.851\,28)$$

$$\simeq 0.212\,82 \text{ m.}$$

(ii) from Simpson's rule:

$$\int_0^2 \frac{v}{9.81 - 0.2v^2}\,dv \simeq \frac{d}{3}[(I + F) + 4(ODD) + 2(EVEN)]$$

$$\simeq \frac{0.25}{3}[0.221\,98 + 4(0.424\,74)$$

$$+ 2(0.315\,55)]$$

$$\simeq 0.083\,33(0.221\,98 + 1.698\,96 + 0.6311)$$

$$\simeq 0.212\,67 \text{ m.}$$

Rates of change

The following simple examples illustrate the use of the differential calculus to determine the rate of change of variables with respect to time.

Example 6.3.31

The motion of a body is modelled by the relationship $s = t^3 - 3t^2 + 3t + 8$, where s is distance in metres and t time in seconds. Determine:

(i) the velocity of the body at the end of 3 seconds;
(ii) the time when the body has zero velocity;
(iii) its acceleration at the end of 2 seconds;
(iv) when its acceleration is zero.

In order to solve this problem you must be aware of the relationship between distance, velocity and acceleration with respect to time.
   The rate of change of distance with respect to time is given by $\dfrac{ds}{dt}$ and is the *velocity*. Similarly the rate of change of

cont.

velocity with respect to time is given by $\dfrac{dv}{dt}$ and is the acceleration of the body.

So $v = \dfrac{ds}{dt}$ and $a = \dfrac{d^2s}{dt^2} = \dfrac{dv}{dt}$.

Now the problem is simple:

$s = t^3 - 3t^2 + 3t + 8$ then

(i) $\dfrac{ds}{dt} = 3t^2 - 6t - 3 = $ velocity,

so at 3 seconds the

velocity $= 3(3)^2 - 6(3) + 3 = 12$ m/s.

(ii) The body will have zero velocity when $\dfrac{ds}{dt} = 0$; that is, when $3t^2 - 6t + 3 = 0$. So we need to solve this quadratic to determine the time t when the velocity is zero.

Then $t^2 - 2t + 1 = 0$

so $(t - 1)(t - 1) = 0$.

Then $\dfrac{ds}{dt} = 0$ when $t = 1$ or body has velocity of zero at $t = 1$ second.

(iii) The acceleration (a) is obtained as stated previously, i.e.

$\dfrac{d^2s}{dt^2} = 6t - 6$ so when $t = 2$

the body has acceleration $\dfrac{d^2s}{dt^2} = a = 6$ m/s^2.

(iv) Acceleration is zero when $\dfrac{d^2s}{dt^2} = 6t - 6 = 0$ so acceleration is zero at time $t = 1$ second.

Example 6.3.32

A particle is subject to harmonic motion given by the relationship $x = A \sin \omega t$. Show that the linear acceleration of the particle is given by $a = \omega^2 x$. Where $\omega = $ angular velocity and $x = $ the linear displacement from the centre of oscillation.

This again relates displacement, velocity and acceleration, so all we need do is differentiate twice.

Then $x = A \sin \omega t$

so $v = \dfrac{dx}{dt} = A\omega \cos \omega t$

and $\dfrac{d^2x}{dt^2} = -A\omega^2 \sin \omega t$, but $x = A \sin \omega t$

so $\dfrac{d^2x}{dt^2} = -\omega^2 x$.

We leave our study of the calculus with a number of problems which have been designed to give you practice in manipulating the calculus as well as applying it. The problems are of varying difficulty, in no particular order.

Questions 6.3

(1) Differentiate the following using function of a function rule (i.e. by substitution) or chain rule.

   (a) $(3x + 1)^2$;   (b) $(2 - 5x)^{3/2}$;   (c) $\dfrac{1}{4x^2 + 3}$;

   (d) $\sin^2 4x$;   (e) $\cos(2 - 5x)$;   (f) $\log_e 9x$;

   (g) Find $\dfrac{d}{dt}\left(\dfrac{1}{\sqrt[3]{1 - 2t}}\right)$;

   (h) Find $f'(t)$ for the function $f(t) = Be^{Kt-b}$.

(2) Differentiate the following using the product rule as required.

   (a) $x \sin x$;   (b) $e^x \tan x$;   (c) $x \ln x$;

   (d) $\dfrac{d}{dt}(\sin t \cos t)$;   (e) Find $\dfrac{d}{dt}[6e^{3t}(t^2 - 1)]$;

   (f) Find $\dfrac{d}{ds}[(s - 3s^2)\log_e s]$.

(3) Differentiate the following:

   (a) $\dfrac{x}{1 - x}$;   (b) $\dfrac{e^x}{3 \sin 2x}$;   (c) $\dfrac{(x^2 - 2)^2}{(3x + 4)^3}$;

   (d) Find: $\dfrac{d}{dt}\left(\dfrac{\cos 2t}{e^{2t}}\right)$;   (e) Find: $\dfrac{d}{d\theta}(\cot \theta)$.

(4) A curve is given in the form: $y = 3 \cos 2\theta - 5 \tan \theta$ where θ is in radians. Find the gradient of the curve at the point where θ has a value equivalent to $34°$.

(5) Find the gradient of the curve $\dfrac{\cos \theta}{\theta}$ at the point where $\theta = 0.25$ (remember radians).

(6) If $y = 4 \log_e(1 - x)$, find the value of $\dfrac{dy}{dx}$ when $x = 0.32$.

(7) If $y = \dfrac{1 + x^2}{x - 2}$ find the value of $\dfrac{dy}{dx}$ if $x = -1.25$.

(8) If $y = 3x^3 + 2x - 7$ find an expression for $\dfrac{d^2y}{dx^2}$ and also its value when $x = -3$.

(9) Given that $y = \dfrac{3t^5 + 2t}{t^2}$ find the value of $\dfrac{d^2y}{dx^2}$ when $t = 0.6$.

(10) Find the value of $\dfrac{d^2u}{dm^2}$ if $u = \frac{1}{2}(e^{3m} - e^{-3m})$ given $m = 1.3$.

(11) A body moves a distance s metres in a time t seconds so that $s = 2t^3 - 9t^2 + 12t + 6$. Find: (a) its velocity after 4 seconds; (b) its acceleration after 4 seconds; (c) when the velocity is zero.

(12) The angular displacement θ radians of the spoke of a wheel is given by the expression $\theta = \frac{1}{2}t^4 - t^3$, where t is the time in seconds. Find: (a) the angular velocity

cont.

after 3 seconds; (b) the angular acceleration after 4 seconds; (c) when the angular acceleration is zero.

(13) Find the maximum and minimum values of y given that $y = x^3 + 3x^2 - 9x + 6$.

(14) An open rectangular tank of height h metres with a square base of side x metres is to be constructed so that it has a capacity of 500 cubic metres. Prove that the surface area of the four walls and the base will be $\dfrac{2000}{x} + x^2$ square metres. Find the value of x for this expression to be a minimum.

(15) A mass of 5000 kg moves along a straight line so that the distance s metres travelled in a time t seconds is given by $s = 3t^2 + 2t + 3$. If v m/s is its velocity and m kg is its mass, then its kinetic energy is given by the formula $\frac{1}{2}mv^2$. Find its kinetic energy at a time $t = 0.5$ seconds. Remember that the joule (J) is the unit of energy.

(16) Find:

(a) $\displaystyle\int 4e^{-3x} - 2\cos 3x - 5\sin 2x\, dx.$

(b) $\displaystyle\int xe^{-x}\, dx.$

(c) $\displaystyle\int \frac{1}{(5x+1)(3-2x)}\, dx.$

(d) $\displaystyle\int_a^b \frac{1}{x}\ln x\, dx.$

(e) $\displaystyle\int \tan x\, dx.$

(f) $\displaystyle\int \frac{1}{\sqrt{1-x^2}}.$

(17) Evaluate:

(a) $\displaystyle\int_1^4 \sqrt[3]{2+3x}.$

(b) $\displaystyle\int_0^{\pi/2} (\sin 2x)(\sin x)\, dx.$

(c) $\displaystyle\int_0^1 x(x^2+2)^6\, d\pi.$

(d) $\displaystyle\int_1^3 \sin^3 x\, dx.$

(18) Evaluate $\displaystyle\int_0^2 x^2 e^{-x^2}\, dx$ using Simpson's rule taking $d = 0.25$, to integrate the function numerically.

cont.

(19) The torque on a clutch plate is given by

$$T = \int_{r_i}^{r_0} 2\pi\rho\mu r^2 \, dr.$$

If $r_0 = 0.15$ m; $r_i = 0.08$ m; $\mu = 0.2$; and $\rho = 2000$ N/m^2, find the value of the torque T in Nm.

(20) Find the mean and rms value of the function $f(x) = \sin\theta + \sin 2\theta$, between the ordinates $\theta = 0$ and $\theta = \dfrac{2\pi}{3}$ rad.

(21) Determine the position of the centroid of the area enclosed between the curve $y = 3x^2 + 1$ and the x-axis, and values at $x = 4$ and $x = 2$.

(22) Prove that the second moment of area of a circle about its Polar axis (zz) is equal to $\dfrac{\pi D^4}{32}$, hence calculate the second moment of area of the section shown below.

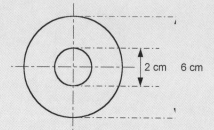

(23) An equation relating the moment M, modulus E, second moment of area I, and deflection y of a beam is:

$$\frac{d^2 y}{dx^2} = \frac{M}{EI}.$$

Find an expression for the deflection y.

(24) If $EI\dfrac{d^2 y}{dx^2} = (Rx - w_1(x - a) - w_2(x - b))$.

Find an expression for y. Then given that when $x = 0$, $y = 0$ and that when $x = L = 2$, $y = 0$. Find the constants of integration given that one constant

$$A = \frac{L^2}{6} - \frac{3(L - 1)^3}{6L} - \frac{4(L - 1.5)^3}{6L}.$$

(25) The emf induced in a secondary circuit is given by:

$e = -M\dfrac{di}{dt}$ V and the emf induced in an inductor is given by $e = -N_2\dfrac{d\Phi}{dt}$ V. Show that the mutual inductance M is given by the expression $M = N_2\dfrac{d\Phi}{di}$. If there are 2000 turns on the secondary (N_2) and the rate of change of flux with current $\left(\dfrac{d\Phi}{di}\right)$ is 20×10^{-4} Wb/amp, find the mutual inductance.

6.4 Advanced topics

In this final section we will look briefly at complex numbers, differential equations, Laplace transforms and harmonic analysis using Fourier coefficients. All the appropriate formulae will be found together, after this introduction. These are followed, as in previous sections, with a review of the methods and then the engineering applications essential for the completion of the subject matter.

Formulae

(1) Complex numbers

(a) $z = x + iy$ where Real $z = x$ and Imaginary $z = y, i = \sqrt{-1}$, $i^2 = -1$.

(b) $\bar{z} = x - iy$ is the *conjugate* of the complex number $z = x + iy$.

(c) $z\bar{z} = x^2 + y^2$.

(d) Modulus $|z| = \sqrt{x^2 + y^2}$.

(e) Distance between two points z_1 and z_2 is $|z_1 - z_2| = |z_2 - z_1|$.

(f) Polar form $x + iy = r(\cos \theta + i \sin \theta) = r\angle\theta$ $r = |z|$ and the argument $\theta = \arg z$ $\tan \theta = x/y$

$$z_1 z_2 = r_1 r_2 [\cos(\theta_1 + \theta_2) + i\sin(\theta_1 + \theta_2)] = r_1 r_2 \angle \theta_1 + \theta_2$$

$$\frac{z_1}{z_2} = \frac{r_1 [\cos(\theta_1 - \theta_2) + i\sin(\theta_1 - \theta_2)]}{r_2} = \frac{r_1}{r_2} \angle (\theta_1 - \theta_2).$$

(g) Exponential form

$$z = re^{i\theta} = \cos\theta + i\sin\theta$$

$$|e^{i\theta}| = 1$$

conjugate of $e^{i\theta}$ is $e^{-i\theta}$.

(h) De Moivre's theorem
If n is any integer, then

$$(\cos\theta + i\sin\theta)^n = \cos n\theta + i\sin\theta.$$

(i) Hyperbolic form

$$\cosh ix = \tfrac{1}{2}(e^{ix} + e^{-ix}) = \cos x$$

$$\sinh ix = \tfrac{1}{2}(e^{ix} - e^{-ix}) = i\sin x.$$

Note: *i* notation often used by mathematicians, while *j* is often used for applications $i = j = \sqrt{-1}$.

(2) Differential equations

(a) Linear first-order:

$$\frac{dy}{dx} + P(x)y = Q(x)$$

Integrating factor $e^{\int P(x)\,dx}$.

(b) Linear second-order with constant coefficients a, b, c :

$$a\frac{d^2y}{dx^2} + b\frac{dy}{dx} + cy = f(x).$$

Auxillary equation $an^2 + bn + c = 0$ roots n_1, n_2.
Complementary function

$Ae^{n_1 x} + Be^{n_2 x}$ (real $n_1 \neq n_2$)

$(Ax + B)e^{n_1 x}$ $(n_1 = n_2)$

$e^{px}(A \cos qx + B \sin qx)$ (complex $n_1 n_2 = p \pm iq$).

General solution $y =$ complementary function plus any particular solution.

Harmonic type $\dfrac{d^2 y}{dx^2} + m^2 y = 0.$

General solution $y = A \cos mx + B \sin mx$.

(3) Vectors

(a) Cartesian co-ordinates:

$$\mathbf{a} = a_x \mathbf{i} + a_y \mathbf{j} + a_z \mathbf{k}$$

$$|\mathbf{a}| = \sqrt{a_x^2 + a_y^2 + a_z^2}$$

where $\mathbf{i}, \mathbf{j}, \mathbf{k}$ are unit vectors orthogonal (at-right-angles to one another).

(b) Dot product (scalar):

$$\mathbf{a} \cdot \mathbf{b} = |\mathbf{a}||\mathbf{b}| \cos \theta = \mathbf{b} \cdot \mathbf{a}$$
$$= a_x b_x + a_y b_y + a_z b_z$$
$$\mathbf{a} \cdot (\mathbf{b} + \mathbf{c}) = (\mathbf{a} \cdot \mathbf{b}) + (\mathbf{a} \cdot \mathbf{c}).$$

(c) Cross product (vector):

$$\mathbf{a} \times \mathbf{b} = |a||b| \sin \theta \mathbf{u} = \mathbf{b} \times \mathbf{a}$$

where $\mathbf{u} =$ any unit vector.

(4) Matrices

(a) Matrix form:

$$\begin{pmatrix} a_{11} & a_{12} & \cdots & a_{1n} \\ a_{21} & a_{22} & \cdots & a_{2n} \\ \vdots & \vdots & & \\ a_{m1} & a_{m2} & \cdots & a_{mn} \end{pmatrix} \quad \begin{array}{l} m \times n \\ (m \text{ rows} \times n \text{ columns}) \end{array}$$

(b) Matrix addition (add each complementary element):

$$\begin{pmatrix} a_{11} & a_{12} & a_{13} \\ a_{21} & a_{22} & a_{23} \end{pmatrix} + \begin{pmatrix} b_{11} & b_{12} & b_{13} \\ b_{21} & b_{22} & b_{23} \end{pmatrix}$$

$$= \begin{pmatrix} a_{11} + b_{11} & a_{12} + b_{12} & a_{13} + b_{13} \\ a_{21} + b_{21} & a_{22} + b_{22} & a_{23} + b_{23} \end{pmatrix}$$

Note: subtraction carried out in a similar way, except complementary elements are subtracted instead of added.

(c) Matrix multiplication:
Columns of first matrix = rows of second matrix.

$$\begin{pmatrix} a_{11} & a_{12} & a_{13} \\ a_{21} & a_{22} & a_{23} \end{pmatrix} \begin{pmatrix} b_{11} & b_{12} \\ b_{21} & b_{22} \\ b_{31} & b_{32} \end{pmatrix}$$

$$= \begin{pmatrix} a_{11}b_{11} + a_{12}b_{21} + a_{13}b_{31} & a_{11}b_{12} + a_{12}b_{22} + a_{13}b_{32} \\ a_{21}b_{11} + a_{22}b_{21} + a_{23}b_{31} & a_{21}b_{12} + a_{22}b_{22} + a_{23}b_{32} \end{pmatrix}.$$

(5) Laplace transforms

The Laplace transform of a function of time, $f(t)$, is found by multiplying the function by e^{-st} and integrating the product between the limits of zero and infinity. The result (if it exists) is known as the Laplace transform of $f(t)$. Thus:

$$F(s) = \mathcal{L}\{f(t)\} = \int_0^\infty e^{-st} f(t)\, dt$$

Table of useful Laplace transforms:

$f(t)$	$F(s) = \mathcal{L}\{f(t)\}$	*Comment*
a	$\dfrac{1}{s}$	Step
t	$\dfrac{1}{s^2}$	Ramp
e^{at}	$\dfrac{1}{s-a}$	Exponential growth
e^{-at}	$\dfrac{1}{s+a}$	Exponential decay
$e^{-at}\sin(\omega t)$	$\dfrac{\omega}{(s+a)^2+\omega^2}$	Decaying sine
$e^{-at}\cos(\omega t)$	$\dfrac{s+a}{(s+a)^2+\omega^2}$	Decaying cosine
$\sin(\omega t + \phi)$	$\dfrac{s\sin\phi + \omega\cos\phi}{s^2+\omega^2}$	Sine plus phase angle
$\sin(\omega t)$	$\dfrac{\omega}{s^2+\omega^2}$	Sine
$\cos(\omega t)$	$\dfrac{s}{s^2+\omega^2}$	Cosine
$\sinh \omega t$	$\dfrac{\omega}{s^2-\omega^2}$	Sinh
$\cosh \omega t$	$\dfrac{s}{s^2-\omega^2}$	Cosh
$e^{at}\sinh \omega t$	$\dfrac{\omega}{(s-a)^2-\omega^2}$	Shift sinh
$e^{at}\cosh \omega t$	$\dfrac{s-a}{(s-a)^2-\omega^2}$	Shift cosh
$t\sin \omega t$	$\dfrac{2\omega s}{(s^2+\omega^2)^2}$	Multiple t
$t\cos \omega t$	$\dfrac{s^2-\omega^2}{(s^2+\omega^2)^2}$	Multiple t

cont.

Table of useful Laplace transforms (cont.)

$f(t)$	$F(s) = \mathcal{L}\{f(t)\}$	Comment
$\dfrac{\mathrm{d}(t)}{\mathrm{d}t}$	$sF(s) - f(0)$	First differential
$\dfrac{\mathrm{d}^2 f(t)}{\mathrm{d}t^2}$	$s^2 F(s) - sf(0) - \dfrac{\mathrm{d}f(0)}{\mathrm{d}t}$	Second differential
$\displaystyle\int f(t)\,\mathrm{d}t$	$\dfrac{1}{s}F(s) + \dfrac{1}{s}F(0)$	Integral

(6) Fourier series

(a) Real form:

$$\frac{1}{2}a_0 + \sum_{n=1}^{\infty} a_n \cos \frac{2n\pi x}{T} + \sum_{n=1}^{\infty} b_n \sin \frac{2n\pi x}{T}.$$

where

$$a_n = \frac{2}{T}\int_{-\frac{1}{2}T}^{\frac{1}{2}T} f(x)\cos\frac{2n\pi x}{T}\,\mathrm{d}x, \quad b_n = \frac{2}{T}\int_{-\frac{1}{2}T}^{\frac{1}{2}T} f(x)\sin\frac{2n\pi x}{T}\,\mathrm{d}x.$$

(b) Harmonic analysis – tabular method:

θ	y	$2\cos\theta$	$2y\cos\theta$	$2s\sin\theta$	$2y\sin\theta$	$2\cos 2\theta$	$2y\cos 2\theta$	$2\sin 2\theta$	$2y\sin 2\theta$	$\ldots$
0										
30										
60										
90										
120										
150										
180										
210										
240										
270										
300										
330										
360										
Total										
Mean	$\frac{1}{2}a_0$		a_1		b_1		a_2		b_2	

$$y = \tfrac{1}{2}a_0 + a_1\cos\theta + b_1\sin\theta + a_2\cos 2\theta + b_2\sin 2\theta + \cdots$$

(c) Fourier series for special waveforms:
Square wave sine series

$$\frac{4K}{\pi}\left[\sin\frac{2\pi x}{T} + \tfrac{1}{3}\sin\frac{6\pi x}{T} + \tfrac{1}{5}\sin\frac{10\pi}{T}x + \cdots\right]$$

Square wave cosine series

$$\frac{4K}{\pi}\left[\cos\frac{2\pi x}{T} - \tfrac{1}{3}\cos\frac{6\pi x}{T} + \tfrac{1}{5}\cos\frac{10\pi x}{T} + \cdots\right]$$

Triangular

$$\frac{8K}{\pi^2}\left[\cos\frac{2\pi x}{T}+\frac{1}{3^2}\cos\frac{6\pi x}{T}+\frac{1}{5^2}\cos\frac{10\pi x}{T}+\cdots\right]$$

Sawtooth

$$\frac{2K}{\pi}\left[\sin\frac{2\pi x}{T}-\tfrac{1}{2}\sin\frac{4\pi x}{T}+\tfrac{1}{3}\sin\frac{6\pi x}{T}-\cdots\right]$$

In the above $f(x)$ is periodic with period T. $\pm K$ is the Amplitude range of waveform.

Methods

Complex numbers

The complex number (z) consists of a real (Re) part $= x$ and an imaginary part $= y$, the imaginary unit (i or j) multiplies the imaginary part y. In normal form, complex numbers are written as:

$$z = x + iy \text{ or } z = x + jy \quad i = j$$

in all respects, j often being used for applications.

Example 6.4.1 shows how we apply the formulae to manipulate complex numbers, that is to add, subtract, multiply and divide them.

Example 6.4.1

Add, subtract, multiply and divide the following complex numbers:

(a) $(3 + 2j)$ and $(4 + 3j)$; (b) general $(a + bj)$ and $(c + dj)$

Addition:

$$(3 + 2j) + (4 + 3j) = 3 + 4 + 2j + 3j$$
$$= 7 + 5j$$
$$(a + bj) + (c + dj) = (a + c) + (b + d)j.$$

Subtraction:

$$(3 + 2j) - (4 + 3j) = -1 - j$$
$$(a + bj) - (c + dj) = (a - c) + (b - d)j.$$

Multiplication:

$$(3 + 2j) \times (4 + 3j) = 3(4 + 3j) + 2j(4 + 3j)$$
$$= 12 + 9j + 8j + 6j^2.$$

Now from definition $j = \sqrt{-1}$, therefore $j^2 = -1$ so,

$$= 12 + 17j + (6)(-1)$$
$$= 6 + 17j$$
$$(a + bj) \times (c + dj) = ac + adj + bcj + bdj^2$$
$$= ac + adj + bcj - bd \text{ (where } j^2 = -1)$$
$$= (ac - bd) + (ad + bc)j$$

cont.

so the result of mutiplication is still a complex number.
Division:

$$\frac{3+2j}{4+3j}$$

here we use an algebraic trick to assist us. We multiply top and bottom by the conjugate of the complex number in the denominator.

So here $z = 4 + 3j$ then $\bar{z} = 4 - 3j$ and we proceed as follows:

$$\left(\frac{3+2j}{4+3j}\right)\left(\frac{4-3j}{4-3j}\right) = \frac{12 - 9j + 8j - 6j^2}{16 + 12j - 12j - 9j^2} = \left(\frac{18 - j}{25}\right)$$

Note denominator becomes real, and in general:

$$\frac{a+bj}{c+dj} = \left(\frac{a+bj}{c+dj}\right)\left(\frac{c-dj}{c-dj}\right)$$

$$= \frac{ac - adj + bcj - bdj^2}{c^2 + cdj - cdj - dj^2}$$

$$= \frac{(ac + bd) + (-adj + bcj)}{c^2 + d^2}.$$

Complex numbers may be transformed from Cartesian to polar form by finding their *modulus* and *argument*, as the next example shows.

Example 6.4.2

Express the complex numbers: (i) $z = 2 + 3j$; (ii) $z = 2 - 5j$ in polar form.

You will remember from your study of co-ordinate systems that polar co-ordinates are represented by an angle θ and a magnitude r (figure example 6.4.2). Complex numbers may be represented in the same way.

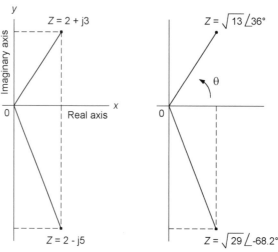

Complex number co-ordinate systems

cont.

To express complex numbers in polar form we first find their modulus and argument. So from formulae 1d, if we have for $z = 2 + 3j$ modulus $= r = \sqrt{2^2 + 3^2} = \sqrt{13}$.   Argument $= \theta$ where $\tan \theta = y/x = 3/2 = 1.5$ $\theta = 56.3°$. Then

$$z = 2 + 3j = \sqrt{13}(\cos 56.3 + j \sin 56.3)$$

$$z = \sqrt{13} \angle 56.3 \text{ (short hand form).}$$

Similarly   for $z = 2 - 5j$ then   modulus $= |z| = r$ $= \sqrt{2^2 + (-5)^2}$ so $r = \sqrt{29}$ and the argument $= \theta = -5/2$ $= -2.5 = -68.2°$. Then

$$z = 2 - 5j = \sqrt{29}\,(\cos(-68.2) + j \sin(-68.2))$$

$$z = \sqrt{29} \angle -68.2.$$

The argument of a complex number in polar form, represents the angle θ measured anticlockwise from the positive x-axis. Its tangent is given by y/x, therefore if we consider the argument θ in radians, it can take on an infinite number of values, which are determined up to 2π radians.

When we consider complex numbers in Cartesian form, then each time we multiply the complex number by $(i = j)$, the complex vector shifts by 90 degrees or $\pi/2$ radians. This fact is used when complex vectors represent *phasors* (electrical vectors), successive multiplication by j, shifts the phase by $\pi/2$, as shown in example 6.4.3. Under these circumstances the imaginary unit j is known as the *j-operator* (see Chapter 3 complex notation).

Example 6.4.3

Multiply the complex number, $z = 2 + 3j$ by the *j*-operator, three times in succession.

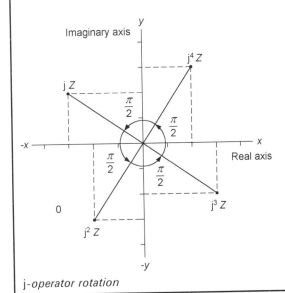

j-operator rotation   *cont.*

Then $jz = j(2 + 3j) = 2j + 3j^2 = 2j - 3$

$j^2z = j(2j - 3) = 2j^2 - 3j = -3j - 2$

$j^3z = j(-3j - 2) = -3j^2 - 2j = -2j + 3$

$j^4z = j(-2j + 3) = -2j^2 + 3j = 2 + 3j.$

Note that $z = j^4z$ we have rotated the vector (phasor) through 2π radians, back to its original position, as shown in the figure on the previous page.

We leave this very short introduction to complex numbers by considering the arithmetic operations of multiplication and division of complex numbers in polar form. Addition and subtraction is not considered because we have to convert the complex number from polar to Cartesian form before we can perform these operations!

When we multiply a complex number in polar form; we multiply their moduli and add their arguments. Conversely for division, we divide their moduli and subtract their arguments.

Example 6.4.4

For the complex numbers given below find the product of z_1 and z_2 also find z_1/z_2.

$z_1 = 3(\cos 120 + j \sin 120)$

$z_2 = 4(\cos -45 + j \sin -45).$

Then $z_1 z_2 =$ multiple of their moduli and the addition of their arguments.

So $z_1 z_2 = (3)(4)[\cos(120 - 45) + j \sin(120 - 45)]$

$\qquad = 12(\cos 75 + j \sin 75).$

and similarly for division $z_1/z_2 =$ divisor of their moduli and the subtraction of their arguments.

So $\dfrac{z_1}{z_2} = \dfrac{3}{4}[\cos(120 + 45) + j \sin(120 + 45)$

$\dfrac{z_1}{z_2} = 0.75(\cos 165 + j \sin 165).$

For the abbreviated version of complex numbers in polar form, we can multiply and divide in a similar manner. Once again they need to be converted into Cartesian form to be added and subtracted.

So in abbreviated form $z_1 = 3 \angle 120$ and $z_2 = 4 \angle 45$. Then to multiply $z_1 z_2$ we multiply their moduli and add their arguments.

So $z_1 z_2 = r_1 r_2 \underline{/\theta_1 + \theta_2} = 12 \angle 120 - 45 = 12 \angle 75°$

Similarly $\dfrac{z_1}{z_2} = \dfrac{r_1}{r_2} \angle \theta_1 - \theta_2 = \dfrac{3}{4} \angle 120 + 45 = 0.75 \angle 165°$

as before.

Differential equations

The theory and solution of differential equations (DEs) is a science in its own right. The methods for their solution include the use of direct integration, separation of variables, integrating factors, D-operators, power series solutions, complex numerical techniques and many others.

We will only be concerned with DEs, which can be solved using elementary techniques. A DE is simply an equation which contains differential coefficients. The order of a differential equation is given by the order of the highest derivative. For example:

$$\frac{dy}{dx} = 6x + 9 \qquad\qquad \text{first order}$$

$$\frac{d^2y}{ds^2} = ms = q \qquad\qquad \text{second order}$$

$$\frac{d^3y}{dx^3} = ky \text{ or } \frac{1}{y}\frac{d^3y}{dx^3} = k \quad \text{third order.}$$

Differential equations are often found in engineering, and express certain properties or laws relating to rates of change. They are formulated as a result of a situation under investigation or as a result of direct differentiation, their solution requires us to reverse these processes. The method of solution will depend on the number of variables and constants present. Methods involving direct integration and *separation of variables* are given below.

Example 6.4.5

Solve the following differential equations:

(a) $\dfrac{dy}{dx} = x^3 - 2$.

(b) $\dfrac{dy}{dx} = 3y - 6$.

(c) $\dfrac{dy}{dx} = \dfrac{y^2}{x}$.

  (a) This equation may be solved by using direct integration.

$$\int \frac{dy}{dx} = \int x^3 - 2 \, dx$$

then $y = \dfrac{x^4}{4} - 2x + c$

(notice the appearance of the constant of integration, we would need to have some information about the function to find a *particular value* for c. This information is referred to as the *boundary conditions* for the DE, as you will see later).

  (b) Again we may integrate directly, after simple rearrangement of the variables.

cont.

$$\frac{dy}{dx} = 3y - 6$$

then $\left(\dfrac{1}{3y - 6}\right) \dfrac{dy}{dx} = 1$

and $\displaystyle\int \left(\dfrac{1}{3y - 6}\right) dy = \int 1 \, dx$

$\frac{1}{3} \ln(3y - 6) = x + c$ (see rules of integration)

$\ln(3y - 6) = 3x + 3c$

$3y - 6 = e^{3x+3c}$ (antiloging!)

$3y = e^{3x+3c} + 6$

$y = \frac{1}{3}e^{3x+3c} + 2.$

(c) $\dfrac{dy}{dx} = \dfrac{y}{x}$

here we need to separate the variables since there are two present x and y. So transposing the equation gives:

$$\frac{dy}{y} = \frac{dx}{x}$$

or $\dfrac{1}{y} d = \dfrac{1}{x} dx$ and integrating

$$\int \frac{1}{y} d = \int \frac{1}{x} dx$$

so $\dfrac{-1}{y} = \ln x + c$

(now since c is a constant let it equal $\ln c$)

then $\dfrac{-1}{y} = \ln x + \ln c$

$\dfrac{-1}{y} = \ln xc$ (laws of logs)

$y = -\dfrac{1}{\ln cx}.$

Linear first order differential equations, which can be put into the form

$$\frac{dy}{dx} + P(x)y = Q(x),$$

where P and Q are functions of x, may be solved using an *integrating factor* (IF).

Where   IF $= e^{\int P \, dx}$

this assumes that we can evaluate the integral of P. If we cannot the method cannot be used.

Now we multiply the whole equation by $e^{\int P \, dx}$

then $\quad e^{\int P \, dx} \cdot \dfrac{dx}{dx} + e^{\int P \, dx} P(x) y = Q(x) e^{\int P \, dx}$.

This can be written as

$$\frac{d}{dx} (e^{\int P \, dx} y) = Q(x) e^{\int P \, dx}.$$

The idea is that the IF will produce a 'perfect differential'. This differential (being a product) may be integrated by parts to solve the equation.

The following example illustrates the integrating factor method for solving linear first order differential equations, if possible.

Example 6.4.6

Solve the equation

$$x^2 \frac{dy}{dx} + xy = x^3$$

We first place the equation in the correct form which is given by

$$\frac{dy}{dx} + P(x) y = Qx$$

so for our equation

$$\frac{dy}{dx} + \frac{y}{x} = x \quad \text{(on division by } x^2\text{)}.$$

We now identify the IF, where we have $P(x) = 1/x$

so $\quad \text{IF} = e^{\int \frac{1}{x} \, dx}$.

Next we multiply each term of our equation by the IF to give:

$$e^{\int \frac{dx}{x}} \frac{dy}{dx} + e^{\int \frac{dx}{x}} \frac{y}{x} = x e^{\int \frac{dx}{x}}.$$

This may be written as:

$$\frac{d}{dx} (e^{\int P dx} \, y) = x e^{\int P dx}. \tag{1}$$

Substituting IF into equation (1) gives:

$$\frac{d}{dx} (xy) = x \cdot x$$

or $\quad xy = \displaystyle\int x^2 \, dx$

and $\quad xy = \dfrac{x^3}{3} + c.$ $\tag{2}$

Now if we are given the boundary conditions we are able to find a *particular* solution. *Let* $y = 1$ when $x = 3$.
  Then from (2)

$$3 = \frac{3^3}{3} + c$$

cont.

giving

$$c = -6$$

and a particular solution is

$$y = \frac{x^2}{3} - \frac{6}{x} \quad \text{(again from equation 2)}.$$

A relatively simple method exists for the solution of second order differential equations with constant coefficients. The method involves finding the roots of the auxiliary equation (the auxiliary equation can be written down directly from the original DE). Then, to find a *general solution*, we find the complementary function (CF) and add to it a particular solution (see formula 2b).

The whole process is described in the following example.

Example 6.4.7

Obtain a general solution for the equation

$$\frac{d^2y}{dx^2} - 5\frac{dy}{dx} + 6y = 2\sin 4x \quad (1)$$

Step 1: Write down the auxiliary equation, which has the form $an^2 + bn + c = 0$. Where n corresponds to dy/dx, then in our case we have $n^2 - 5n + 6 = 0$.

Step 2: Find the roots of the auxiliary equation. Factorising gives $n = 3, n = 2$.

Step 3: Write down the CF. From formula 2b with values of n being real and not equal we get

$$y = Ae^{3x} + Be^{2x} \quad \text{(where } n_1 = 3, n_2 = 2\text{)}.$$

Step 4: Find a particular solution (often called the Particular Integral, PI).

Here we need to do a little 'guess work'. When $f(x)$ in equation (1) involves $a\sin bx$ or $a\cos bx$, try the solution

$$y = C\sin 4x + D\sin 4x.$$

Then $\quad \dfrac{dy}{dx} = 4C\cos 4x - 4D\sin 4x$

and $\quad \dfrac{d^2y}{dx^2} = -16C\sin 4x - 16D\cos 4x$

so substituting these values into equation (1):

$$\frac{d^2y}{dx^2} - 5\frac{dy}{dx} + 6y = 2\sin 4x$$

gives

$$[-16C\sin 4x - 16D\cos 4x] - 5[4C\cos 4x - 4D\sin 4x]+$$

$$\ldots 6[C\sin 4x + D\cos 4x] = 2\sin 4x$$

and on simplifying this expression we get:

cont.

$-10C \sin 4x - 10D \cos 4x - 20C \cos 4x$

$+20D \sin 4x = 2 \sin 4x.$

Now equating coefficients! (see partial fractions) we obtain:

$-10C + 20D = 2$ (1)

$-20C - 10D = 0$ (2)

multiplying equation (2) by 2 and subtracting from (1):

$$-10C + 20D = 2$$
$$-40C - 20D = 0$$
$$-50C \quad\quad = 2 \quad \text{and} \quad C = -\tfrac{1}{25},$$

and substitution into (1)

gives $(-10)(-\tfrac{1}{25}) + 20D = 2$

$$20D = 2 - \tfrac{2}{5} \quad \text{and} \quad D = \tfrac{2}{25}.$$

So our particular solution is

$$y = -\frac{\sin 4x}{25} + \frac{2 \sin 4x}{25}$$

and the *general solution* to the equation:

$$\frac{d^2 y}{dx^2} - 5\frac{dy}{dx} + 6y = 2 \sin 4x$$

is $= Ae^{3x} + Be^{2x} - \dfrac{\sin 4x}{25} + \dfrac{2 \sin x}{25}.$

The engineering application of DEs given in the next section limits their solution to that of direct integration and the determination of the constants of integration for given boundary conditions. Never the less the techniques illustrated here will prove useful for your future studies.

Vectors and matrices

The techniques required for vector addition and the use of the dot and cross-product will be looked at in the next section, when we consider their application to frameworks. We concentrate here on one or two techniques which will enable you to multiply matrices and represent a system of linear equations in matrix form.

The process of matrix multiplication is detailed in formula 4c. In order to multiply two matrices the number of columns of the first matrix must equal the number of rows of the second matrix. Matrix multiplication requires each matrix to be taken in order. So for two matrices, A and B, the product AB does not equal BA. Matrix multiplication methods are illustrated in Example 6.4.8.

Example 6.4.8

Find AB and BA for the following matrices

$$A = \begin{bmatrix} -1 & 0 \\ 2 & 3 \end{bmatrix} \qquad B = \begin{bmatrix} 1 & 2 \\ 3 & 0 \end{bmatrix}.$$

Both are 2×2 matrices, that is both have 2 rows and 2 columns, so multiplication may be carried out.

$$\text{so} \quad AB = \begin{bmatrix} -1 & 0 \\ 2 & 3 \end{bmatrix}\begin{bmatrix} 1 & 2 \\ 3 & 0 \end{bmatrix} = \begin{bmatrix} a_{11} & a_{12} \\ a_{21} & a_{22} \end{bmatrix}\begin{bmatrix} b_{11} & b_{12} \\ b_{21} & b_{22} \end{bmatrix}$$

$$= \begin{bmatrix} (-1)(1) + (0)(3) \\ (2)(1) + (3)(3) \end{bmatrix} + \begin{bmatrix} (-1)(2) + (0)(0) \\ (2)(2) + (3)(0) \end{bmatrix}$$

$$= \frac{(a_{11}b_{11} + a_{12}b_{21}) + (a_{11}b_{12} + a_{12}b_{22})}{(a_{21}b_{11} + a_{22}b_{21}) + (a_{21}b_{12} + a_{22}b_{22})}$$

$$= \frac{(-1 + 0) + (-2 + 0)}{(2 + 9) + (4 + 0)}$$

$$= \begin{bmatrix} -1 & -2 \\ 11 & 4 \end{bmatrix}$$

$$\text{and} \quad BA = \begin{bmatrix} 1 & 2 \\ 3 & 0 \end{bmatrix}\begin{bmatrix} -1 & 0 \\ 2 & 3 \end{bmatrix} = \begin{bmatrix} b_{11} & b_{12} \\ b_{21} & b_{22} \end{bmatrix}\begin{bmatrix} a_{11} & a_{12} \\ a_{21} & a_{22} \end{bmatrix}$$

$$= \begin{bmatrix} (1)(-1) + (2)(2) \\ (3)(-1) + (0)(2) \end{bmatrix}\begin{bmatrix} (1)(0) + (2)(3) \\ (3)(0) + (0)(3) \end{bmatrix}$$

$$= \frac{(b_{11}a_{11} + b_{12}a_{21}) + (b_{11}a_{12} + b_{12}a_{22})}{(b_{21}a_{11} + b_{22}a_{21}) + (b_{21}a_{12} + b_{22}a_{22})}$$

$$= \begin{bmatrix} (-1 + 4) + (0 + 6) \\ (-3 + 0) + (0 + 0) \end{bmatrix}$$

$$= \begin{bmatrix} 3 & 6 \\ -3 & 0 \end{bmatrix}.$$

Note $AB \neq BA$.

Square matrices with 1s on the lead diagonal are of special interest, such as:

$$\begin{bmatrix} 1 & 0 \\ 0 & 1 \end{bmatrix}, \begin{bmatrix} 1 & 0 & 0 \\ 0 & 1 & 0 \\ 0 & 0 & 1 \end{bmatrix} \ldots, \text{etc.}$$

A matrix of this form is called an *identity matrix* and is denoted by the symbol I. Multiplying a matrix A by the identity matrix I we get $IA = A$. In other words the identity matrix acts like the number 1 in arithmetic. Also, if A is any square matrix, and if a matrix B can be found such that $AB = I$ then A is said to be *invertible* and B is called an *inverse* of A.

Example 6.4.9

Show that the matrix $B = \begin{bmatrix} 3 & 5 \\ 1 & 2 \end{bmatrix}$ is an inverse of $A = \begin{bmatrix} 2 & -5 \\ -1 & 3 \end{bmatrix}$. Then we need to show that $AB = I$

$$AB = \begin{bmatrix} 2 & -5 \\ -1 & 3 \end{bmatrix} \begin{bmatrix} 3 & 5 \\ 1 & 2 \end{bmatrix} = \begin{bmatrix} (6-5)+(10-10) \\ (-3+3)+(-5+6) \end{bmatrix}$$

$$= \begin{bmatrix} 1 & 0 \\ 0 & 1 \end{bmatrix} = I.$$

If A is invertible as above then the *inverse* of A is represented as A^{-1} and so $AA^{-1} = I$.

One more example of matrix multiplication is given in example 6.4.10, make sure you can obtain the resulting identity matrix.

Example 6.4.10

For matrix A and matrix B given below, show that matrix B is the inverse of A.

$$A = \begin{bmatrix} 2 & 1 & 0 \\ 1 & -1 & 5 \\ -1 & -1 & 2 \end{bmatrix} \qquad B = \begin{bmatrix} -3 & 2 & -5 \\ 7 & -4 & 10 \\ 2 & -1 & 3 \end{bmatrix}.$$

Then we need to show that $AB = I$, where $B = A^{-1}$.

$$AB = \begin{bmatrix} 2 & 1 & 0 \\ 1 & -1 & 5 \\ -1 & -1 & 2 \end{bmatrix} \begin{bmatrix} -3 & 2 & -5 \\ 7 & -4 & 10 \\ 2 & -1 & 3 \end{bmatrix}$$

$$= \begin{bmatrix} (-6+7)+(4-4)+(-10+10) \\ (-3-7+10)+(2+4-5)+(-5-10+15) \\ (3-7+4)+(-2+4-2)+(5-10+6) \end{bmatrix}$$

'0' products omitted for clarity

$$= \begin{bmatrix} 1 & 0 & 0 \\ 0 & 1 & 0 \\ 0 & 0 & 1 \end{bmatrix} = I$$

So $AB = I$, or $AA^{-1} = I$ where $B = A^{-1}$.

Systems of linear equations may be put into matrix form and solved using *elementary row operations*. The complete process is illustrated in our final example on matrix methods, given below.

Example 6.4.11

Solve the system of linear equations:

$$-x + y + 2z = 2$$

$$3x - y + z = 6$$

$$-x + 3y + 4z = 4.$$

The first step in the process is to write down the *augmented matrix*. All we do is represent the coefficients of the variable and the numbers on the right-hand side of the equality signs, in the form of a matrix array.

Then $A = \begin{bmatrix} -1 & 1 & 2 & 2 \\ 3 & -1 & 1 & 6 \\ ⊝{-1} & 3 & 4 & 4 \end{bmatrix}$ note we remove the $=$ sign.

Next we carry out elementary row operations as necessary to produce a matrix of the form

$$ax + by + cz = d \qquad (1)$$

$$0 + ey + fz = g \qquad (2)$$

$$0 \quad 0 \quad hz = i \qquad (3)$$

then we use backward substitution to find a value for z, from equation (3), placing value for z into equation (2) and solving for y and so on.

So our first set of row operations need to eliminate $⊝{-1}$, from matrix A as indicated. The row operations we are allowed to perform, correspond to the operations we can perform on the systems of equations. These are:

(1) multiply a row through by a non-zero constant;
(2) interchange two rows;
(3) add a multiple of one row to another.

Then for our augmented matrix we eliminate x from row 3, ie. -1 by adding ($-1 \times$ row 1) to row 3, then adding ($-1 \times$ row 1) to row 3 gives

$$\begin{bmatrix} -1 & 1 & 2 & 2 \\ 3 & -1 & 1 & 6 \\ 0 & 2 & 2 & 2 \end{bmatrix}$$

(Notice only the row being operated on changes).

Adding ($3 \times$ row 1) to row 2 gives

$$\begin{bmatrix} -1 & 1 & 2 & 2 \\ 0 & 2 & 7 & 12 \\ 0 & 2 & 2 & 2 \end{bmatrix}.$$

Adding ($-1 \times$ row 2) to row 3 gives

cont.

$$\begin{bmatrix} -1 & 1 & 2 & 2 \\ 0 & 2 & 7 & 12 \\ 0 & 0 & -5 & -10 \end{bmatrix}.$$

Now our system of equations looks like:

$$-x + y + 2z = \quad 2 \tag{1}$$

$$2y + 7z = \quad 12 \tag{2}$$

$$-5z = -10 \tag{3}$$

and using *back substitution* we find that from equation (3) $z = 2$ and from equation (2) $y = -1$ and finally from equation (1) $x = 1$.

Laplace transforms

Differential equations are difficult to formulate and to solve as well as being difficult to manipulate. To help us overcome these problems we can revert to the use of *operators*. So, for example, if we find multiplication of numbers difficult we use the logarithm operator to convert multiplication into the simpler process of addition. We can adopt this idea by using the *s-operator*, or *Laplace transform* (LT). This enables us to take our differential equation or other complex function which is in the time domain, convert it into the s domain (a complex domain where $s = j\omega$) manipulate the simpler result, then convert it back to the time domain using Inverse Laplace Transforms to obtain our desired solution.

The definition of the Laplace Transform is given at the beginning of the table in the formula sheet, it is included again here with some further comments.

The Laplace Transform of a function $f(t)$ is given by:

$$\mathcal{L}[f(t)] = \int_0^\infty e^{-st} f(t)\, dt = F(s)$$

whenever the integral exists.

Note that the Laplace Transform (LT) is denoted by the capital letter corresponding to the original function and is a function of s rather than t.

We will assume that any required transforms do actually exist. We will also assume that we are considering $t \geq 0$ only.

The above definition may be used to evaluate a Laplace Transform of a function. For example to find the LT of $f(t) = t$, then

$$\mathcal{L}[t] = \lim_{k \to \infty} \int_0^k e^{-st} \cdot t\, dt.$$

This integral is a product, so integrating by parts we get:

$$= \left[-\frac{1}{s^2} e^{-st}(st + 1) \right]_0^k$$

$$= -\frac{1}{s^2} e^{-sk}(sk + 1) + \frac{1}{s^2}$$

and as $k \to \infty$ the first term $\to 0$ (true if $s > 0$). Then:

$$\mathcal{L}[t] = \frac{1}{s^2}.$$

As you can see, obtaining Laplace transforms from first principles can be a tedious and often difficult process, fortunately there exist tables of standard LT to which we may refer. In practice we nearly always look up Laplace transforms in tables and, you should always try this first. The following rules may then help you to evaluate LTs of more complicated functions using the results for elementary functions, which may be read directly from the tables.

(1) $\mathcal{L}[f(t) + g(t)] = \mathcal{L}[f(t)] + \mathcal{L}[g(t)]$

(2) $\mathcal{L}[af(t)] = a\mathcal{L}[f(t)] \quad a = \text{constant}.$

The above rule indicates that $\mathcal{L}$ is a *linear operator* (the s-operator).

(3) If $\qquad \mathcal{L}[f(t)] = F(s)$

then $\quad \mathcal{L}[f(at)] = \dfrac{1}{a}F\left(\dfrac{s}{a}\right) \quad$ the scale rule.

(4) If $\qquad \mathcal{L}[f(t)] = F(s)$

then $\quad \mathcal{L}[e^{at}f(t)] = F(s - a).$

This rule is the *shift rule*, so called because the transform function $F(s)$ is 'shifted' a distance 'a' along the s-axis by the presence of e^{at}.

(5) If $a > 0 \quad f(t) = 0 \quad t < 0$ and $g(t) = f(t - a)$ then

$$\mathcal{L}[g(t)] = e^{-as}\mathcal{L}[f(t)].$$

This rule is equivalent to:

$$\mathcal{L}[u(t - a)f(t - a)] = e^{-as}\mathcal{L}[f(t)]$$

where $u(t)$ is the limit step function

$$u(t) = \begin{cases} 0 & t < 0 \\ 1 & t \geq 0 \end{cases}.$$

(6) $\mathcal{L}[t^n] = F\left(\dfrac{n!}{s^{n+1}}\right) \quad$ for integer $n \geq 0.$

The above rule enables us to apply LTs to powers.

The following example shows the first stage of the LT process, finding the LT for a given function using tables.

Example 6.4.12

Find the Laplace Transform of the following functions

(a) $\mathcal{L}[t^3 - 8t^2 + 1].$

(b) $\mathcal{L}[e^{2t}\sin 3t].$

(c) $\mathcal{L}[3t^2 + 2e^{-t} - 4\cos t].$

(d) $\mathcal{L}[e^{-3t}\sin 2t].$

cont.

(e) $\mathcal{L}[t \cos 3t]$.

(a) $\mathcal{L}[t^3 - 8t^2 + 1] = \mathcal{L}(t^3) - 8\mathcal{L}t^2 + \mathcal{L}1$

Tables and rule 6 $= \dfrac{6}{s^4} - \dfrac{16}{s^3} + \dfrac{1}{s}$.

(b) $\mathcal{L}[e^{2t} \sin 3t] = \dfrac{3}{(s-2)^2 + 9}$ (shift rule 4).

(c) $\mathcal{L}[3t^2 + 2e^{-t} - 4 \cos t] = 3\mathcal{L}(t^2) + 2\mathcal{L}(e^{-t}) - 4\mathcal{L}(\cos t)$

Tables and rule 6 $= \dfrac{6}{s^3} + \dfrac{2}{s+1} - \dfrac{4s}{s^2+1}$.

(d) $\mathcal{L}[e^{-3t} \sin 2t] = \dfrac{2}{(s+3)^2 + 4} = \dfrac{2}{s^2 + 6s + 13}$ decaying sine.

Note that you are given the shift rule in the tables for this function. Otherwise it requires you to apply it to $\mathcal{L}[\sin 2t] = 2/(s^2 + 4)$, then by shift rule 4 with $a = -3$, the above result follows.

(e) $\mathcal{L}[t \cos 3t] = \dfrac{s^2 - 9}{(s^2 + 9)^2}$ for multiple t.

Note the meaning of the multiple in the LT. The LT for $\cos 3t$ is

$$\mathcal{L}[\cos 3t] = \frac{s}{s^2 + 9}$$

on multiplication by 't' when transformed we get

$$\mathcal{L}[t \cos 3t] = -\frac{d}{ds}\left(\frac{s^2}{s^2 + 9}\right) \quad \text{differential of transfom so}$$

$$= -\frac{9 - s^2}{(s^2 + 9)^2} = \frac{s^2 - 9}{(s^2 + 9)^2}$$

that is the result from tables.

We now turn our attention to the techniques required for finding Inverse Laplace transforms. This requires us to use our rules in reverse and to apply the odd algebraic trick to get the function into the form we require. The important *reverse rules* for finding the inverse LTs are given below.

Rules for finding the Inverse Laplace Transform:

(1) $\mathcal{L}^{-1}[F(s) + G(s)] = \mathcal{L}^{-1}[F(s)] + \mathcal{L}^{-1}[G(s)]$.

(2) $\mathcal{L}^{-1}[aF(s)] = a\mathcal{L}^{-1}[F(s)]$.

(3) $\mathcal{L}^{-1}[F(s - a)] = e^{at}\mathcal{L}^{-1}[F(s)]$.

(4) $\mathcal{L}^{-1}[e^{-as}F(s)] = u(t - a)\mathcal{L}^{-1}[F(s)]$

$u(t)$ is the step function.

Note the symmetry between the Laplace transforms and their Inverses.

The following example shows how we put these rules into practice.

Example 6.4.13

Find the Inverse LTs for the following functions:

(a) $\mathcal{L}^{-1}\left(\dfrac{4}{s^2+22}+\dfrac{12}{s-5}\right)$.

(b) $\mathcal{L}^{-1}\left(\dfrac{2}{s^2+2s+5}\right)$.

(c) $\mathcal{L}^{-1}\left(\dfrac{s+1}{s(s^2+1)}\right)$.

(a) $\mathcal{L}^{-1}\left(\dfrac{4}{s^2+22}+\dfrac{12}{s-5}\right)$.

The only difficulty here is being able to put the constants into a form which enables us to read the Inverse directly from the tables. This means that the constant in the numerator needs to be the square root of the constant in the denominator. Then:

$$\mathcal{L}^{-1}\frac{4}{s^2+22}+\mathcal{L}^{-1}\frac{12}{s-5}=4\mathcal{L}^{-1}\frac{1}{s^2+22}+12\mathcal{L}^{-1}\frac{1}{s-5}$$

$$=4\mathcal{L}^{-1}\frac{1}{s^2+22}+12e^{5t}$$

(direct from tables).

The trick for the left-hand inverse is to multiply top and bottom by the square root of 22, i.e. $\sqrt{22}/\sqrt{22}$.
   So we get:

$$4\mathcal{L}^{-1}\frac{\sqrt{22}}{(s^2+22)\sqrt{22}}$$

which gives

$$\frac{4}{\sqrt{22}}\mathcal{L}^{-1}\frac{\sqrt{22}}{s^2+22}$$

this is the form required from the tables

$$\frac{4}{\sqrt{22}}\mathcal{L}^{-1}\frac{\sqrt{22}}{s^2+22}=\left(\frac{4}{\sqrt{22}}\right)\sin\sqrt{22}t+12e^{5t}$$

(b) $\mathcal{L}^{-1}\left(\dfrac{2}{s^2+2s+5}\right)$.

We need this inverse LT in shift form so the denominator looks like $(s+a)^2+\omega^2$. We achieve this by 'completing the square', the quadratic expression in the denominator.
   To complete the square we use the following rule:

$$s^2+as+b=\left(s+\frac{a}{2}\right)^2+b-\frac{a^2}{4}.$$

cont.

Then using this rule we get:

$$s^2 + 2s + 5 = (s+1)^2 + 5 - \tfrac{4}{4}.$$

So $\mathcal{L}^{-1}\left(\dfrac{2}{s^2 + 2s + 5}\right) = \left(\dfrac{2}{(s+1)^2 + 4}\right) = e^{-t}\sin 2t$

direct from tables.

(c) Find $\mathcal{L}^{-1}\left(\dfrac{s+1}{s(s^2+1)}\right)$.

This requires us to use partial fractions (PFs) which you met when studying the algebra in this chapter.

so $\mathcal{L}^{-1}\left(\dfrac{s+1}{s(s^2+1)}\right) = \mathcal{L}^{-1}\left[\dfrac{A}{s} + \dfrac{Bs+c}{s^2+1}\right]$.

Then the PF's are given from:

$$s + 1 = A(s^2 + 1) + (Bs + c)s$$

and equating coefficients:

s^0 gives $\quad 1 = A$

s^1 gives $\quad 1 = C$

s^2 gives $\quad 0 = A + B$

from which $A = 1, B = -1, C = 1$.

$$\mathcal{L}^{-1}\left(\dfrac{s+1}{s(s^2+1)}\right) = \mathcal{L}^{-1}\dfrac{1}{s} + \left(\dfrac{-s+1}{s^2+1}\right)$$

$$= \mathcal{L}^{-1}\dfrac{1}{s} - \dfrac{s}{s^2+1} + \dfrac{1}{s^2+1}$$

$$= 1 - \cos t + \sin t.$$

Our final technique involving the use of LTs is concerned with their use in the solution of differential equations (DEs). In our Laplace transform table (formula 5) we have transforms for the first and second derivatives. The extension to these formulae is given below, where we are able to find the LT for nth order differential equations.

The Laplace Transform for nth derivative ordinary differential equations, may be found by using:

$$\mathcal{L}[f^{(n)}(t)] = s^n \mathcal{L}[f(t)] - s^{n-1}f(0) - s^{n-2}f'(0) \dots$$

$$-s^{n-3}f''(0) \dots - sf^{(n-2)}_{(0)} \dots - f^{(n-1)}_{(0)}.$$

where $f^n(t) = \dfrac{d^n f}{dt^n}$ is the nth derivative.

In the case of ordinary differential equation problems where the function (f) and its derivatives are specified at time $t = 0$ (as in the above rule), they are referred to as *initial value problems*. As opposed to problems where (f) and its derivatives are specified at other values of (t) which are called *boundary value problems*.

The procedure for applying LTs to ordinary differential equations is as follows:

(1)   Initial value problem given in the form of an ordinary differential equation (ODE).

(2)   The Laplace transform is applied to the ODE to produce an algebraic equation.

(3)   The algebraic equation is solved.

(4)   The Inverse Laplace transform is applied, giving the solution of the initial value problem.

Example 6.4.14

Solve $y'' - 4y = t$ given $y(0) = 1, y'(0) = -2$.
(1)   We apply the LT to the ODE using the rules:

$$\mathcal{L}[y''] - 4\mathcal{L}[y] = \mathcal{L}[t]$$

$$\mathcal{L}[y''] = s^2\mathcal{L}[y] - sy(0) - y'(0)$$

$$\text{(applying differential rule)}$$

$$= s^2\mathcal{L}[y] - s + 2.$$

Since the ODE is second order (note the use of the initial value conditions),

and the $\mathcal{L}[t] = \dfrac{1}{s^2}$.

So the transformed equation is:

$$s^2\mathcal{L}[y] - s + 2 - 4\mathcal{L}[y] = \frac{1}{s^2}.$$

(2) Rearranging and manipulating the algebraic equation gives:

$$(s^2 - 4)\mathcal{L}[y] = \frac{1}{s^2} + s - 2.$$

So $\mathcal{L}[y] = \dfrac{1}{s^2(s^2 - 4)} + \dfrac{s - 2}{s^2 - 4}.$

(3)   Now applying Inverse Laplace Transform:

$$y = \mathcal{L}^{-1}\left[\frac{1}{s^2(s^2 - 4)} + \frac{s - 2}{s^2 - 4}\right].$$

Again we require the use of PFs for the term:

$$\frac{1}{s^2(s^2 - 4)} = \frac{A + Bs}{s^2} + \frac{s + D}{s^2 - 4}$$

where $A = -\frac{1}{4}, D = \frac{1}{4}, B = C = 0$, (make sure you are able to find these coefficients).

Then: $y = \mathcal{L}^{-1}\left[-\dfrac{1}{4s^2} + \dfrac{1}{4(s^2 - 4)} + \dfrac{1}{s + 2}\right]$

since $\dfrac{s - 2}{s^2 - 4} = \dfrac{\cancel{(s - 2)}}{\cancel{(s - 2)}(s + 2)}$

$$y = -\tfrac{1}{4}\mathcal{L}^{-1}\frac{1}{s^2} + \tfrac{1}{4}\mathcal{L}^{-1}\frac{1}{s^2 - 4} + \mathcal{L}\frac{1}{s + 2}.$$

cont.

One last piece of algebraic manipulation on the middle term gives:

$$y = -\tfrac{1}{4}\mathcal{L}^{-1}\frac{1}{s^2} + \tfrac{1}{8}\mathcal{L}^{-1}\frac{2}{s^2-4} + \mathcal{L}\frac{1}{s+2}$$

and $\;\; y = -\tfrac{1}{4}t + \tfrac{1}{8}\sinh 2t + e^{-2t} \;\;$ (from tables).

Perhaps you can now see why manipulating algebra is so fundamental to all of your analytical study!

This concludes our study on advanced methods. The methods used for studying waveforms using Harmonic analysis is treated briefly in the next section, as an application.

Engineering applications

We begin our study of advanced applications, with a brief look at the use of harmonic analysis for determining the nature of waveforms, which is a skill needed by those wishing to become Electrical Engineers. We then return to look at the application of complex numbers, differential equations, vectors, and finally Laplace transforms.

Harmonic analysis

In Chapter 4.3 Fourier analysis is used in the study of complex waves. All waveforms whether continuous (like sinusoidal waves) or discontinuous (sawtooth waves), can be expressed in terms of a *convergent series*. You will remember from your study of series, that convergence guarantees the series is *bounded*.

The technique used for harmonic analysis, allows us to produce a better and better approximation to the shape of any waveform by successive addition of a series of sine waves, until the parent wave is replicated to the desired degree of accuracy. The more terms of the series we use, then the greater the number of waves which may be added to produce the parent wave shape. This process is illustrated for the synthesis of a rectangular impulse wave in figure 6.4.1.

For an illustration of the application of Fourier and harmonic analysis to a typical engineering application, we consider a triangular wave form.

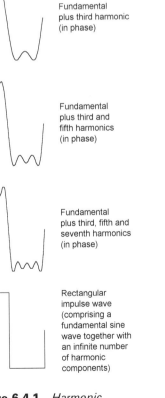

Fundamental sine wave

Fundamental plus third harmonic (in phase)

Fundamental plus third and fifth harmonics (in phase)

Fundamental plus third, fifth and seventh harmonics (in phase)

Rectangular impulse wave (comprising a fundamental sine wave together with an infinite number of harmonic components)

Figure 6.4.1 *Harmonic synthesis of rectangular impulse wave*

Example 6.4.15

Find the Fourier series for the triangular waveform shown below.

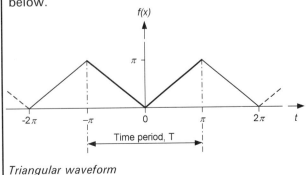

Triangular waveform   *cont.*

From the formula given in 6a, we know that the Fourier series for our triangular wave may be found using:

$$f(t) = \tfrac{1}{2}a_0 + \sum_{n=1}^{\infty} a_n \cos \frac{2\pi nt}{T} + \sum_{n=1}^{\infty} b_n \sin \frac{2\pi nt}{T}. \tag{1}$$

Now for our problem the time period $T = 2\pi$, so the above equation becomes:

$$f(t) = \tfrac{1}{2}a_0 + \sum_{n=1}^{\infty} a_n \cos nt + \sum_{n=1}^{\infty} b_n \sin nt.$$

Note also that $\omega = 2\pi/T$ which may be used in the Series if desired

We now need to find the Fourier coefficients a_0, a_n and b_n for the series. If we were given tabular data (or more numerical information about the triangular wave) we could use the tabular methods given in Example 4.3.3.

Even so, knowing whether the function is EVEN or ODD enables us to *halve* our work, since the integrals involving sine *or* cosine will equal zero. This can be determined by inspection of the waveform under consideration. If the integral of the function between the limits $-\pi$ to $\pi = 0$ (in other words the areas under the graph are both positive and negative) the function is ODD. The formal definition for ODD functions is that $f(x) = -f(-x)$, under these circumstances all the coefficients a_n vanish since $f(x) \cos nx$ is ODD.

If between the limits $-\pi$ to π the function has areas that produce a value the function is EVEN. Or formally if $f(x) = f(-x)$ giving symmetry about the y-ordinate the function is EVEN. For similar reasons as before, the coefficients b_n vanish.

The result of all this is, when the function is EVEN we need only consider the Fourier coefficients a_0, a_n. If the function is ODD then we need only consider the Fourier coefficient b_n.

I hope you can see that the *triangular function* given above is EVEN.

So coefficients $b_n = 0$ $(n = 1, 2, \ldots)$.

$$\text{Coefficients } a_n = \frac{2}{T} \int_{-\frac{1}{2}T}^{\frac{1}{2}T} f(x) \cos \frac{2n\pi x}{T} \, dx \quad \text{(formula 6a)}$$

and since $T = 2\pi$ then:

$$a_n = \frac{1}{\pi} \int_{-\pi}^{\pi} f(t) \cos nt \, dt$$

or $\quad a_n = \dfrac{2}{\pi} \displaystyle\int_0^{\pi} f(t) \cos nt \, dt \quad$ (notice the limits of integration!).

Now $f(t)$ for the interval $-\pi$ to π gives

$$f(t) = \begin{cases} -t & \text{if } -\pi \leq t \leq 0 \\ t & \text{if } 0 \leq t \leq \pi. \end{cases}$$

cont.

This is the *definition for the triangle function*.

So, $a_n = \dfrac{2}{\pi} \displaystyle\int_0^\pi t \cos nt\, dt$ (since $f(t) = t$ from 0 to π).

Now to find a_n we will need to use integration by parts! For this example we will show the working in full.

Then using $\displaystyle\int uv'\, dx = uv - \int u'v\, dx$

where $\qquad\qquad\qquad u = t, u' = 1$

$$v' = \cos nt \quad\text{so}\quad v = \frac{1}{n}\sin t$$

then $\dfrac{2}{\pi}\displaystyle\int_0^\pi t \cos nt\, dt = \dfrac{t}{n}\sin t - \int \dfrac{1}{n}\sin nt\, dt$

$$= \frac{t}{n}\sin t - -\frac{1}{n^2}\cos nt$$

$$= \left[\frac{t}{n}\sin t + \frac{1}{n^2}\cos nt\right]_0^\pi$$

$$= \left[\frac{\pi}{n}\sin \pi + \frac{1}{n^2}\cos n\pi\right] - \left[0 + \frac{1}{n^2}\cos n0\right]$$

$$= \left[0 + \frac{1}{n^2}\cos n\pi\right] - \left[0 + \frac{1}{n^2}\right]$$

$$= \frac{1}{n^2}\cos n\pi - \frac{1}{n^2}$$

$$a_n = \frac{2}{\pi}\int_0^\pi t \cos nt\, dt$$

$$= \frac{2}{\pi n^2}[\cos n\pi - 1].$$

Now if integer n is EVEN $\cos n\pi = +1$ hence $a_n = 0$. If integer n is ODD $\cos n\pi = -1$ hence $a_n = -4/(\pi n^2)$.
We still need to find

$$a_0 = \frac{2}{\pi}\int_0^\pi t \cos 0\, dt = \frac{2}{\pi}\int_0^\pi t\, dt$$

$a_0 = \pi$ (solving integral between limits).

So finally collecting the coefficients and putting them into equation (1) yields:

$$f(t) = \tfrac{1}{2}(\pi) + \sum_{n=1}^{\infty} -\frac{4}{\pi n^2}\cos nt \quad \text{(for ODD 'n')}$$

and expanding

$$f(t) = \frac{\pi}{2} - \frac{4}{\pi}\left[\frac{\cos t}{1^2} + \frac{\cos 3t}{3^2} + \frac{\cos 5t}{5^2} + \cdots\right].$$

Note that the Fourier series expansion does not include EVEN values of 'n' within the bracket, remembering that when n is EVEN $a_n = 0$.

If you have managed to plough your way through the last example, you will realise that to find the coefficients using integration, requires care, skill and extreme patience! This is why, whenever possible you should resort to the tabular method given in Example 4.3.3. Even the tabular calculations associated with this method can be eliminated with the aid of specially designed computer software (see Appendix).

Complex numbers

You may have already applied complex numbers in your study of a.c. We will complement this work by looking at one or two examples, similar to those given in Chapter 3.3.

Example 6.4.16

Find the resultant impedance in a circuit where the impedances are connected in parallel and are defined as follows:

Z_1 Comprises a $20\,\Omega$ resistance connected in series with an inductive reactance (at $+90°$) of $8\,\Omega$.

Z_2 Comprises a 15Ω resistance connected in series with a capacitive reactance (at $-90°$) of $10\,\Omega$.

Complex numbers are required to find an expression for the resultant (total) impedance Z_T of the circuit. Before this can be done we need to write down Z_1 and Z_2 remembering that the j-operator shifts the complex variable by $90°$.

Then $Z_1 = 20 + j8$

$Z_2 = 15 - j10.$

Now impedances (from Chapter 3) in parallel are found using

$$\frac{1}{Z_T} = \frac{1}{Z_1} + \frac{1}{Z_2} + \cdots$$

so we add $\dfrac{1}{Z_1} + \dfrac{1}{Z_2}$ using fractions!

$$\frac{1}{Z_T} = \frac{1}{Z_1} + \frac{1}{Z_2}$$

$$\frac{1}{Z_T} = \frac{Z_2 + Z_1}{Z_1 Z_2} \quad \text{(Now we can turn upside down)}$$

$$Z_T = \frac{Z_1 Z_2}{Z_2 + Z_1}.$$

Now $Z_T = \dfrac{(20 + j8)(15 - j10)}{(20 + j8) + (15 - j10)}.$

Applying complex multiplication and addition gives:

cont.

$$Z_T = \frac{300 + j120 - j200 - j^2 80}{(35 - j2)}$$

$$Z_T = \frac{380 - 80j}{35 - 2j} \quad \text{(now multiplying top and bottom by the conjugate of denominator)}$$

then $\quad Z_T = \frac{(380 - 80j)(35 + 2j)}{(35 - 2j)(35 + 2j)}$

$$= \frac{13\,300 + 760j - 2800j - 160j^2}{1225 + 70j - 70j - 4j^2}$$

$$= \frac{13\,460 - 2040j}{1229}$$

so $\quad Z_T = 10.95 - 1.66j.$

Example 6.4.17

An impedance $Z = (200 + j100)\,\Omega$ is connected to a 50 V supply. Find an expression for the current in complex form given that $I = V/Z$ and I, V and Z are complex variables.

Then $I = V/Z$ simply requires complex division

$$I = \frac{50}{200 + j100}$$

$$= \frac{50(200 - j100)}{(200 + j100)(200 - j100)}$$

$$= \frac{10\,000 - j5000}{40\,000 - j^2 10\,000}$$

$$I = \frac{10000 - j5000}{50\,000}$$

$$= 0.2 - j0.1$$

so $\quad I = (0.2 - j0.1)$ Amps.

Vectors and matrices

You should be familiar with the concept of a vector and vector addition of forces. Look at the formulae on vectors at the beginning of this section. The Cartesian co-ordinate system in three dimensions is represented by the position vectors i, j, and k which act at right-angles to each other, they are *orthogonal*.

A *position vector* system is directly comparable with the Cartesian vector system and uses the same direction vectors i, j and k. The position vector r is defined as a fixed vector which locates a point in space relative to another point. The position vector can be expressed in Cartesian form as $r = xi + yj + zk$ (Figure 6.4.2). Let us consider an example, which will illustrate the power of the position vector system.

Figure 6.4.2 *Position vector system*

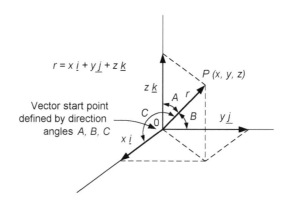

Example 6.4.18

Determine the magnitude and direction of the pole r, shown in figure example 6.4.18.

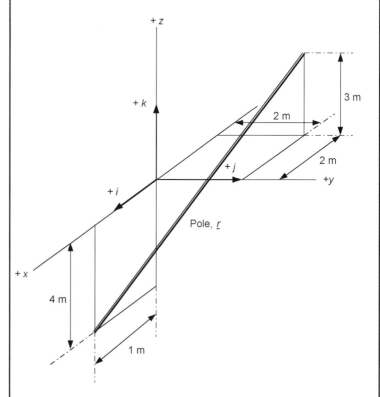

The vector representing the pole can be found by determining the co-ordinates of its start and finish points A and B.

$A(1, 0, -4), B(-2, 2, 3)$.

Now $r = r_B - r_A$ (in other words position at finish minus position at start, defines size and position of r).

so $r = (-2 - 1)i + (2 - 0)j + (3 - -4)k$

$r = -3i + 2j + 7k$.

cont.

Then the magnitude of $|r| = \sqrt{(-3)^2 + (2)^2 + (7)^2} = 7.9$ m. The unit vector u is defined as $u = r/r$ that is the unit vector in the direction of r is formed by dividing the individual components of r by the size of $|r|$.

This is identical to dividing a vector by its scalar multiplier to obtain the unit vector. Then the unit vector

$$u = \frac{t}{r} = \frac{-3}{7.9}i + \frac{2}{7.9}j + \frac{7}{7.9}k.$$

Now the components of the unit vector give the cosines of the coordinate direction angles, of the strut from its start point. So these angles position the strut in its fixed direction as it leaves A for B.

Then $\cos \theta_i = -0.3797$ so $\theta_i = 112.3°$

$\cos \theta_j = 0.2538$ $\theta_j = 75.3°$

$\cos \theta_k = 0.866$ $\theta_k = 27.6°$

So we have determined both the size of the strut and its direction in space.

The important points to note in this example are

(i)   the technique for finding the vector representing a *line in space*, considering its start and finish points that is
$r = r_B - r_A$

(ii) Finding the *direction cosines* using the unit vector u (this always works).

(iii) Noting that the magnitude of the vector $|r|$ was found in the normal way (formula 3a).

If you find difficulty in following this application of vectors you will need to ensure that you review the theory associated with vector geometry.

The next example shows the use of the vector dot product. The dot product has two very useful applications. It enables us to find the angle formed between two vectors or intersecting lines and, when used with the unit vector it enables us to find the component of a vector parallel to a line. These applications are illustrated for the force system shown in the next example.

Example 6.4.19

An overflow pipe is subject to a force $F = 300\,N$ at its end B. Determine the angle θ between the force and the pipe section AB and, find the magnitudes of the components of the force F, which act parallel and perpendicular to B. Figure example 6.4.19 shows the situation.

To find the angle θ we use the dot product as given in the formula sheet.

cont.

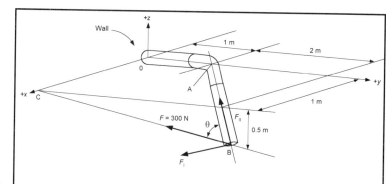

Overflow pipe assembly

$$A \cdot B = |A||B|\cos\theta$$

$$\text{or} \quad \cos\theta = \frac{A \cdot B}{|A||B|}.$$

To locate the point of contact of force vector **F** we first find the position vector along BA and BC. Thus by inspection

$$r_{BA} = (-2i - 2j + 0.5k)m.$$

Note that to get from B to the point A, we travel back 2 m along the x-axis, 2 m back along the y-axis and 0.5 m up the z-axis.

Similarly: $\tau_{BC} = (0i - 3j + 0.5k)$. That is from B to C we travel 3 m along the y-axis and up 0.5 m along the z-axis, C and B are the same distance along the x-axis.

We are now in a position to find $\cos\theta$ using the dot product

$$A \cdot B = [(-2)(0) + (-2)(-3) + (0.5)(0.5)]$$

$$= (0 + 6 + 0.25)$$

and $\quad |r_{BA}| = \sqrt{(-2)^2 + (-2)^2 + (0.5)^2} = \sqrt{8.25}$

$$|r_{BC}| = \sqrt{(0)^2 + (-3)^2 + (0.5)^2} = \sqrt{9.25}$$

then $\quad \cos\theta = \dfrac{A \cdot B}{|r_{BA}||r_{BC}|} = \dfrac{6.25}{\sqrt{8.25}\sqrt{9.25}} = \dfrac{6.25}{8.736} = 0.7155$

giving $\quad \theta = 44.3°.$

Now we have θ then force parallel to pipe section F_{11} is given by $F_{11} = F\cos\theta = (300)(0.7155) = 214.7$ N and $F_1 = F\sin\theta = (300)(0.6984) = 209.5$ N.

We now look at an application for elementary row operations, for the solution of linear equations. This method is sometimes referred to as *Gaussian elimination*.

Example 6.4.20

The equations shown below were obtained by modelling an electric circuit having two emfs of 24 V and 12 V, using Kirchoff's laws. Using Gaussian elimination find the values of the currents I_1, I_2, I_3.

$$12I_1 + 4(I_1, -I_2) = 24$$
$$4(I_2 - I_1) + 3(I_2 - I_3) = 0$$
$$3I_3 + 3(I_3 - I_2) = 12.$$

We first simplify the equations to give:

$$16I_1 - 4I_2 = 24$$
$$-4I_1 + 7I_2 - 3I_3 = 0$$
$$-3I_2 + 6I_3 = 12.$$

Now write down the augmented matrix

16	−4	0	24	(row 1)
−4	7	−3	0	(row 2)
0	−3	6	12	(row 3)

Now using elementary row operations we get the matrix into *row-echelon* form, that is zeros to the left and below the lead diagonal. We need to eliminate the −4 term from row 2. This can be achieved by adding a quarter of row 1 to row 2.

So Adding ($\frac{1}{4} \times$ row 1) to row 2 gives:

16	− 4	0	24	(1)
0	6	− 3	6	(2)
0	− 3	6	12	(3)

Adding ($\frac{1}{2} \times$ row 2) to row 3 gives:

16	− 4	0	24	(1)
0	6	− 3	6	(2)
0	0	4.5	15	(3)

We are now in row-echelon form, so replacing matrix with original equations gives:

$$16I_1 - 4I_2 = 24$$
$$6I_2 - 3I_3 = 6$$
$$4.5I_3 = 15.$$

Now using backward substitution we get:

cont.

$$I_3 = \frac{15}{4.5} = 3\tfrac{1}{3},$$

$6I_2 - (3)(3\tfrac{1}{3}) = 6$ so $I_2 = 2\tfrac{2}{3}$.

and finally: $16I_1 - (4)(2\tfrac{2}{3}) = 24$ so $I_1 = 2\tfrac{1}{6}$.

Then required current values are:

$$I_1 = 2\tfrac{1}{6}A, \quad I_2 = 2\tfrac{2}{3}A, \quad I_3 = 3\tfrac{1}{3}A.$$

Differential equations

We know from our previous work that a first order differential equation will have one constant of integration, a second order will have two, and so on. In order to solve these equations completely we need to be given boundary conditions. In Chapter 5, we considered the use of the Macaulay method for the bending of beams, this required the formulation of differential equations, for different situations. To solve these equations the boundary conditions were considered, as the next example shows.

Example 6.4.21

The figure shows a simply supported beam subject to a uniformly distributed load, causing a deflection y.

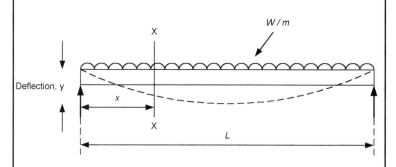

Simply supported beam

If the differential equation modelling the situation is given by

$$EI\frac{d^2 y}{dx^2} - M = 0$$

where the bending moment M at section x–x is $\dfrac{\omega Lx}{2} - \dfrac{\omega x^2}{2}$.

The product EI may be treated as a constant. Solve the DE given the boundary conditions:

$x = 0, \quad y = 0$

$x = L, \quad y = 0$.

Then we have:

cont.

$$EI\frac{d^2y}{dx^2} = \frac{\omega Lx}{2} - \frac{\omega x^2}{2} \quad \left(M = \frac{\omega Lx}{2} - \frac{\omega x^2}{2}\right).$$

This is a second order differential equation which can be solved by direct integration.

So integrating both sides once gives:

$$EI\frac{dy}{dx} = \frac{\omega Lx^2}{4} - \frac{\omega x^3}{6} + A \quad (A \text{ is constant of integration}).$$

and integrating again gives:

$$EI_y = \frac{\omega Lx^3}{12} - \frac{\omega x^4}{24} + Ax + B. \tag{1}$$

Now using the boundary conditions to find constant A and B. Then when $x = 0, y = 0$ and substituting these values into (1) gives $B = 0$. Also when $x = L, y = 0$ and from equation (1) again we get:

$$0 = \frac{\omega L^4}{12} - \frac{\omega L^4}{24} + AL$$

$$-\frac{\omega L^4}{24} = AL$$

and $\quad A = -\dfrac{\omega L^3}{24}.$

Therefore $\quad EI_y = \dfrac{\omega Lx^3}{12} - \dfrac{\omega x^4}{24} - \dfrac{\omega xL^3}{24}$

hence $\quad y = \dfrac{1}{EI}\left[\dfrac{\omega Lx^3}{12} - \dfrac{\omega x^4}{24} - \dfrac{\omega xL^3}{24}\right].$

The next example involves a differential equation of first order and relies on knowledge of Newton's second law of motion for its solution.

Example 6.4.22

From Newton's second law $F = ma$ it can be shown that for a body falling freely in a vacuum then:

$$m\frac{dv}{dt} = m\boldsymbol{g}$$

where $\quad m =$ mass of body,

$\qquad \boldsymbol{g} =$ acceleration due to gravity (9.81 m/s^2),

and $\quad v =$ velocity of body (m/s).

Given that $v = 0$, when $t = 0$, solve the differential equation and find the velocity of the body when $t = 3.5$ seconds.

cont.

The differential equation $m\dfrac{dv}{dt} = m\boldsymbol{g}$ is easily solved by separation of variables.

Rearranging gives $dv = \boldsymbol{g}\, dt$

so $\displaystyle\int 1 \cdot dv = \boldsymbol{g} \int dt$ \hfill (1)

$$= v = \boldsymbol{g}t + A.$$

Now using boundary condition $v = 0$ when $t = 0$ in equation (1) gives $A = 0$.

so $v = \boldsymbol{g}t$ and at time $3.5\,\text{s}$

$$v = (9.81)(3.5) = 34.3 \text{ m/s}.$$

Laplace transforms

Laplace transforms play a particularly important role in the modelling of engineering control systems. A control system transfer function G gives the ratio of the output signal to the input signal for the system. Since the modelling of engineering systems often involves the formulation of DEs, then it becomes necessary to manipulate these equations in order to modify system behaviour. This difficult process is best achieved by using Laplace transforms and carrying out the desired manipulation in the *s-plane*.

The analysis of a.c. electrical circuits involves the representation of system variables such as capacitive reactance, as complex functions. Again, the DEs formed using these variables are best manipulated in the s-plane, after the Laplace operator has been applied.

The final examples in this section, illustrate the above uses.

Example 6.4.23

The closed loop transfer function for a control system is given below. Determine the time domain response to a unit-step input.

$$G = \frac{2}{s(s-5)}.$$

Note first that G is already in the s-domain, that is, the Laplace transform has been performed on G.

To convert to the time domain we are required to find the $\mathcal{L}^{-1}$.

Before we do this we are told that the system is subject to a unit step response. From our Laplace transform table a unit step response in the 's' domain is given as $\dfrac{1}{s}$. Therefore, the system transfer function G, is now:

$$F(G) = \frac{1}{s}\left(\frac{2}{s(s-5)}\right) = \frac{2}{s^2(s-5)}$$

therefore we require

cont.

$$\mathcal{L}^{-1}\frac{2}{s^2(s-5)}$$

this cannot be found directly from the tables, so we need to simplify G, by finding the PFs.

Then $\dfrac{2}{s^2(s-5)} = \dfrac{A}{s^2} + \dfrac{B}{s} + \dfrac{C}{s-5}.$

So: $2 = As - 5A + Bs^2 - 5Bs + Cs^2.$

If we equate coefficients then:

$s^2 \quad 0 = B + C$

$s^1 \quad 0 = A - 5B$

$s^0 \quad 2 = -5A.$

So: $A = -\dfrac{2}{5}, \quad B = -\dfrac{2}{25}, \quad C = \dfrac{2}{25}.$

Then we require $\mathcal{L}^{-1}\left[\dfrac{2}{25(s-5)} - \dfrac{2}{25s} - \dfrac{2}{5s^2}\right]$

or $\dfrac{2}{25}\mathcal{L}^{-1}\dfrac{1}{s-5} - \dfrac{2}{25}\mathcal{L}^{-1}\dfrac{1}{s} - \dfrac{2}{5}\mathcal{L}^{-1}\dfrac{1}{s^2}.$

These may now be read directly from the tables to give our function $f(G)$ in the time domain as:

$$f(G) = \frac{2}{25}e^{5t} - \frac{2}{25} - \frac{2t}{5}.$$

Example 6.4.24

An electric circuit is modelled by the DE:

$$V = Ri(t) + L\frac{di(t)}{dt} + \frac{1}{C}\int i(t)\,dt.$$

Solve the equation *in terms of the current* (i) using Laplace Transforms, given that $i = 0$ at $t = 0$ and $v = 12, L = 2, R = 8$ and $C = 0.025$. Applying the Laplace Transform gives:

$$\mathcal{L}[v] = R\mathcal{L}[i(t)] + L\mathcal{L}\left[\frac{di(t)}{dt}\right] + \mathcal{L}\left[\frac{1}{C}\int i(t)\,dt\right].$$

Using our table then,

$$\frac{V}{s} = RI(s) + sL[I(s) - i(0)] + \frac{1}{sC}[Is + I(0)]$$

and applying initial conditions $i = 0$, at $t = 0$ we get $i(0) = 0$ and $I(0) = 0$.

Therefore: $\dfrac{V}{s} = RI(s) + sL[I(s) - 0] + \dfrac{1}{sC}[I(s) + I(0)]$

$\dfrac{V}{s} = RI(s) + sLI(s) + \dfrac{1}{sC}Is$

cont.

$$\frac{V}{s} = I(s)\left(R + sL + \frac{1}{sC}\right).$$

So in terms of $I(s)$;

$$I(s) = \frac{V}{s\left(R + sL + \frac{1}{sC}\right)} = \frac{V}{s^2 L + sR + \frac{1}{C}}$$

and on substitution of given values:

$$I(s) = \frac{12}{2s^2 + 8s + 40} = \frac{6}{s^2 + 4s + 20}.$$

Now we need to find the *inverse LT*

$$\mathcal{L}^{-1} \frac{6}{s^2 + 5s + 20}$$

We cannot obtain this inverse directly from the tables but, using the method of completing the square gives:

$$\mathcal{L}^{-1} \frac{6}{\left(s + \dfrac{5}{2}\right)^2 + 20 - \dfrac{16}{4}} \quad \text{or}$$

$$\mathcal{L}^{-1} \frac{6}{(s + 2.5)^2 + 16} \quad \text{and}$$

$$6\mathcal{L}^{-1} \frac{1}{(s + 2.5)^2 + 16} \quad \text{and on multiplication by} \quad \frac{\sqrt{16}}{\sqrt{16}}$$

$$6\mathcal{L}^{-1} \frac{\sqrt{16}}{[(s + 2.5)^2 + 16]\sqrt{16}} \quad \text{or}$$

$$\frac{6}{\sqrt{16}} \mathcal{L}^{-1} \frac{\sqrt{16}}{(s + 2.5)^2 + 16} \quad \text{which is required form}$$

for direct use of tables where $F(s) = \dfrac{\omega}{(s + a)^2 + \omega^2}$

then: $I(s) = \left(\dfrac{6}{\sqrt{16}}\right) \dfrac{\sqrt{16}}{(s + 2.5)^2 + 16}$

so $i(t) = \dfrac{6}{\sqrt{16}} e^{-2.5t} \sin \sqrt{16}\, t.$

Compare with example 4.4.3!

We end this section and our study of analytical methods, with a number of problems of varying difficulty.

Questions 6.4.1

(1) Perform the required calculation and express the answer in the form $a + ib$ for:

   (a) $(3 - 2i) - i(4 + 5i)$;   (b) $(7 - 3i)(5i + 3)$;

   (c) i^5;   (d) $\dfrac{1 + 2i}{3 - 4i}$;

   (e) $\dfrac{(4 - i)(1 - 3i)}{-1 + 2i}$.

(2) Represent the following complex numbers in polar form:

   (a) $6 - 6i$;   (b) $3 + 4i$;   (c) $(4 + 5i)^2$;

   (d) $\dfrac{1}{(1 - 2i)^2}$.

(3) Express the following complex numbers in Cartesian form:

   (a) $\sqrt{30} \angle 60°$;   (b) $\sqrt{13} \angle \dfrac{\pi}{4}$;   (c) $2 \cos \dfrac{\pi}{6} + j\, 2 \sin \dfrac{\pi}{6}$.

(4) $Z_1 = 20 + 10j$; $Z_2 = 15 - 25j$; $Z_3 = 30 + 5j$.

   Find: (a) $|Z_1||Z_2|$;   (b) $Z_1 Z_2 Z_3$;   (c) $\dfrac{Z_1 Z_2}{Z_3}$;

   (d) $\dfrac{Z_1 Z_2}{Z_1 Z_3}$.

(5) An impedance Z of $(200 + j\, 80)\,\Omega$ is connected to a 100 V R.C. supply. Given that $I = \dfrac{V}{Z}$, determine the current flowing and express your answer in complex form.

(6) Solve the following differential equations:

   (a) $\dfrac{dy}{dx} = \dfrac{2x}{2x + y}$;

   (b) $x\dfrac{dy}{dx} = -y + 2x$ given $y = 0$ at $x = 2$.

(7) Obtain the general solution of the differential equation:

   $$\dfrac{d^2 y}{dt^2} = -\omega^2 \cos \omega t.$$

(8) Obtain the particular integral for the equation:

   $$\dfrac{d^2 y}{dx^2} + 4\dfrac{dy}{dx} + 2y = 2x + 3.$$

(9) Solve $y'' + 4y = 0$,   given   that $y(0) = 0$ and $y'(0) = 1$.

(10) The bending moment for a beam is related by the equation:

cont.

$$M = EI\frac{d^2y}{dx^2} \quad \text{where} \quad M = \frac{WL}{4} - \frac{Wx}{2}.$$

Find an equation for y given that when $x = 0$, $y' = 0$ and when $x = \frac{6}{2}$, $y = 0$.

(11) The second order differential equation

$$\frac{d^2q}{dt^2} + \frac{6\,dq}{dt} + 40q = 115$$

models a CRL series circuit with an emf of 115 V. Find an equation for q (the charge) in terms of t.

(12) Solve the system of linear equations:

$$3x - 7y + 4z = 10$$
$$x + y + 2z = 8$$
$$-x - 2y + 3z = 1$$

using Gaussian elimination.

(13) Determine the length of the crankshaft AB by first formulating a Cartesian position vector from A to B and then determining its magnitude.

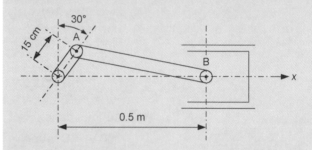

Crank and piston assembly

(14)

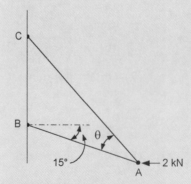

Strut assembly

Determine the angle θ (acute) between the two struts so that the horizontal force (2 kN) has a component of 2.3 kN directed from A towards C.

cont.

(15) Using an analytical method show that the Fourier series for a fully-rectified sine wave is given by:

$$\frac{4x}{\pi}\left[\frac{1}{2} - \frac{1}{1.3}\cos\frac{4\pi x}{T} - \frac{1}{3.5}\cos\frac{8\pi x}{T}\right.$$
$$\left. - \frac{1}{5.7}\cos\frac{12\pi x}{T} - \cdots\right]$$

for limits $-\frac{1}{2}T \leqslant x \leqslant \frac{1}{2}T$, amplitude k.

(16) Use Harmonic analysis, with tabulated values at $30°$ steps to find the Fourier series up to the fourth harmonic for the waveform given by the following data.

θ	0	30	60	90	120	150	180	210	240	270	300	330	360
y	10	9.5	8.5	7.5	7.0	6.5	6.0	5.0	4.5	3.0	1.5	1.0	0

(17) Find the inverse Laplace transforms for the following system transfer functions:

(a) $G(s) = \dfrac{100}{s^2 + 5s + 200}$

(b) $G(s) = \dfrac{200}{s(s + 5)(s - 4)}$.

(18) A control system can be described by the second order differential equation:

$$\frac{d^2\theta_0}{dt^2} + 2\xi\omega\frac{d\theta_0}{dt} + \omega^2\theta_0 = k\omega^2\theta_i.$$

Solve the equation for $\dfrac{\theta_0}{\theta_i}$ using Laplace transforms, given that $\xi = 0.7$ (damping ratio), the angular velocity (ω) applied to the system is 15 rad/s and $k = 4$.

$\dfrac{\theta_0}{\theta_i}$ = the gain of system,

where θ_0 = magnitude of output

θ_i = magnitude of input.

Mapping of contents to units and reference material

Business Management Techniques

BTEC unit code: 21716P (9497M)

Outcome	Topic	Pages	Reference material
1	Costing systems and techniques	1–14 (see also pages 53–36)	Henderson, Illige and McHardy, *Management for Engineers* (Butterworth-Heinemann, 1994) – chapter 8.
			Curtis T., *Business and Marketing for Engineers and Scientists* (McGraw-Hill, 1994) – chapter 9.
			Microsoft Excel spreadsheet models for various costing exercises can be downloaded from: http://members.aol.com/highereng
2	Financial planning and control	14–26	Henderson, Illige and McHardy, *Management for Engineers* (Butterworth-Heinemann, 1994) – chapters 6, 7 and 9.
			Curtis T., *Business and Marketing for Engineers and Scientists* (McGraw-Hill, 1994) – chapter 9.
			Lock, *Handbook of Engineering Management* (Newnes 1992), chapters 19 and 20, information on costing and budgeting processes.
			Microsoft Excel spreadsheet models for various budgetary planning exercises can be downloaded from: http://members.aol.com/highereng
3	Project planning and scheduling	26–35 (see also pages 70–72)	Curtis T., *Business and Marketing for Engineers and Scientists* (McGraw-Hill, 1994) – chapter 9.
			Microsoft Excel spreadsheet models for various budgetary planning exercises can be downloaded from: http://members.aol.com/highereng
			A Visio stencil containing symbols for constructing network diagrams can be downloaded from: http://members.aol.com/highereng

Engineering Design

BTEC unit code: 21719P (9498M)

Outcome	Topic	Pages	Reference material
1	Design specification	36–44	Pugh, *Total Design* (Addison Wesley, 1991) – chapter 3.
			Pahl and Beitz, *Engineering Design* (Springer, 1995) – information on general design process and formulating specification, chapters 3 to 5.
			BS 7373, *Guide to the preparation of specifications* (British Standards Institute).
2	Design report	45–57	Pahl and Beitz, *Engineering Design* (Springer, 1996) – chapters 4 and 5.
			BS 4811, *The presentation of research and development reports* (British Standards Institute).
			Norman, *Advanced Design and Technology* (Addison Wesley Longman, 2nd ed., 1995), chapter 2.
			Lock, *Handbook of Engineering Management* (Newnes 1992), chapters 19 and 20, information on costing and budgeting processes.
3	Computer aided design	58–76	Groover and Zimmers, *Computer Aided Design and Manufacturing* (Prentice Hall 1984) – Part II
			Lock, D., *Handbook of Engineering Management* (Newnes 1992) – chapter 31 (Applying Computers to Engineering).
			Grimston, C. *PC Maintenance* (Arnold 1996) – chapters 1 to 3, provides a useful introduction to computer hardware and microprocessor architecture.
			Moeler, Phillips and Davis, *Project Management with CAM, PERT and Precedence Diagramming.* chapter 2 provides an introduction to networks.
			Access to CAD and analytical packages for engineering, such as AutoCAD, CadKey, MathCAD, and CODAS.
			Software packages for the computer analysis of structural systems, such as Analysis by Ing. Frank Crylaerts, can be downloaded from: http://www.club.innet.be/~year1335
			The student's version of QuickField, a software package for field analysis, can be downloaded from: http://www.tor.ru/quickfield and also from Simtelnet at: http://www.simtel.net/pub/simtelnet/Win95/engin
			A Visio stencil containing symbols for constructing network diagrams can be downloaded from: http://members.aol.com/highereng

Engineering Science

BTEC unit code: 21718P (9499M)

Outcome	Topic	Pages	Reference material
1	Static and dynamic engineering systems (torsion and bending)	88–111 (see also 291–301 and 311–313)	Hearn E.J., *Mechanics of Materials – Volume 1* (Butterworth-Heinemann, 3rd ed., 1997) – chapters 3 and 4 provide comprehensive coverage of shear force, bending moments and bending theory. chapter 8 provides coverage of torsion.
			Bacon and Stephens, *Mechanical Technology* (Butterworth-Heinemann, 3rd ed.,1998) – chapters 1 to 5, 10 and 18.
			Bolton, W., *Engineering Science* (Newnes, 3rd ed., 1998), chapters 6 and 7.
			Bolton W., *Mechanical Science* (Blackwell 1993), chapters 3 and 4 cover bending and deflection of beams; chapters 10, 11 and 16 cover dynamics.
			Drabble G., *Dynamics* (Macmillan 1990) – programme 6.
2	Energy transfer in thermal and fluid systems	112–143	Eastop and McConkey, *Applied Thermodynamics* (Addison Wesley Longman) – chapter 17 provides good coverage of all aspects of heat transfer.
			Rogers and Mayhew, *Engineering Thermodynamics, Work and Heat Transfer* (Addison Wesley Longman 1986) – part IV, heat transfer.
			Kreith and Bohn, *Principles of Heat Transfer* (PWS, 5th ed., 1996).
			Mott, R., *Applied Fluid Mechanics* (Macmillan 1993) – chapters 1 to 3 and 6 to 9 contain excellent information on viscosity, the energy equation, and energy losses.
			Douglas, Gasiorek and Swaffield, *Fluid Mechanics* (Addison Wesley Longman 1991) – provides a general coverage of fluid systems at an advanced level.
			Bolton, W., *Engineering Science* (Newnes, 3rd ed., 1998) – chapter 12.
			Daugherty et al., *Fluid Mechanics with Engineering Applications* (McGraw-Hill, 9th ed., 1997) – chapters 1 and 4.
3	Single-phase a.c. theory	144–180 (see also 455 to 458)	Bird, J. *Higher Electrical Technology* (Newnes, 2nd ed., 1996) – chapters 1 to 4.

Hughes, *Electrical Technology* (Addison Wesley Longman, 7th ed., 1995) – chapters 10 to 14.

Silvester, P., *Electric Circuits* (Macmillan, 1993) – programmes 4 and 5.

Bolton, W., *Engineering Science* (Newnes, 3rd ed., 1998) – chapter 16 provides a useful introduction to a.c. circuits; chapter 18 develops this further.

4	Information and energy control systems	180–214	Hughes, *Electrical Technology* (Addison Wesley Longman, 7th ed., 1995) – chapters 17 to 19 and 29.

Mazda, F., *Power Electronics Handbook* (Newnes, 3rd ed., 1997) – this book provides numerous examples of electrical power and energy controllers. It also contains a useful introduction to a variety of semiconductor power control devices.

Tooley, M., *PC-based Instrumentation and Control* (Newnes, 2nd ed., 1995) – this book provides numerous examples of interfacing sensor, transducers and other control devices to personal computers.

Bolton, W., *Engineering Science* (Newnes, 3rd ed., 1998) – chapter 1 provides a general introduction to engineering systems.

Electrical and Electronic Principles

BTEC unit code: 21759P, (9502M)

Outcome	*Topic*	*Pages*	*Reference material*
1	Circuit theory	215–259	Bird, J., *Higher Electrical Technology* (Newnes 2nd ed., 1996) – chapter 8.
			Bird, J., *Electrical Circuit Theory and Technology* (Newnes, 1997) – chapters 11, 13 and 27 to 32.
			Hughes, *Electrical Technology* (Addison Wesley Longman, 7th ed., 1995) – chapters 4, 13 and 33.
			Silvester, P., *Electric Circuits* (Macmillan, 1993) – programmes 1, 2, 3 and 5.
2	Networks	259–270	Bird, J., *Higher Electrical Technology* (Newnes 2nd ed., 1996) – chapters 7, 9 and 13.
			Bird, J., *Electrical Circuit Theory and Technology* (Newnes, 1997) – chapters 27 and 38.
			Hughes, *Electrical Technology* (Addison Wesley Longman, 7th ed., 1995) – chapters 4, 15 and 30.

3	Complex waves	270–282 (see also 473 to 476)	Bird, J., *Higher Electrical Technology* (Newnes 2nd ed., 1996) – chapter 10.

Outcome	Topic	Pages	Reference material
			Bird, J. *Electrical Circuit Theory and Technology* (Newnes, 1997) – chapters 23, 33, and 34.
			A Microsoft Excel spreadsheet model for Fourier analysis can be downloaded from: http://members.aol.com/highereng
4	Transients in R-L-C circuits	282–289 (see also 467 to 473 and 484 to 487)	Silvester, P. *Electric Circuits* (Macmillan, 1993) – Programme 7.
			Bird, J. *Electrical Circuit Theory and Technology* (Newnes, 1997) – chapter 42.

Mechanical Principles
BTEC unit codes: 21722P, 21793P, (9503M)

Outcome	Topic	Pages	Reference material
1	Complex loading systems	291–301	Bolton, W., *Mechanical Science* (Blackwell, 1995) – chapter 8 has information on complex stress and strain.
			Hearn E.J., *Mechanics of Materials – Volume 1* (Butterworth-Heinemann, 3rd ed., 1997) – chapters 2 and 4 cover all the requirements of this outcome.
			Gere and Timoshenko, *Mechanics of Materials* (Chapman and Hall, 3rd ed., 1993) – chapters 1 and 6.
2	Loaded cylinders and beams	301–334	Hearn E.J., *Mechanics of Materials – Volume 1* (Butterworth-Heinemann, 3rd ed., 1997) – chapters 9 and 10.
			Benham and Crewford, *Mechanics of Engineering Materials* (Addison Wesley Longman, 2nd ed., 1996), chapters 2 and 15.
3	Power transmission and rotational systems	334–366	Bolton, W., *Mechanical Science* (Blackwell, 1995) – chapters 12 to 14.
			Drabble, G., *Work Out Dynamics* (Macmillan 1987) – chapter 4 provides good coverage of coupled and geared systems.
			Juvinall and Marshek *Fundamentals of Machine*

Component Design (Wiley, 2nd ed., 1991) – chapters 15 to 19 provide good coverage of belt, clutch and geared systems.

Outcome	Topic	Pages	Reference material
			Mott *Machine Elements in Mechanical Design* (Merrill, 2nd ed., 1994) – chapters 11 to 16 provide good coverage of gears and clutch systems.
			Ryder and Bennett, *Mechanics of Machines* (Macmillan, 2nd ed., 1990) – the balancing of machines is well covered in chapters 10 and 12.

Analytical Methods for Engineers

BTEC unit code: 21717P

Outcome	Topic	Pages	Reference material
1	Algebra	368–395	Bird, J., *Early Engineering Mathematics* (Newnes, 1995) – chapters 4 to 8 provide an introduction for those with a limited mathematics background.
			Bolton, W., *Essential Mathematics for Engineering* (Newnes, 1997) – chapter 2 provides a useful introduction to polynomials and partial fractions.
2	Trigonometry	395–413	Bird, J., *Early Engineering Mathematics* (Newnes, 1995) – chapters 19 to 22 provide an introduction to trigonometry for those with a limited mathematics background.
			Bird, J., *Higher Engineering Mathematics* (Newnes, 3rd ed., 1998) – chapter 4 provides an introduction to hyperbolic functions.
3	Calculus	413–451	Bird, J., *Higher Engineering Mathematics* (Newnes, 3rd ed., 1998) – chapters 15 to 24.
			Bolton, W., *Essential Mathematics for Engineering* (Newnes, 1997) – chapters 22 to 28.
			Yates, J. C., *National Engineering Mathematics – Volume 3* (Macmillan, 1996) – chapters 4 to 10.
4	Statistics and probability	Not covered in this book	Bird, J., *Higher Engineering Mathematics* (Newnes, 3rd ed., 1998) – chapters 33 to 36 deal with binomial, Poisson and normal distributions, linear correlation and linear regression.
			Bolton, W., *Essential Mathematics for Engineering* (Newnes, 1997) – chapters 42 to 45 cover probability and statistics.

Answers

Chapter 1

Question 1.1.1
(page 5)

Hardware, £2.85; Semiconductors, £3.33;
Passive components, £3.51; Miscellaneous, £2.41;
Total cost, £11.83 (overall reduction in cost = 16p).

Question 1.1.2
(page 8)

Manual data switch, £37; automatic data switch, £85.

Question 1.2.1
(page 19)

12 month figures are as follows:

(a) Battery charger, £66,000
Trolley jack, £466,500
Warning triangle, £88,200
Total, £620,700

(b) Battery charger, £69,000
Trolley jack, £464,250
Warning triangle, £82,000
Total, £615,250

(c) Battery charger, £40,000
Trolley jack, £390,000
Warning triangle, £73,000
Total, £503,000

(d) Battery charger, £58,500
Trolley jack, £431,750
Warning triangle, £79,625
Total, £569,875

Question 1.2.2
(page 20)

Totals are as follows:

(a) Battery charger, £22,456.5
Trolley jack, £161,540
Warning triangle, £32,227.5
Total, £216,224

(b) Battery charger, £19,906.5
Trolley jack, £152,150
Warning triangle, £32,227.5
Total, £204,284

(c) Battery charger, £21,337.5
Trolley jack, £152,150
Warning triangle, £25,770
Total, £199,257.1

(d) Battery charger, £18,787.13
Trolley jack, £161,540
Warning triangle, £25,770
Total, £206,097.1

Question 1.2.3

(page 21)

Total labour costs are as follows:

(a) Battery charger, £15,675
   Trolley jack, £113,281.25
   Warning triangle, £11,531.25

(b) Battery charger, £13,537.5
   Trolley jack, £113,281.25
   Warning triangle, £10,890.63

(c) Battery charger, £15,675
   Trolley jack, £122,343.75
   Warning triangle, £10,890.63

(d) Battery charger, £13,537.5
   Trolley jack, £122,343.75
   Warning triangle, £11,531.25.

Question 1.2.4

(page 23)

Total overheads are as follows:

(a) £98,094.38
(b) £98,894.38
(c) £97,623.81
(d) £101,823.81.

Question 1.2.5

(page 23)

Additional semi-variable cost; cost of plant and equipment maintenance.

Additional fixed cost; energy cost for factory and office lighting and heating.

Problems 1.3.1

(page 30)

1. Critical path; B1, B2, C, D, F.
2. Critical path; C, D, H, J, $t_e = 15$ days.

(Please see the Tutor Resource Pack for a model answer with critical path diagram.)

Question 1.3.2–1.3.4

(pages 32 and 34)

Please see the Tutor Resource Pack for model answers.

Chapter 3

Problems 3.1.1

(page 81)

1. $151.2 \, \text{MN/m}^2$.
2. $201.6 \, \text{MN/m}^2$.

Problems 3.1.2

(page 87)

1. $0.227 \, \text{mm}$.

2. 50.1 kN.
3. 23.8 MN/m^2 compression.

Problems 3.1.3

(page 94)

1. $\tau_{max} = 49.73$ MN/m^2, $\theta = 0.9°$ per unit length.
2. $\tau_{max} = 40.0$ MN/m^2, $\tau_{min} = 28$ MN/m^2, $\theta = 1.22°$.

Problems 3.1.4

(page 103)

1. 53.3 rad/s.
2. 35.07 kg m^2.
3. Retardation = 2.2 rad/s^2, braking torque = 2.475 Nm.

Problems 3.1.5

(page 111)

1. Time period = 0.513 s, velocity = 0.92 m/s.
3. Velocity = 0.105 m/s.
4. Velocity = 1.57 m/s, acceleration = 9.87 m/s^2.
5. Velocity of piston = 2.45 m/s, acceleration of piston = 53.5 m/s^2.

Problems 3.2.1

(page 126)

2. 37.5 W/m^2.
3. 25.5 kW.
4. (i) 13.94 W/m^2; (ii) 12.5 kW.

Problems 3.2.2

(page 143)

2. 3.3×10^{-4} m^2/s.
3. Turbulent.
4. (i) 15.3 m/s; (ii) 0.0769 m^3/s; (iii) 63.8 kg/s.
5. (i) 62.5 kph; (ii) 103840 Pa; (iii) 19 mm.
6. f = 0.04.
7. 163 Nm.
8. f = 0.0123 and power loss = 348.8 W.

Problems 3.3.1

(page 153)

1. (a) 250 mA; (b) 176.75 mA; (c) 100 Hz; (d) 10 ms;
   (e) 237 V.
2. $v = 160 \sin(377t)$, 144.8 V.
3. $i = 4.713 \sin(2513t)$, 3.33 A.
4. (a) 585.43 Ω; (b) 11.709 Ω.
5. (a) 7.536 Ω; (b) 1,507.2 Ω.
6. 233.24 Ω, 0.857 A.
7. 85.32 Ω, 1.29 A.
8. $V_C = 21.44$ V, $V_R = 10.8$ V.
9. 159 kHz.
10. 13.63 µH.

Question 3.3.1
(page 154)

(a and b) Please see Tutor Resource Pack for model answers.
(c) $v = 15.14 \, \text{V}$, $i = 0.22 \, \text{A}$.

Question 3.3.2
(page 154)

(a) $V_L = 1.036 \, \text{V}$;   (b) $V_C = 3.388 \, \text{V}$;   (c) $V_R = 1.65 \, \text{V}$;
(d) $Z = 57.46 \, \Omega$;   (e) $V = 2.873 \, \text{V}$;   (f) $\phi = -55°$.

Problems 3.3.2
(page 166)

1. $V = 200 + \text{j}240 \, \text{V}$.
2. $I = 2 + \text{j}5 \, \text{A}$.
3. $V = 0.4 + \text{j}0.75 \, \text{V}$.
4. $I = 0.04 + \text{j}0.0126 \, \text{A}$.
5. $V_L = 40 \, \text{V}$, $V_C = 40 \, \text{V}$, $V_R = 0.3 \, \text{V}$.
6. $f_0 = 47.7 \, \text{kHz}$, $Z_d = 1,000 \, \Omega$.
7. $C_{max} = 1.25 \, \text{nF}$, $C_{min} = 312.5 \, \text{pF}$.
8. $f_{max} = 711 \, \text{kHz}$, $Q_{max} = 89.35$.
   $f_{min} = 224 \, \text{kHz}$, $Q_{min} = 28.15$.

Problems 3.3.3
(page 173)

1. Reactive component $= 1.75 \, \text{A}$, active component $= 1.785 \, \text{A}$.
2. Power factor $= 0.6$, apparent power $= 1.148 \, \text{kW}$.
3. True power $662.9 \, \text{W}$, reactive power $= 499.5 \, \text{W}$.
4. Power factor $= 0.193$, capacitance $= 24.4 \, \mu\text{F}$.
5. Power factor $= 0.178$, current $= 0.459 \, \text{A}$.
6. Power factor $= 0.606$, phase angle $= 52.7°$.
7. Power factor $= 0.598$, current $= 6.313 \, \text{A}$.
8. Capacitance $= 116.7 \, \mu\text{F}$.

Chapter 4

Questions 4.1.1
(page 230)

1. $0.4 \, \Omega$, $20 \, \text{V}$.
2. $-18 \, \text{V}$ in series with $3 \, \Omega$.
3. $3.692 \, \text{A}$ in parallel with $4.333 \, \Omega$.
4. $0.98 \, \text{A}$ (Thévenin equivalent $2.5 \, \text{V}$ in series with $1.55 \, \Omega$).
5. $4.62 \, \text{V}$ (Norton equivalent $2.7 \, \text{A}$ in parallel with $10.91 \, \Omega$).
6. $8.076 \, \text{V}$.
7. $R_A = 206.7 \, \Omega$, $R_B = 310 \, \Omega$, $R_C = 124 \, \Omega$.
8. $R_1 = 40 \, \Omega$, $R_2 = 10 \, \Omega$, $R_3 = 8 \, \Omega$.
9. $R_1 = 6.67 \, \Omega$, $R_2 = 3.33 \, \Omega$, $R_3 = 4.44 \, \Omega$.
10. $R_A = 37 \, \Omega$, $R_B = 29.6 \, \Omega$, $R_C = 24.67 \, \Omega$.

Questions 4.1.2

(page 236)

1. (a) $33.54 \angle 63.44° \, \Omega$;   (b) $12 \angle 90° \, \Omega$;   (c) $0.5 \angle -53.13° \, \Omega$;
   (d) $1,000 \angle -90° \, \Omega$.
2. (a) $12.5 + j21.65 \, \Omega$;   (b) $88.39 - j88.39 \, \Omega$;
   (c) $56.72 + j647.52 \, \Omega$.
3. $0.618 - j0.379 \, A$ or $0.725 \angle -31.52° \, A$.
4. (a) $1.118 \angle 26.57° \, A$;   (b) $0.988 \angle -18.44° \, A$.
   (c) $0.771 \angle 40.6° \, A$.
5. $Z_1 = Z_2 = -j120$, $Z_3 = j160$.
6. $Z_A = Z_B = Z_C = 30 + j30$.
7. $1.212 + j0.308 \, A$.

Questions 4.1.3

(page 241)

1. Turns ratio, $22.4 : 1$; Primary voltage $= 141.1 \, V$.
2. $8 - j2 \, \Omega$.

Questions 4.1.4

(page 245)

1. $I_1 = 0.118 - j0.344 \, A$, $I_2 = -0.123 - j0.615 \, A$,
   $I_3 = 0.241 + j0.271 \, A$.
2. $V_A = 5.05 - j3.21 \, V$, $V_B = 0.08 + j0.72 \, V$.

Questions 4.1.5

(page 253)

1. $18 \, \mu Wb$.
2. $-40 \, V$.
3. 0.133.
4. 0.26.
5. $0.2 \, A$.
6. $1.146 \, mWb$, 20% increase.
7. (a) $55 \, V$;   (b) $5 \, A$;   (c) $275 \, W$.

Questions 4.1.6

(page 259)

1. (a) $237 \, kHz$;   (b) 22.3;   (c) $10.62 \, kHz$.
2. $I = 0.05 - j1.825 \, mA$,
   $V_L = 2578 + j71 \, mV$,
   $V_C = -580 - j16 \, mV$,
   $V_R = 1.5 - j54.75 \, mV$.
3. (a) $12 \, mA$;   (b) $40.5 \, kHz$.
4. 22.75.
5. (a) $425.6 \, kHz$, 140.3, $3.03 \, kHz$;   (b) $45.5 \, kHz$.

Questions 4.2.1

(page 269)

1. (a) $32.25 \, \Omega$;   (b) $53.3 \, \Omega$;   (c) $35.8 + j27.9 \, \Omega$.
2. $R_1 = 40.9 \, \Omega$, $R_2 = 10.1 \, \Omega$.
3. $R_1 = 1.273 \, k\Omega$, $R_2 = 945.8 \, \Omega$.

4. $\gamma = 0.672 + \text{j}0.087$.
5. 2.08 Nepers, 18.06 dB.

Question 4.3.1
(page 275)

(b) $y = 1 - \dfrac{2}{\pi} \sin \omega t - \dfrac{1}{\pi} \sin 2\omega t - \dfrac{2}{3\pi} \sin 3\omega t - \ldots$

(d) $y = \dfrac{2}{\pi} - \dfrac{4}{3\pi} \cos 2\omega t - \dfrac{4}{15\pi} \cos 4\omega t - \ldots + \sin \omega t$

(e) $y = \dfrac{20}{\pi} \cos \omega t - \dfrac{20}{3\pi} \cos 3\omega t + \dfrac{20}{5\pi} \cos 5\omega t - \ldots$

Question 4.3.2
(page 278)

Fourier constants are as follows: $A_0 = 4.55$, $A_1 = -4.79$, $A_2 = -1.85$, $B_1 = 4.79$, $B_2 = 1.79$, etc.

Question 4.3.3
(page 281)

1. (a) 100 W, 16 W, 4 W; (b) $i = 2 \sin \omega t + 0.8 \sin(2\omega t - \pi/2) + 0.4 \sin(4\omega t - \pi/2)$; (c) $V_{\text{RMS}} = 77.46$ V, $I_{\text{RMS}} = 1.55$ A; (d) 120 W.
2. (a) 21.87 V; (b) 0.715 A; (c) 15.28 W; (d) 15.63 W; (e) 0.977.
3. (a) $i = 100 \sin(314t - 0.64) + 0.51 \sin(942t - 1.15) - 0.26 \sin(1570t - 1.31)$ A. (b) 118.74 V; (c) 1.47 A; (d) 87.93 W; (e) 174.5 W; (f) 0.504.

Question 4.4.1
(page 289)

1. $I(s) = \dfrac{0.354s + 70.7}{s^2 + 10^4}$.
2. $i = \text{e}^{-40t}$, 0.607 mA.
3. $i = 0.6(1 - \text{e}^{-5t})$.
4. $i = \text{e}^{-2t} \sin(14t)$.

Chapter 5

Questions 5.1.1
(page 293)

1. Change in length $= 0.05$ mm (extension),
   Change in dimension on 60 mm surface $= 0.0036$ mm (compression),
   Change in dimension on 25 mm surface $= 0.0015$ mm (compression).
2. Change in length $= 0.0714$ mm.
3. $\sigma_1 = 35.32$ MPa; $\sigma_2 = 17.96$ MPa.

Questions 5.1.2
(page 300)

1. $\sigma_1 = 95.3$ MPa; $\sigma_2 = 72.69$ MPa.

2. $\nu = 0.212$; K = 46.3 GPa.
3. $\nu = 0.303$, E = 57.87 GPa, G = 22.2 GPa, K = 48.96 GPa.

Questions 5.2.1

(page 311)

1. Change in diameter = 17.7 μm; change in length = 16.67 μm.
2. Maximum safe gas pressure = 1375 kPa.
3. (i) P = 427.9 kPa.
   (ii) σ_h = 34.23 MPa, σ_1 = 17.12 MPa (compression).
4. σ_h = 28.4 MPa, σ_r = 15 MPa (compression).
5. (i) At inner surface: σ_h = 38 MPa, σ_r = 70 MPa (compression).
       At outer surface: σ_h = 8 MPa, σ_r = 40 MPa (compression).
   (ii) σ_L = 16 MPa.

Questions 5.2.2

(page 319)

2. R_A = 120 kN R_B = 100 kN.
   From R_A; SF = 120, 95, −25, −80 (kN),
   BM at left-hand support = 268.75 kNm,
   BM at right-hand support = − 20 (kNm).

Questions 5.2.3

(page 322)

1. (i) 162 kNm;   (ii) 1350 cm^3.
2. (iii) BM = 450 kNm   (iv) σ_{max} = 128.6 MPa.

Questions 5.2.4

(page 333)

1. (a) 12 kNm;   (b) 210 kNm.
2. 3.188 mm.
3. Maximum deflection = 5.7 mm at 1.92 m from left-hand end.
4. (i) −15 Nm;   (ii) 144 MPa;   (iii) 3.47 m.
5. Deflection at centre of beam = 15.5 mm.

Questions 5.3.1

(page 365)

1. (i) θ_1 = 168.52°, θ_2 = 191.48°;   (ii) 2.95 kW.
2. (i) T_1 = 962 N, T_2 = 143 N;   (ii) 6.1 MPa.
3. 112 Nm.
4. (i) t_A = 5;   (ii) 1466 W;   (iii) 1173; W   (iv) 1454 N.
5. T_1 = 1769 N; T_2 = 977 N and L = 4.48 m.

Chapter 6

Questions 6.1

(page 393)

1. $A = 1.63 \times 10^{-3}$
2. 4.49 (see worked solution).
3. $P_2 = 374.28$.
4. $P_{noise} = 0.02$ (see worked solution).

5. $r = 0.247$.

6. $x = 2, y = 4$.

7. Law of graph is $H = 0.05V^3$

8. (i) $\dfrac{1}{(x+2)(x-3)} = \dfrac{-1}{5(x+2)} + \dfrac{1}{5(x-3)}$;

   (ii) $\dfrac{x}{(x+1)(x^2+2x+6)} = \dfrac{-1}{5(x+1)} + \dfrac{x+6}{5(x^2+2x+6)}$;

   (iii) $\dfrac{x^2}{(x+1)(x+2)^2} = \dfrac{1}{9(x-1)} + \dfrac{8}{9(x+2)} - \dfrac{4}{3(x+2)^2}$.

9. $A = 2, b = 2.772$ (to 4 s.f.).

10. We know that the general condition for convergence is that in the limit as $n \rightarrow \infty$ the modulus of:
    $$\dfrac{U_n + 1}{U_n} < 1 \quad \text{where} \quad U_n \quad \text{is the } n\text{th term.}$$
    Then by expanding each series and applying the above criteria, we are able to establish convergence or other wise.

    Thus both of these series are *convergent*.

11. Coefficients are 1, 6, 15, 20, 15, 6, 1 (see worked solution).

12. Q increases by 6% (see worked solution).

13. Using the exponential series for e^x we get:

    $$x^2 e^{-x} = x^2 - x^3 + \dfrac{x^4}{2!} - \dfrac{x^5}{3!} + \dfrac{x^6}{4!} - \ldots$$

    also $\quad -e^{-x} = -1 + x - \dfrac{x^2}{2!} + \dfrac{x^3}{3!} - \dfrac{x^4}{4!} \ldots$

    then $\quad x^2 e^{-x} - e^{-x} = -1 + x + \dfrac{x^2}{2} - \dfrac{5x^3}{6} + \dfrac{11x^4}{24} - \ldots$

    are the first five terms of required series.

14.

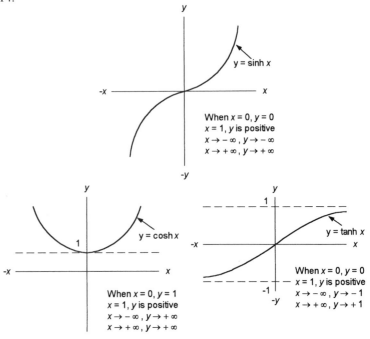

When $x = 0, y = 0$
$x = 1, y$ is positive
$x \rightarrow -\infty, y \rightarrow -\infty$
$x \rightarrow +\infty, y \rightarrow +\infty$

$y = \sinh x$

$y = \cosh x$

When $x = 0, y = 1$
$x = 1, y$ is positive
$x \rightarrow -\infty, y \rightarrow +\infty$
$x \rightarrow +\infty, y \rightarrow +\infty$

$y = \tanh x$

When $x = 0, y = 0$
$x = 1, y$ is positive
$x \rightarrow -\infty, y \rightarrow -1$
$x \rightarrow +\infty, y \rightarrow +1$

15. (i) $\sinh x = 1.5$ then $x = 1.195$ (correct to 4 s.f.);

    (ii) $\cosh x = 1.875$ then $x = 1.242$ (correct to 4 s.f.);

(iii) $\sinh^{-1} 1.375 = x$ then $x = 1,123$ (correct to 4 s.f.);

(iv) $\tanh x = 0.32$ then $x = 0.3316$ (correct to 4 d.p.).

Questions 6.2

(page 411)

1. (i) $s = 9.43$, $A = 84.8$;   (ii) $s = 56.5$, $A = 679$.
2. (i) $5, 306.9°$;   (ii) $1.5, 56.3°$.
3. (i) $x = 2.828$, $y = 2,828$;   (ii) $x = \sqrt{3}$, $y = -1$.
5. (i) Amplitude $= 6$, phase angle $= 30°$ lagging;
   (ii) Amplitude $= 4$, phase angle $= 60°$ lagging.
6. (i) Amplitude $= 3$ A
      Frequency $= 100$ Hz
      Periodic time $= 1/f = 1/100$ sec
      Phase angle $= 45°$ lagging
      Time to first maximum $t = 3/800$ sec.
   (ii) Amplitude $= 0.7$ V
      Frequency $= 200$ Hz
      Period time $= 1/200$ sec.
      Phase angle $= 60°$ leading
      Time to first maximum $= 1/2400$ sec or 0.416 ms.
8. (i) Amplitude $= 40$ Amps
      Periodic time $= 1/50$ sec.
      Frequency $= 50$ Hz
      Phase angle $= 18.3°$ lagging.
   (ii) $i = -12.55$ A.
   (iii) $t = 1.13$ ms.
9.

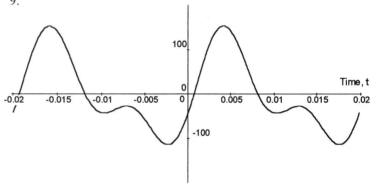

      Peak–peak voltage ≈ 260 V.
12. $0 < \mu < 0.67$.
13. 443.4 kN m.
15. (i) $\sin 6\theta$;   (ii) $\cos 7t$.

Questions 6.3

(page 448)

1. (a) $6(3x + 1)$;   (b) $-\frac{15}{2}(2 - 5x)^{1/2}$;   (c) $-8x(4x^2 + 3)^{-2}$;
   (d) $8 \sin 4x \, ws \, 4x$;   (e) $5 \sin(2 - 5x)$;   (f) $\frac{1}{x}$;   (g) $\frac{2}{3}(1 - 2t)^{-4/3}$.
2. (a) $\sin x + x \cos x$;   (b) $e^x \sec^2 x + e^x \tan x$;   (c) $1 + \ln x$;
   (d) $\cos^2 t - \sin^2 t$;   (e) $12te^{3t} + (t^2 - 1)18e^{3t}$;
   (f) $1 - 3s + \log_e s - 6s \log_e s$.

3. (a) $\dfrac{1}{(1-x)^2}$;

   (b) $\dfrac{3e^x \sin 2x - 6e^x \cos 2x}{9 \sin^2 x}$;

   (c) $\dfrac{4x(x^2-2)(3x+4) - 6(x^2-2)^2}{(3x+4)^3}$;

   (d) $\dfrac{-e^{2t}(2 \sin 2t + e^{2t} \cos 2t)}{e^{4t}}$;

   (e) $-\operatorname{cosec}^2 \theta$.

4. -12.836.

5. -16.49.

6. 5.88.

7. -0.3254.

8. -54.

9. 12.96.

10. 90.16.

11. (a) $36\,\text{m/s}$; (b) $30\,\text{m/s}^2$; (c) $t = 1$ and 2 seconds.

12. (a) $27\,\text{rad/s}$; (b) $72\,\text{rad/s}^2$; (c) at time $t = 0$ and $t = 1$ second.

13. Minimum at $y = 1$, maximum at $y = 33$.

14. $x = 12.5992\,\text{m}$.

15. Kinetic energy $= 62.5\,\text{kJ}$.

16. (a) $-\frac{4}{3} e^{-3x} - \frac{2}{3} \sin 3x + \frac{5}{2} \cos 2x + c$;

   (b) $-xe^{-x} - e^{-x} + c$;

   (c) $\frac{1}{17} \ln(5x+1) - \frac{1}{17} \ln(-2x+3) + c$;

   (d) $\frac{1}{2}[(\ln b)^2 - (\ln a)^2]$;

   (e) $-\ln(\cos x) + c$;

   (f) $\arcsin x + c$.

17. (a) 6.298; (b) $\frac{2}{3}$; (c) 147.07; (d) 1.15.

18. 0.423 to 3 d.p.

19. $T = 2.399\,\text{Nm}$.

20. Mean value $= 1.074$.

21. $\bar{x} = 3.2$, $\bar{y} = 16.38$.

22. $40\pi\,\text{cm}^4$.

23. $y = \dfrac{M}{2EI} x^2 + Ax + B$.

24. A is given, $B = O$.

25. 4 Henry.

Questions 6.4.1

(page 487)

1. (a) $8 - 6i$; (b) $36 + 26i$; (c) i; (d) $-\frac{1}{5} + \frac{2}{5}i$; (e) $-\frac{27}{5} + \frac{11}{5}i$.

2. (a) $\sqrt{72} \angle -45°$; (b) $\sqrt{25} \angle 36.9°$; (c) $\sqrt{1681} \angle -12.7°$;

   (d) $\sqrt{\frac{1}{25}} \angle -36.9°$.

3. (a) $2.74 + 4.74i$; (b) $\dfrac{\sqrt{13}}{\sqrt{2}} + \dfrac{\sqrt{13}}{\sqrt{2}}i$; (c) $\sqrt{3} + i$.

4. (a) 651.92 (b) $21{,}250 - 7250j$; (c) $\dfrac{710}{37} - \dfrac{550}{37}i$;

   (d) $\dfrac{87}{185} - \dfrac{181}{185}i$.

5. $I = 0.431 - 0.172j$ Amps.

6. (a) $(2x - y)^2 = Ax^3$; (b) $xy = x^2 + c$ and $y = x - \dfrac{4}{x}$.

7. $y = \cos \omega t + ct + D$.

8. $y = x - 0.5$.

9. $y = -\frac{1}{4}e^{-4x}$ with respect to x.

10. $y = \dfrac{I}{EI}\left[\dfrac{WLx^2}{8} - \dfrac{Wx^3}{12} + \dfrac{27\,W}{12} - \dfrac{9\,WL}{8}\right].$

11. $q = Ae^{4t} + Be^{-10t} + c.$

12. $x = 3,\ y = 1,\ z = 2.$

13. 44.4 cm.

14. –

15. –

16. See model answer.

17. (a) $G(t) = \dfrac{100}{193.75}\,e^{-2.5t}\sin\sqrt{193.75}\,t;$

    (b) $G(t) = -10 + 4.44\,e^{-5t} + 5.55\,e^{4t}.$

18. $G(t) = \dfrac{900}{\sqrt{4.5}}\,e^{-10.5t}\sin\sqrt{4.5}\,t.$

Index